F

H.C

Fundamentals of Friction:
Macroscopic and Microscopic Processes

NATO ASI Series

Advanced Science Institutes Series

A Series presenting the results of activities sponsored by the NATO Science Committee, which aims at the dissemination of advanced scientific and technological knowledge, with a view to strengthening links between scientific communities.

The Series is published by an international board of publishers in conjunction with the NATO Scientific Affairs Division

A Life Sciences	Plenum Publishing Corporation
B Physics	London and New York
C Mathematical and Physical Sciences	Kluwer Academic Publishers Dordrecht, Boston and London
D Behavioural and Social Sciences	
E Applied Sciences	
F Computer and Systems Sciences	Springer-Verlag
G Ecological Sciences	Berlin, Heidelberg, New York, London,
H Cell Biology	Paris and Tokyo
I Global Environmental Change	

NATO-PCO-DATA BASE

The electronic index to the NATO ASI Series provides full bibliographical references (with keywords and/or abstracts) to more than 30000 contributions from international scientists published in all sections of the NATO ASI Series.
Access to the NATO-PCO-DATA BASE is possible in two ways:

– via online FILE 128 (NATO-PCO-DATA BASE) hosted by ESRIN,
Via Galileo Galilei, I-00044 Frascati, Italy.

– via CD-ROM "NATO-PCO-DATA BASE" with user-friendly retrieval software in English, French and German (© WTV GmbH and DATAWARE Technologies Inc. 1989).

The CD-ROM can be ordered through any member of the Board of Publishers or through NATO-PCO, Overijse, Belgium.

Series E: Applied Sciences - Vol. 220

Fundamentals of Friction: Macroscopic and Microscopic Processes

edited by

I. L. Singer

Tribology Section – Code 6170,
U.S. Naval Research Laboratory,
Washington, DC, U.S.A.

and

H. M. Pollock

School of Physics and Materials,
University of Lancaster,
Lancaster, U.K.

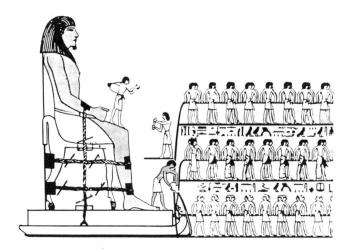

Kluwer Academic Publishers

Dordrecht / Boston / London

Published in cooperation with NATO Scientific Affairs Division

Proceedings of the NATO Advanced Study Institute on
Fundamentals of Friction
Braunlage, Harz, Germany
July 29–August 9, 1991

Library of Congress Cataloging-in-Publication Data

Fundamentals of friction : macroscopic and microscopic processes /
   edited by I.L. Singer, H.M. Pollock.
      p.   cm. -- (NATO ASI series. Series E, Applied sciences ; no.
   220)
      "Published in cooperation with NATO Scientific Affairs Division."
      Synthesis of material from lectures, discussions, and workshops
   from the NATO Advanced Study Institute on the Fundamentals of
   Friction, held July/August 1991 at Braunlage, Germany.
      Includes bibliographical references and index.
      ISBN 0-7923-1912-5 (alk. paper)
      1. Friction--Congresses.   2. Tribology--Congresses.   I. Singer,
   I. L. (Irwin L.) II. Pollock, H. M. (Hubert M.)   III. NATO Advanced
   Study Institute on the Fundamentals of Friction (1991 : Braunlage,
   Germany) IV. North Atlantic Treaty Organization. Scientific
   Affairs Division. V. Series.
   TA418.72.F85   1992
   620.1'1292--dc20   92-24977

ISBN 0-7923-1912-5

Published by Kluwer Academic Publishers,
P.O. Box 17, 3300 AA Dordrecht, The Netherlands.

Kluwer Academic Publishers incorporates the publishing programrnes of
D. Reidel, Martinus Nijhoff, Dr W. Junk and MTP Press.

Sold and distributed in the U.S.A. and Canada
by Kluwer Academic Publishers,
101 Philip Drive, Norwell, MA 02061, U.S.A.

In all other countries, sold and distributed
by Kluwer Academic Publishers Group,
P.O. Box 322, 3300 AH Dordrecht, The Netherlands.

Printed on acid-free paper

Printed in the Netherlands

CONTENTS

IV. LUBRICATION BY SOLIDS AND TRIBOCHEMICAL FILMS

V. LUBRICATION BY LIQUIDS AND MOLECULARLY-THIN LAYERS

VI. NEW APPROACHES AT THE NANO- AND ATOMIC SCALE

VII. MACHINES AND MEASUREMENTS

VIII. APPENDIX

PREFACE AND ACKNOWLEDGEMENTS: THE BRAUNLAGE MEETING

The NATO Advanced Study Institute on the fundamentals of friction, held in July/August 1991 at Braunlage in the Harz mountains of Germany, was set up in order to take advantage of a particular opportunity - the convergence of two hitherto separate disciplines. Classical tribologists have used concepts of surface contact, adhesion, deformation and fracture with increasing success in recent decades in order to model macroscopic friction behaviour. However, gaps in understanding remain, in part because surface-mechanical behaviour (including friction) has only recently attracted the full attention of experts within a second discipline, that of modern surface science. These include theorists, molecular dynamicists, and users of surface proximity devices (surface force apparatus, single-asperity probes, atomic force/scanning tunnelling microscopes) who now have the tools to carry out fundamental studies of sliding interfaces at the atomic level.

The objective of the ASI was to bring together experts in these two fields, in order to improve their understanding of friction processes and to pass on this knowledge to the ASI students and to the scientific/engineering community in general. We hope that this book will show that friction has come of age as an interdisciplinary subject involving theoretical physics and chemistry as well as engineering.

We organised the ASI by inviting outstanding investigators from the two fields, conferring with them on the "burning issues" needing to be addressed, and then structuring lectures, discussions and workshops accordingly. Sixty-five students, mainly from NATO countries, joined the twenty-one lecturers for ten days of morning and evening sessions (the participants are listed in section 8). Lengthy question-and-answer sessions followed the lectures, and many lecturers posted copies of their presentations after their talks as an aid to informal discussion. Students had the opportunity to present their work at either of two poster sessions, beginning with a three-minute overview from each contributor. In response to individual requests, lecturers and students gave afternoon tutorials on special topics. Workshops on "Computer simulations and modelling" and on "Friction: machines and measurements" included short contributions and video presentations by students. Finally the penultimate morning was devoted to topical discussion groups of five to fifteen participants or "wandering scholars", and this resulted in a typed collection of important issues. These served as focal items for the final day's overview on "Future issues in microscopic and macroscopic friction".

The directors are very grateful for the hard and time-consuming oral and written work which lecturers and discussion contributors devoted to this meeting, and which led to its quite special atmosphere and success. We hope that this book reflects some of this atmosphere; that it will contribute to the integration of the two disciplines; and that tribologists will gain new insights into friction processes, while proximity-probe scientists will recognise the challenges (and rewards) of solving tribological problems.

The major support for the Institute was provided by NATO Scientific Affairs Division. We thank also the home institutions of the directors for support, both direct and indirect: the U S Naval Research Laboratory and the School of Physics and Materials of Lancaster University, in particular Miss H Coates for her secretarial work. Additional co-sponsors were:

US Naval Weapons Support Center - Crane
US Air Force Office of Scientific Research, AR & D
US Army Research Office, ERO
US National Science Foundation
US Office of Naval Research, ERO
Unilever Research and Engineering Division
Mobil Research and Development Corporation
Exxon Research and Engineering Company.

We are especially grateful to the ASI Administrator, Mr M G de St V Atkins, who was largely responsible for the smooth running of the months of preparation, the meeting itself and the follow-up period. We warmly thank also the following: Mr Jonathan Singer for recording over fifty hours of lectures, discussions, workshops and panels on videotape and transcribing them to audiotape; Mrs Barbara Kester of International Transfer of Science and Technology, Overijse, for helpful advice and assistance with planning; Frau Silke Harling and the staff of the Maritim Hotel Braunlage, whose facilities, food, comfort and friendly atmosphere were outstanding; members of the organising committee, especially Dr B J Briscoe for help with the planning of the scientific programme; Professor D Dowson, for his evening presentation on "Friction in everyday life", which informed and entertained a mixed scientific and non-technical audience; and Prof Dr-Ing J Holland, Dipl-Ing A Linnenbrügger and Dipl-Ing M Tychsen of the Clausthal Technical University's Institut für Reibungstechnik und Maschinenkinetik, who organised transport, arranged highly successful excursions, provided a computer and made arrangements for the use of electronic mail. We are also grateful to Longman Group (UK) Ltd. for permission to reproduce seven figures, shown here on pages 2, 136, 236, 324, 404, 522, and 568, from *History of Tribology* by D Dowson (1979).

IRWIN L SINGER
Surface Chemistry Branch (6176)
Naval Research Laboratory
Washington DC 20375, USA

HUBERT M POLLOCK
School of Physics and Materials
Lancaster University
Lancaster LA1 4YB, England

INTRODUCTION AND SUMMARY

This book describes what is known about friction from models and experiments on a macroscopic scale, and what is being learned at the microscopic level. It has been 40 years since the last comparable publication appeared,[1] and it is unusual for tribology conference proceedings to focus on this specific topic. The 1952 publication followed a gathering of scientists in London at the Royal Society in April 1951. At that time, questions were raised about fundamental mechanisms of friction, such as energy dissipation, but the techniques needed to probe friction at microscopic scales (of both space and time) did not evolve until the 1980s. By 1991 the time was ripe for presenting and discussing the "burning issues" of friction with new insights gained with the help of these techniques, and the NATO advanced study institute, whose aims are spelled out in the Preface, was the chosen forum.

The book is a synthesis of the material from lectures, discussions and workshops that took place at Braunlage. The texts of most of the lectures are included. Although other relevant aspects of tribology, such as contact mechanics, surface treatments, and wear behavior, are reported, the emphasis here is on defining the state of knowledge and the gaps in understanding of friction processes. It will be for the reader to judge how successfully the various issues have been clarified. In this respect, we hope that the transcribed discussions following many of the chapters will prove useful. In addition, an *epilogue* found in the Appendix (section 8) summarizes issues and recommendations for scientists and engineers seeking to contribute to future understanding of the fundamentals of friction. The rest of the book is divided into seven sections:

1. SCIENTIFIC AND ENGINEERING PERSPECTIVES. Current understanding of friction involves concepts of adhesion, elastic and plastic deformation, boundary lubrication by surface films, and surface forces. In addition to reviewing some of the general ideas that emerged during his fifty-five years of the study of friction, D. Tabor treats the question: how is energy dissipated when friction occurs? In many cases the mechanism is clear, but recently, there has been increasing interest in the intriguing physical problem of energy dissipation for interfacial sliding under elastic or near-elastic conditions.[2]

On the engineering side, E. Rabinowicz documents the fallacy that friction coefficients of materials are constant. Many of the parameters that contribute to friction fluctuations (as distinct from variations) are described, and the size and nature of surface asperities and junctions are suggested as especially important

---

[1] Proc. Royal Soc. **A 212** (1952) pp. 439-520

[2] This question is especially relevant to the sliding of diamond and other hard ceramics - a topic that has always proved difficult to model, hence the absence of a specific chapter on the friction of such materials. However, it will be clear from discussions transcribed here that to some extent this question is being answered through an understanding of atomic-scale processes.

factors.

Sections 2 - 6 address the more scientific aspects of friction, and section 7, the engineering issues:

2. CONTACT MECHANICS, SURFACES AND ADHESION. This section contains background material needed for an understanding of models that have been used with increasing success to model macroscopic friction behaviour. (For those needing a review of the fundamentals of contact mechanics, K.L.Johnson's summary of the principal formulae is presented in an appendix). Of particular interest are the topics where gaps in understanding remain. Current theories of roughness are reasonable as regards the number and size of the micro-contacts between surfaces, but as one of the originators of models of surface roughness, J.A. Greenwood, sees it, we know little more now of how to describe it than we did in 1933 when the profilometer was invented. Will the fractal approach prove more effective than the spectral density approach? Surface forces can be measured by means of various types of "adhesion" experiment - but, as H.M. Pollock points out, the influence of surface forces on "adhesive friction", or on any other variety of friction, is complex. Here, as described by A. R. Thölén, electron microscopy is proving to be a powerful tool, being able to reveal adhesion stress fields, with elastic and plastic deformation as well as material transfer, at the asperity scale. The usual pragmatic approach to friction regards the friction coefficient as an intrinsic material constant: one consequence of shirking the question of the physical aspects is that predictability is largely lost when we are outside empirically tested limits. The fracture mechanics approach, presented by A.R. Savkoor, can explain why an interface is much weaker in the normal than in the tangential direction, and promises to clarify the differences between peeling and slipping. At last, it is becoming possible to say to what extent kinetic friction (an energy-dissipative process) may be considered as a series of sequential processes of static friction (a conservative process!)

3. FRACTURE, DEFORMATION AND INTERFACE SHEAR. While fracture processes may play an important role in friction, B.R. Lawn illustrates the converse part played by friction in the fracture of brittle ceramics. This can be deleterious, as regards strength, abrasion and wear, or beneficial, by helping to dissipate energy and thus to increase toughness. Recent development of the two-term, non-interaction model of friction, with adhesion and ploughing components, owes much to studies of the friction of organic polymers, thanks partly to their wide spectrum of mechanical response. According to B.J. Briscoe, the model's predicative capability is good, and there is confidence in the validity of concepts such as intrinsic interface shear stress. The friction of flowing powders, according to M.J. Adams, can also be understood in terms of the same two-term model. While the friction of powders against smooth walls requires only relatively simple multi-asperity contact models, the bulk deformation of compacted powders

involves mechanical interlocking (Coulombic friction), and computer simulation is required to model friction at the shear planes. The two-term model of friction also provides an intuitive starting point for a more detailed description of the deformation of rough, elasto-plastic surfaces in the presence of adhesion. From continuum models for elastic and plastic deformation, T.H.C. Childs develops flow maps that delineate regions of elastic contact from wave, wedge, or chip-forming flow as a function of surface roughness and interfacial shear strength. The section concludes with a "speculative argument" by K.L. Johnson that one may include the possibility of mode II fracture and take the relevant critical stress intensity factor into account when constructing a friction map. This approach promises to be valuable in attempts to establish the conditions for bulk shearing, intrinsic stick-slip, or elastic contact with steady sliding in metals, polymers and ceramics.

4. LUBRICATION BY SOLIDS AND TRIBOCHEMICAL FILMS. Lubrication "cheats" Mother Nature of the damage to which sliding surfaces are predisposed. The mechanisms by which thin solid films reduce friction are reviewed by I.L. Singer from both a macroscopic and a microscopic viewpoint. Interfacial films that form during sliding contact are investigated, and models of film generation and the chemistry of formation are proposed. Boundary lubrication, when successful, produces a thin film responsible for reduced friction. J-M. Georges et al. analyze the different physicochemical processes induced by a boundary lubricant and describe mechanical properties of boundary films developed over a wide range of pressures, from 1 kPa to 10 GPa. Boundary lubricants must also react with the solid surface if they are to lubricate effectively. R.S. Timsit demonstrates that the shear strength and other tribological properties of molecular layers of stearic acid on glass, aluminum and gold surfaces correlate with the chemical reactivity of stearic acid to the solid surface. Chemical reactivity often controls both the friction and wear behavior of tribomaterials in many practical situations. Tribochemical reactions that affect a variety of engineering materials, from oxidative wear of steels to embrittlement of ceramics, are described by T.E. Fischer. Interactions of sliding solid surfaces with gases or liquids lead to surface modification and enhance reaction rates. Tribochemical reactions involved in extreme pressure (EP) lubrication of steel motivate J.T. Yates et al. to investigate surface chemical reactions between Fe(100) surfaces and CCl_4 in UHV. The influence of surface defects, chemisorbed oxygen and temperature on the stability of $FeCl_2$ EP layers is described.

5. LUBRICATION BY LIQUIDS AND MOLECULARLY-THIN LAYERS. Liquid lubrication can provide the most effective means of separating surfaces in relative motion. Presenting the subject from an historical perspective, D. Dowson reviews the past century's developments in our understanding of bearing lubrication, from fluid films through elasto-hydrodynamic contacts to suggestions of fluid solidification. An instrument that has enabled scientists to observe

directly the behavior of liquids confined to molecularly-thin volumes, the Surface Force Apparatus (SFA), has provided much of the material for the next two chapters. The principal developer of the SFA, J.N. Israelachvili, begins by reviewing recent advances in experimental and theoretical techniques for probing adhesion, interfacial friction and lubrication at the Ångstrom level. His very comprehensive chapter describes experiments that have evaluated a range of theories, from contact mechanics (e.g., JKR vs Hertz) to friction mechanisms (e.g., Amontons vs. Bowden-Tabor), and presents results of computer simulations that account for "quantized jumps" that occur during shear of molecularly-thin fluids. S. Granick also examines the rheological behavior of fluids a few molecular diameters thick, and provides further evidence that the properties of fluids at interfaces differ considerably from those in the bulk. Exotic fluid behaviors such as shear-rate dependent viscosity and time-dependent yield strength of confined ultrathin liquid films are presented, and some of the engineering implications of liquids behaving like solids are discussed.

6. NEW APPROACHES AT THE NANO- AND ATOMIC SCALE. This section introduces a variety of proximal probe and computational techniques for studying the fundamental behavior of contacting and sliding interfaces at the atomic level. Two experimental techniques, the atomic force and scanning tunnelling microscopes (AFM/STM), offer the potential to investigate adhesion and friction at a scale of single-atom contact. G.M. McClelland and J.N.Glosli review simple theoretical models and recent experimental advances concerned with frictional processes at the atomic scale. Models for wearless (*and even frictionless!*) interfacial sliding are discussed and compared with molecular dynamics calculations of frictional forces and energy dissipation between close-packed alkane monolayers. AFM measurements of atomic-scale friction are but one of many applications of "proximal probes" techniques in the field of nanotribology. E. Meyer et al. demonstrate that another technique, friction force microscopy (FFM), enhances the capability of depicting surface topographical features from the atomic to the micron scale. FFM images reveal features such as monolayer steps in 2-layer LB films and periodic track spacing of 1.6 μm in magneto-optical discs. Mechanisms of failure of boundary lubricant layers are also investigated by intentionally damaging LB films by overloading them, then imaging the damage sites.

The final three chapters present several tools of surface physics theory that have recently been introduced into tribology. J. Ferrante and G. Bozzolo describe a variety of computational techniques, ranging from first-principles to semi-empirical methods, that are available for studying tribology at the atomic level. In addition to discussing the techniques, their accuracy and their limitations, they show that surface energy calculations using semi-empirical techniques can be relatively simple to perform, and then apply these methods to solve problems of adhesion and friction. Molecular dynamics provides not only the computational techniques to simulate contacts between solid-solid and solid-liquid interfaces, but also visually exciting portrayals of these interactions. U. Landman et al. give a comprehensive review of their molecular dynamics investigations of atomistic mechanisms involved in adhesive contact formation, friction, and wear processes. Results are presented for several tip and substrate materials (metallic, ionic and covalent), and thin alkane films. Visual presentations are replete with stress and strain contour plots and stunning photo-sequences of atomic configurations during jump-to-contact, elastic and plastic deformation, neck formation and surface reconstruction, wetting, molecular re-ordering, material transfer and atomic-scale stick-slip. J. Belak and I.F. Stowers take a similar approach in their study of the indentation and scraping of a clean metal surface by a hard diamond tool. While calculated stress fields give excellent agreement with continuum models during the initial elastic indentation, the onset of plastic deformation occurs at a much higher yield stress in the atomistic simulations. This enhanced hardness at shallow indentations is discussed in terms of the theoretical yield stress required to create dislocations a few lattice spacings deep.

7. MACHINES AND MEASUREMENTS. This section elaborates on the theme introduced in the second chapter by E. Rabinowicz, the engineering concerns about friction fluctuations and variations. P.J. Blau describes a set of sliding friction experiments designed to address the issue of whether friction is a fundamental property of two contacting materials or a property of the larger tribosystem. The studies show that frictional behavior (steady state plus variations) depends on the tribosystem and confirm, therefore, that a single friction coefficient cannot adequately characterize the behavior of laboratory or engineering tribosystems. Fluctuations in friction associated with vibrations is the topic of the last two papers. Beginning with a tutorial overview of dynamical friction during continuous sliding, D.P. Hess and A. Soom predict the contact and surface roughness conditions that lead to unsteady friction in dry sliding. Although many of the results still await experimental verification, the models warn us that while vibrations may not always change the average friction coefficient, complex nonlinear dynamic behavior can lead to unstable and even chaotic behavior in seemingly simple sliding systems. One of the undesirable by-products of friction between sliding surfaces is friction-induced vibrations (stick-slip). M.T. Bengisu and A.Akay present a stability analysis of friction-induced stick-slip vibrations to demonstrate the significance of the number of degrees of freedom. Finally, systems with multiple degrees of freedom are shown to exhibit bifurcations in their response characteristics, and may respond chaotically.

SCIENTIFIC AND ENGINEERING PERSPECTIVES

..... What is happening at the interface between solids during sliding? Nowadays, when so many physicists are devoting their time to studying the complexities of nuclear disintegration, this may seem a humdrum topic, but there are times when it is well to remember that the world in which we live is not entirely made up of disintegrating nuclei .....

F P BOWDEN, in "A discussion on friction", Proc. Roy. Soc. **A 212**, 439 (1952).

..... If an understanding of the nature of surfaces calls for such sophisticated physical, chemical, mathematical, materials and engineering studies in both macro- and molecular terms, how much more challenging is the subject of ".... interacting surfaces in relative motion"?

D DOWSON, "History of Tribology", Longman 1979, p 3.

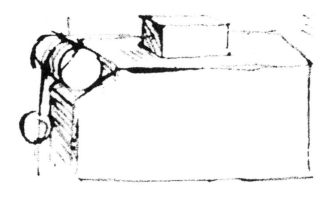

Leonardo da Vinci's studies of friction. Sketches showing experiments to determine: (top) the force of friction on a horizontal plane by means of a pulley; (bottom) the friction torque on a roller and half bearing. [From D. Dowson, <u>History of Tribology</u> (Longman, London, 1979) p. 98, with permission.]

FRICTION AS A DISSIPATIVE PROCESS

DAVID TABOR
University of Cambridge
Cavendish Laboratory
Madingley Road
Cambridge CB3 0HE
United Kingdom

ABSTRACT This paper consists of two interwoven strands. One strand describes selected experimental studies of friction carried out by the author over the last 55 years and indicates some of the rather more general ideas that emerged from that work.

The second strand deals with the question: how is energy dissipated when friction occurs? It is clear that if friction involves permanent damage to the sliding surfaces, say plastic flow, fracture, tearing or fragmentation the energy expended should be explicable in terms of the strength properties of the materials. If it involves viscous flow or inelastic deformation (hysteresis) there may be little visible damage and one is tempted to treat the energy loss as an example of internal friction (this is another problem). There are, however, some apparently puzzling cases where none of these explanations are plausible. The paper will discuss these and other borderline cases.

As a starting point it takes up the ideas of plastic shear in a crystalline solid where it is known that atoms in the shear plane are displaced from their equilibrium position until they reach an unstable configuration: at this point they flick back to another equilibrium position and slip occurs by a single atomic spacing. The strain energy is lost by vibrations in the lattice and these in turn are degraded into heat. (This also applies to crystals containing dislocations).

These ideas are certainly applicable to frictional processes involving plastic deformation. It is suggested that a similar mechanism applies, in many cases, to elastic or near-elastic conditions, particularly if sliding occurs truly in the contact interface. Thus it would explain the friction observed when oriented fatty acid molecules on a solid substrate slide over a similar surface without producing any permanent damage. It is probable that this also applies to experiments with the Atomic Force Microscope where it has been found that riders sliding over a surface experience a finite frictional force without damaging the surface (like passing a finger nail over the teeth of a comb).

In this paper this idea is applied to the sliding of oxide-coated metals, to polymers and tentatively to ceramics. (The behaviour of rubber is left to Dr Savkoor). The distortion produced by the frictional force and hence the energy dissipated depends on the strength of the bonds between the surfaces and may be expressed in terms of surface forces or in terms of surface energies.

In those cases where the friction is speed and/or temperature dependent the behaviour may be understood in terms of the effect of these variables on the elastic constants of the bodies. It may be that this is equivalent to certain rate-theories developed by Eyring and his school many years ago: but the Eyring approach gives little guidance as to the modes of energy dissipation.

The basic idea developed in this paper is not original: it is to be found in a 1929 paper by Tomlinson though, for some reason, he does not emphasise the important role of atomic vibrations as a means of dissipating energy.

3

I. L. Singer and H. M. Pollock (eds.), Fundamentals of Friction: Macroscopic and Microscopic Processes, 3–24.
© 1992 *Kluwer Academic Publishers. Printed in the Netherlands.*

1. Introduction

When one body is placed on another a finite force is required to move it. If the movement involves sliding this is called sliding friction. If the body can be moved by rolling, especially if it is spherical or cylindrical the force may be called rolling friction. We first deal with sliding.

The early history of the subject will be dealt with by other speakers: here I refer only to the classical work by the French scientists and engineers Amontons and Coulomb. It is interesting to recall that Coulomb was an engineer in the French army and that much of his work was stimulated by military and industrial factors. His main observation (for our purposes) is that for extended surfaces the sliding frictional force F is proportional to the normal load N and independent of the size of the bodies: that is to say the coefficient of friction μ, defined as

$$\mu = F/N \tag{1}$$

is a constant for a given pair of materials. Most of his experiments were carried out with wooden bodies which he recognised as being rough. He suggested that the friction corresponded to the force required to slide one set of asperities over the other. If the average slope of the roughnesses is θ, simple mechanics shows that

$$\mu = \tan \theta \tag{2}$$

No other factor is involved and thus the experimental laws of friction are fulfilled.

In 1804 Leslie wrote a book on Heat in which he showed that this model was unsatisfactory because as much energy is gained in sliding down the slopes as is expended in climbing up them (like an ideal roller-coaster or switchback). He suggested that energy could be expended by deforming the asperities. Other workers (as recently as 1980) have suggested that in sliding down the asperities there may be impact and energy may be lost either by impact deformation or by the generation of waves. The crucial issue in friction is to understand its dissipative nature. This problem is still with us.

2. An early theory of metallic friction

When I began my research under the late Professor Bowden 55 years ago, we were very conscious of the fact that real surfaces are rough. The area of real contact depends on the way the asperities interact. It seemed to us that we needed to know the real area of contact since this is where the frictional (in these days we would say tribological) action takes place. There were several approaches. One was to use molecularly smooth surfaces - more of this later. Another was to use metal surfaces and measure the electrical conductivity: not very successful since oxide films complicate the interpretation. Furthermore with extended surfaces making contact over numerous asperities the electrical conductance of each asperity contact is not proportional to the contact area but proportional to its diameter so that one needs to know the **number** of asperity contacts. The same problem applies to thermal conductance and acoustic impedance. Even so our experiments showed that the real area of contact for extended surfaces is a minute fraction of the nominal area [2] and could not be quantified with any precision.

The third approach was to study the behaviour of a "concentrated" contact - a sphere in contact with a flat [2,3]. If the surfaces are "smooth" the geometric area of contact will be a fair approximation to the real area. One surface was hard, the other soft. The most striking result was obtained with the metal indium which is not only very soft (one quarter the hardness of lead) but is also relatively inert so that oxide films are not important. Our loads were of order of Kgf so that the deformations were relatively large. With an indium sphere pressing on a smooth flat surface we made three observations. First the area of geometric contact was proportional to the load. The deformation was clearly plastic. This at once indicates that for extended surfaces, if each asperity deforms plastically under the load that it bears, the total area of contact is proportional to the total load. This was a gratifying conclusion. The second was that the indium actually stuck to the other surface and a finite normal force was required to pull it off. Clearly strong interfacial bonds were formed. The third observation was that in sliding there was plastic flow as the indium sheared away from its adhered region. The friction was thus due to the shearing of adhered junctions. If the system was inverted so that a hard sphere slid over indium there was clearly additional resistance to sliding due to the work of plastically grooving the indium surface.

Similar results were obtained with other metals such as lead, copper and platinum. If the sphere-flat configuration is considered to represent the behaviour of a single asperity the behaviour of extended surfaces could be explained. It seemed that we could generalise. We proposed that friction is due to two main factors [2]:-
  (i) the adhesion of one surface to the other
  (ii) the ploughing of the softer surface by asperities on the harder
  Hence F = Fad + Fdef.   (3)

In broad terms a whole range of frictional phenomena could be explained including surface damage, transfer of one metal to the other and certain types of wear. There were of course complications. Harder surfaces showed no normal adhesion although transfer and plastic shearing occurred during sliding. Explanations were given particular by K L Johnson 20 years later involving released elastic stresses and surface forces [4]. And the general validity of equation [3] was confirmed by Oxley in a whole series of 2D experiments in which wedge-shaped sliders transversed softer metals [5]. Using slip-line field models he showed that for moderate interfacial adhesion one could indeed add the shearing and the deformation terms to give the total friction.

3. Early experiments on adhesion and asperity contact

Many experiments were carried out to study the mechanism of adhesion. Forty years or more ago experiments were carried out in high vacuum (Bowden and Young) and they showed that most clean metals stick very strongly and that a monolayer of contaminant greatly reduces the adhesion [2]. These results have since been confirmed and extended with far more sophisticated apparatus by Buckley [6] and others. This work also confirmed the role of released elastic stresses in weakening the interfacial junctions. In addition it emphasised the importance of ductility in achieving strong normal adhesion. I may add that the role of ductility has, on the whole, been neglected in the general study of friction.

Suppose a single asperity is loaded on to a clean surface so that plastic flow occurs. If a tangential force is applied in an attempt to produce sliding, the plasticity conditions in the asperity are exceeded, because of the additional shear stress applied, and further plastic flow will occur. The area of contact and the adhesion both increase (Figure 1). This effect which we call junction growth was first described by Courtney-Pratt and Eisner 35 years ago [2,3]. The growth is limited by the ductility of the metal and by the presence of contaminant films. Adhesion is clearly reduced if the surfaces are dirty and if the metals are lacking in ductility. The latter is one reason why very little adhesion is observed between ceramics even if they are clean.

Another line of research involved the contact between a flat specimen and a very fine point to simulate more realistically the behaviour of a single asperity. Experiments along these lines were carried out over 20 years ago by Gane & Skinner [7] in one of the first scanning electron microscopes, and by Andarelli, Maugis and Courtel [8], using transmission electron microscopy. The deformation of the flat could be directly observed and constitutes a very early example of microindentation. This line was taken up 10 years later by Pethica and Pashley working with surfaces in ultrahigh vacuum and with such delicate controls that they were able to observe the effect of surface forces in producing plastic flow at the contact zone [9]. It is significant that Pethica has become an expert in nano indentation studies, Pashley in tunnelling microscopy. These observations are in the nature of a diversion. I revert to the general idea that so long as the deformation processes involve plasticity there is no problem in explaining energy dissipation.

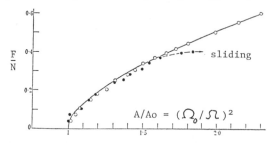

Figure 1. "Concentrated" contact between a smooth hemisphere and a flat of platinum in air. The normal load N produces plastic deformation at the contact region. Electrical resistance (Ω) is used to estimate the area of contact (A) assuming a single contact region and metallic contact. Ao is the initial area of contact. As the tangential force F is applied further plastic deformation occurs and there is a steady increase in A before sliding occurs. The figure plots F/N against A/Ao where the latter is taken as equal to $(\Omega o/\Omega)^2$. With clean surfaces junction growth is sustained and limited only by ductility. With contaminated surfaces there is an earlier limit to junction growth and sliding occurs when F/N (in this experiment) exceeds about 0.4.

4. The mechanism of plastic deformation

When a metal is deformed plastically slip occurs across various planes in the material and energy is almost totally converted into heat. It is for this reason as Philip Bowden pointed out many years ago that Joule observed a very close correlation between frictional work and heat generated. Apparently very little energy is expended in stored strain energy. This raises an interesting question. There are some "memory" alloys which undergo major phase transformations under shear and a large amount of stored energy is involved. Does the Joulian equivalent of friction and heat no longer apply?

How is energy dissipated by slip over crystallographic planes? A simple model is shown in Figure 2. In (a) the atoms in one plane are in equilibrium with those in the neighbouring plane. In (b) a shear has been applied to the top part: the attractive forces between the atoms deforms the lattice elastically. If the shear is increased an instability is reached where the atoms in both planes flick back to a new equilibrium position (c) and vibrate until all the displacement energy has been dissipated.

If there are dislocations there is again a distortion, an instability and vibration as the dislocation moves along by one atomic spacing. Displacement, instability and a flickingback to another equilibrium position involving vibrations in the lattice are the key elements to energy dissipation.

At this point it may be relevant to point out that for a typical single crystal the shear stress τ required to produce slip over the slip plane in the absence of dislocations is of order

$$\tau = G/30 \qquad\qquad (4)$$

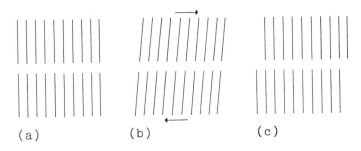

Figure 2. Slip over a crystallographic plane. In (a) the atoms are in equilibrium. In (b) a shear has been applied to the top part: the lattice deforms elastically. If the shear is further increased an instability is reached and the atoms flick back to a new equilibrium position and vibrate until all the strain energy has been dissipated. Slip has occurred by one atomic spacing and all the elastic energy has been converted to heat [10].

where G is the shear modulus of the metal. If a dislocation is present the shear stress to cause it to move through the slip plane is a thousand times less.

If we know the bulk deformation properties of a solid i.e. its stress - strain curve it is in principle possible to estimate the amount of plastic energy expended in deforming it by a harder slider (or an asperity). Since plasticity involves a shear stress we describe the macroscopic properties in terms of shear-stress τ vs shear strain γ. Figure 3a shows a typical curve and we consider a two dimensional system in which a bar of uniform cross section is subjected to shear. We draw a square element in the bar and study its change in shape.
 At O it is undeformed. As the stress is applied the element first deforms elastically along OA. At A plastic deformation occurs and the material work hardens. At B the stress is removed and a slight amount of elastic recovery occurs, BC. (We ignore the Bauschinger effect). The shape of the element at C is shown. The change in shape of the element is a measure of the plastic work done. Quantitatively the plastic work per unit volume of material is given by the area under the curve OABC. In the 2 D deformation of one surface by say a hard cylindrical slider it should be possible to study the shape change of every element under the slider and so estimate the integrated plastic work. There is a danger here. If the elements have undergone complex stress-strain cycles, it is quite misleading simply to compare the shape of each element before and after the traversal of the slider. A simple example is shown in Figure 3(b). A square element A may be sheared to shape B : it may then be subjected to shear in the opposite direction to bring it back almost to its original shape C. Comparing A and C suggests negligible plastic work whereas in fact work has been done in shearing from A to B and from B to C. In the specific example of a hard cylinder passing over a softer material in the complete absence of interfacial adhesion a rough picture of the deformation of a square element is shown in Figure 4. It is seen that as passage takes place the element A is first sheared in one direction (B), then compressed i.e. the shear is now turned through 45° (C), then sheared in the opposite direction (D) and finally released at E with a very small amount of change in the shape of the element*. Plastic work is expended in all these parts of the deformation process and this must be accounted for in any valid calculation of the plastic work expended. Of course if adhesion occurs the deformation is far more severe and more complicated.

——————————————————————————————————————

* A similar problem arises in the deformation losses produced in the deformation of a viscoelastic solid. Similar processes also occur in rolling (see Appendix).

8

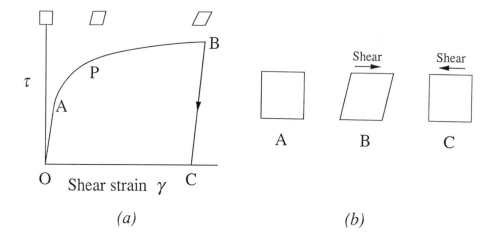

(a)   *(b)*

Figure 3(a). Shear stress τ versus shear strain γ for a typical elastic-plastic solid. OA represents elastic deformation, PB plastic deformation involving work hardening. The area under OAPBC represents the plastic work performed per unit volume of solid. For a 2D specimen, an initial square (above O) is deformed to a parallelogram (above B). The change in shape provides a measure of the plastic work expended. (3b) A square element (A) is sheared to shape B and then sheared back again to shape A at C. Although A and C have virtually the same shape, plastic work has been expended in the two processes A to B and B to C.

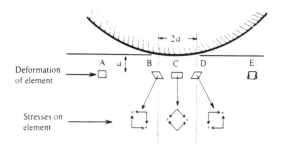

Figure 4. A hard cylinder traverses a softer plastic solid in the **absence** of interfacial adhesion. The deformation below the surface of a typical element is shown. The element is undeformed at A. As it enters the contact zone it is first sheared (B) then compressed (C) resheared in the opposite sense (D) and finally emerges at E. The element has been subjected to three complete stain cycles involving considerable plastic work although its final change in shape is quite small. This indicates how misleading it may be to assess plastic work simply from the difference between the initial and final shape of an element.

If strong adhesion occurs the material at the adhered region must be sheared if sliding is to take place. This again involves plastic work. Evidently both the deformation and adhesive terms are dissipative processes. There is no energy problem here.

In the Preface to the 1986 paperback edition of our Monograph on the Friction and Lubrication of Solids [2] originally published in 1950 and 1954 I wrote that in our early work "the experiments and the interpretation tended to overemphasise the role of plastic deformation" (page vii). In many practical situations, particularly if the surfaces are contaminated, a "shake-down" situation may be achieved where very little plastic deformation occurs and where the ploughing term becomes negligible. The friction is then mainly due to interfacial effects. As I wrote on page (viii) of the Preface "the detailed behaviour of systems in which plastic deformation is unimportant raises many problems still unresolved concerning the nature of the interfacial shear process".

In what follows I would like to discuss this problem in relation to metals, ceramics and polymers. First I will digress and talk about the shear of oriented boundary films.

5. Mica surfaces, surface forces, boundary lubrication

In attempting to come to grips with the problem of the contact between rough surfaces Prof Bowden had the bright idea of getting rid of the asperities altogether. The cleavage surface of mica is molecularly smooth and areas of several sq. cm. can be prepared without a single cleavage step. The contact between such surfaces thus gives a molecular area of contact identical with the geometric area. There was a further trick. If both faces of the mica sheet are molecularly smooth the sheet is of uniform thickness. One face may be silvered to make it highly reflecting. If the unsilvered face is placed in contact with a similarly prepared specimen the contact may be studied using multiple beam optical interference. The interference occurs between the silvered faces so that there is no difficulty in measuring the separation between the mica even though it may be as little as 2 or 3 Å. A typical arrangement is that of crossed cylindrical sheets shown in Figure 5a: the contact zone and the shape of the surfaces can be precisely determined. These and many other experiments with mica by Courtney-Pratt and Anita Bailey almost 40 years ago [2,3], provided information about surface forces, surface energies and surface films. It also consolidated the interference technique in the laboratory. The main difficulty with mica sheets is that they are very thin and floppy and there is a limitation to the experiments that can be done with them.

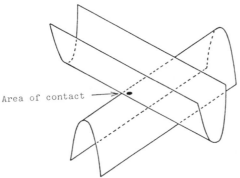

Figure 5a. Contact between molecularly smooth crossed sheets of mica silvered on the back face. Optical interference provides precise data of area of molecular contact.

There was a time gap of nearly 15 years when a new approach, which resolved this problem, was provided by Richard Winterton who found that it is possible to glue the mica sheets on to the surface of a glass cylinder without producing wrinkles in the mica. He was thus able to produce mica surfaces (silvered on the back face) which are molecularly smooth and relatively rigid (Figure 5b).

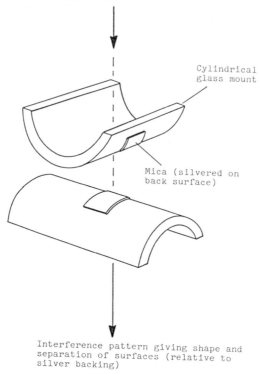

Figure 5b. Mica sheets glued to glass cylindrical mounts. The surfaces are molecularly smooth, of well defined curvature and relatively rigid. Multiple beam interferometry provides the shape of the gap between the surfaces and their separation to an accuracy of 2 angstroms.

With these specimens he studied the van der Waals forces between two such mica surfaces as a function of distance. The technique was greatly improved and extended by Jacob Israelachvili [12] and their combined results are shown in Figure 6. This activity has led to a whole new area of research in surface forces, in particular, the influence of liquid and surface films. This provides a fundamental approach to the study of colloids and colloid stability.

I mention here one type of experiment relevant to the study of friction first undertaken by Jacob Israelachvili [13]. The mica surfaces may be covered with one or several monolayers of fatty acid, or soap, from a Langmuir trough, the CH_3 groups being exposed to the air. It is then possible to slide one such surface over a similarly coated surface as indicated in Figures 7a and 7b.

These beautiful experiments provide a means of measuring (i) the thickness of the fatty acid sandwich (ii) the area of molecular contact (iii) the shear force to produce sliding. From these data the effective shear strength τ of the interface can be determined.

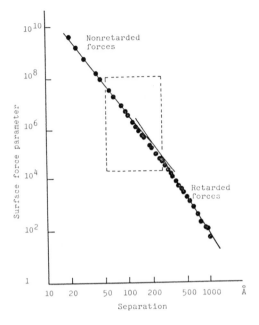

Figure 6   Force between crossed cylinders of mica in air as function of separation as determined by Israelachvili [12].   For separations less than about 200 Å the forces are normal van der Waals forces: for large separations, retarded.   The law of force is different in these two regimes. Data in the dotted rectangle are from Winterton [11].

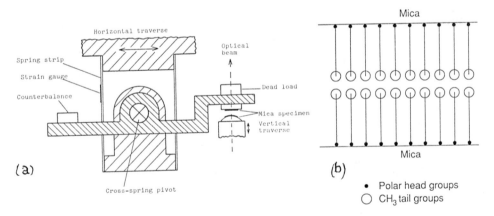

Figure 7(a) Arrangement for studying the shear strength of monolayers of stearic acid. Length of beam about 8cm.   (b) Structure of monolayers deposited on the cylindrically shaped mica surfaces.   The polar head groups are adsorbed on the mica: shear occurs between the CH3 tail groups.

These experiments and later ones by Briscoe and Evans [14] show that the shear strength increases with the pressure p on the sandwich according to a relation of the form

$$\tau = \tau_0 + \alpha p \tag{5}$$

where α is about 0.04 (see Figure 8).

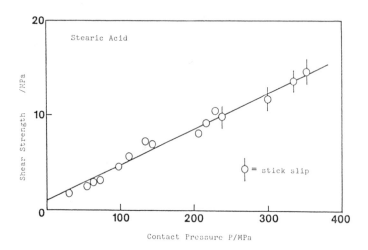

Contact Pressure P/MPa

Figure 8. Interfacial shear strength τ of monolayers shown in Figure 7b, as function of contact pressure p. The results show that $\tau = \tau_0 + \alpha p$ where $\tau_0 \approx 0.6$ MPa and $\alpha \approx 0.04$.

From the point of view of our present discussion, however, there was a far more important observation: the films can retain their integrity, that is, sliding occurs between the CH_3 groups of each surface and in some cases there is no evidence of any damage to the films. How then is frictional energy dissipated. It is tempting to believe that the layers are distorted elastically in a manner resembling the model in Figure 2 and that the energy is lost by elastic vibrations.

The same comment may be true of the beautiful experiments recently described by Meyer and Frommer using the atomic force microscope [15]. In this work the displacement of a single atom in the surface of one specimen by the tangential movement of a probe a few angstroms away could be directly observed. If only one could incorporate an ultra high speed camera with atomic resolution it might be possible to observe the vibration of the disturbed atom as it returns to an equilibrium position! The absence of permanent damage to the surface, although a finite frictional force is involved, may be likened to passing a finger nail over the teeth of a comb.

6. Elastic deformation of asperities

Amongst the many seminal ideas that Jack Archard [16] introduced into Tribology was his model of multi-asperity contacts in which he assumed that each asperity had a spherical shape. Assuming **elastic** deformation each asperity would give an area of contact proportional to $\omega^{2/3}$ where ω is the load born by the asperity. If however the asperities are distributed in height above some datum line the total area of contract of all the asperities is very nearly proportional to the total load $W = \Sigma\omega$. This model was elegantly extended by Greenwood and Williamson to cover the situation where some of the asperities may exceed their elastic limit and yield plastically. A plasticity index was derived. Here again it was shown that the total area of contact is very nearly proportional to the total applied load. We thus have two very good models which show that we do not need to assume that all the asperities undergo plastic deformation in order to give linear proportionality between load and real area of contact. We have already dealt with the plastic case. What is the situation if all the asperities deform elastically?

The most direct study of this for metals was carried out by Pashley and Pethica in Cambridge [9], Hubert Montagu–Pollock in Lancaster [17] and Courtel's group in Paris [8]. In the Cambridge work a very fine point could be brought into contact with the smooth face of a single crystal of the same metal. The experiment was carried out in an ultrahigh vacuum and the state of the surfaces could be characterised using Auger spectroscopy. The loads were of order μN. The electrical conductance could be measured accurately and, assuming a single region of contact, this provides a measure of the area of contact. The results show that as the load is applied the area of contact increases and the behaviour suggests elastic deformation. When, however, the load is reduced the area does not decrease in a reversible manner; the area of contact remains constant until a negative load is applied (Figure 9).

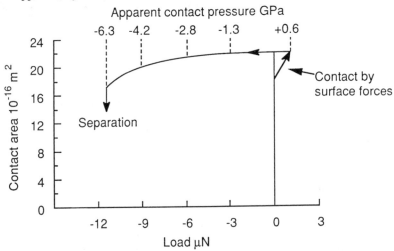

Figure 9. Contact between a clean tungsten tip (R = 1μm) and a clean nickel single crystal in U.H.V. The area of contact is deduced from the electrical contact resistance. Initial load less than 1 μN. Under tension the area of contact remains virtually unchanged until ductile failure occurs.

Clearly strong metal-metal bonds have been formed. The pull off force divided by the area of contact corresponds roughly to the tensile strength of the metal. Thus elastic contact leads to a condition where the interface must be deformed plastically if the surfaces are to be separated. We are back to plastic deformation. (There is also some evidence that surface forces may contribute to plastic deformation as the surfaces are first brought together [17]).

With clean metals it seems impossible to avoid plastic effects. The only way elastic deformation can be dominant is if the surfaces are coated with a protective film such as oxide, lubricant or polymer. The major frictional process is then largely confined to the surface film itself.

7. Oxide coated metals

In some situations oxide films can protect the substrate and prevent metal-metal contact. In that case even in the absence of lubricants, the surfaces may endure repeated traversals without seizure or metallic transfer. The substrate may, after initial plastic deformation, achieve a condition of plastic shake-down and subsequent deformation of the substrate may be primarily elastic. The friction is then determined by the interaction between the oxide and the other surface. The tangential traction will depend on the strength of the bond i.e. of the surface forces and on the physical properties of the oxide. The deformation of the oxide may involve plastic yielding or it may behave elastically.

Some oxides, as Bowden and Whitehead showed many years ago [2,3], are remarkably ductile - can undergo appreciable deformation without exposing the underlying metal. The oxide then absorbs most of the energy in plastic flow. With repeated traversals it may also reach some type of shakedown condition though it will also work harden so that the stress during sliding may reach some steady state. The whole process will be much gentler if adhesion between oxide and counter surface is small. The oxide may then have a very prolonged life even if further oxide growth does not take place. I believe this occurs with some oxide covered metals particularly in the presence of contaminant films. If the adhesion is strong, failure occurs below the contact interface: oxide is removed during sliding and the life will be shortened. There is also the possibility of fatigue of the oxide.

If the oxide deforms elastically the deformation will resemble the deformation and recovery of the surface as in Figure 2. The energy dissipation may be expected to be small. But such oxides will invariably be brittle: this raises very complicated consequences which we shall not discuss further.

8. Real elastic contact

Consider the contact between a strip of clean rubber and a clean hard surface (I exclude rubber-rubber contact since chain diffusion across the interface introduces too many complications). The adhesion of the system has been studied by Kendall in U.K. [18], by Maugis in France [19], by Savkoor in Holland [20], and by Gent in U.S.A.[21]. The adhesion can be explained in terms of surface energy or in terms of van der Waals forces. To peel the surfaces apart these forces must be overcome. In the abstract world of thermodynamic reversibility this process is perfectly reversible. The only work done is in creating free surfaces which have a higher potential energy than atoms in the bulk. In such an ideal system all elastic strain energy produced in the bodies as the surfaces are separated will be conserved. This will be discussed in the appendix.

To model a single asperity consider a rubber sphere in contact with a clean flat. The adhesion in this system has been studied by Johnson, Kendall and Roberts [22], by Maugis [19] and by Savkoor [20]. It is well understood in terms of surface energy and the enhancement of that energy by hysteresis within the rubber. The **sliding** condition is complex and waves of detachment are generated [Schallamach, 23]. Apparently the rear portion of the contact zone peels away but the mode of sliding of the rest of the contact is difficult to explain: it will be dealt with by Dr Savkoor.

9. Polymers

With polymers there are two extreme types of frictional behaviour and many intermediate types. If a PTFE hemisphere is slid over a clean glass surface a thin film of polymer is drawn out of the slider and remains weakly attached by van der Waals forces to the glass [24]. The frictional work is expended primarily in extracting the polymer chains out of the bulk. One may regard this as molecular friction within the polymer: it must involve distortion of the chains and their neighbours and probably resembles the shear processes observed in the irreversible shear deformation of other solids.

The second type is exemplified by UHDPE. If a slider of this polymer is slid over a clean glass surface there is no detectable transfer and consequently the wear rate is phenomenally low. The friction is not particularly low ($\mu \approx 0.2$) and the interfacial shear strength τ i.e. the frictional force divided by the area of contact is comparable with, though smaller than, the bulk shear strength of the polymer. Incidentally both these shear strengths are of order

$$\tau = (\frac{1}{30} - \frac{1}{50}) \, G \tag{6}$$

where G is the elastic shear modulus of the polymer. The relation between τ and G is representative of solids in which there are no free dislocations (though with crystalline solids dislocations may be formed by the application of the stress itself).

We may also note that for many polymers G is a function of the pressure p and can be expressed as

$$G = Go + \alpha p \tag{7}$$

The connection between τ and pressure is thus clearly shown. When sliding occurs we assume that the polymer is attached to the glass surface by surface forces and that shearing occurs at the interface in a manner resembling the examples quoted above. The energy is dissipated as vibrations as individual atomic contacts are displaced to an unstable position and then flick back to a new equilibrium position. If the sliding is smooth this must imply that as bonds are snapped at the rear of the contact region new bonds are formed at the front. In some ways this resembles the passage of a dislocation through the interface. Alternatively all the bonds of a particular asperity contact are broken and this is replaced by new asperity contacts.
We may note that if a soap is used as a lubricant the interfacial shear strength is essentially the same whether the molecules are oriented as in Figure 1 or completely disordered. The shear strength is comparable with that observed in the friction of polyethylene. We conclude that the detailed structure does not appear to be critical. The only common feature is that all these materials have approximately the same surface energy, 25 to 40 mJ m^{-2}. For this reason I once suggested a simple model which connected the shear strength with the surface energy and the mean spacing between atoms in the chain [25]. I am not so pleased with this model now as I was at the time mainly because it gives no indication of the mode of energy dissipation.

10. Brittle solids, ceramics

K.L.Johnson has proposed a very interesting model of the sliding of hard elastic solids. Each asperity contact is assumed to fail by a type II fracture mode [26]. On this view sliding constitutes the propagation of a crack through the interface. The energy dissipated can be explained by fracture mechanics. However there is one difficulty. In type II fracture the crack is not open since it still bears a normal load which tends to press the crack surfaces together. Does this mean that there is friction acting on the crack faces? i.e. are we back to square one? One possible way out is provided by a suggestion put forward by Gittus in 1974 [27]. He suggested that a dislocation travels through the interface so that the load is always borne by the asperity over the whole of its area of contact except where the dislocation is located. This will be discussed in greater detail by K.L.Johnson in a later lecture. My own impression is that failure will rarely occur in the interface itself, except in the presence of some contaminant film.

11. Viscous and viscoelastic losses

Frictional losses in liquids are due to their viscosity. There are several approaches to this but one that has been particularly fruitful has been that due to Eyring. It suggests that molecules are continuously vibrating with a frequency of about 10^{12} s^{-1}. If they have enough thermal energy they will be able to break bonds with their neighbours and jump into another position. If the energy required for this is ε, the fraction of molecules with this energy will be

$$\exp(\frac{-\varepsilon}{\kappa T}) \quad \text{or} \quad \exp(\frac{-Q}{RT}) \tag{9}$$

If Q is of order 10kJ/mol this fraction is of order of 1/100 so that swapping will occur at a rate of about 10^{-10} s^{-1}. If a shear stress is applied the mechanical work done by the stress on the molecule reduces the thermal energy required to produce jumping in the shear direction (ε_1) and increases it in the opposite direction (ε_2) as indicated in Figure 10. As a result jumping is more frequent in the shear direction and less frequent in the opposite direction.

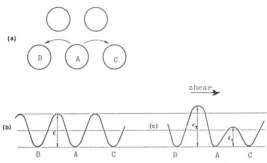

Figure 10. The Eyring model of viscous flow for simple molecules. The activation energy for hopping to a neighbouring site is ε. A shear stress decreases the thermal activation energy in the direction of shear (ε_1) and increases it in the opposite direction (ε_2).

There is thus a net movement of molecules in the shear direction. The full development of this idea leads to an equation for viscosity in terms of temperature, pressure and shear rate [10]. The equation works very well.

There are however two difficulties. First, the pioneer molecular dynamics of Alder and Wainwright thirty years ago [28] showed that the molecules in a simple liquid do not jump from site to site but shuffle around as a collective group gradually changing their locations. Second, it is not clear on the Eyring model how the viscous energy is dissipated. A simple approach is to regard each group or cluster of molecules as constituting a cage that at any instant appears to resemble a solid. Displacing a molecule is not unlike the process observed in the shear of solids. There is a shake-up of the structure and added vibration which degrades as heat.

Viscoelastic losses in rubbers and polymers can often be expressed by Eyring - like equations and many workers describe the behaviour in terms of activated processes. The problem remains of identifying the various processes in physical terms. Alternatively the behaviour may be associated with the way in which the shear modulus G depends on temperature, pressure and rate of deformation. Whichever approach is adopted it is clear that the energy loss is ultimately linked with vibrations of constituent parts of the molecule: these appear as heat.

12. Conclusion

The broad conclusion of this paper is very simple. It takes up the ideas of plastic shear in a crystalline solid where it is known that the atoms in the shear plane are displaced from their equilibrium position until they reach an unstable configuration: at this point they flick back to another equilibrium position. The strain energy is lost in the form of vibrations and these in turn are degraded into heat.

These ideas are certainly applicable to frictional processes involving plastic deformation. It is suggested that a similar mechanism applies, in many cases, to elastic or near elastic conditions particularly if sliding occurs truly in the contact interface. Thus it would explain the friction observed when oriented fatty acid molecules on one solid substrate slide over a similar surface without producing any permanent damage. It is probable that this applies to the AFM studies of riders sliding over a surface without damaging it.

In this paper the idea is applied also to the sliding of oxide-coated metals, to polymers and tentatively to ceramics. In these materials the distortion of the surface layers depends on the strength of the bond between the surfaces, that is, on surface forces or it may be expressed in terms of surface energies. The shear distortion is dissipated as vibrations.

In those cases where the friction is speed and/or temperature dependent the behaviour may be understood in terms of the effect of these variables on the elastic constants of the bodies. This is equivalent to the stress-aided rate theories developed by Eyring and his school: but gives a slightly different slant on the mode of energy dissipation.

The basic idea expressed here is not original; it is to be found in a slightly different form in a 1929 paper by Tomlinson [29] where, for some reason, he does not emphasise the important intermediate role of atomic vibrations.

As a final remark I wish to express my thanks to a referee of this paper who brought to my attention two important developments of Tomlinson's work. One is by John Skinner in an unpublished internal report issued in 1974 [30] which in many ways is well ahead of its time. The other is a more recent powerful analytical study by Gary McClelland [31] published in 1989: it shows how Tomlinson-type dissipation for body A sliding on body B depends on intermolecular forces and lattice spacings of the two bodies. In particular it shows that for incommensurate spacings and weak interfacial forces the displacement of atoms at the interface is slow and reversible and, in principle, there is no energy dissipation. For other systems the displacements are irreversible and energy is dissipated. Although these papers carry some aspects of the subject further than my own contribution I have left my paper in the form in which it was originally delivered "warts and all".

13. APPENDIX

Energy balance in formation of free surfaces: surface energy

We consider here the energy balance where a new pair of surfaces is formed by peeling i.e. by normal separation. The situation is very different from that which applies when surfaces are separated by shear, as in sliding.

Consider first the force-displacement curve for parallel surfaces as in Figure A1. The area under the curve corresponds to the work of forming two new unit areas of surface so that it equals 2γ. Now consider an idealised crystal with uniformly spaced crystal planes as shown in Figure A2. We may regard this crystal as two bodies I and II in atomic contact. We now specify that XY is marginally weaker than all the other interplanes so that the bodies will ultimately separate along XY. As the applied force is increased all the planes are pulled further apart. When the maximum force is reached, corresponding to point A on Figure X, the maximum dilatations have occurred. As the separation of the bodies along XY increases, the forces gradually decrease (portion ABC of Figure X). The planes in the crystal gradually recover their original separation. Thus no elastic energy is dissipated in the solids as the new surfaces on each side of XY are created (but see also page 580).

18

(This is in marked contrast to the shear process described in Figure 2 where instabilities are unavoidable and are bound to lead to energy loss). The only work done, 2γ, is that involved in creating new surfaces: the atoms in the surface have a higher energy than the atoms in the bulk.

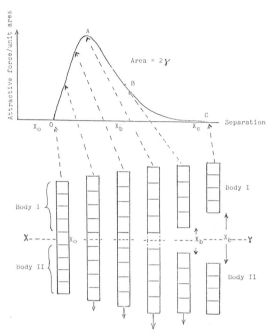

We note that if the bodies I and II are separated faster than the speed at which the solid can relax (this equals the velocity of longitudinal elastic waves), some of the elastic energy in the solids will appear as waves travelling to and fro in the bodies. If the solid is considered to be the ideal structureless continuum sometimes assumed by elasticians (could such a material have surface energy?!) these waves will persist indefinitely. But because the bodies consist of atoms with their own thermal movements, these vibrations will scatter the elastic waves and the latter will gradually merge with the thermal motion of the atoms. They will in fact be degraded into heat. For many crystalline solids the velocity of longitudinal waves is of order 1000 ms^{-1} but for soft rubbers it is of order 20 m s^{-1}. If the interface between the soft rubber specimens is separated more rapidly than this some energy will be expended and lost as elastic waves. Consequently the work done will be more than 2γ per unit area. This might be relevant to rolling friction on an ideally elastic rubber. At low rolling speeds the surface energy expended on the peeling side will be exactly equal to the surface energy gained on the joining side, and the rolling friction will be zero. However at very high rolling speeds the work expended on the peeling side will be greater than 2γ while on the joining side part of the surface energy gain (2γ) will be lost as compressional elastic waves in the rubber. Consequently there will be a finite resistance to rolling.

Of course if rolling is carried out on a rubber which shows appreciable elastic hysteresis the rolling friction will be dominated by the hysteretic losses even at low rolling speeds [32].

REFERENCES

1.Dowson, Duncan (1979) History of Tribology, Longmans, London.

2.Bowden, F P and Tabor, D (1950) Friction and Lubrication of Solids Part I, Oxford University Press; Revised edition (1954); Paperback edition, (1986).

3.Bowden, F P and Tabor, D (1964) Friction and Lubrication of Solids, Part II, Oxford University Press.

4.Johnson, K L (1976) "Adhesion at the contact of solids", Theoretical and Applied Mechanics, Proceedings 4th IUTAM Congress, Koiter,W T (ed.) Amsterdam; North Holland, pp.133-143.

5.Challen, J M and Oxley, P L B (1979) "Different regimes of friction and wear using asperity deformation models", *Wear*, **53**, 229-241.

6.Buckley, Donald H (1981) Surface Effects in Adhesion, Friction, Wear and Lubrication, Elsevier, Amsterdam.

7.Gane, N and Skinner, J (1973) "The friction and scratch deformation of metals on a micro-scale", *Wear*, **24**, 381-384.

8.Maugis, D, Desalos-Andarelli, G, Heurtel, A and Courtel, R (1978) "Adhesion and friction of Al thin foils related to observed dislocation density", ASLE Transactions, **21**, 1-19.

9.Pashley, M D, Pethica, J B and Tabor, D (1984) "Adhesion and micromechanical properties of metal surfaces", *Wear*, **100**, 7-31.

10.Tabor, David (1991) Gases, Liquids and Solids, and Other States of Matter, 3rd edition, Cambridge University Press.

11.Tabor, D and Winterton, R (1969) "The direct measurement of normal and retarded van der Waals forces", *Proceedings of the Royal Society, London*, **A312**, 435-450.

12.Israelachvili, J N and Tabor, D (1972). "The measurement of van der Waals dispersion forces in the range 1.5 to 130 nm", *Proceedings of the Royal Society, London*, **A331**, 19-38.

13.Israelachvili, J N and Tabor, D (1973) "The shear properties of molecular films", *Wear*, **24**, 386-390.

14.Briscoe, B J and Evans, D C B (1982) "The shear properties of Langmuir-Blodgett layers", *Proceedings of the Royal Society, London*, **A380**, 389-407.

15.Meyer, E and Frommer, J (1991) "Forcing surface issues", *Physics World 4* **(4)**, 46-49.

16.Archard, J F (1959) "Elastic deformation and the laws of friction", *Proceedings of the Royal Society, London*, **A243**, 190-205.

17.Pollock, H M and Chowdhury, S K R (1982) "A study of metallic deformation,

adhesion and friction at the thousand-angstrom level", in
Microscopic Aspects of Adhesion and Lubrication, J M Georges (ed.),
Elsevier, Amsterdam, pp253-262.

18.Kendall, K (1971) "The adhesion and surface energy of elastic solids",
J. Phys. D: Appl. Phys. **4**, 1186-1193.

19.Maugis, D and Barquins, M (1978) "Fracture mechanics and the
adherence of viscoelastic bodies", *J. Phys. D: Appl. Phys.* **11**,
1989-2023.

20.Savkoor, A R (1982) "The mechanics and physics of adhesion of
elastic solids" in Microscopic Aspects of Adhesion and Lubrication,
J M Georges (ed.) Elsevier, Amsterdam.

21.Gent, A N and Petrich, R P (1969) "Adhesion of viscoelastic materials
to rigid substrates", *Proceedings of the Royal Society, London,* **A310**,
433-438.

22.Johnson, K L, Kendall, K and Roberts, A D (1971) "Surface energy
and the contact of elastic solids", *Proceedings of the Royal Society,
London,* **A324**, 301-313.

23.Schallamach, A (1971) "How does rubber slide?", *Wear,* **17**, 301-312.

24.Pooley, C M and Tabor, D (1972) "Friction and molecular structure: the
behaviour of some thermoplastics", *Proceedings of the Royal
Society, London,* **A329**, 251-274.

25.Tabor, D (1982) "The role of surface and intermolecular forces in
thin film lubrication", in Microscopic Aspects of Adhesion and
Lubrication, J M Georges (ed.), Elsevier, Amsterdam, pp 651-679.

26.Johnson, K L "Aspects of Friction", paper delivered at Institute of
Physics Conference "Frontiers of Tribology", Stratford-upon-Avon, April,
1991 ; this volume, page **227**.

27.Gittus, J H (1974) "Dislocations in sliding interfaces: interfaceons",
Wear, **30**, 393-394.

28.Alder, B J and Wainwright, T E (1959) "Studies in molecular dynamics
I. Journal of Chemical Physics, **31**, 459-466.

29.Tomlinson, G A (1929) "A molecular theory of friction", *Phil. Mag.*
7, 905-939.

30.Skinner, J (1974) "An atomic model of friction", Central Electricity
Generating Board (U.K.), Report No. RD/B/N3137 Berkeley, England.

31.McClelland, G M (1989) "Friction between weakly interacting smooth
surfaces" in Adhesion and Friction, N J Kreuzev and M Grunze (eds)
Springer, Series in Surface Science, Springer Verlag Berlin, pp. **1-16**.

32.Greenwood, J A, Minshall, J and Tabor, D (1961) "Hysteresis losses
in rolling and sliding friction", *Proceedings of the Royal Society,
London,* **A259**, 480-507.

Discussion *following the lecture by D Tabor on "Friction as a dissipative process":*

J N ISRAELACHVILI. I would like to discuss a point raised at the end of Professor Tabor's lecture when suggested that there is a reversible energy change (surface energy) if surfaces are separated slowly enough. In fact on separating an interface to create two new surfaces, you can never separate them slowly enough - the surfaces will always jump apart. This mechanical instability implies on the atomic scale that the separating atomic bonds go pop, pop, pop and dissipate energy.

H M POLLOCK. Work by John Ferrante on atom-atom interactions which we shall hear about in next week's sessions shows that there is an intrinsic instability between flat parallel surfaces when they are brought together or separated. This effect seems to be quite distinct from the macroscopic jump-to-contact predicted by the JKRS model for a sphere approaching a flat.

D TABOR. Suppose we think of two surfaces held together by van der Waals forces where no individual bonds are involved. Surely as the surfaces begin to separate the forces between them fall off more slowly than the speed of the elastic waves in the solid bodies, so that elastic strains are gradually released and no elastic energy is lost. I accept that in some of the experiments described the surfaces jump together or jump apart. But I am not sure if this is fundamental or an artefact of the experimental procedure.

U LANDMAN. Our simulation experiments show that if two surfaces are moved towards one another, at a separation of a couple of angstroms they will suddenly jump together involving energy loss. When they are in contact it is possible to calculate the stresses below the interface to determine when yielding will occur. I prefer the von Mises approach because it deals with energy, namely the elastic shear strain energy available for producing plastic deformation. I find this more meaningful than the Tresca criterion which involves the difference between principal stresses. If we consider what happens when the surfaces are pulled apart it is not necessary for the interface to separate instantly over its whole area. The surfaces may separate bit by bit and may indeed form a neck.

M J ADAMS. The nature of the separation process between two bodies depends primarily on their ductility. A plastically-deforming contact deforms stably by strain concentration or necking, although often the ultimate mechanism is net section rupture involving void coalescence. Brittle materials fail by crack propagatopn as a result of a stress concentration. Crack stability depends on the geometry; for example it is possible to grow stable cracks in highly brittle materials by employing a chevron notch for which the failure surface area increases with crack growth. In the absence of kinetic energy contributions, the criterion for instability is that the rate of increase in the energy release rate with respect to the fracture surface area must exceed the rate of increase in the crack resistance.

B BRISCOE. I see no difficulty in Professor Tabor's presentation. I think it is correct. In principle there are two ways of doing the experiment. You can impose a displacement and measure the force, or apply a force and measure the displacement. Ultimately you will rely on some spring to sense the force or the displacement. In practise it will be difficult, if not impossible, to find a spring system stiff enough to avoid the instability that some of the speakers have observed.

B R LAWN. The basis of Professor Tabor's model is related to the mechanism of brittle fracture and is discussed in a very useful paper written by J Rice in 1978. In brittle materials the bonds are strongly localised. As fracture occurs, a row of bonds at the crack tip goes pop as the crack progresses. At elevated temperatures, because of thermal motion the bonds may not be so localised, so that the bonds near the crack tip may be spread out. If the activation energy is rather small the fracture process may therefore be gentler: you can run the crack forward and remain in a state of thermal equilibrium. The larger the energy barrier, the slower you need to do this (depending on the temperature).

D MAUGIS. I revert to the question of friction. If the coefficient of friction is defined as the ratio of the frictional force to the applied normal load, we are not allowing for the role of attractive forces between the surfaces. For example if we consider a magnet sliding on a steel surface, the coefficient of friction as defined above will be extremely high (in the limit it could approach infinity). By studying the frictional force as a function of the applied load we could deduce, by extrapolation, the magnitude of the interfacial attractive forces.

K L JOHNSON. I would like to raise two points: first the von Mises yield criterion. This is attractive because of its mathematical symmetry, but plastic flow in metals occurs along crystal planes so that a critical shear stress (the Tresca criterion) is physically more appealing. Secondly, fracture mechanics can be applied to non-brittle materials but only if the plastic zone at the tip of the crack is small compared with the size of the specimen.

D TABOR. This is the first time I have heard an account of Professor Landman's atomic simulations and I hope to learn more about this during the course of the conference. My description of the separation of a body across a single plane by the application of a tensile force was an attempt to persuade myself that all the energy goes into creating new surfaces, and that if it is done very slowly, all the elastic strain energy is restored reversibly and no elastic energy is lost - that is to say, it was an attempt to validate the thermodynamic concept of surface energy. This has proved very successful in explaining peeling. Of course if the system contains spring supports, elastic energy in these parts may be lost. I may be mistaken but it is my impression that the many examples of instability that we have heard may be due to the support system or the geometry [see also page 580].

But the real point of my paper was that there is a very real contrast between adhesion normal to the interface and the shearing of the interface as in an idealised friction experiment. In the latter case, the shear produces distortion of the structure leading to an instability in which atoms flick back from an unstable to a stable position. This involves atomic vibrations: the elastic shear-strain energy is thus dissipated and ultimately the frictional work appears as heat.

B J BRISCOE. Following on a point raised earlier in the discussion I should like to consider in a slightly different way the role of adhesion in friction. We have studied the adhesion and friction between two crossed fibres (20μm diameter) of polyethylene terephthalate. We can measure the normal force required to overcome the van der Waals attraction, that is the adhesion between the fibres. We then observe that even if a negative load is applied (less than the adhesional force - so that the fibres are still in contact) - a finite friction is observed. If we study the frictional force over a range of loads, we find that by adding the adhesive force to the applied load all the results of frictional force versus total load lie on a single curve extending into the full region of negative loads.

H M POLLOCK (written question). My question is in connection with frictional energy dissipation without adhesion or transfer of material. I'd be very interested to hear your comments on how your "combs" model, with its sets of engaging teeth, relates to the slightly larger-scale engagements between asperities that you described in your published work on sliding diamond surfaces; also to the two mechanisms put forward by B Samuels and J Wilks, J. Mater. Sci. 23, 2846-64 (1988) - which they termed (a) a "Ratchet (ride-over)" mechanism, (b) an "elastic push-aside" mechanism.

D TABOR. I am familiar with the paper by Samuels and Wilks. It contains a large body of experimental results and some helpful and thoughtful ideas on the mechanism of diamond-friction in air. Their basic assumption is that interfacial adhesion plays no part, or a negligible part, in the frictional process and they attribute the whole of the friction to asperity interaction. This is rather persuasive, since the principal cleavage plane of diamond is (111) and this implies that the polished surface of diamond always exposes small projections or ledges: indeed the frictional anisotropy observed can be explained in terms of these features. For example the (100) face of diamond is covered with small pyramidal projections about $50\text{\AA}$ in height. The friction in the <110> direction, where one can imagine that sliding occurs between the asperity peaks, is about one half of the friction in the <100> direction where sliding involves riding over the peaks. The anisotropy is particularly marked at higher loads. The authors discuss two models:

(i) a ratchet model which resembles the Coulomb mechanism. However, they suggest that contacting asperities are forced apart elastically at those regions where they bear part of the load. Because the asperities are rough and irregular, the transfer of load to different sets of asperities will be abrupt and irreversible. If 10% of the elastic energy is lost, this could account for the observed friction. In my view elastic recovery is too rapid. Further, a loss of 10% seems very high for diamond.

(ii) "elastic push-aside". It assumes that an asperity on one surface deforms an asperity on the other by pushing it aside. "The contact between two such asperities will generally be lost abruptly, leaving them with stored elastic energy which will cause them to vibrate". Ultimately this energy will be dissipated as heat. The model therefore closely resembles the flicking of the engaging teeth of two combs as they pass over one another. This is an original idea, and can explain many of the features of diamond-diamond friction. There is however one difficulty. The geometry of engaging teeth clearly involves mechanical instability. (That is why the comb model resembles the instability of atoms in the slip plane of metals when they are sheared). With the pyramidal projections on the diamond surface the geometry is quite different. They are relatively shallow, with a base length more than twice their height (unlike the teeth of a comb). Their elastic deformation is very small compared with their dimensions and it seems more likely that their elasticity will enable them to conform to the changing shape of the contact zone as contact is made and released. Elasticians would express this differently: they would say that elastic deformation is quasi-static unless the speed of deformation exceeds that of elastic waves (Rayleigh waves). I therefore regard this model as very attractive but possibly flawed. In addition, as Feng and Field have recently shown (Z Feng and J E Field, paper presented at Institute of Physics, International Conference on Frontiers of Tribology at Stratford-Upon-Avon, UK, April 1991, to be published in J. Phys. D. (1992), it is possible to reduce the friction of diamond-on-diamond in air merely by wetting the surfaces with water. On this view some part of the friction of diamond in air must be due to interfacial adhesion.

There is little doubt that surface asperities play an important part in the frictional behaviour, but our detailed knowledge is still imperfect. For example we are still unable to specify precisely the mode of contact, the deformation of asperities, the scratching of surfaces on a fine scale produced by sliding, the change in behaviour at some critical load which may produce plastic grooving or cracking or both: and finally the fact that in the industrial polishing of diamond, where the surfaces may become almost red hot, the frictional anisotropy is still observed. Some of these issues are discussed in an article by D Tabor and J E Field in "Diamond", edited by J E Field, to be published in 1992.

G M McCLELLAND. Samuels' and Wilks' models are indeed based on the same principle as the so-called independent oscillator model [see p.405], with asperities taking the role of oscillators. As I have emphasized in my lecture, it is essential in such models to treat properly the release of the strain energy stored in intrasolid bonds or in the displacement of the asperities. Is this energy released irreversibly into heat or back into translational motion?

FRICTION FLUCTUATIONS

ERNEST RABINOWICZ
Department of Mechanical Engineering
Massachusetts Institute of Technology
77 Massachusetts Avenue, Room 35-010
Cambridge, MA   02139, USA.

ABSTRACT.   Although the friction coefficient has often been assumed to
be a constant, and indeed tabulated values are widely available, it is
well established that friction values vary widely, being functions
of time of stick, sliding velocity, load, surface configuration, surface
roughness, surface temperature, lubrication, contamination, surface
films, etc.   Many of these parameters vary quasi-randomly with time and
distance slid; the nature of the friction measuring apparatus often
changes the dependences.
    Good explanations for the dependences of friction on other
variables are available in some cases, but not in others.   Generally,
the size and nature of the surface asperities and junctions, at a
scale of about 10 microns, are regarded as especially important factors.

1.   INTRODUCTION

1.1   Early Research

The earliest researchers into friction thought that the friction
coefficient for any pair of materials was a constant, independent of
the normal load and the apparent area of contact, and dependent but
little on the sliding velocity (1-3).   On examining Morin's friction
table of 1833 (4), it becomes clear that Morin assigns friction
coefficient values independently of the above variables and of
roughness, temperature, etc.   Indeed, the fact that friction is
independent of the important variables forms the basis of the well-
known laws of friction.   Modern research has rather extended this
notion of friction independence to the point where one may ask, if
indeed the friction coefficient does not depend on this or that or the
other, what does it depend on?   Clearly, friction depends primarily on
the nature of the sliding surfaces and the lubricants and contaminants
at the interface.

1.2   Recent Findings

I. L. Singer and H. M. Pollock (eds.), Fundamentals of Friction: Macroscopic and Microscopic Processes, 25–34.
© 1992 Kluwer Academic Publishers. Printed in the Netherlands.

26

As soon as high quality friction tests were carried out, and this takes us to the early 20th century, (5), it became recognized that the friction coefficient was not quite constant, and my own way of quantifying these variations is by a "factor of 10" rule. If you change any parameter like load or sliding speed by a factor of 10, then the friction coefficient changes up or down by 10% or less. The variation of friction with surface roughness obeys the same rule (Figure 1, taken from reference 6). Pin on disk geometry.

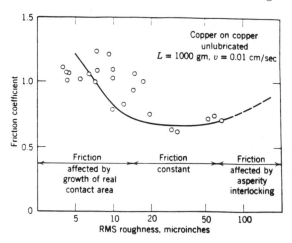

Figure 1. Plot of friction coefficient against surface roughness. For roughnesses in the middle range, from 10 to 100 microinches, the friction is independent of roughness.

This is of course a modest variation. If we have a log-log plot in which y varies by 10% while x varies by a factor of 10, then the slope variation is only 4%, and in fact it is often hard to detect a slope of this modest magnitude unless the x parameter is changed by several orders of magnitude. However, even in those cases where the friction coefficient variations can be detected and measured, this is generally something of no particular importance to anyone.

2. FRICTIONAL OSCILLATIONS

2.1 Variation of friction with velocity.

Although I have stated the general rule that friction variations don't interest anybody there is only one exception, that involving velocity. If we have a sliding system in which the friction goes down when the velocity goes up, we have a situation which can lead to harmonic oscillations, one of the manifestations of the stick-slip phenomenon. It does not require careful experimentation to detect these harmonic oscillations, as the squeaks and squeals are readily noted by the human ear.

2.2 Friction variations and fluctuations.

However, although few workers in tribology have thought that friction variations were worth studying, every one realizes that friction values are subject to great fluctuations, (see Figure 2 taken from reference 7). Note: A variation is a systematic change, while a fluctuation is a random one . Fluctuations are observed when all the variables are kept constant but the system undergoes repeat testing, or else more simply, when we measure the friction continuously while sliding under constant conditions.

The fluctuations are of importance to two kinds of users. On the one hand, it is important for a design engineer to know how extreme are the fluctuations which can be obtained in a sliding system (8). On the other hand, by having a research scientist studying friction fluctuations, he may gain information on the nature of the junctions at the interface of the two sliding bodies, and thus improve our knowledge of the friction phenomenon.

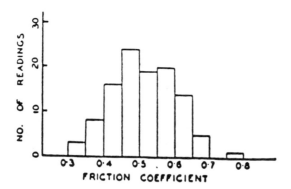

110 separate values—standard deviation is .099.

Figure 2. Hysteresis plot of 110 static friction coefficient values for steel on steel surfaces. The standard deviation is about 20% of the mean friction.

3. FRICTION FLUCTUATIONS AS INDICATORS OF JUNCTION SIZE

3.1 Early studies.

The idea that friction fluctuations could be used to deduce junction size dates back to the early 1950's, as by -product of an enquiry into the transition between static and kinetic friction (9). At that time the nature of the transition was not clear. Was it instantaneous or was it a function of the relevant variables? What were these variables? If, as seemed plausible, it was assumed that a characteristic distance of sliding was required to go from the static to the kinetic friction coefficient (figure 3), what was this distance?

In order to resolve this question, an apparatus was used in which an impacting ball imparted momentum to a block on an inclined plane, and the quantity measured was whether or not gross sliding down the plane was initiated (see figure 4). It was found that whether or not gross sliding was reached was determined by whether or not microslip at the interface produced as a consequence of the impact was or was not large compared to the size of the junctions at the interface. Typical junction sizes were computed to be 10 μm. (i.e., microslip large compared to 10 μm would trigger full scale slippage, while microslip less than 10 μm would not).

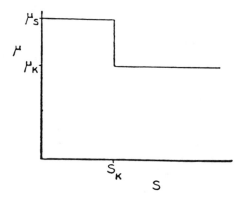

Figure 3. Plot of the friction coefficient as a function of distance slid. The friction will not drop to its smaller, kinetic value unless the distance of microslip exceeds the junction size.

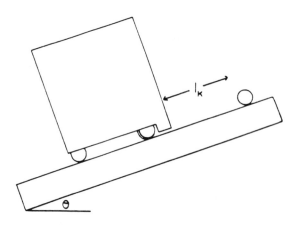

Figure 4. Apparatus for studying microslip. If the distance slid as result of the ball impact exceeds the junction size, then macroslip is initiated.

There has been later work in this area. For example, studies have been carried out using the autocorrelation coefficient values of the friction - displacement plot (10). The assumption is that the auto-correlation becomes zero when the sliding distance becomes great enough so that a new set of junctions is involved. The results indicate that autocorrelation falls to zero when the sliding distance becomes equal to about 10 μm (see figure 5), which is in good agreement with other estimates of the size of the junctions.

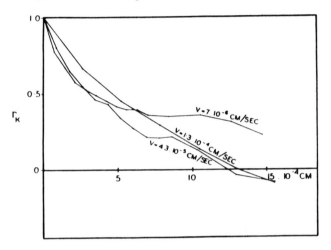

Figure 5. Correlation for three friction-distance traces obtained at various sliding speeds. The correlation reaches zero when the distance of correlation reaches the junction sizes.

3.2 Critical distance effects control onset of stick-slip.

Incidentally, the observation that the occurrence of stick-slip is related to a critical slip distance forms the basis of a method for eliminating stick-slip (11). If a sliding system is made so stiff that the slip distance involved in going from static conditions to full speed sliding to static conditions again is less than the junction size (approximately 10 μm), then stick-slip is suppressed, since one of the conditions associated with the onset of stick-slip (i.e., that the slip distance is greater than the junction size) is not obeyed.

3.3 Use of critical distance effects in super conducting magnets and geophysics.

The concept that the study of variations and fluctuations in the fric-tion coefficient could have importance in various contexts, especially the very mundane one of estimating the onset of slippage in a stick-slip process, has been used and extended by various specialized groups of scholars, in particular in the superconducting magnet and geophysi-cal communities. In the superconducting magnet case, it has been found

that there is a critical distance and if slippage of the magnet
windings during magnet activation is greater than the critical value,
enough energy dissipation occurs to drive the system outside the superconducting range (12). In the geophysical case, if slip during displacement of rocks exceeds a critical distance, then large-scale
relative motion occurs and an earthquake is initiated (13). If the
slip distance is less than the critical distance, then no slippage
occurs. It is perhaps too early to say whether this model after further
improvement can indeed be used to preduct the occurrence of earthquakes.

4. SIGNIFICANCE OF STICK-SLIP

One of the questions which is raised from time has to do with the
significance of stick-slip and other frictional fluctuation phenomena.
What is the 'meaning' of stick-slip, or, putting it another way, how
can we use the knowledge that a certain sliding system does or does
not oscillate to tell us something we did not know before.

4.1 Disagreement between English and Russian schools.

This line of questioning was the basis of a lively disagreement
(around 1940) between the English school of Bowden and associates (14)
and the Russian school of Haykin and associates (15). Bowden argued
that the occurrence of stick-slip indicated that, with prolonged time
of stick, juctions were becoming stronger, and thus stick-slip indicates
junction strengthening. Haykin however, maintained that the presence
of stick-slip indicates merely that the friction-velocity function had
a negative slope, for whatever reason.
    My own vote goes with Haykin in this disagreement, and I can
readily cite examples to support my position. For example, I have
encountered stick-slip during rolling contact with rough surfaces, a
situation in which strong welding of junctions is highly unlikely.
However the up and down motion of the rolling surfaces can provide a
mechanism for a negative friction-velocity curve, since at low speeds
one rolling surface rolls down into the crevices of the other, and must
be dug out, so to speak. At high speeds the two specimens do not penetrate each other to the same extent, so the friction is lower.

5. VARIATION OF FRICTION WITH LOAD

5.1 Dependence of friction on load.

Over the past two or three centruies, there was relatively little
interest in the magnitude of the friction coefficient. If an engine or
motor had a friction coefficient which was too big, then more input
power would have to be applied. Similarly, if a brake or clutch gave
a friction coefficient which was too small, a larger normal load would
have to be provided. Recently a few devices have indeed been produced
which will only operate well if the friction is as low as possible, in

particular battery- powered equipment.

    An interesting finding of the recent few years has been that, for
a wide range of material combinations, as the loads increase, the friction coefficients decrease somewhat (figure 6). This behavior has been
attributed to a transition from elastic to plastic loading, the influence of surface energy, and various other factors. Fortunately, this
effect is quite moderate, within the "factor of ten" rule.

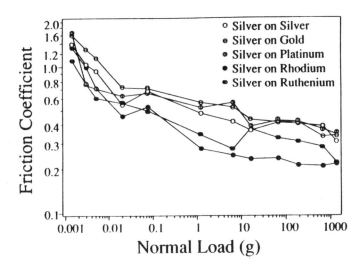

Figure 6.   Static friction of various noble metal pairs.

5.2   Great friction variations at very light loads.

In addition, in the past ten years a couple of specialized friction
problems have arisen, in which the values of the friction coefficient
cannot be readily explained by present theory, but it seems clear that
in the absence of good theories difficult practical problems will surely
arise. One problem relates to the friction of micromechanisms (motors,
gears and other engineering components with weights in the microgram
range). The other relates to the performance of atomic force microscopes, in which sliding surfaces operate at loads in the nanogram
range. The discrepancy of the results obtained by these devices is
perhaps most easily seen if we note that at very light normal loads
micromechanisms tend to stall, indicating infinite friction, (17), while
atomic force microscopes often give friction coefficient values tending
to zero (16). No comprehensive explanation for these discordant friction coefficient values is yet available. In any case, even in the
absence of good theories the practical results are of great interest,
especially the finding that at light loads the friction tends to zero.
This find might eventually be used as the basis for a zero friction
bearing.

32

6. DISCUSSION

6.1 Current status of friction.

In the above summary of our knowledge of friction I have tried to give a flavor of the way friction research has developed. On the one hand friction represents a low level scientific problem which has been studied for some 300 years, and although a lot of empirical information is available, and we are able to use such aspects as friction fluctuations to deduce some reasons for some friction effects, we seem to be well short of a comprehensive friction theory.

6.2 Current problem areas.

Currently, interest in friction is at an all time high, both because of the connection between friction and such phenomena as earthquakes, but also because there have been a large number of modern devices which only perform if the friction can be closely controlled, and here I should mention magnetic recording devices, atomic force microscopes, micromechanisms, nanogram hardness testers.

6.3 Future prospects.

Perhaps what friction research needs now is better integration of the various phenomena known to be involved in the friction process. In particular we must combine our theoretical knowledge of such effects as elastic and plastic deformation, the nature and thickness of surface films, surface energy, surface roughness, and doubtless others.

7. REFERENCES

1. Amontons, G., On the resistance originating in machines, (in French), Mem. Acad. Roy., 206-222, 1699.
2. Coulomb, C. A., The theory of simple machines (in French) Mem. Math. Phys. Acad. Sci., 10, 161-331, 1785.
3. Morin, A., New Friction experiments carried out at Metz in 1831-1833, (in French), Mem. Acad. Sci., 4, 1-128, 591-696, 1833.
4. Rankin, W.J.M., Table 338 - Friction, Smithsonian Physical Tables, 9th Edition, Smithsonian Institution, Washington, 1954.
5. Bowden, F. P. and Tabor, D., Friction and Lubrication of Solids, Clarendon Press, Oxford, 1950.
6. Rabinowicz, E., Friction and Wear of Materials, John Wiley and Sons, NY, Section 4.2, 1965.
7. Rabinowicz, E., The determination of the compatibility of metals through static friction tests, ASLE Trans., 14, 198-205, 1971.
8. Rabinowicz, E., Rightmire, B. G., Tedholm, C. E. and Williams, R. E., The statistical nature of friction, Trans. ASME, 77, 981-4, 1955.

9. Rabinowicz, E., The nature of the static and kinetic coefficients of friction, J. Appl. Phys., 22, 1373-79, 1951.
10. Rabinowicz, E., Autocorrelation analysis of the sliding process, J. Appl. Phys., 27, 131-135, 1956.
11. Rabinowicz, E., The intrinsic variables affecting the stick-slip process, Proc. Phys. Soc. (London) 71, 668-675, 1958.
12. Kensley, R. S. and Iwasa, Y., Frictional properties of metal insulator surfaces at cryogenic temperatures, Cryogenics, 20, 25-36, 1980.
13. Gu, J.-C., Rice, J. R., Ruina, A. L. and Tse, S. T., Slip motion and stability of a single degree of freedom elastic systems with rate and static dependent friction, J. Mech. Phys. solids, 32, 167-196, 1984.
14. Bowden, F. P. and Leben, L., The nature of sliding and the analysis of friction, Proc. Roy. Soc. A, 169, 371-9, 1939.
15. Haykin, S., Lissovsky, L. and Solomonovich, A., On the jerky character of the friction force, J. Phys. USSR, 2, 253-8, 1940.
16. McClelland, G. M. et al., Atomic scale friction measured with an Atomic Force Microscope, pp. 11-12 of 'Engineered Materials for Advanced Friction and Wear Applications', ed. F. A. Smidt and P. J. Blau, ASM International, 1988.
17. Lim, M. G., et al., Polysilicon microstructures to characterize static friction, Proc. 3rd IEEE Workshop on Micro Electron Mechanical Systems, Napa Valley, CA, 1990.

Discussion *following the lecture by E Rabinowicz on "Friction fluctuations":*

P BLAU. I am glad you brought up the effect of time on the friction coefficient. This leads us to the question of running-in or breaking-in. On this question of friction force versus time, there is a great deal of information that could be extracted from running-in data obtained during the past 15 years or more. We have competition between various processes as they approach steady states, allowing us to identify different friction mechanisms which could operate simultaneously, so it would be worthwhile doing more work on the analysis of the initial transitions.

E RABINOWICZ. The area of analysis of transitions is one of the most difficult in the whole of tribology; I take off my hat to Dr Blau who has done this (see his book "Friction and wear transitions of materials", Noyes, 1989), and to anyone else who is able to do this. I quite agree with you that in many applications it is very important to study what happens before equilibrium is established.

H M POLLOCK. (written question). I wonder if you would care to comment on one or two ideas on the subject of stick-slip, that were published many years ago?

1. J N Sampson et al (J. Appl. Phys. 14, 689-700, 1943!) proposed that

"*Time* is required for local adhesive junctions to form, because plastic deformation and rearrangement of surface atoms must first take place. The friction force will not return to its *static* value (higher than the kinetic value) unless the surfaces remain at rest - long enough for new kinetic junctions to be formed"

If so, what implications does this have for the variation of stick-slip amplitude with sliding speed?

2. The old question on how fundamental is the distinction between static friction, and kinetic friction with stick-slip, was tackled by T E Simkins, Lubrication Engineering Journal of ASLE 23 (1967) 26-31, and by W E Campbell and J Aronstein, ASLE Trans. 16 (1973) 223-232. Simkins claimed that Sampson's time-dependency hypothesis is unnecessary and that there is no fundamental distinction between static and kinetic friction: the so-called static friction force at the stick peak is merely a "local maximum". Campbell and Aronstein, however, distinguish between (a) microslip resulting from junction growth, (b) irregular slip during which individual junctions within the contact area are at different stages of contact formation and local failure, and (c) smooth, gross slip.

E RABINOWICZ. Sampson's fine pioneering paper was one of my inspirations in undertaking research in stick-slip. What I got out of it was the following:

1. Low sliding speeds give more severe stick-slip,
2. Weak springs in the friction apparatus give more severe stick-slip.

My own work lead to a third statement:

3. If the distance slid during the sliding part of the stick-slip cycle is less than the junction size, then the tendency to stick-slip is suppressed.

CONTACT MECHANICS, SURFACES AND ADHESION

.... Can an interface offer resistance to a shear stress if adhesion forces are zero? If so, it is incredible, all properties being known (surface energy, elastic properties, loss properties), that a friction coefficient cannot be found by *a priori* calculation .....

D MAUGIS, in "Microscopic aspects of adhesion and lubrication" (J-M Georges, ed), Elsevier 1982, p 221.

Three of the following chapters (by J A Greenwood, H M Pollock and A R Savkoor) treat certain introductory topics which underlie the macroscopic approach to friction. As an essential foundation, a lecture was given on:

AN INTRODUCTION TO CONTACT MECHANICS

by

K L JOHNSON
Department of Engineering
University of Cambridge
Cambridge CB2 1PZ
England

This outlined the mechanics of frictionless normal contact of non-conforming solid surfaces. In particular it dealt with the following topics.

(1.) Geometry of non-conforming surfaces.
(2.) Hertz theory of elastic contact.
(3.) The elastic stress field induced by Hertz contact.
(4.) Initiation of plastic yield.
(5.) Plastic indentation.
(6.) Non-dimensional plot of contact pressure.

Professor Johnson has covered the subject in full, in his book entitled "Contact mechanics", Cambridge University Press, 1985. For the reader's convenience, the main points of his lecture are listed in appendix 1, page 589.

CONTACT OF ROUGH SURFACES

J. A. GREENWOOD
Department of Engineering
University of Cambridge
Trumpington Street
Cambridge CB2 1PZ, U.K.

Introduction

Engineering surfaces are rough. To the finger, there seems to be little difference between a flake of cleared mica or the surface of a ball bearing; but when the bearing is examined by a profilometer – traditionally a hi-fi system but with a chart recorder instead of a loudspeaker: soon perhaps a compact disk player – it looks like Fig. 1. Note immediately what has led to the "Talysurf delusion" named after one of the more popular profilometers: the vertical magnification is typically 100× greater than the horizontal magnification (here 10,000×, 100×) – and the slopes are not really all that steep.

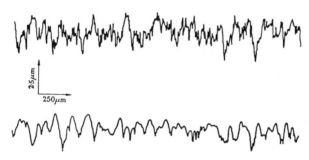

Figure 1. Profiles of mild steel specimen after three surface treatments. (a) Surface ground only. (b) Surface ground and then lightly polished. (c) Surface ground and then lightly abraded on 600 papers.

I. L. Singer and H. M. Pollock (eds.), Fundamentals of Friction: Macroscopic and Microscopic Processes, 37–56.
© 1992 Kluwer Academic Publishers. Printed in the Netherlands.

The profilometer was invented in 1933. The first one used a broken razor blade [1] rather than a diamond stylus, so the technology has certainly improved – but we have no more idea now how to describe surface roughness than we had then. We do, however, know a lot more about the difficulties. The invention by Abbott was immediately followed by a tribological application: if we slice the profile at a series of levels and find the area of the slice, we obtain the Abbott 'bearing-area curve'. The idea was that wear would rapidly remove the high points to leave a series of plateaux, but once the total area of the plateaux was, say, 50% of the nominal area, wear would become slow: so the 'slop' in a bearing, or a piston, would be related to the height difference between the original highest points – or more conveniently, the level for 1% bearing area – and the level for 5% bearing area: the 'peak roughness'. Abbott also offered a 'valley roughness' and a 'medial roughness' – very perceptively recognising the need to describe the form of the height distribution (Fig. 2). But of course he was then working in a university. By 1938 he formed a company to market his "Profilometer", and registered that it would only sell if it produced a <u>single</u> magic number: it took until 1962 for the science to recover.

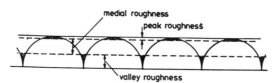

Figure 2a. Abbott's parameters.

Figure 2b. Centre-line-average
$R_a \equiv$ (total shaded area)/length is the same for both surfaces.

For the metrologists and standards committees had a field day – or quarter century. As a quality control tool, the only requirement is that a parameter should be <u>measurable</u>: there is no requirement that it be <u>significant</u>. Is the production process running today as it did yesterday? So we define a datum as a line (or with a round component as a circle) with equal areas above and below; then the cla roughness (centre-line-average: the proper symbol now is R_a) is defined as the average (absolute) deviation from this datum. Unlike Abbott's roughness triplet, it cannot distinguish between the two surfaces in Fig. 2, on only one of which one can slide!

Times have changed, and now no profilometer is complete without its built-in computer giving 10 different surface roughness parameters (but not including Abbott's). It is not clear whether the driving force was the computer salesman or the need to understand surface contact: perhaps both.

What does determine surface contact? Undoubtedly surface roughness: in Bowden's memorable words: [2]

> Putting two solids together is rather like turning Switzerland upside down and standing it on Austria – the area of intimate contact will be small.

Yes: so the area will consist of localised spots, and the total area will be small – but how many spots and how small? For many years tribologists were content with knowing the total area, found by arguing that each contact resembled a Brinell hardness impression, in which a hard steel ball is loaded against the specimen and the contact area measured. The hardness H, the ratio of load/area is approximately constant (and as we heard in the Contact Mechanics talk, approximately 3Y): it follows that the total area of contact A between two solids is simply

$$A = W/H.$$

Amontons' laws of friction follow immediately – and we didn't need to specify the surface roughness.

Having been brought up by Professor Tabor in a Physics laboratory, it took me a long time to bring myself to use the word pressure – particularly since there is a bad habit in the U.S. of using 'pressure' to mean 'load'. In dry contact it really is the <u>load</u> which matters: and whether it is spread over 1 cm^2 or 100 cm^2 usually doesn't matter. If the load is 10 Kg and the hardness 200 (kg/mm^2) then the area of contact is the same: 0.05 mm^2 – and the only difference between the two cases is that the distances between the individual contacts are greater. Despite the objections to A = W/H, it remains extremely valuable for setting the scene: contacts are not quite as sparse as planets in the solar system, but they are small and well-separated.

[In lubricated contacts, of course, pressure comes in to its own; and even in a dry contact, the thermal behaviour depends on the area: engineers are never complete fools.]

But what are the objections to A = W/H? Primarily, it gives a value of A which is too small: so let me show my open-mindedness with a counter-example: who says A = W/H in a hardness indentation? Fig. 3 shows a profilometer trace across a Brinell indentation into bead-blasted aluminium: clearly the area of contact is closer to ½(W/H) than to W/H. One can get the same result using gold, which is said not to have a surface film; or through a lubricant, so it is not a frictional effect: calculation suggests that we are not seeing elastic recovery (contact having been complete when the load was present) and the strain seems too small to produce appreciable work hardening. The best we can do is to attribute it to interacting deformation fields around adjacent asperity contacts: which is convenient because this means that it will not occur in normal tribological contacts.

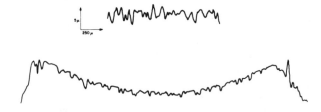

Figure 3. Talysurf profile across a Brinell dent. (Using the
radius attachment to remove most of the curvature.)

A more serious objection to A = W/H comes from the Contact Mechanics
lecture:

Load for first yield $\quad P_y = 21 \; R^2 Y^3 / E^{*2}$ $\qquad$ (Eqn 3.12)

Load for full plasticity $\quad P \approx 8000 \; R^2 Y^3 / E^{*2}$

(Elastic/plastic contact master curve)

There is no doubt that asperity loads often exceed the yield load –
but do they reach the fully plastic value? We can only answer this by
knowing something about asperity radii of curvature, and about the
number of asperities taking part in contact.

In fact Archard [3] raised the fundamental objection to A = W/H. It
is believable if what we do is prepare a surface, slide over it, and
throw it away. But Engineers like to use their machine next day as
well – and preferably next year as well – with the same surface. How
can we have an area of contact determined by plastic, <u>irreversible</u>
deformation, the millionth time we make contact? But how can we obtain
Amontons' laws unless we can show that the area of contact is
proportional to the load?

Elastic contact of nominally flat rough surfaces [4]

Suppose we regard our rough surface as being produced by scattering
asperities on to a plane. A reasonable first approximation is to take
them to have spherical caps of the same radius of curvature R: if they
also have the same height the analysis is simple but gives an unwelcome
answer: for with N asperities each will carry a load $W_1 = W/N$, and so
by the Hertz equations governing contact between a ball – or a
spherical cap – and a plane, each contact will have an area $A_1 = \pi a_1^2$

where

$$a_1{}^3 = \frac{3}{4} \frac{W_1 R}{E^*}$$

(Eqn 2.12)

Then the total area is

$$A = NA_1 = N\pi \left[\frac{3}{4} \frac{WR}{NE^*}\right]^{2/3} = 2.6 \ (WR/E^*)^{2/3} \ N^{1/3}$$

and is <u>not</u> proportional to the load.

But of course our profilometer does not suggest that the asperities have the same height: so assume instead that there are $N\phi(z)dz$ with heights z to z+dz. $\phi(z)$ is going to become a height probability function, but for now we can take it to be the explicit frequency function of the surface considered.

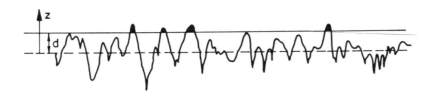

Figure 4. Contact between a rough surface and a rigid plane. All asperities with height z greater than the separation d make contact.

For convenience, suppose we load it against a rigid plane: and let the separation between the rigid plane and the datum plane of the rough surface be d (Fig. 4). All asperities with height greater than d will make contact, and will be compressed a distance $\delta = (z-d)$. From equation §2.13 this will lead to an asperity contact area $\pi a_1{}^2 = \pi R\delta = \pi R(z-d)$, and an asperity load $W_1 = \frac{4}{3} E^* R^{1/2} \delta^{3/2}$: so that

number of contacts $\qquad n = \displaystyle\int_d^\infty N \ \phi(z) \ dz$

total area of contact $A = \displaystyle\int_d^\infty N\pi R(z-d)\phi(z)dz$

$$\text{total load} \qquad W = \int_{}^{\infty} N \cdot \frac{4}{3} E^* R^{\frac{1}{2}} (z-d)^{\frac{3}{2}} \phi(z) dz$$

This tells us what we need to measure – the height distribution (the derivative of the Abbott bearing area curve rather than the curve itself!), the number of asperities, and their mean radius of curvature.

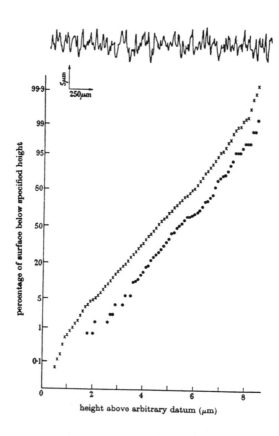

Figure 5.   Cumulative height distribution of bead–blasted aluminium.   Both the distributions of all heights (×) and of peak heights (•) are Gaussian, at least in the range ± 2 standard deviations.

With an analogue–digital converter and a computer connected to a profilometer this is straightforward.   Fig. 5 shows a typical height distribution: a grit–blasted surface with an almost–Gaussian ("normal") height distribution.   Ground surfaces also have Gaussian height distributions to a good approximation.   It does not particularly matter to contact theory whether or not the distribution is Gaussian, though it is convenient to have a specific distribution which we can specify by a single parameter, the standard deviation (R_q) – in other words, the r.m.s. roughness – and perform our integrations once only. The assumption of a single radius of curvature is rather drastic (Fig. 6); but it seems that taking the mean radius is satisfactory.

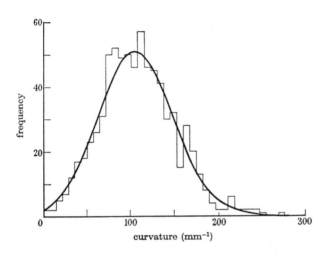

Figure 6.   Histogram of curvatures of 749 peaks on a specimen of bead-blasted gold.   The continuous curve is the Gaussian with the same mean and standard deviation.

In fact, the important feature of the height distribution is that the number of asperities of a given height should fall off rapidly with height.   In particular, if the number falls off exponentially, $\phi(z) = e^{-\lambda d}$, then our three integrals become

$$n = \frac{N}{\lambda} \, e^{-\lambda d}$$

$$A = \frac{N\pi R}{\lambda^2} \cdot e^{-\lambda d}$$

$$W = \frac{NE^* R^{\frac{1}{2}}}{\lambda^{5/2}} \sqrt{\pi} e^{-\lambda d}$$

and the area of contact is directly proportional to the load.

The average size of a micro-contact (A/n) and the average contact pressure W/A both become constants.   This is surprising, since we started with Hertz theory according to which, as the surfaces approach, the size and mean pressure of each contact steadily increase. The answer is, of course, that new contacts are continually forming, initially with zero size and zero pressure: and with this height distribution the growth and formation exactly balance.

The results for a Gaussian height distribution, or indeed any height distribution approximating to an exponential, approximate to this. For a Gaussian at normal nominal pressures, the planes tend to be 2-3 standard deviations apart so the _relevant_ height distribution is near-exponential: the mean pressure increases slowly as the planes draw together, and the average contact size increases sightly; but the general picture remains the same.

Plasticity index

Can we now settle the question about the mode of deformation of the asperities? We need to rewrite the contact mechanics equations, for the independent variable is now the asperity compression δ: we find that $p_0 = (2/\pi)E^*\sqrt{R\delta}$. First yield occurs when $p_0 = 1.6(H/3)$ i.e. when $\delta = 0.8R(H/E^*)^2$.

Usually we raise this slightly on the grounds that the very first yield is unimportant, and take

$$\delta_Y = R(H/E^*)^2 \ .$$

Deciding exactly when we reach full plasticity is harder. The Hertz relation $\delta = a^2/R$ will no longer hold; but it is not clear that the geometrical relation $\delta = a^2/2R$ holds either. However, the difference is not too important. From the contact mechanics master curve we have $E^*a/YR \sim 30$ for full plasticity, so

$$\delta_F = 50 \ R(H/E^*)^2$$

What does this mean? Suppose we have a grit-blasted steel surface with a Gaussian height distribution and $\sigma = 0.4$ μm, loaded so that the mean planes are 2.5 σ apart. Then the proportion of the asperities making contact is

$$\int_{2.5\sigma}^{\infty} \phi(z)dz \qquad \text{i.e.} \qquad \Phi(2.5) \equiv 0.0062$$

How many will be plastic? Suppose $R = 0.4$ mm: $H = 1$ kN/mm^2, $E^* = 100$ kN/mm^2 so that $\delta_Y = 0.04$ μm: then all asperities higher than $2.5\sigma + \delta_Y = 2.6\sigma$ will have yielded: i.e. a fraction $\Phi(2.6) = 0.0047$.

All those higher than $2.5\sigma + \phi_F$ will be fully plastic, i.e. $\Phi(7.5)$ – well, my tables stop at $\Phi(5) = 3 \times 10^{-7}$: shall we just say 'not many'?

It is clear that the proportion of contacts which have yielded depends on the ratio $\delta_Y/\sigma \equiv R/\sigma(H/E^*)^2$. We define the plasticity index ψ as $\sqrt{(\sigma/\delta_Y)}$, i.e.

$$\psi = \frac{E^*}{H} \sqrt{\frac{\sigma}{R}}$$

and it turns out that for an exponential height distribution, the proportion of non-elastic contacts depends only on ψ – it does not depend on the load, for the same reason as before. Of course increasing the load brings each contact nearer to yield, or nearer to full plasticity: but at the same time new elastic contacts are formed, and the proportion does not change. As usual, for the Gaussian this holds approximately: increasing the load leads to a very slight increase in the proportion of non-elastic contacts.

$\psi = 1$ corresponds, rather conveniently, to 1% of the total contact area at yielded contacts, which, given that on this theory we can never avoid yield completely, is a reasonable definition of 'elastic contact'. Correspondingly, $\psi = 7$, or thereabouts, corresponds to 1% of the total contact area being 'fully plastic'.

These are important matters for tribology. Elastic contacts may, ultimately, fail by fatigue; but can be expected to give useful service before they do. Elastic strains are unlikely to crack the surface oxide, the fundamental protective layer of a metal. We really do need to be able to assess the contact state.

Other contact models

An immediate corollary of the model just described applies to the contact of lubricated surfaces: provided we can relate the film thickness which is predicted by smooth-surface lubrication theory to the separation between our reference planes. Christenson's theory of stochastic lubrication [5] supports the idea that, at least at large separations, lubrication theory gives the distance between the mean planes: so we need to add the distance between the mean surface height and the mean asperity height (Fig. 7). Then we throw away the equation for the load, and the theory is complete. As before, the plasticity index tells us how serious the asperity contacts are. The result depends somewhat on the film thickness, which because it is determined hydrodynamically, is rather more variable than for dry contacts (it can vary from $h = 3\sigma$ to ∞ instead of from $h = 2.5\sigma$ to 4σ). But the general conclusion is the same: the number of contacts, and the 'no-contact' time, depend on h/σ; but when $\psi < 1$, even when contacts do occur, they will be elastic and, largely, harmless.

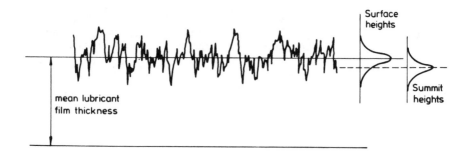

Figure 7. The distinction between the mean plane and the mean
summit height is important in a lubricated contact.

When studying the contact of plane surfaces, the micro-contacts will
be well-separated, and it is reasonable to treat them as independent:
that is, to assume the load at one contact does not affect the
compression at another. But what happens if our nominal geometry is a
Hertzian contact between a ball and a plane [6]? Presumably we still
have asperity contacts, just as with the Brinell indentation of Fig. 3.
We can study such contacts approximately by assuming there are so many
micro-contacts within the Hertzian contact, that there are many even
within an annulus so small that the separation between the two surfaces
can be taken as constant. Our basic model relating load to separation
now gives the local <u>nominal</u> pressure over that annulus, and we can
solve the contact problem treating the roughness as a compliant layer
(with a highly non-linear pressure/compliance relation) separating the
two elastic bodies. The nominal pressure is of course composed of
individual contact loads at the micro-contacts.

The results depend [7] primarily on a parameter $\alpha \equiv \sigma_s R/a^2$ where R is
now the <u>ball</u> radius, and 'a' the Hertzian contact radius for smooth
surfaces: for $\alpha > 1$ the micro-contacts extend over a region larger
than the Hertz area, and the nominal pressures may be much lower than
Hertz: but for $\alpha < 0.1$, although there is always a small 'fringe'
outside the Hertz area, the nominal pressures revert to almost
Hertzian. In both cases the real area of contact is nearly
proportional to the load.

Advances in computing technique now permit this problem to be solved
'exactly'. Webster & Sayles [8] have analysed the 2-D Hertz contact
between a cylinder and a plane, taking the surface of the plane to be a
measured surface profile, with the height recorded at 2 μm intervals.
Fig. 8 shows their results for a load giving a smooth-surface Hertz
width (2b) of 0.96 mm: once again, the overall geometry is close to
Hertz, but they find the real area of contact (with elastic
micro-contacts) to be closely proportional to load.

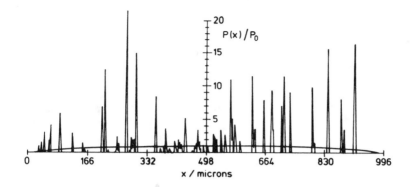

Figure 8. Pressure distribution and contact geometry of a smooth
cylinder of radius 40 mm loaded against a rough surface
($\sigma = 0.276$ µm, $\sigma_m = 0.071$) under a load of 0.5 kN/mm, and with

$E^* = 110$ kN/mm^2. (After Webster & Sayles).

This analysis is related to a very much simpler model of dry contact
of plane surfaces: the 'profilometric' theory [9] – essentially a
modification of Abbott's bearing area curve. We postulate, not that
all material above a given height is worn away, but that it is
"absorbed" by some form of local deformation without affecting the
shape of the surface in the non-contact regions. This is certainly not
true for elastic contacts (Fig. 9a); but for plastic contacts, despite
the conservation of volume in plastic deformation, it is reasonably
accurate – to the extent that there is no agreement on whether we get
'piling-up' or 'sinking-in' around the contact. (Fig. 9b).

Figure 9.
a) Elastic contact: contact area less than the geometrical area 2Rδ.
b) Plastic contact with "piling-up": contact area greater than
geometrical area.

Thus, the Abbott curve gives us immediately the area of contact for a given separation. If we also study the number of 'level crossings' of our profile, we can deduce the number and characteristic size of the contact areas. The problem is to introduce the load: and here we have to rely on the results of our earlier theory: that the contact pressure varies very little with asperity compression, provided we are beyond the initial elastic range when it builds up as $\sqrt{\delta}$. It will be rare for the average contact pressure to lie outside the range Y/3 to 3Y: so we can estimate the load as the bearing-area multiplied by the yield stress.

This isn't as bad as at first sight appears, because of the strong variation of bearing-area with separation. For example, with a Gaussian height distribution and a nominal pressure of Y/1000, the separation is given by $Y\Phi(h/\sigma) = p_{nom}$ so that $h/\sigma = 3.09$: if we were wrong and the pressure is 2Y, then $\Phi = 0.002$ and $h/\sigma = 2.88$.

Thus, we get the area (and the contact pressure) wrong: but we shall get reasonable estimates of the number and size of the micro-contacts, of the electrical and thermal contact resistance, and of the mechanical stiffness of the contact.

The real virtue of this model only appears when we begin to look more closely into the nature of surface roughness. At that point, we find that the foundations of the cleverer models begin to crumble, and we are relieved to have a theory left.

Surface roughness

We need to consider in more detail the procedure for and the results of analysing surface roughness. The profilometer stylus makes a traverse across the surface, and the resulting height variations are converted into a voltage signal. We may digitise this at regular (horizontal) intervals – the sampling interval h – to get a sequence of heights z_1, z_2, z_3 .... We define a peak as any point higher than its immediate neighbours ($z_i > z_{i-1}$, z_{i+1}), and we calculate the curvature there by fitting a parabola through the three points to get

$$\kappa = (2z_i - z_{i-1} - z_{i+1})/h^2$$

This gives the basic pieces of information: the peak height distribution, the peak curvature distribution, and the peak density. A convenient way of displaying the height distribution is to use probability paper, where we plot the cumulative number of heights (the Abbott bearing-area curve) on a non-linear scale arranged to straighten the Gaussian ogive into a straight line. Fig. 10 shows the results of a study of a wear process by Williamson [10]: the overall distribution is not Gaussian: but we can happily fit a straight line to the upper parts of the curves and so use our Gaussian integrals. Note that

Abbott was wrong: it seems that we do not remove all the material above a given height to produce plateaux.

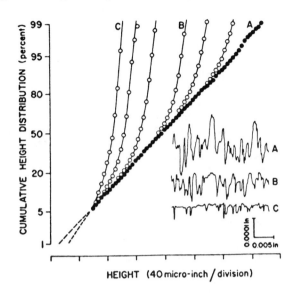

Figure 10. Transitional topography due to wear of a band-blasted surface. As wear continues, the curve moves steadily to the left. (From Williamson.)

Evidence suggests that if we sample larger numbers of points, the data still lie on the same curve: and in particular the highest point found becomes steadily higher. Earlier theories of contact taking the first point of contact as the datum were choosing a rather obscure starting point. It seems that surface heights really do form a random population – and that the rather small number we study with a profilometer should be regarded as a sample from which to estimate the population rather than as the height distribution.

We have already seen that peak curvatures show a comparable spread; but more complicated calculations suggest that the mean peak curvature can be used, even when determining the amount of elastic contact. We note that the stylus trace is unlikely to pass over the tops of the asperities: but one can show that if the asperities were spherical caps, the peak curvature found is only a little too high. Nayak demonstrated that this is nonsense. The typical mountain is not a single spherical cap; it has multiple tops, with connecting ridges; and what the profilometer sees as a peak is much more likely to be a nondescript point on a ridge than a near-miss of the top. He introduced the term 'summit' to distinguish tops from 'peaks'; and showed that just as the peak height distribution is higher than the surface height distribution, so the summit heights are higher still. This is qualitatively obvious: and can be measured experimentally once one can overcome problems of making multiple parallel traces while retaining the datum and registration. Nayak, however, established the use of random field theory as a model of surface roughness. Earlier Tallian and colleagues [11] at SKF had used Rice's random signal theory [12] (of noise down telephone lines) to study the frequency of metallic contacts in a lubricated ball-bearing: Nayak [13] took the two-dimensional form of this, formulated by Longuett-Higgins [14] to describe the roughness of the ocean (a _moving_ random field!) and developed it to answer the tribologists questions.

It should be pointed out that there is a basic requirement which must be fulfilled before random field theory can be applied: the overall height distribution must be Gaussian: and it usually is not – but the difference between engineering and mathematics is that we don't expect our basic assumptions to be exactly true. Anyway, according to Nayak, the heights must be Gaussian: peak and summit heights are then almost Gaussian, with means shifted upwards and standard deviations slightly reduced – all tying in with observation. Peak curvatures follow a Rayleigh distribution

$$p(\kappa) = \lambda^2 \kappa e^{-\frac{1}{2}\lambda^2\kappa^2}$$

which again agrees with observation – and had indeed already been suggested as an empirical deduction. This has some interesting implications: the mean curvature is $\sqrt{(\pi/2)}/\lambda$, while the mean radius of curvature is $\lambda\sqrt{\pi/2}$: so the relation between the mean radius of curvature and the mean curvature is

$$\bar{\kappa} \, \bar{R} = \frac{\pi}{2}$$

and one should not simply write $\bar{R} = 1/\bar{\kappa}$: but more seriously, while the standard deviation of κ is $0.523\bar{\kappa}$, that of R is _infinite_: so one really must not attempt to estimate $\bar{R}$ directly from measurements of R.

But, of course, we are not interested in peaks, but in summits. How do we estimate the number of asperities, given the _linear_ peak density on a profile? Naively we originally said

$$D_s = D_p^2$$

– and since we couldn't count summits we had no problem. But consider Figure 11. With the simple asperity of Fig. 11a, each profilometer trace detects one peak, and the above result is correct. With the realistic mountain of Figure 11b, many of the traces detect two peaks: so presumably $D_s < D_p^2$. Now modify the hill slightly to make subsidiary summits: the peak count is the same, but the mountain now has _three_ summits: $D_s > D_p^2$.

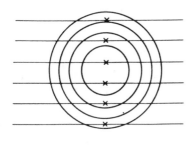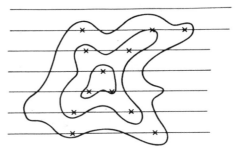

Figure 11a. Simple asperity: an intersecting traverse detects one peak.

Figure 11b. Real asperity with shoulders. Intersecting traverses detect multiple peaks – the asperity may or may not have multiple summits.

Random field theory gives the formal answer:

$$D_s = (2\pi/3\sqrt{3})D_p^{\,2} = 1.209D_p^{\,2}$$

– but in the course of doing so, perhaps eliminates the entire model of individual, isolated asperities from serious consideration.

Random field theory makes some inconvenient predictions. According to the theory – and experiment supports this – the mean peak curvature steadily increases with peak height: so it is no use finding the overall average peak curvature: one should study the high peaks which will be involved in contact. Or rather, the high <u>summits</u>. The theory tells us that the overall mean summit curvature $(\frac{1}{2}\overline{(\kappa_x + \kappa_y)} \equiv \overline{\kappa}_s)$ is 1.2 $\overline{\kappa}_p$: but varies even more strongly with height than the mean peak curvature (limited experimental work contradicts this – fortunately) – so we ought to measure directly the curvatures of the relevant summits – an uninviting prospect, particularly as a suggestion for a routine process for someone not primarily devoted to the analysis of surface roughness.

I see that I have forgotten to mention the most impressive of all the predictions of random field theory: that there are only three parameters governing all these different problems – or perhaps just one parameter and two scale-factors. If we measure the rms profile height σ, the rms profile slope σ_m, and the rms profile curvature σ_κ, random field theory tells us everything else. The one parameter is $\alpha \equiv (\sigma\sigma_\kappa/\sigma_m^2)^2$: then if we scale all heights by σ, and all curvatures by σ_κ, an isotropic surface is characterised by this single parameter.

Table I. Predictions of the Rice/Longuett-Higgins/
Nayak theory

	Peaks	Summits
Density	$\dfrac{1}{2\pi}\dfrac{\sigma_\kappa}{\sigma_m}$	$\dfrac{1}{6\pi\sqrt{3}}\dfrac{\sigma_\kappa^2}{\sigma_m^2}$
Mean height	$\sqrt{\dfrac{\pi}{2\alpha}} = 1.2533\ \sigma_m^2/\sigma_\kappa$	$\dfrac{4\sigma}{\sqrt{\pi\alpha}} = 2.2568\ \sigma_m^2/\sigma_\kappa$
Variance	$(1-.5708/\alpha)\sigma^2$	$(1-.8468/\alpha)\sigma^2$
Mean curvature	$\sigma_\kappa\sqrt{\dfrac{\pi}{2}} = 1.2533\sigma_\kappa$	$\dfrac{8\sigma_\kappa}{3\sqrt{\pi}} = 1.5045\sigma_\kappa$
Variance	$.4242\sigma_\kappa^2$	$.5178\sigma_\kappa^2$

In the study of hydrodynamic lubrication of rough surfaces, it is often assumed that anisotropic surfaces are represented by corresponding anisotropic asperities, and that an isotropic surface has isotropic asperities – as indeed did our dry contact model. With dry contact there is some justification, for as the Contact Mechanics lectures showed, there is little difference in behaviour between an elliptical contact and the equivalent elliptical one (the mean curvature should be taken as $\sqrt{\kappa_1\kappa_2}$, not $\frac{1}{2}(\kappa_1+\kappa_2)$ as in Nayak's work). But it should be emphasised that quoting a 'mean curvature' disguises the fact that the asperities on an isotropic surface are not axisymmetric: the most common ratio of summit principal curvatures is 3:1, and isotropy is achieved by random orientation.

McCool [15] has proposed that the surface contact model described should have these equations incorporated in it, so that, for example, the plasticity index will become

$$\frac{E^*}{H}\sqrt{\sigma_s\bar{\kappa}_s} = 1.226\,\frac{E^*}{H}\cdot\sqrt{\sigma_s\bar{\kappa}_s}\,(1-.8908/\alpha)^{\frac{1}{4}}$$

The advantage, of course, is that it is straightforward to determine the necessary parameters, since they are now overall properties of a profile, and do not require us to hunt for high summits. He then compares the predictions of different contact models, and concludes that one may as well stick to the simplest.

Actually, he overlooks the simplest, and so have we. Perhaps it isn't a model at all. But the whole analysis of surface contact goes back to Archard, who said all that matters:

> if the primary result of increasing the load is to cause existing contact areas to grow, then area and load will not be proportional: but if the primary result is to form new areas of contact, then area and load will be proportional.

Perhaps we might add another sentence:

> the property that brick, wood, and copper have in common is not ideal plastic flow, but roughness.

We are, of course, here to talk about the fundamentals of friction; and I have discussed only normal contact. There is no reason why the same considerations should not hold during sliding: so we still expect the area of contact to be proportional to the load: if some clever person would explain why friction exists, and is proportional to the area of contact, our problem would be solved.

References

[1]   Abbott, E.J. and Firestone, F.A. (1933), Mech. Eng., 55, p. 569.
[2]   Bowden, F.P. and Tabor, D. (195) 'Friction and Lubrication of Solids', O.U.P.
[3]   Archard, J.F. (1961) 'Single contacts and multiple encounters', J. Appl. Phys., 32, p.1420-1425.
[4]   Greenwood, J.A. and Williamson, J.B.P. (1966) 'Contact of nominally flat surfaces', Proc. Roy. Soc., A295, p.300-319.
[5]   Christensen, H. (1970) 'Stochastic models for hydrodynamic lubrication of rough surfaces', Proc. I. Mech. E, 184, p.1013-1026.
[6]   Greenwood, J.A. and Tripp, J.H. (1967) 'The elastic contact of rough spheres', J. Appl. Mech., 89, p.153-159.
[7]   Greenwood, J.A., Johnson, K.L. and Matsubara, E. (1984) 'A surface roughness parameter in Hertz contact', Wear, 100, p.47-57.
[8]   Webster, M.N. and Sayles, R.S. (1986) 'A numerical model for the elastic frictionless contact of real rough surfaces', J. Tribology (ASME), 106, p.314-320.
[9]   Greenwood, J.A. (1967) 'On the area of contact between rough surfaces and flats', J. Lub. Tech. (ASME), 1, p.81.

54

[10] Williamson, J.B.P., Pullen, J. and Hunt, R.T. (1970), 'The shape of solid surfaces', ASME, Surface Mechanics, p.24.

[11] Tallian, T.E., Chin, Y.P., Kamenshine, J.A., Sibley, L.B., Sindlinger, N.E. and Huttenlocher, D.F. (1964) 'Lubricant films in rolling contact of rough surfaces', ASLE Trans., 7, p.109-126.

[12] Rice, S.O. (1944, 1945) 'Mathematical analysis of random noise', Bell System Technical Journal, 23, p.282, 24, p.46.

[13] Nayak, P.R. (1971) 'Random process model of rough surfaces', J. Lub. Tech (ASME), 93, p.398-407

[14] Longuett-Higgins, M.S. (1957) 'Statistical analysis of a random, moving surface', Phil. Trans. Roy. Soc., A249, p.321-387, A250, p.157-174.

[15] McCool, J.I. (1986) 'Comparison of models for the contact of rough surfaces', Wear, 107, p.37-60.

Discussion *following the lecture by J A Greenwood on "Contact of rough surfaces"*

U LANDMAN. Observations of surfaces in an electron microscope or a scanning microscope show faceting: an asperity can never be an ideal sphere since it grows preferentially with crystal planes exposed.

J A GREENWOOD. The mechanical engineer cannot get down to the level at which faceting might be observed. But there is a remarkable absence of the effects of the individual grains forming a polycrystalline metal: I recall some German work showing a connection some 25 years ago, but I have seen nothing since. This must be due to the presence of a worked layer on top of the normal grain structure in most practical surfaces.

ANON. The chemically interesting sites are where the facets meet: there we find steps and defects, and the chemistry is as different from the chemistry of a flat as between one element and another.

ANON. Surely we all look at the same materials: the material doesn't know our background - whether we are engineers or chemists or physicists. Dr Greenwood says he meets only worked surfaces - what is his picture of this worked layer? Surely it has a structure with facets and periodic potentials etc, just like the ones we meet in a force microscope. It may be a much more complex system, but not by any means a "goo".

J A GREENWOOD. I don't believe we all look at the same surfaces - or at least, that we don't *see* the same ones. Engineers forget adsorbed films: surface chemists forget oxides. Both ignore the difference between ferrite and pearlite. Sticking four oxygen atoms on a metal surface is not the same as having a 0.5 μm oxide layer - not oxygen on metal but oxygen on iron on oxygen on iron ... - a brute of its own sort.

An engineer asked to make a surface is going to scrape it: scrape it with a cutting tool, scrape it with a grinding wheel, with a hone, with carborundum paper ... The grain structure the metallurgist tends to look at is the uniform structure of oxide and metal forming the very top layers - and when he does he often finds a "white-etching layer" he can't make any sense of.

B J BRISCOE. There certainly is a highly disordered, probably amorphous layer on most of the surfaces that engineers use. It is often called the "Beilby" layer: at the surface there is a highly dislocated region: below that there is a less dislocated region and finally - after perhaps 5μm - we reach the normal microstructure. [Editor's comment: for a *critical* review of this controversial topic, see J F Archard, "Mechanical polishing of metals", Physics Bulletin **36**, 212-4 (1985).]

H M POLLOCK. My question relates to high surface energy contacts where the amount of plastic deformation of a rough surface may be significantly increased at small or even zero externally-applied load. A simple energy argument[see p. 87] predicts that if the work of adhesion w exceeds $H\sigma$, there should be a dramatic increase in friction. The values that you quoted for steel were 2kN/mm^2 and $\sigma = 0.4$ μm. If we consider a softer material, with H reduced by a factor of 10, and with a much smoother surface (σ down by a factor of 100), then for w of order unity, $H\sigma$ is of the same order as w. Do you think this could occur in practice?

I would also make the point that in contrast to what happens when smooth surfaces touch, surface roughness allows surface energy to be dissipated at *all* stages of increase in overall contact area, i.e. as each individual asperity makes contact. [This effect is mentioned on p. 90]. But I have no idea whether or not this is significant when frictional energy is being dissipated.

J A GREENWOOD. It is of course well-known that with soft metals - indium and to some extent tin and lead - we can obtain adhesion without the high-vacuum techniques needed with harder metals. No-one at the time this work was done attributed the adhesion or the junction growth which occurs when sliding starts, to surface energy: but in retrospect perhaps this should have been done. The numbers do not look impossible.

M AKKURT. It is very important when estimating the size of the area of contact that you use the hardness of the asperities and not the bulk hardness of the material. One can readily find a bulk harness of 200 H_v, but asperities with a hardness of 600.

J A GREENWOOD. I certainly agree that one must make microhardness measurements and so use the hardness of the surface layer. But I may have confused you in my talk: there are no such things as asperities. A metal has the mildest of ups and downs on its surface - grossly exaggerated by the profilometer with its 100:1 distortion - but what we call an asperity is only an ordinary bit of the surface, just slightly higher than the rest. The whole surface tends to be much harder than the bulk, not just the "asperities".

U LANDMAN. Thin films of liquid may behave as if much thinner than they really are, because a liquid near a solid surface behaves quite differently from a liquid in bulk. So one should beware of applying hydrodynamic theory to the apparent thickness.

J A GREENWOOD. This worry has been raised from time to time when dealing with thin films - particularly thin films of long-chain molecules - (lubricating oils!). The evidence of traction measurements in elastohydrodynamic contacts, with surface polished to 0.025 μm and film thicknesses of 0.5μm, is that the lubricant is behaving normally, but that is hardly the right range.

PROBLEMS WITH SURFACE ROUGHNESS

J.A. GREENWOOD
Department of Engineering
University of Cambridge
Trumpington Street
Cambridge CB2 1PZ, UK

INTRODUCTION. The difficulty about regarding a rough surface as an assembly of asperities is perhaps that facing a friend who works at the Royal Geographical Society on cataloguing the Himalayas: is she to accept that the summit reached by the 1973 Ruritanian expedition is actually a separate mountain ? Of course it was summit – those concerned were all honourable climbers – but shouldn't it really be counted as a subsidiary of K10 ? If we walk along this connecting ridge, how far must we descend to make it a separate mountain ?

More specifically, we saw in the earlier lecture that if we sample a profile of a rough surface at a regular spacing h – the sampling interval – we can define a peak as any point higher than its immediate neighbours: $z_0 > z_1$, z_{-1}; and define the curvature of the peak as

$$- \kappa_p = (z_{-1} - 2z_0 + z_1)/h^2 .$$

On a ground steel surface, this is what Whitehouse & Archard found (Fig. 1).   Different sampling intervals gave similar distributions of peak curvatures – but the mean curvature varied from 300 mm^{-1} for a sampling interval of 1 μm to 6 mm^{-1} for h = 15 μm. The difference is connected to the differing numbers of peaks identified: Fig. 2 shows how the peak density varied. Whitehouse & Archard actually quote the proportion of ordinates found to be peaks: this rose from 1 in 10 at h = 0.5 μm to 1 in 3 at h = 15 μm.

Can we understand these values – as a first step towards choosing the correct ones ? To do so, we need to investigate random field theory more seriously.

I. L. Singer and H. M. Pollock (eds.), Fundamentals of Friction: Macroscopic and Microscopic Processes, 57–76.
© 1992 Kluwer Academic Publishers. Printed in the Netherlands.

58

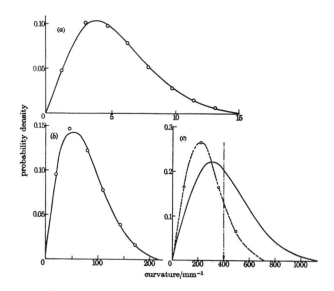

Figure 1. Probability densities of an ordinate being a peak of given curvature. Experimental points are from digital analysis of a ground surface (Aachen 64-13, $\sigma = 0.5$ μm, $\beta* = 6.5$ μm, for three different values of the sampling interval a) h = 15 μm, b) h = 3 μm, c) h = 1 μm. The arrow indicates the nominal stylus curvature. Full lines are from the theory for an exponential autocorrelation function. (from Whitehouse & Archard)

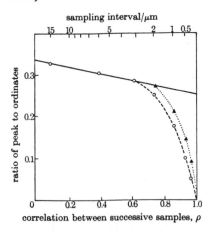

Figure 2. Ratio of peaks to ordinates as a function of sampling interval. Full line is exponential theory: experimental points for Aachen 64-13. O Normal stylus, nominal tip dimension 2-5 μm; ▲ special stylus (0.25 μm). (from Whitehouse & Archard).

Random Field Theory

The basic postulate is that the surface consists of a superposition of an infinite number of waves of arbitrary wave number (i.e. reciprocal wavelength) and phase:

$$z = \Sigma \, a_{j,k} \cos(jx + hy + \epsilon_{j,k})$$

(j,k are not necessarily integers). The Central Limit Theorem of statistics guarantees that the height z will then be a Gaussian random variable: and that two points separated by a distance s will be correlated random variables with some correlation coefficient $\rho(s)$. The correlation coefficient falls off with distance (from 1 to 0; but it may well go negative in between): but if we consider only one particular sampling interval h and an isotropic surface, we need only the two values $\rho(h) = \rho_1$ and $\rho(2h) = \rho_2$.

Then the rms slope of the surface is

$$\sigma_m^{\,2} = E\left\{[(z_1 - z_0)/h]^2\right\} = 2\sigma^2 (1 - \rho_1)/h^2$$

and the rms curvature is

$$\sigma_\kappa^{\,2} = E\{[z_1 - 2z_0 + z_{-1})/h^2]^2\} = 2\sigma^4 (3 - 4\rho_1 + \rho_2)/h^4$$

where $E\{\ \}$ denotes the expectation of the quantity inside the braces $\{\ \}$. It will be convenient to regard σ_m, σ_κ and the rms height σ as the three basic variables and drop ρ_1 and ρ_2: but we note here that Whitehouse & Archard found that the correlation coefficient decayed exponentially ($\rho_1 = e^{-\beta h}$, $\rho_2 = \rho_1^{\,2} = e^{-2\beta h}$); so that as h tends to zero,

$$\sigma_m^{\,2} = 2\sigma^2 (1 - e^{-\beta h})/h^2 \to \infty$$

$$\sigma_\kappa^{\,2} = 2\sigma^4 (3 - 4e^{-\beta h} + 3e^{-2\beta h})/h^4 \to \infty$$

If the autocorrelation function really decreases exponentially, two of the basic parameters, the rms slope of the profile and the rms curvature of the profile, are both infinite, and it is only because we use a finite sampling interval that we obtain finite values.

Probability theory tells us the joint statistical distribution of three adjacent heights on the profile; or more conveniently the distribution of the three variables central height z, slope m_2 and curvature. We have

$$P(z,m_2,\kappa) =$$

$$\frac{1}{(2\pi)^{3/2}\sigma_2\sigma_\kappa\sqrt{1-r^2}} \exp\left[\frac{-1}{2\sigma^2\sigma_\kappa^2(1-r^2)}(\sigma_\kappa^2 z^2+\sigma^2\kappa^2+2\sigma_m^2 z\kappa)\right] \exp\left[\frac{-m_2^2}{2\sigma_2^2}\right]$$

(where $m_2 = (z_1 - z_{-1})/2h$ and $\sigma_2^2 = E\{m_2^2\}$; and $r = \sigma_m^2/\sigma\sigma_\kappa$).

This is not as bad as it looks. The factor

$$\frac{1}{\sigma_2\sqrt{2\pi}} \exp\left[\frac{-m_2^2}{2\sigma_2^2}\right]$$

states that the slope m_2 is a Gaussian random variable, with a distribution independent of height or curvature. The remaining factor

$$\frac{1}{2\pi\sigma\sigma_\kappa\sqrt{1-r^2}} \exp\left[\frac{-1}{2(1-r^2)}(\frac{z^2}{\sigma^2} + \frac{\kappa^2}{\sigma_\kappa^2} + \frac{2\ r z\kappa}{\sigma\sigma_\kappa})\right]$$

states that central height and curvature are correlated Gaussian variables, with standard deviations σ and σ_κ respectively, and a correlation coefficient $(-r)$.

Note that we have introduced two different definitions of slope: $m_2 = (z_1 - z_{-1})/2h$ and $m = (z_1 - z_0)/h$. They have different standard deviations: in fact

$$\sigma_m^2 = \sigma_2^2 + h^2\sigma_\kappa^2/4$$

which to anyone brought up on Pythagoras suggests writing

$$\sigma_2 = \sigma_m \cos\theta: \qquad \frac{h\sigma_\kappa}{2} = \sigma_m \sin\theta.$$

It is now straightforward to find the probability that the central ordinate is higher than its neighbours and so is a peak; or the probability that it is a peak of given height or of given curvature: we find for example

$$P(\text{peak}) = \theta/\pi$$

$$\text{mean peak curvature } \bar{\kappa}_p/\sigma_\kappa = \sqrt{\pi/2} \, \sin\theta/\theta$$

$$\text{mean peak height } \bar{z}_p/\sigma = r\sqrt{\pi/2} \, \sin\theta/\theta$$

θ turns out to be a rather important parameter! What values does it take ?

From the definition, $\cos\theta = \dfrac{1}{2}\sqrt{\dfrac{1-\rho_2}{1-\rho_1}}$, so that if the correlation

coefficient decreases with distance $(\rho_2 < \rho_1)$, $\cos\theta > \dfrac{1}{2}$ and $\theta < \pi/3$. If the correlation coefficient is convex

$$1 - \rho_1 < \rho_1 - \rho_2$$

then $\theta < \pi/4$. An exponential satisfies the first but not the second: it follows that

$$\frac{1}{4} < \frac{\theta}{\pi} < \frac{1}{3}$$

and between one third and one quarter of all points sampled will be peaks!

The assumption made by Rice, Longuett-Higgins and Nayak – and indeed by most users of random field theory e.g. Newland "Random Vibrations" – is that the "signal" has a derivative – i.e. that the limiting rms slope and curvature exist as $h \to 0$. This is equivalent to saying that the auto-correlation function possesses derivatives at the origin, and has a Taylor expansion

$$\rho(h) = 1 + \frac{1}{2!} \, h^2 \rho^{ii}(0) + \frac{1}{4!} \, h^4 \rho^{iv}(0) + \ldots$$

– which of course an exponential does not.

We shall refer to such a surface as a "Rice surface", although Rice's pioneering work was restricted to the analysis of random signals – profiles. The classical theory which applies to such a surface deals with real slopes $\dfrac{\partial z}{\partial x}$ and curvatures $\dfrac{\partial^2 z}{\partial x^2}$: there is no need to introduce finite difference approximations or a sampling interval. The results for a profile are exactly what we find from the present theory by noting that for a Rice surface σ_m and σ_κ tend to definite limits as $h \to 0$:

$$\sigma_m^{\;2} = -\sigma^2 \rho^{ii}(0) \;\; ; \;\; \sigma_\kappa^{\;2} = \sigma^4 \rho^{iv}(0) \;.$$

As a result, the sampling interval parameter $\theta \equiv \sin^{-1}(\dfrac{h\sigma_\kappa}{2\sigma_m})$ tends to zero, and we now obtain a peak $\underline{\text{density}}$:

$$\text{peak density} \equiv \frac{\theta}{\pi h} = \frac{\sigma_\kappa}{2\pi\sigma_m} \cdot \frac{\sin\theta}{\theta} \approx \frac{\sigma_\kappa}{2\pi\sigma_m}$$

and we obtain a definite value for all small sampling intervals falling off only by the factor $(\sin\theta/\theta)$; qualitatively as we should expect. (Note that for a non-Rice surface, it is not the factor $\sin\theta/\theta$ which causes difficulty, but the fact that σ_κ and σ_m do not tend to definite limits).

The effect of the sampling interval parameter on the other profile properties is small. It enters only as this same factor $(\sin\theta/\theta)$ (or $\sin2\theta/2\theta$ when variances are considered) – so that as θ increases from 0 to $\pi/3$, the mean peak height falls only by a factor of 0.827, as does the non-dimensionalised peak curvature $\bar\kappa_p/\sigma_\kappa$. The limiting results as $\theta \to 0$:

$$\text{mean peak curvature} \;\; \bar\kappa_p = \sigma_\kappa\sqrt{\pi/2}$$
$$\text{mean peak height} \;\;\;\;\;\; \bar z_p = \sigma r\sqrt{\pi/2}$$

are just the answers found by Nayak directly.

Figures 3 and 4 show typical peak curvature and height distributions, and confirm that the value of the sampling interval parameter is unimportant. Since it is the only parameter affecting curvatures, we deduce that there is effectively a universal peak curvature distribution, which can be regarded as the Rayleigh distribution. Peak heights are almost Gaussian, with much the same standard deviation; but here the second parameter $r \equiv \sigma_m^{\;2}/\sigma\sigma_\kappa$ appears also and has a distinct influence, especially on the mean peak height. This parameter appears equally in the classical theory of Rice surfaces, where it is constant (and denoted by $\alpha^{-1/2}$); it seems to have exactly the same effect in both theories. It may be shown that for a profile derived from an isotropic surface, $\alpha > 3/2$: correspondingly in the

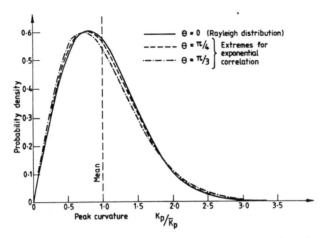

Figure 3.   Peak curvature distributions for varying θ.   θ = 0 is the
Rayleigh distribution $P(t) = te^{-\frac{1}{2}t^2}$

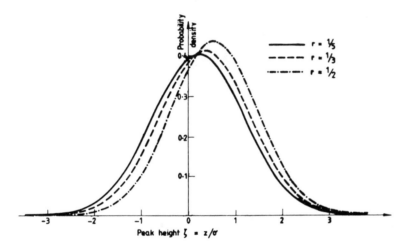

Figure 4.   Peak height distributions for θ = π/3: r = 1/5, 1/3, 1/2.

present theory r<c where c ~ $\sqrt{2/3}$.   The values α = 5 (r = .447) or
α = 10 (r = .316) are often taken as typical, giving $\bar{z}_p$ = 0.5600 or
0.3960 respectively as the height difference between the mean plane and
the mean level of the peaks – as is found experimentally.

Note the unexpectedly close relation between the mean peak height and
the mean peak curvature:

$$\text{mean peak height} = (\sigma_m/\sigma_\kappa)^2 \times \text{mean peak curvature}$$

SURFACES

The theory has been extended to two dimensions in two different ways.
Longuett-Higgins and Nayak analysed a Rice surface, and studied, among
other things, the properties of true summits. Whitehouse & Phillips,
and Greenwood, extended the finite-interval theory, but only by
following the lead of the experimentalists, and defining a summit as
any point higher than its four nearest neighbours (on a square grid).
It can be shown that this introduces an appreciable number of false
summits: in the limiting case of a Rice surface a surprising 30% are
spurious. However, the mean summit curvature and mean summit height
are reasonably correct in this case, so can perhaps be trusted.

True summits or not, the predictions of random field theory certainly
agree with experiment. Figures 5 and 6 show comparisons with
measurements by Sayles & Thomas on a grit-blasted surface: that is, the
direct measurement of the property is compared with its predicted value
based on the measured profile properties σ, σ_m and σ_κ.

Why is this agreement so distressing ? Note first that it is the
probability that a point is a summit which we are plotting, not the
summit density. Thus, Fig. 5 shows that increasing the sampling
interval by a factor of 7 decreases the summit density from $540/\text{mm}^2$ (at
$20~\mu\text{m}$) to $7/\text{mm}^2$ (at $141~\mu\text{m}$). Fig. 6 shows that the mean peak
curvature falls by a factor of 10 as the sampling interval increases
from $10~\mu\text{m}$ to $100~\mu\text{m}$: the mean summit curvature falls similarly.

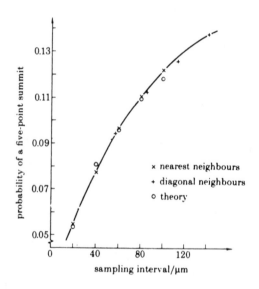

Figure 5. Experimental and theoretical summit densities for a
grit-blasted surface (courtesy R.S. Sayles).

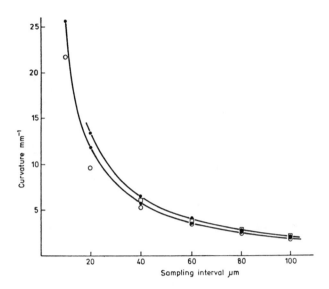

Figure 6.   Comparison of measured peak, ○, and summit, □, properties for a grit-blasted surface with predicted values, ●.

Quite generally, the non-dimensional curvatures κ/σ_κ vary very little; and random field theory correctly predicts the values.   But σ_κ itself: the rms profile curvature, has an enormous variation (Fig. 7). We have no evidence about what happens below 1 μm: but we know quite definitely that in the engineering range of sampling intervals, the variation is enormous - in effect, we can get any value we like.

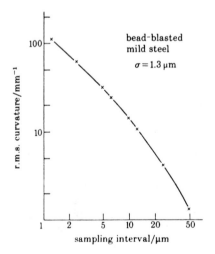

Figure 7.   Variation of σ_κ with sampling interval.

Surface Specification

How then can we describe the properties of a surface ? If the auto-correlation _were_ exponential ($R(x) = \sigma^2\exp(-\beta x)$), then, although we might dislike the implications, the surface would at least be easily described: two numbers, σ and β would say all. (Archard & Onions indeed developed a plasticity index $\psi_A = \dfrac{E*}{H} \cdot \dfrac{\sigma}{\beta}$, which is certainly more realistic than the original Greenwood & Williamson one $\dfrac{E*}{H} \sqrt{\sigma_s \kappa_s}$). But while an exponential is a good approximation over some range of sampling intervals, the more sophisticated test of plotting θ reveals that it normally strays from the range $\dfrac{\pi}{4} < \theta < \dfrac{\pi}{3}$ at shorter sampling intervals (Fig. 8): there is no suggestion that θ tends to zero, but it certainly falls below $\pi/4$.

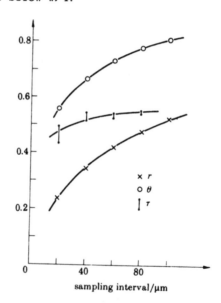

Figure 8. Variation of r and θ with sampling interval for the grit-blasted surface of Figs. 5 and 6.

SPECTRAL DENSITY

There seems to be no obvious alternative for the auto-correlation function, so what then can we do ? Sayles & Thomas suggest that the missing overall description can be found by considering, instead of the auto-correlation function $R(x)$, the profile spectral density $G(k)$, which is the Fourier transform of $R(x)$:

$$G(k) = \frac{2}{\pi} \int_0^\infty R(x) \cos kx \, dx$$

$$R(x) = \int_0^\infty G(k) \cos kx \, dk.$$

This seems to be the approved definition of the spectral density $G(k)$; but I feel it misses the point. Let's try again: if the profile is

$$z(x) = \int_0^\infty f(k) \cos(kx + \epsilon(k)) \, dk$$

then the mean square value of $z(x)$ is

$$\overline{z^2} \equiv \sigma^2 = \frac{1}{2} \int_0^\infty f^2(k) \, dk$$

and the mean value of $z(x) \, z(x+s)$ is

$$E\{z(x)z(x+s)\} = R(s) = \frac{1}{2} \int_0^\infty f^2(k) \cos ks \, dk$$

and comparing, we say that $G(k) \equiv \frac{1}{2} f^2(k)$ is the spectral density (strictly, the "power spectral density", of the profile).
   In terms of the spectral density, we have

$$\sigma^2 = \int_0^\infty G(k) \, dk$$

$$\sigma_m^2 = 2\int_0^\infty G(k) \left(\frac{1-\cos kh}{h^2}\right) dk$$

$$\sigma_\kappa^2 = 4\int_0^\infty G(k) \left(\frac{1-\cos kh}{h^2}\right)^2 dk \quad .$$

If we let h tend to zero, these integrals become the moments of the power spectrum:

$$m_{2m} = \int_0^\infty k^{2m} G(h) \, dk \quad (k = 0,1,2)$$

and classical random field theory uses m_0, m_2 and m_4 where we have used σ^2, σ_m^2, σ_κ^2. Our belief that σ_m and σ_κ increase indefinitely as $h \to 0$ is equivalent to claiming that the moments of the power spectrum are infinite.
   Suppose now that the auto-correlation function were exponential

$(R(x) = \sigma^2 \exp(-\beta x))$.   Then

$$G(k) = \frac{2}{\pi} \cdot \frac{\beta \sigma^2}{\beta^2 + k^2}$$

so we are expecting the spectral density to be something like this. Sayles & Thomas suggest that it follows the simple law

$$G(k) = C/k^2$$

- which is equivalent to saying that all <u>wavelengths</u> are equally represented in the profile. The moments m_2 and m_4 will then indeed be infinite: and so also will the rms height σ. Sayles & Thomas produce evidence that roughness is indeed a non-stationary random process, and that the longer the sample, or the sample <u>length</u>, (not the sampling <u>interval</u> now) the larger is the measured roughness. Because in practice the sample length is finite, the spectral density necessarily has a lower cut-off (at $k_1 = 2\pi/L$ where L is the sample length: k can then only take discrete values k_1, $2k_1$, $3k_1$ ... but for high wave-numbers these are indistinguishable from a continuous spectrum) and so in practice $m_o = \int_{k_1}^{\infty} C/k^2 \, dk$ is finite.

An upper cut-off is also in practice imposed on the spectrum, this time by the finite size of the profilometer stylus. So finite values of all three parameters are necessarily obtained - but only because of considerations unlikely to be relevant to any intended application: the finite values are more misleading than infinite ones would be ! Clearly, the experimenter should not accept these imposed values. Instead, he should identify the range of wave-numbers relevant to his application (hopefully all lying within the measurable range), and impose his own filtering in the spectrum.

REAL FILTERING PROCESSES

The simplest way of obtaining finite values of the spectral moments m_o, m_2 and m_4 is to truncate the spectral density: to ignore all wave-numbers except those in a range $k_1 < k < k_2$. For $G(k) = C/k^2$ this gives (for $k_2 \gg k_1$)

$$m_o \sim C/k_1 \; : \; m_2 \sim Ck_2 \; : \; m_4 \sim \frac{1}{3} Ck_2^3$$

However, it is worthwhile examining more closely the likely effects of filtering on the measured profile. A reasonable postulate for the effect of the stylus is that it responds to the average height of the profile over a distance 2b - the stylus "size". It is easily seen that this replaces

$$z = \int f(k) \cos(kx+\epsilon(k))\ dk \qquad \text{by} \quad y = \int f(k) \frac{\sin kb}{kb} \cdot \cos(kx+\epsilon(k))\ dk$$

so the spectral density becomes $G(k) \cdot (\frac{\sin kb}{kb})^2$.

A common technique for eliminating longer wavelengths is to subtract a moving average over a suitable distance (2ℓ) from the measured height. This replaces $f(k)$ by $f(k)(1 - \frac{\sin k\ell}{k\ell})$: so the combined effect of stylus size and moving average is to give

$$G^f(k) = G(k)\ (\frac{\sin kb}{kb})^2 (1 - \frac{\sin k\ell}{k\ell})^2 \ .$$

Figure 9 shows this filtered spectral density when the original density is $G(k) = C/k^2$, taking $\ell = 25$ b. Clearly using a sharp cut-off is a

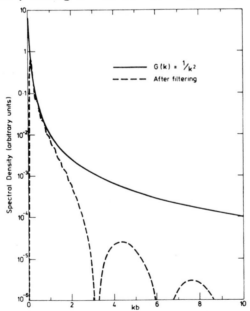

Figure 9. Filtered spectral density for a stylus size 2b and a moving average over $2\ell = 50b$.

drastic simplification. Figure 10 shows the rms slope and curvature predicted by this model - there is a general resemblance to the behaviour found experimentally (Fig. 7) but the agreement is not impressive. However it seems at least possible that the measured properties of profiles can be explained in this manner: that surface

70

roughness has an intrinsic spectral density C/k^2, and what we observe is the result of modifying this by our instrumentation. And perhaps understanding what we see is the first step to understanding what to do about it.

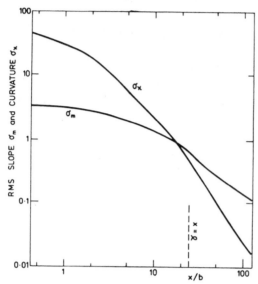

Figure 10. Predicted σ_m and σ_k variation with the spectral density of Fig. 9.

FRACTALS

It takes some time to become accustomed to the idea that ordinary, engineering, surfaces should be such difficult phenomena: that to represent a profile correctly, we should draw a curve which does not possess a derivative at any point: that however much we magnify the surface, it will still look rough. My instinct on first recognising these implications of Whitehouse & Archard's work was to recall distant memories of mathematical analysis, and examples chosen to destroy a naive faith in commonsense - in particular that one could indeed construct a continuous function which had no derivative - and to reject such idiotic concepts as part of my past. They are now beginning to look like the future too.

A major role in making such things respectable has been played by Mandelbrot, and his assertion that smooth, measurable curves are a mathematical invention, and play no part in the real world. By his classic question 'How long is the coastline of Britain ?' and his

examples of snowflakes, Brownian motion, fracture surfaces, river flows
- by examples from every branch of science, Mandelbrot has obliged us
to accept that surface roughness is normal - ie unmeasurable !

Mandelbrot is somewhat reluctant to define 'fractals' or 'fractal
dimension' preferring to offer examples: if for example, we measure
the length L of a curve using measuring rods of varying lengths ℓ, and
find that

$$L \sim \ell^{1-D}$$

then we say that the curve has dimension (*fractal dimension*) D. (The
coast of Britain is of dimension 1.24)

A curve is said to be 'self-similar' if it appears the same after
magnification. A fractal curve need only be 'self-affine' - the
magnifications along and normal to the curve need not be equal. The
Weierstrass function of my analysis lectures is

$$z(x) = \sum_{n=1}^{\infty} \frac{\cos 2\pi \gamma^n x}{\gamma^{(2-D)n}} \quad .$$

(Unlike a Fourier series, the wave numbers are $1, \gamma, \gamma^2 \ldots$ instead of
$1, 2, 3 \ldots$). Mandelbrot's ideal fractal can be scaled upwards or
downwards, so he modifies this to form the Weierstrass-Mandelbrot (W-M)
function

$$z_1(x) = \sum_{n=-\infty}^{+\infty} \frac{1-\cos 2\pi\gamma^n x}{\gamma^{(2-D)n}} \quad .$$

Berry & Lewis published plots of W-M functions, calculated using
$\gamma = 1.5$ but with different fractal dimensions D. They also show that
the discrete frequencies of the W-M function approximate to a
(continuous) spectral density

$$G(k) = \frac{1}{2 \ln \gamma} \cdot \frac{1}{k^{(5-2D)}} \quad .$$

This suggests that similar curves might be obtained by taking

$$z(x) = \sum_{n=1}^{\infty} \frac{1}{n^{2.5-D}} \cos(2\pi nx + \epsilon_n)$$

provided that ϵ_n is taken to be a random phase angle. Fig. 11 shows
the results: they do indeed resemble the W-M function. We note that it
is possible to estimate the fractal dimension by eye - and that a
typical profilometer trace has $D \sim 1.5$. Which brings us back to the
Sayles & Thomas spectral density $G(k) \sim 1/k^2$.

72

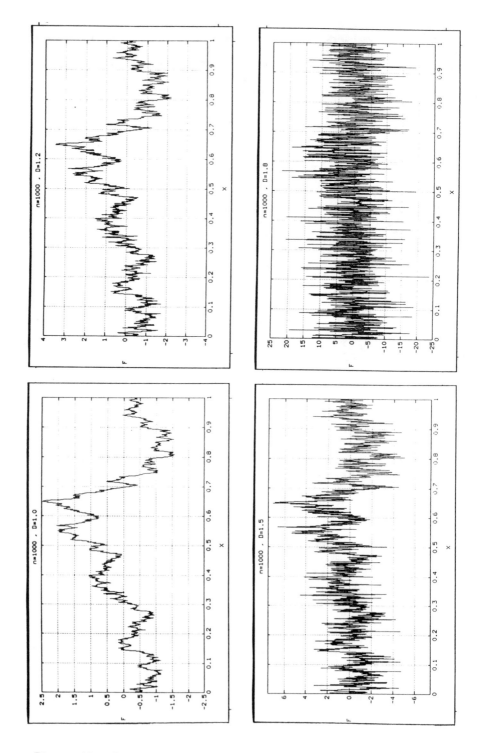

Figure 11.   Surface profiles of different fractal dimension D.

It is not yet clear whether the fractal approach will prove more effective than the spectral density approach, or even whether the two are essentially the same. What is clear, is that the study of surface roughness continues to get more complicated, and that we are a long way from understanding it.

REFERENCES

[1]   Bowden, F.B. and Tabor, D. (1950) Friction and Lubrication of
        Solids, Oxford University Press.
[2]   British Standard 1134:1950. The assessment of surface texture by
        the centre-line-average height method.
[3]   Archard, J.F. (1961) Single contacts and multiple encounters ,
        J. Appl. Phys. 32, 1420-1425.
[4]   Greenwood, J.A. and Williamson, J.B.P. (1966) Contact of
        nominally flat surfaces , Proc. Roy. Soc. A295, 300-319.
[5]   Nayak, P.R. (1971) Random process model of rough surfaces , ASME
        J. Lub. Tech. 93, 398-407.
[6]   Longuett-Higgins, M.S. (1957) Statistical analysis of a random,
        moving surface , Phil. Trans. Roy. Soc. A249, 321-387.
        Statistical analysis of an isotropic random surface , Phil.
        Trans. Roy. Soc. A250, 157-174.
[7]   Rice, S.O. (1944) Mathematical analysis of random noise , Bell
        System Technical Journal 23, 282; 24, 46.
[8]   Whitehouse, D.J. and Archard, J.F. (1970) The properties of
        random surfaces of significance in their contact , Proc. Roy.
        Soc. A316, 97-121.
[9]   Sayles, R.S. and Thomas, T.R. (1979) Measurements of the
        statistical microgeometry of engineering surfaces , ASME J.
        Lub. Tech. 101F, 409-418.
[10]   Whitehouse, D.J. and Phillips, M.J. (1978, 1982)
        Discrete properties of random surfaces , Phil. Trans. Roy.
        Soc. A290, 267-248.
        Two-dimensional discrete properties of random surfaces , Phil.
        Trans. Roy. Soc. A305, 441-468.
[11]   Greenwood, J.A. (1984) Unified theory of surface roughness ,
        Proc. Roy. Soc. A393, 133-157.
[12]   Mandelbrot, B. (1982) The Fractal Geometry of Nature,
        W.H. Freeman, New York.
[13]   Majumdar, A. and Bhushan, B. (1990) Role of fractal geometry in
        roughness characterisation and contact mechanics of surfaces ,
        ASME J. of Tribology 112, 205-216.

Discussion *following the lecture by J A Greenwood on "Problems with surface roughness"*

M O ROBBINS. Have spectral densities actually been measured, or do you only get them by "eyeballing" surface profiles?

J A GREENWOOD. Yes, I have measured them, and others, notably Sayles and Thomas before me. Recently Majumdar in the USA has reported many spectra: not always with an exponent of 2, but usually centered round there.

J KRIM. When we do a spectral analysis of our surfaces, we always go back and look again at the surface to check that we have it right. With the surfaces that we examine in our force microscope - deposited films with an area of 1 cm² - the upper wavelength cut-off is not the specimen size as found by Dr Greenwood, but something very much less, perhaps 1μm.

J A GREENWOOD. I have never found anything like that: with the surfaces I use, the sample length is the limitation: and if the sample length is increased, then the roughness increases.

E VANCOILLE. Two questions: (1) can you say how the plasticity index varies with sliding distance? (2) If I make an emergency stop with my car, what happens to the fractal dimension of the tyres?

J A GREENWOOD. Under benevolent sliding conditions, the roughness decreases - though with a smooth surface against a rough it almost certainly increases - and more importantly, the mean asperity curvatures decrease, so that the plasticity index goes down. With a sufficient degree of optimism it is even possible to claim that sometimes it appears to tend to 1.

I would guess that the fractal dimension of the types does not change: but I'm afraid it's an experiment I haven't done.

I L SINGER. When you take two surfaces that are fairly rough and rub them together so that some wear takes place, while they may end up being rough when you take a profile trace over either one of them, there must be some degree to which the pair is smooth because they have been run in to conformity. To what extent should we ignore the roughness of one surface, or use the sum of the squares of the roughnesses to characterise the interface - always assuming that you maintain that same track and that you really do have two rather well-conforming surfaces?

J A GREENWOOD. If you rub them so that they don't mate on the same points, then of course there is no problem: you have two random signals, and you can simply add together the variances to form an equivalent surface: this is well-established as a method of anal ysing the interface. But if, as you say, we arrange that a point on the first surface always contacts the same point on the second surface, and no other point, then as you say, one acts as a die and stamps its form on the other, or in some way they rub each other away until they mate. There is then nothing you can learn about the interface by studying one surface alone. I think the problem is well-recognised: but I don't know of any work going on to quantify it. Alan Dyson did some work at Leicester a few years ago, and showed that the roughness on the two disks in a disk machine became correlated if the angular velocities were simply related, but I don't believe he decided what the next step was. Of course the advantage of not having simply related numbers of teeth in gear technology is well-known, and one uses an extra tooth - a hunting tooth - to ensure that each tooth of one gear in time contacts every tenth of the second gear. And the well-known lapping principle: that to get *two* flat surfaces you need to make *three*, lapping each against both of the others - is really the same idea.

H M STANLEY. We have seen the strong influence of sampling interval. What is it about the contact situation which helps us determine the "proper" sampling rate (interval)?

J A GREENWOOD. I am sorry, but this was a talk on current "problems" in surface roughness, and I'm afraid that currently I don't have any answers. One well-known proposal is that if the application involves a Hertzian contact, then the diameter of the Hertz region should be taken as the low-wavelength cut-off. The difficulty is that an alternative proposal is that it should be taken to be the high-wavelength cut-off - that sums up the state of the art.

M O ROBBINS. Professor Rabinowicz told us that the typical size of contacts, and his critical distance, were both of order 1μm. However, he said that this was really the size of the ten largest contacts and that there might be smaller ones. If surfaces are self-affine fractals, there should be contacts of all sizes. The largest will always be comparable to the size of the total contact region. Could it be that the reason people often cite 1μm as a contact size is that typical contact regions are 3 to 10μm?

J A GREENWOOD. I find it very difficult to make the first step in assessing the typical contact size. What is normally regarded as a single contact region can be very much longer than a few microns: for many purposes the Hertzian contact between a locomotive wheel and track is a single region, but is typically 10 mm across: whether it is subdivided I don't know. The slide I showed of a Brinell indentation (Contact of Rough Surfaces, Fig 3) is a single contact region 1 mm across, and I am reasonably certain that the roughness after unloading does correspond to separate contact areas during loading: that it is not all elastic recovery.

Counting level-crossings on a surface profile at a sampling interval of 1.7 μm - before we knew any better - suggests that at characteristic separations of 2σ-3σ, typical contact areas would be a few microns (5-10) across: but we don't know whether this is really the size of a contact area of the width of an outlying shoulder (see Contact of Rough Surfaces, Fig 11b).

What we now know is that although level-crossings with an intervening elevation of reasonable height have this sort of spacing, actually each level-crossing is a multiple crossing because of the "fuzz" on the profile - and the evidence is that, as usual, the density of level-crossings increases indefinitely as the sampling interval is reduced. We haven't as yet found a criterion for distinguishing between "significant" level-crossings and "insignificant" ones - and on the fractal hypothesis, we shall find it very difficult to do so: when we scale up the picture it can look exactly the same. Perhaps an assessment based like Professor Rabinowicz's on the sizes of the larger wear particles is more helpful than all this surface topography - except then we get involved in questions of the mechanism of the wear process: how many contacts are involved in making a single wear particle, and exactly how? In particular, it is possible that the typical contact size during sliding is appreciably larger than the typical contact size under normal loading - as shown in two beautiful papers by Cocks (J. Appl. Phys. **33** (1962) p 2152 and **35** (1964) p 1807) - even without the increase in the total area of contact ("junction growth") shown by Professor Tabor.

H M POLLOCK. I'll just mention one possible criterion for a lower cut-off to the range of individual asperity contact radii. Adhesion maps confirm that the smaller the relevant dimension (radius of curvature, or contact radius), the more the surface energy will contribute to deformation, either elastic or plastic, thereby preventing this dimension from falling below some cutoff value. From the non-dimensional quantity plotted along the x-axis [of the map on p. 85], we see that in your notation, the relevant natural length is $E^{*2}w/Y^3$. Do you think that this relates to the problem of selecting the "correct" sampling interval or low-wavelength cut off?

J A GREENWOOD. I suspect not: or at least not directly. For a single contact you (and independently K L Johnson) have argued that surface energy will induce plastic deformation even without an applied load if $E^{*2}w/(RY^3) > 3$. I don't think we can directly convert $E^{*2}w/Y^3$ into a sampling interval: but perhaps one could argue that no asperity curvature greater than $3Y^3/(E^{*2}w)$ could persist - at least, not for long enough to take part in a frictional contact - and hence that we should never choose a sampling interval which would give such curvatures.

As discussed earlier [p 55], you have shown also that contact areas and hence friction and wear would be increased - perhaps dramatically - for a rough surface if $w/H\sigma$ approaches unity: but I think it is fundamental that σ is a *vertical* distance - i.e. normal to the surface - not a horizontal one, and thus does not relate to this question of sampling interval.

SURFACE FORCES AND ADHESION

H.M. POLLOCK
School of Physics and Materials
Lancaster University
Lancaster LA1 4YB
England

ABSTRACT. This chapter is concerned, not with the controversial topic of the adhesion component of friction (covered later in this book), but with how surface forces may influence a wide range of friction processes. This influence is either direct, or as a result of changes in the real area of contact.

First it is necessary to clarify some confusing terminology associated with the word "adhesion", and to list current techniques used to measure surface forces. The chapter then describes the analysis of the elastic contact between   a sphere and a flat, as a first step in understanding the adhesion of solids. Useful approximations are given by the "fracture mechanics" or "energy balance" model described by Sperling and independently by Johnson, Kendall and Roberts (JKRS), and the "deformed profile" model due to Derjaguin, Muller and Toporov (DMT). The JKRS approximation assumes a value of the work of adhesion or Dupré adhesion energy. It takes into account the additional deformation near the contact periphery resulting from surface forces, over and above the Hertzian deformation that would be given by the external load alone, and is valid in cases of strong adhesion, large radius and low elastic modulus. In the opposite situation, the DMT approximation applies: an arbitrary deformed profile (e.g. Hertzian), and the appropriate intermolecular force law, are assumed, and the force of attraction outside the contact zone is obtained by integration.

The effective force of attraction between the surfaces (not the same as the pull-off force required to separate them) can give rise to plastic as well as elastic increases in contact area. A size effect operates here, in that for smaller radii of curvature, the force *required* to initiate plastic deformation decreases faster than the force of attraction: this is opposed by a compressive reaction which therefore will produce plastic deformation if the scale of the contact region is small enough. The conditions for this "adhesion-induced plastic deformation", and for ductile or brittle failure of an adhesive contact, have been summarized in the form of maps.

There are two at first sight contradictory effects of roughness on adhesion. For two surfaces glued together, roughness increases the force needed to peel them apart. However, for two solids placed in contact, roughness reduces the pull-off force. The classical analysis of this effect, for elastic and plastic contact, is reviewed: recent theoretical and experimental work has described an additional phenomenon known

I. L. Singer and H. M. Pollock (eds.), Fundamentals of Friction: Macroscopic and Microscopic Processes, 77–94.
© 1992 *Kluwer Academic Publishers. Printed in the Netherlands.*

as avalanching, in which enlarged asperity junctions are formed when adhesion energy is released. This leads to the topic of energy dissipation in static contact, the associated hysteresis involving mechanical, chemical, or bulk effects (plasticity, viscoelasticity). As will be evident from later chapters, corresponding processes are involved in sliding contact.

1. Introduction: what is meant by adhesion?

Whatever type of friction we are observing, a satisfactory model must account for data of two kinds: the magnitude of the frictional force transmitted across the sliding interface, and the rate at which mechanical work is dissipated through friction. Both quantities may or may not depend upon the presence and magnitude of some kind of adhesive force, such as intermolecular attraction across the interface. Here we meet an initial difficulty — the loose terminology often used when the topic of adhesion is discussed. This is perhaps not surprising when we consider the variety of different units in which adhesion of various kinds has been measured or described. For example,

(a) the well-known phenomenon of adsorption, in which a solid surface, for example charcoal, attracts individual gas molecules onto itself. *Adsorption energy* is measured in electron-volts (or joules) per molecule and in effect quantifies the adhesion of a molecule to a surface;

(b) the macroscopic *force of attraction between two surfaces* in contact, in newtons. Although this is generally a difficult quantity to measure, it plays a vital part in many everyday processes. Examples have been quoted [1] ranging from the coalescence of latex particles in emulsion paint to the hard brown stain that appears under your dripping water tap. Ceramics technology depends on the ability of inter-particle forces to promote sintering;

(c) the adherence or externally-applied *pull-off force*, in newtons, which is not the same quantity as the force of attraction. Although Newton himself detected adhesion between polished marbles, it is relatively seldom that we are aware of having to exert a force in order to separate two dry solids placed in contact;

(d) the thermodynamic work or free *energy of adhesion*, in joules per square metre. In principle this is a material parameter associated with a particular solid-solid interface, by analogy with the surface energy of a single solid;

(e) various quantities used to quantify the performance of coatings and glued joints. These have included normal or transverse forces, "critical loads" in certain specific types of test, and tensile or shear "strengths". But as emphasized by Kendall[1], "the force required to break a lap joint, for example, does not depend simply upon the area of overlap. How can the strength of a body be defined if the breaking force does not increase in proportion to area? Such difficulties, unbelievably, have not prevented engineers from presenting their glue failure test results in the form of shear strengths". In many cases the logical interpretation involves a fracture mechanics analysis, the relevant quantities including work of adhesion (Jm^{-2}) or critical *stress intensity factor* $(N. \, m^{-3/2})$;

(f) as a final example of yet another use of the word "adhesion" — possibly the most important — we have, of course, the *adhesion component of friction* itself.

Strong support for the adhesion theory of metallic sliding friction is given by a widely-observed and progressive change in behaviour that has been summarized by Spurr[3] as follows:

— ultra-clean metal surfaces adhere to one another when placed in contact;

— then as the surfaces are progressively filmed, sliding is needed to cause seizure;

— next, gross seizure can no longer occur, but friction is still high and some plucking (transfer of material across the interface) occurs;

— then as the surface becomes more and more contaminated, friction is reduced until eventually boundary lubrication takes place.

Critics have doubted the significance of adhesion, pointing to the general absence of a detectable normal pull-off force required to separate dry surfaces under normal laboratory conditions (unless cold-welding or seizure has occurred). This particular argument is not convincing, since for the attraction between the surfaces to be significant we must apply a normal load to achieve the necessary intimate contact as will be discussed later. However, this does imply that terms such as "strength of the adhesion", in units of shear stress ($N\ m^{-2}$), are misleading. Even when true shear sliding at the interface occurs, the resistance to sliding that intermolecular forces provide is a different (and largely unexplored) phenomenon to the "mode I" bond-breaking involved when we measure adhesion. In fact, the question posed ten years ago [42] by Maugis[1] has still not been fully resolved: "when adhesion forces are zero, can an interface offer resistance to a shear stress, and is it possible to calculate the resulting coefficient of friction?"

Thus the adhesion component of friction is controversial, as emphasized by Johnson[4] who has used the term "tangential interaction" to include both true interfacial shear sliding [46] and the shearing or "tearing" of the intermetallic micro-junctions that result in the first place from the gripping action of intermolecular forces. In the second case, of course, any values of shear strength that are derived from analysis of friction data will bear *no relation* to any adhesion parameters associated with forces across the interface. This is because the mechanism by which work is converted into frictional heat involves plastic deformation of metal near the junctions, or other non-interfacial mechanisms.

In any case, we see that "adhesion" tends to mean different things to different authors. Apart from this question of the adhesion component of friction, attractive forces can modify and even dominate frictional processes in other ways. How this occurs is the theme of this chapter. Meanwhile, some of the main types of sliding interaction are listed, in an over-simplified and schematic way, along the three axes shown in figure 1. We shall see that an important consequence of intermolecular attraction is an increase in the real area of contact, which in principle will affect friction values governed by any of the mechanisms listed in the figure. In addition, certain types of friction, in particular those grouped along the B-axis, depend directly on the strength of the attractive force.

---

[1]For Maugis' question in full, see page 233.

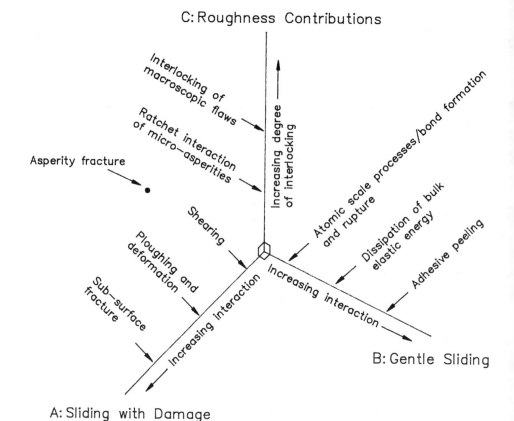

Figure 1 Some mechanisms of friction (from [30]). "Friction coefficient" is not an intrinsic material property. It is sometimes helpful to over-simplify the main types of sliding interaction by grouping them as follows:
A. Processes involving *damage* to one or both of the surfaces (shearing, ploughing and deformation, sub-surface fracture);
B. *gentle sliding*, where neither surface is permanently altered by the sliding event (atomic-scale bond formation and rupture, dissipation of bulk elastic energy, adhesive peeling);
C. processes that depend on *roughness* (ratchet interaction of micro-asperities, interlocking of macroscopic flaws). Note, however, that "smooth" does not mean "frictionless"!

2. Forces involved in adhesion

These include the same types of force as are responsible for the cohesion of solids. The coulomb force between charges gives rise to purely electrostatic interactions between permanent dipoles and quadrupoles as well as charges themselves. Forces that are quantum-mechanical in nature give rise to covalent or metallic bonding and to the repulsive exchange interactions that balance the attractive forces at very small interatomic separations. Polarization forces arise from the dipole moments induced in atoms and molecules by the electric fields of nearby charges and dipoles. The whole subject of intermolecular and surface forces has been outlined by Israelachvili[5] who describes, for example, how the term van der Waals force includes the Keesom "orientation" interaction between rotating permanent dipoles, the Debye "induction" contribution, and the normally dominant "dispersion" force involving dipole-dipole interactions between non-polar molecules. He emphasizes an important distinction. The properties of gases and condensed phases, apart from ionic crystals, are determined mainly by the strength of the operative forces at or close to molecular contact. However, when we sum the relevant pair potentials in order to calculate the interaction between two *macroscopic* particles, the result proves to be still appreciable at long range (separations of up to 100 nm or more). The interaction energy is greater the larger the size of the particle, and at contact typically exceeds kT by a large factor, so that for complex potential/separation functions, the ultimate thermodynamic equilibrium separation will not be reached. Given the shapes of two bodies in contact, it is possible to integrate the appropriate potentials or forces, so that, for example, the omnipresent van der Waals force varies as the inverse square of the separation in the case of sphere-against-flat geometry. Only in the case of strong covalent or metallic bonding do we find that short-range attractions are dominant.

On the experimental side, three techniques in particular have been developed to explore the major complications resulting from the presence of adsorbed layers, oxide films, etc., on real surfaces. With the Surface Forces Apparatus [5,6], force/separation curves are obtained for smooth surfaces (mica) which may be coated, or immersed in an electrolyte solution. The sharply-curved specimen used in the "Mechanical Microprobe" [7] allows a single area of contact at the atomic level to be obtained even with surfaces whose roughness normally results in multi-asperity contact : load/area curves and values of pull-off force, together with field emission monitoring, are used to elucidate adhesion mechanisms. Most recently, Atomic Force Microscopes have been used in non- imaging, point contact mode [44], thereby allowing several features of the other two techniques to be combined. We note in passing that no-one so far has succeeded in using point force microscopy to verify the inverse square dependence on distance expected for purely van der Waals forces. Instead, measured attractive forces seem to be of much longer range than expected, an effect which could indicate that under ordinary laboratory atmosphere ambient conditions, the surfaces are covered with a layer of dielectric material or of permanent dipoles, or that electrostatic forces between patch charges [45] may be significant for sharp asperities.

3. Mechanics and mapping of adhesion

Correct analysis of the elastic contact between a sphere and a flat is a first step in the understanding of the adhesion of solids. The radius of contact a is determined by the radius of the sphere R^*, the appropriate elastic constant K, the applied normal load P, and the forces of attraction acting between the surfaces. A suitable force law may be assumed, such as the normalised interaction

$$f(z) = \frac{12w}{5z_0}\left[\left(\frac{z_0}{z}\right)^3 - \left(\frac{z_0}{z}\right)^{13}\right].$$ (1)

Here z_0 is the equilibrium interatomic separation and w is the Dupré energy of adhesion, equal to $\gamma_1 + \gamma_2 - \gamma_{12}$ where γ_1 and γ_2 are the surface energies and γ_{12} the interfacial energy; $f(z)$ is the force per unit area as a function of separation z that results when the attractive term in the 6-12 Lennard-Jones interaction energy expression is modified to describe the interaction between unit area of one surface and an effectively infinite area of the opposing surface, and the whole expression is then differentiated to give the force. Muller et al[8] showed that a rigorous analysis is possible on the basis of such an interaction, together with the equations of elasticity, but a simpler approach is useful if we wish to predict the effect of different types of contact geometry or attractive force. This proves to be of great importance, given that pull-off force, for example, depends not only on attractive forces but also on geometric factors such as roughness, energy-dissipative processes such as ductility or viscoelasticity, and elastic properties such as the stiffness of the measuring apparatus. Consequently, values of adhesion energy calculated from first principles, typically of the order of up to 2 J m^{-2}, are related only indirectly to forces measured in practice: to use the value of (pull-off force) $\times z_0$ as an estimate of w × (contact area) would be highly misleading!

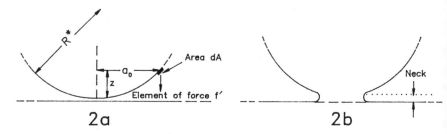

Figure 2. Cross-sections of sphere/plane contacts.
(a): ideally rigid case. The force of attraction is given by: $\Pi = \Sigma f' = \int f(z)\,dA$ (z varying along the profile). Substitution using the approximation $dA \approx 2\pi R^* dz$ gives: $\Pi = 2\pi R^* \int f(z)\,dz$, the integral being equal to w by definition (if z is now considered to vary vertically). Equation (2) then follows, whatever the function $f(z)$.
(b): elastic case.

We consider first the "deformed profile" model due to Derjaguin et al[9], the basis of the so-called DMT approximation. An arbitrary deformed profile (eg Hertzian), and the appropriate intermolecular force law, are assumed, and the force of attraction outside the contact zone is obtained by integration. *Additional deformation of the profile, resulting from this force, is neglected.* Interestingly, the force of attraction, Π, is independent of the elastic properties of the solids and of the form of the force law assumed, and is given by the very simple expression

$$\Pi = 2\pi R^* w$$ (2)

Figure 2a illustrates a derivation of the same formula given nearly sixty years ago by Bradley[10], valid for the case of ideally rigid solids (i.e. zero contact area!). The DMT approximation is particularly useful in predicting the effect of complex surface geometries. Its range of validity has been discussed in detail by Maugis [11]: when two elastic solids touch, a "neck" of pulled-out material is formed (fig. 2b), and the DMT approximation is valid only when the length of this neck is small compared with z_0. This condition is satisfied[12] in cases of low w, small R^* or large K.

In the opposite situation (large adhesion energy, large radius, low elastic modulus) the "JKRS approximation" applies. This is based on a fracture mechanics or energy balance model described by Sperling[13] and independently by Johnson, Kendall and Roberts[14]. The reduction in contact area, as separation is approached, may be considered as resulting from the propagation of a peripheral crack, inwards towards the centre of the contact. The JKRS approximation takes into account the additional deformation near the contact periphery resulting from attractive forces, over and above the Hertzian deformation that would be given by the external load alone. The attractive forces are taken to act at the periphery itself (crack tip), and *attraction outside the contact zone is neglected.* Maugis[15] has recently reviewed the fracture mechanics treatment of adhesion, and we next discuss various general points that result from this analysis, some of which are perhaps surprising at first sight:

1. *The force of attraction is not proportional to the contact area.* Let us define an effective peripheral attractive force, valid for all axisymmetric geometries: at equilibrium,

$$\Pi = -\left(\frac{\partial U_s}{\partial \delta}\right)_P \tag{3}$$

where U_s is the surface energy (equal to $-\pi a^2 w$) and δ is the distance moved by the load P. The value of Π is given by

$$\Pi = (1.5\pi w K a^3)^{1/2} \tag{4}$$

The full proof is given by Pollock et al[16]. Here we note simply that as we expect, this force is proportional neither to the area nor to the circumference of the contact region, since its value is determined by a Griffith fracture criterion. Thus here, the stress intensity factor is proportional to $(wK)^{1/2}$, the stresses involved vary as $(wK/a)^{1/2}$, and the force is proportional to stress × area or to $(wKa^3)^{1/2}$, as given by equation (4).

2. *Attractive forces produce an increase in elastic contact area. The smaller the radius of curvature, the more important are the adhesive forces in comparison with a given external load.* To calculate a, we can show[16] that its value in the presence of surface forces is equal to the calculated value that would result from an external load of $P + 2\Pi$ (not $P + \Pi$!) in the absence of surface forces:

$$P + 2\Pi = Ka^3/R^* \tag{5}$$

If we eliminate Π from equations (4) and (5), we obtain the rather complex expression for a that Johnson et al[14] first derived by minimising the sum of the surface, elastic and potential energies involved.[2] As an extreme example of the increased contact radius resulting from the action of attractive forces, consider Kendall's analysis[1] of the coalescence of latex particles involved when emulsion paint hardens. How small do the particles have to be? Suppose that we require that

$$a = R^*/2 \qquad (6)$$

for a hexagonally packed sheet to be formed, with no pores. With zero externally-applied load P, equations (4) and (5) give

$$Ka^3 = 6\pi w R^{*2} \qquad (7)$$

(equivalent to equation 2 of the following chapter by Thölén, who presents another example), so that from (6), we require that

$$R^* \leq 48\pi w/K \qquad (8)$$

Solid spheres can pull themselves together through the action of surface forces alone, provided that they are small enough for equation (8) to be satisfied. For smooth elastomers in contact, several research groups have verified the predicted elastic increase in contact area, and there is some indirect evidence that the JKRS theory may apply to contacts between stiff solids also, in the case of assemblies of unsintered ceramic particles[17].

3. Let us imagine a friction experiment in which the adhesion or "tangential" interaction" model of friction, as mentioned earlier, applies in its simplest form: namely, in which the frictional force F is proportional to the real area of contact. Now for extended *rough* surfaces as distinct from smooth spheres in contact, as is shown in a later chapter the true area of contact, and therefore F also in this case, is independent of the size of the bodies and is *directly proportional to the load*, in both elastic and plastic regimes of contact (see chapter by J.A. Greenwood, also [18, 19]). But what do we mean by load? Kendall[20] suggests that F should be proportional to the apparent load Ka^3/R^* (equation 5), which is not proportional to P unless Π is negligible. To be able to employ a constant quantity, μ, as the "coefficient of friction", it is necessary to define it by the equation $F = \mu \times Ka^3/R^*$. Eliminating a from equation 5 he thus obtains

$$F = 3\pi R^* w\mu + \mu P + \mu.(6\pi R^* wP + 9\pi^2 R^{*2} w^2)^{1/2} \qquad (9)$$

Thus frictional force may vary non-linearly with load, and it is not enough to write $F = \mu \times$ *(external load + attractive force)*. In equation (9), the first term represents surface attraction, the second is the load term, and the third quantifies the fact that the attractive force itself varies with contact area and hence with load. This result from friction theory corresponds to an important general point in adhesion theory: adhesion modifies contact mechanics in two ways[1]. Not only is a finite pull-off force needed to separate two solids, but the intermolecular attraction also pulls the solids together, thereby increasing the contact area. For a polymer of low modulus (≈ 1 GPa) and high adhesion energy ($w \approx 1$ Jm^{-2}), equation (8) suggests that the

---

[2]The relevant equation is: $a^3 = \frac{PR^*}{K}[1 + \frac{3\pi w R^*}{P} + (2 \times \frac{3\pi w R^*}{P} + [\frac{3\pi w R^*}{P}]^2)^{1/2}]$.

first and third terms on the right-hand side of equation (9) will be important for R^* as large as a few hundred nm. Data from friction experiments involving polymer fibres[21] and silica powder[20] have been quoted [20] in support of equation (9), but a modified theory has been presented by Thornton [46] (see also the discussion following chapter , friction of granular non-metals).

4. *Surface forces alone can induce plastic deformation, even in the absence of an externally-applied load* ("adhesion-induced plastic deformation"). The increase in contact area produced by attractive forces need not remain elastic. This follows from a simple consideration of the size effect involved. Whereas Π varies as $a^{3/2}$ (equation 4), the force *required* to initiate plastic deformation tends to vary as the area, i.e. as a^2. Thus, at small enough values of a, the former will exceed the latter.

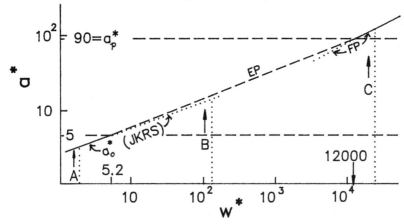

Figure 3 (from [24]). Theoretical equations relating contact radius a and adhesion energy w in reduced coordinates, *at zero applied load*. For points on the curve "JKRS", we have elastic and adhesive contact. For values of w^* between 5.2 and 12 000, elasto-plastic deformation ("EP") is predicted. Curve "FP" (full plasticity) represents equation (12) with $P = 0$. Equilibrium is attained as shown by dotted lines such as A, B or C, depending on the value of w^*.

Of course, in practice a second, material-dependent size effect may complicate the issue : the yield stress may vary with the volume of the deforming region. The stress tensor for adhering spheres has been calculated[23] and shown to be that of fracture mechanics in plain strain near the edge of the contact. It was shown [24] that for a wide range of values of Poisson's ratio and w, the following approximation predicts a load for plastic *initiation* on the axis that agrees to within 12% with exact calculations following the von Mises criterion:

$$P + 1.5\Pi = 1.1\pi a^2 Y \tag{10}$$

where Y is the elastic limit (yield stress). From equations 4, 5 and 10, it follows that at $P = 0, R^*$ would need to be less than ca. $0.7\ K^2 w/Y^3$. During the subsequent elasto-plastic stage of indentation, the mean pressure at first increases linearly with $ln(Ea/Yr)$, where E is Young's modulus. When the plastically-deformed zone reaches the surface (full plasticity), it is a reasonable approximation to assume that

a is given by

$$a_p \approx 60R^*Y/E \tag{11}$$

as is typically found for non-adhesive contact; however, we must replace the usual "non-adhesive" relation between P, a, Y, and H (hardness), by adding a surface term to the load:

$$P + 2\pi wR^* = \pi a^2 H \tag{12}$$

as shown some years ago [25,26]. Thus in theory, fully plastic deformation can be reached if both equations (11) and (12) are satisfied, even at $P = 0$.

Figure 3 is an example of an adhesion map, showing what should happen as regards adhesion-induced plastic deformation in the special case of zero load, for different values of w and R^*. It is convenient to use the dimensionless quantities $w^* \equiv wK^2/(R^*Y^3)$; $a^* \equiv aK/(R^*Y)$. At low values of w, the value of the equilibrium zero-load radius of contact, a_0, is given by $a_0^* = (6\pi w^*)^{1/3}$ (according to the JKRS theory, equation (5)). For plastic deformation to be initiated, equation (10) must also be satisfied, giving $w^* = 1.13a_0^*$. From these two relations we predicted that for the particular value $w^* \approx 5$ or 6, a_0 just reaches the value required for *initiation* of plastic deformation ($a^* \approx 5$) on the axis at the "Hertz point" (in fact a peripheral plastic zone is already well developed, as at any crack tip). For example, this would be the case for two gold spheres of radius 0.24 mm in air $(w \approx 0.1 \text{ J m}^{-2})$ or of radius 5.8 mm in UHV $(w \approx 2.4 \text{ J m}^{-2})$. At zero load again, the contact radius given by equation (12) *(full* plasticity) is $a_0^{*\prime} = (2w^*/3)^{1/2}$. For the particular value $w^* \approx 12,000$, a_0' just reaches the value required for full plasticity, equation (11) $(a*_p \approx 90)$. The evidence for adhesion-induced plastic deformation consists largely of data from experiments on contact between metallic microparticles, as discussed in the next chapter, and on measurements performed with mechanical microprobes in which a single area of contact at the atomic level is achieved (for a review, see ref. 7). Quantitative comparison with theory[24] suffers from the limitations imposed by the use of macroscopic concepts such as hardness, elasto-plasticity and full plasticity, in a situation where the physics is that of the nucleation and multiplication of dislocations.

5. *Surface roughness appears to have two contradictory consequences.* For two surfaces glued together, roughness greatly *increases* the force needed to peel them apart — that is why, if your cycle has a flat tyre, you roughen the punctured tube with sandpaper before sticking on the repair patch. On the other hand, for two surfaces placed in contact, roughness normally *reduces* the pull-off force — thus, the patched tube is much less likely to stick to the inside of the outer tyre if you roughen it by dusting it with chalk beforehand. The topic of multi-asperity contact is discussed in another chapter, and was extended to include adhesion, by Fuller and Tabor[27]. They showed that when $K^2 \gg rw^2/\sigma^3$, where σ is the standard deviation of the asperity heights, for a wide range of loads the surface energy becomes negligible compared to the elastic strain energy stored in the deformed asperities. When compressed, the higher asperities act as springs pushing the surfaces apart, so that the adhesion forces between the lower asperities that are trying to hold the surfaces together are swamped. However, Kendall et al[17] have pointed out that this model will not be valid for particles less than a micron in diameter. Moreover, for materials such as very soft rubbers that exhibit creep, the stored strain energy is reduced and both adhesion and rolling friction may

even be increased by slight roughening [41]. This effect is enhanced if some of the strain energy builds up, during peeling, within isolated contact points and is then irrecoverably lost when the final contacts are broken.

We extended an analysis due to Johnson[28] to describe the situation in which there are both brittle and ductile contributions to the pull-off force, as a consequence of the spread of asperity heights. If N, a' and r_a are the number of asperities per unit area, the contact radius of an individual asperity, and the radius of curvature (taken as constant), then, from equation (12), the total load per unit area to compress the asperities plastically is[26]

$$P = N \int_d^\infty (\pi a'^2 H - 2\pi r_a w)\phi(z)dz = 2\pi r_a n(H\sigma - w) \tag{13}$$

where n is the number of asperities actually making contact per unit area, d is the separation between the mean plane and the flat surface during loading and $\phi/(z) = (N/\sigma) \exp(-z/\sigma)$ is the exponential distribution of the asperity height z above the mean plane. Since the total real area of contact, a_r, per unit area is $N \int_d^\infty \pi a'^2 \phi(z)dz = 2\pi r_a n\sigma$, the ratio of apparent pressure to hardness is given by the simple expression

$$P/(A_r H) = 1 - w/(H\sigma) \tag{14}$$

It is interesting to note that some asperities will be plastically deformed even at zero load if $w \geq H\sigma$, in which case there will be important consequences as regards friction and wear[26]. With the help of a more elaborate elastic-plastic contact model, Chang et al[29] have concluded that for metals, the adhesion should in theory be quite large even for relatively rough surfaces.

Such models may be useful for elastomers or ductile metals, but are probably of little relevance to contact between hard ceramic materials. With very hard elastic solids and reasonable values of r and w, the adhesion vanishes for roughnesses (σ) of the order of a few nanometers, i.e. in theory *hard ceramic particles should not naturally adhere*. In practice, fine particles of alumina, TiN and other hard materials display strong adhesion. Ross et al[30] have described evidence that here, we need to consider a particular type of nanometre-scale stepped topography, since the distribution of atomic steps near the contact periphery can have the effect of greatly altering the magnitude of the effective force of attraction, as compared with its value for smooth profiles. In the extreme case of ideally rigid solids the force will be reduced, because the stepped profile holds them apart at the immediate periphery of the contact, at a separation greater than is obtained with a smooth profile. With highly elastic solids, the peripheral gap is everywhere narrower than in the case of the smooth Hertzian profile, so that the attractive force is greater.

6. It is possible that with two surfaces in static or sliding contact, *enlarged single-crystal junctions may be formed across the interface as a result of what has been termed "avalanching"*[31]. This is most likely to occur when a coherent, single-crystal junction is formed thereby, though incoherent boundaries with lattice mismatch could also result. At least four groups have described experimental data that appear to confirm the possibility of an abrupt establishment of finite contact area at the instant of contact at zero load. The mechanical microprobe technique,

88

mentioned earlier, has been used to detect this effect under certain conditions[32,30]. Gimzewski and Möller[33] used scanning tunnelling microscopy to observe a sharp transition from tunnelling to atomic contact. Burnham et al [34], and Cohen et al[35], used atomic force microscopes to detect "snapping" into contact, although as pointed out by Landman et al [36], it is important to eliminate the effect of the well-known cantilever instability artefact in order to verify the inherent jump-to-contact phenomenon. There are at least two reasons why this phenomenon, in a sense "the inverse of a fracture process" [31], is likely to be important in friction. First, during sliding the newly-formed crystalline junction will not necessarily crack at the original interface. Moreover, the process of energy dissipation in sliding will be modified (we return to this point later).

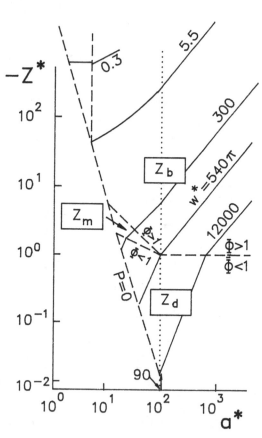

Figure 4 (from [24]). Theoretical equations relating a (contact radius formed plastically or elasto-plastically at maximum load) and Z (subsequent pull-off force), for different contours of w, in reduced coordinates.

Zones, bounded by dashed lines, correspond to different modes of separation (Z_d ductile, Z_b and Z_m brittle). ϕ is a strain energy index.

What are the principal causes of such phenomena? This is not yet clear, but contributing factors may include:

(i) competition between adhesive energy for interaction across the interface, and cohesive energy for interaction between surface and sub-surface layers. Theoretically it has been shown[31] that this will give a tendency for the solids to avalanche together at a critical separation, especially in the case of planar surfaces, regardless of the stiffness of external supports. Landman et al use the term "jump to contact", and describe its analysis by molecular dynamics simulations[36]. They show that the effect is driven by the tendency of atoms near the interface to optimise the energy needed to embed each atom into the local electron density provided by neighbouring atoms. Also the jump to contact itself produces strain which in turn affects the relevant energies;

(ii) reduced local rigidity of the solids outside the contact region, arising from non-planar geometries such as hemisphere on flat. Theoretically[37] this again limits the proximity at which two solids can be held apart;

(iii) partial conversion of surface energy, released on contact, into energy of plastic deformation, in the case of contact between an asperity and a flat surface[24]; an extreme example of such adhesion-induced plastic deformation could bring the material far into the full plasticity regime at zero load, with the radius of the contact region approaching R^*, giving rise to an almost punch-like geometry. The presence of nanometre-scale stepped topography will affect subsequent discontinuities in contact area at finite loads[30].

7. *To explain strong adhesion, equilibrium fracture mechanics theory is not enough — one must invoke hysteresis[1].* In this chapter we are not primarily concerned with how values of pull- off force are calculated, as this topic is not directly relevant to friction. The details have been reviewed elsewhere[7], and here we need only say that in principle there are two extreme modes of separation of the two solids: gross ductile flow within the bulk of one of the samples, and brittle separation by crack propagation along the interface. The effects of stored elastic strain energy, plastic deformation, and variations in K, Y, w, R^* and P, are summarised in the form of an "adherence map" (fig. 4), which complements the zero-load adhesion map discussed earlier. The equilibrium theory is useful in accounting for two of the most important variables, the geometry and the elasticity, but in practice the large values of w needed for agreement with experiment in practical situations are a sign that other energy-dissipative mechanisms are important also.

Chen et al [43] have recently presented a useful classification of relevant mechanisms that can give rise to hysteresis, including

(a) mechanical hysteresis, arising from intrinsic mechanical irreversibility of many adhesion/decohesion processes, while not implying that the surfaces become damaged or changed in any way,

(b) chemical hysteresis, involving rearrangement of chemical groups, and

(c) bulk plastic or viscoelastic effects.

Under (a) it would be possible to list various types of cyclic variation of contact area with load, with loss of surface energy in the form of phonons giving rise to hysteresis. For smooth adhering bodies following the JKRS behaviour, energy is dissipated mainly upon initial contact (and to a lesser extent during any subsequent decrease in load and area, as discussed earlier by Professor Tabor). If however the surfaces are rough, then at all stages of increase in overall contact area, significant surface energy is lost as each individual asperity makes contact. Any avalanching will tend to amplify this effect. Fig. 5 is a schematic illustration of hysteresis found in the case of contact between a flat surface and a sphere exhibiting stepped nano-topography as suggested by observations of "jumps" (discontinuities) in contact

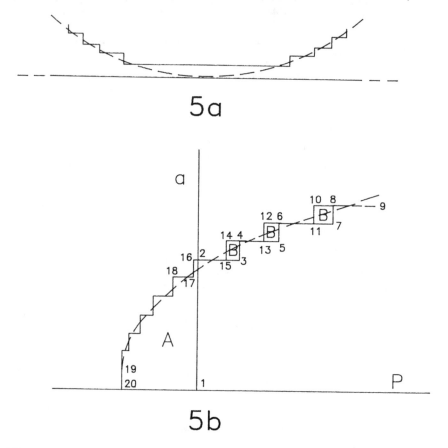

Figure 5. Adhesive contact between plane and asperity whose topography is not perfectly spherical [30]. (a) Profile, showing stepped nano-topography; (b) Corresponding schematic plot of contact radius a as a function of load P, in the order 1 2 3 4 ....Hysteresis areas A, B correspond to dissipation of surface energy. For an ideally smooth surface (dashed line), areas B would not be seen.

area at finite loads. The load at which each flat portion of the asperity parts from the counterface is greater than the load at which it made contact.

An important example of a bulk energy dissipation mechanism is the hysteresis in cyclic variation of stress undergone by an element of volume near the tip of a moving crack. As shown by Barenblatt[38] and discussed by Greenwood and Johnson[39] and by Maugis[15], cohesive forces act across the peripheral crack over a finite length. These deform the crack elastically until the stress singularity predicted by Griffith crack theory is removed, so that the crack tip profile becomes cusp-shaped. At the crack tip the stress now reaches the theoretical strength which is proportional to w, and consequently the drag on the moving crack also varies as w, which will determine the limiting crack speed. Adsorbed gases may screen the adhesion forces, reduce w, and hence increase the crack speed.

For sliding contacts the corresponding processes will necessarily be more complex. However, at this point we may like to note a satisfying analogy, emphasized in Johnson's paper on Aspects of Friction[4]. Generally the adhesive (more precisely, "tangential interaction") term is far too small to account for observed friction forces: however, adhesion can "trigger" a large increase in the plastic strains involved in ploughing, and indeed the principal energy dissipation mechanism is through deformation losses within the bulk of the solids. Correspondingly, the adhesion energy w can trigger bulk energy losses in a *viscoelastic* solid, thus producing a large increase in pull-off force. Quantitatively, this phenomenon is expressed by the master curve of Maugis and Barquins[40]:

$$G - w = w(a_T v)^{0.6} \qquad (15)$$

Here G is the strain energy release rate which describes how the elastic energy varies with contact area at constant δ; $G - w$ is like a crack extension force, given that the simple Griffith criterion is just $G = w$; and the right-hand term represents the drag, which (as discussed above) is proportional to w. It also depends upon the crack speed as shown, a_T being the William-Landel-Ferry shift factor for frequency-temperature equivalence (the exponent 0.6 is empirical - the theoretical prediction simply includes an arbitrary dimensionless function of $a_T v$). Note the multiplicative or "triggering" effect of w on the right-hand side of the equation. If, for example, w is reduced by adsorption of water, then to separate the solids at a given peripheral crack speed, the required strain energy release rate, and thus the required pull-off force, will be lower. The details will, of course, depend upon the experimental geometry, but the master curve is valid for all geometries, loading conditions and elastic properties[15], and completely decouples these from the surface properties (w) and viscoelastic properties (function of $a_T v$).

While on the topic of energy loss mechanisms, it is appropriate to mention the vexed question of *liquid bridges*[5]. Don't we all know that we can build sandcastles with moist sand and not with dry sand, because water has condensed in the narrow gaps between the solid particles? This can happen to the extent that the liquid fills in the gaps between contacting asperities, thereby reducing the effect of surface roughness. But as emphasised by Kendall[1], this effect is often outweighed by the fact that if liquid molecules are present on the solid surface through intermolecular attraction, then the overall value of w must decrease. The presence of a liquid near a good contact between smooth solids *reduces* the adhesion. The meniscus radius r is determined by the surface energy, temperature, molar volume and relative vapour pressure according to the classical Kelvin equation. The force of attraction between the solids is then determined by the contact angle, the wetted area and the Laplace pressure γ_L / r, where γ_L is the surface energy of the liquid. (The additional force arising from the resolved surface tension around the circumference is always small in comparison, unless the contact angle is close to 90°). There is evidence from experiments with the surface forces apparatus[5] that for molecules that interact mainly via a simple Lennard-Jones potential, their bulk surface energy is already manifest at meniscus radii as small as the size of a molecule. By means of the force microscope technique mentioned earlier, the presence of liquid bridges may be detected through a characteristic shape of attractive force curve, which on approach is relatively short-range and shows a discontinuity, whereas on retraction, long-range behaviour is seen, again with a discontinuity. Maugis [15] treats the liquid bridges topic by considering the Laplace term as a restraining pressure acting at the tip of an external crack, and is thus able to evaluate the pull-off force by means of fracture mechanics.

4. Other processes

It will be obvious that this survey of surface forces and adhesion and their relevance to friction has been selective rather than comprehensive - no mention has been made, for example, of the evidence for thermal activation of bonding between ceramic surfaces in contact [7]. The use of adhesive peeling models of stick-slip friction, an important topic [46], is covered in the chapters by Savkoor and Briscoe.

5. Conclusions

1. When considering the role of surface forces in friction, it is important to distinguish the concepts of force of attraction, pull-off force, and adhesion energy. The relation between these parameters is complex, as is the controversial concept of the adhesion or "tangential interaction" component of friction.

2. The main predictions of the contact mechanics of adhesion have been summarised in the form of maps, and the experimental work appears to have entered a new phase.

3. Particular points covered here have included the increase in contact area, elastic or plastic, that is produced by attractive forces; the various effects of roughness on adhesion; the phenomenon of avalanching; and hysteresis processes, involving either surface-mechanical or bulk viscoelastic energy dissipation.

6. Acknowledgment

I thank the Science and Engineering Research Council for research funding granted through its specially promoted programme for particulate technology.

7. References

1. Kendall, K. (1980) Contemp. Phys. **21**, 277.

2. Dowson, D., Taylor, C.M., Godet M and Berthe, D. eds., 'Friction and Traction', Westbury House, Guildford 1981.

3. Spurr, R.T. ref. (2), 34.

4. Johnson, K.L. ref. (2), 3.

5. Israelachvili, J.N. (1985) 'Intermolecular and surface forces', Academic Press, London.

6. Israelachvili, J.N. and McGuiggan, P.M. (1990) J. Mater. Res. **5**, 2223.

7. Guo, Q., Ross J.D.J. and Pollock, H.M. (1989) in 'New materials approaches to

tribology: theory and applications', (MRS Fall Meeting Symposium Proceedings 140), L E Pope et al. (eds.), Materials Research Society, Pittsburgh.

8. Muller, V.M., Yushchenko, V.S. and Derjaguin, B.C. (1980) J. Colloid Interface Sci. **77**, 91.

9. Derjaguin, B.V., Muller V.M. and Toporov, Y.P. (1975) J. Colloid Interface Sci. **53**, 314.

10. Bradley, R.S. (1932) Phil. Mag. **13**, 853.

11. Pashley, M.D. (1984) Colloids and Surfaces **12**, 69.

12. Tabor, D. J. (1977) J. Colloid Interface Sci. **58**, 2.

13. Sperling, G. (1964) Dissertation, Karlsruhe Technical High School.

14. Johnson, K.L., Kendall, K. and Roberts, A.D. (1971) Proc. R. Soc. **A324**, 301.

15. Maugis, D. in 'Adhesive bonding' (L-H Lee, ed.), Plenum 1991, p.303; see also [11].

16. Pollock, H.M., Maugis D. and Barquins, M. (1978) App. Phys. Lett. **33**, 798.

17. Kendall, K., McN Alford N. and Birchall, J.D. (1987) Proc. R. Soc. **A412**, 269.

18. Tabor, D. (1975) in: 'Surface Physics of Materials 2' (J M Blakely, ed.), Academic Press.

19. Tabor, D. (1987) in 'Tribology - 50 years on' (I Mech E, ed.), Mechanical Engineering Publications.

20. Kendall, K. (1986) Nature **319**, 203.

21. Briscoe, B.J. and Kremnitzer, S.L. (1979) J. Phys. D: Appl. Phys **12**, 505.

22. Georges, J-M (ed.), 'Microscopic aspects of adhesion and lubrication', Elsevier (1982).

23. Barquins, M and Maugis, D (1982) J. Mécanique Theor. Appl. **1**, 331.

24. Maugis, D. and Pollock, H.M. (1984) Acta Metall. **32**, 1323.

25. Roy Chowdhury, S.K., Hartley, N.E.W. Pollock H M and Wilkins, M.A. (1980) J. Phys. D: Appl. Phys. **13**, 1761.

26. Roy Chowdhury, S.K. and Pollock, H.M. (1981) Wear **66**, 307.

27. Fuller, K.N.G. and Tabor, D. (1975) Proc. R. Soc. **A345**, 327.

28. Johnson, K.L. in 'Theoretical and applied mechanics' (W T Koiter, ed.), North Holland 1977, p 133.

29. Chang, W.R., Etsion, I. and Bogy, D.B. (1988) Trans. ASME J Tribol. **110**, 50.

30. Ross, J.D.J., Pollock H.M. (1991) and Guo, Q. Powder Technol. **65**, 21.

31. Smith, J.R., Bozzolo, G., Banerjea A. and Ferrante, J. (1989) Phys. Rev. Lett. 63, 1269.

32. Pollock H.M. and Roy Chowdhury S.K. in ref. 22, p253.

33. Gimzewski, J.K. and Möller, R. (1987) Phys. Rev. B26, 1284.

34. Burnham, N.A., Dominguez, D.D., Mowery R.L., and Colton, R.J. (1990) Phys. Rev. Let. 64, 1931.

35. Cohen, S.R., Neubauer G. and McClelland, G.M. (1990) J. Vac. Sci. Technol. A8, 3449.

36. Landman, U., Luedtke, W.D., Burnham N.A. and Colton, R.J. (1990) Science 248, 454.

37. Pethica, J.B. and Sutton, A.P. (1988) J. Vac. Sci. Technol. A6, 2490.

38. Barenblatt, G.I. (1962) Adv. Appl. Mech. 7, 55.

39. Greenwood, J.A. and Johnson, K.L. (1981) Phil. Mag. 43, 697.

40. Maugis, D. and Barquins, M. (1978) J. Phys. D 11, 1989.

41. Fuller, K.N.G. and Roberts, A.D. (1981) J. Phys. D: Appl. Phys. 14, 221.

42. Maugis, D. in ref. 22, p.221.

43. Chen, V.L., Helm, C.A. and Israelachvili, J.N., J. Phys. Chem. to be published.

44. Burnham, N.A., Colton, R. and Pollock, H.M. (1991) J. Vac. Sci. Technol. A9, 2548.

45. Burnham, N.A., Colton, R. and Pollock, H.M. (1992) Phys. Rev. Lett., to be published.

46. Thornton, C. (1992) J. Phys. D : Appl. Phys. 24, 1942.

PARTICLE-PARTICLE INTERACTION IN GRANULAR MATERIAL

A. R . THÖLÉN
Laboratory of Applied Physics
Technical University of Denmark
Building 307
2800 Lyngby, Denmark

ABSTRACT. Transmission electron microscopy is used to study adhesion contacts between small (20-100 nm) Fe-Ni particles as well as scanning tunneling microscope (STM) tips before and after use. The finer details of adhesion contact are revealed by computer simulation of electron microscope contrast. The electron microscope is a very powerful tool in friction studies being able to reveal elastic and plastic deformation at contact as well as material transfer.

1. INTRODUCTION

Electron microscopy can help in elucidating adhesion contact between flat surfaces and/or small asperities which are of great importance in friction. Strain fields from adhesion contacts between oxide-free small metal particles (20 - 100 nm) in contact have earlier been observed (1 - 4) and explained with the JKR theory (5). Further results on particle-particle or tip-surface interaction can be found in (6-10).

We will here present results on the adhesion contact between Fe-Ni particles, which have been produced with a new method. Improved computer simulations of electron microscope contrast, which help in gaining more information from the displacement field near contact, have also been performed.

Furthermore,some results from transmission electron microscopy of STM - tips will be shown .This represents an area with intentional or unintentional contact between two bodies and has obvious parallels in friction . Electron microscopy is here of great help in revealing more about the geometry and possible problems at contact. It is also a

I. L. Singer and H. M. Pollock (eds.), Fundamentals of Friction: Macroscopic and Microscopic Processes, 95–110.
© 1992 *Kluwer Academic Publishers. Printed in the Netherlands.*

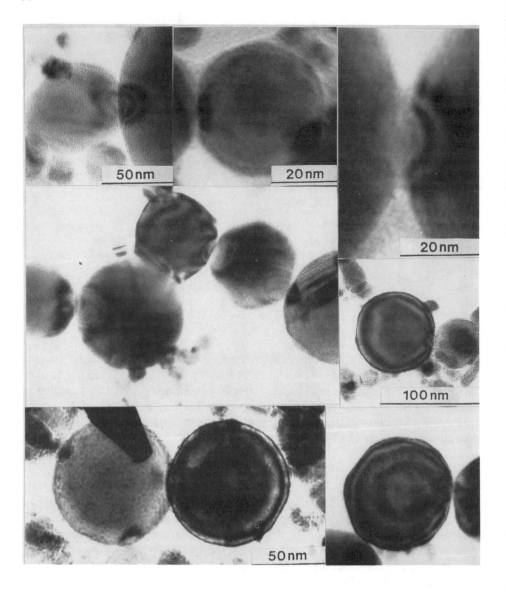

Fig.1 Adhesion strain fields between Fe-Ni particles. In general the contrast is visible in only one of the particles, but a correct tilting can render it visible in the other particle instead. The contrast is thus very dependent on the orientation of lattice planes in the two particles.

convenient tool to study local deformation and transfer of material.

2. EXPERIMENTAL

In order to observe contrast from metal contacts in the electron microscope it is essential that two clean (oxide and contamination free) surfaces meet. This prerequisite can be fulfilled in different ways. In Ref.1 it was described how coherent Fe-Ni particles were extracted from a copper matrix by means of an etching-evaporation-etching process. The particles first touch below a liquid surface and are not oxidized at contact. An alternative

method was gently to press fine aluminium particles against each other, which caused the aluminium oxide film to break and this resulted in a pure metal-metal contact (1).
Particles and adhering particle contacts can also be produced by evaporating metal in an inert gas (2-4). The particle size depends on gas pressure and evaporation temperature and increases with these variables. The particles in the adhesion studies are 20-100 nm and were typically obtained in argon at a pressure of 10 - 20 torr. The enormous surface area when evaporating ensures that the ensemble of particles acts as a pump and that many particles thus meet with very clean surfaces. They grow by repeated coalescence and reach a final size essentially depending on gas pressure.
The Fe-Ni particles presented here were made by sintering 5 - 10 nm particles, which had first been produced by simultaneous evaporation of iron and nickel in helium (3 torr), at 591 K for 48 hours in a reducing atmosphere (11). In this way particles with diameter 50 - 100 nm were obtained which could interact with one another having oxide-free surfaces.
The STM tips of tungsten were made by electropolishing a thin wire of tungsten (12,13), while those of platinum-iridium simply were cut from a wire with a pair of tongs. The particles and the STM-tips were observed in a Philips 430 transmission electron microscope operating at 300 kV.

3. EXPERIMENTAL OBSERVATIONS AND DISCUSSION

Most particle contacts do not show any special electron microscope contrast. However, at some Fe-Ni particle contacts stress fringes were observed (Fig 1). These were usually visible in only one of the particles, but could sometimes be observed in both particles. If only visible in one of the particles a proper tilting in the electron microscope would render them visible in the other particle. The orientation of lattice planes relative to the incoming beam is thus essential for the contrast. In a contact with no visible contrast, stress fringes may appear if the particles are properly tilted. It is on the other hand found that most contacts between evaporated particles do not show any strain. This could be part of the history of forming larger particles form from smaller ones or it could be due to sintering in stressed necks. A further observation is that smaller particles, with a size less than about 20 nm, never showed contrast at contact. It can be argued that the contrast is weaker and more difficult to detect for smaller particles and eventually impossible to register due to the diffraction conditions. A more likely explanation is, however, that the stressed particle necks rapidly sinter, which of course would exclude any possibility to observe a strain contrast.

4. THE GEOMETRY OF CONTACT

The JKR-model describes how spherical (or ellipsoidal) particles are squeezed into intimate contact due to adhesion forces. The adhesion makes the particles meet across an area with a certain radius **a** rather than at a point. The contact area is stressed compressively in the centre and the stress is tensile further out. Along the rim the normal stress, σ, in fact reaches infinity in this model. The predicted values are given by the following expressions.

$$\sigma = \frac{2*E*a}{3*\pi*(1-v^2)*R} * \frac{2-3*\dfrac{r^2}{a^2}}{\sqrt{1-\dfrac{r^2}{a^2}}} \tag{1}$$

$$a = \sqrt[3]{\frac{9*\pi*(1-v^2)*R^2*\gamma_{eff}}{4*E}} \tag{2}$$

$$\gamma_{eff} = 2*\gamma_s - \gamma_{gb} \tag{3}$$

σ=normal stress across particle contact
R=particle radius
a=radius of contact
r=radius in the plane of contact
E=modulus of elasticity ($=2.1*10^{11}$ N/m² here)
v=Poisson's number (=0.33 here)
γ_s =surface energy
γ_{gb}=grain boundary energy ($=1/3*\gamma_s$ here)
γ_{eff}=effective grain boundary energy
ξ_g=extinction distance ($3*10^8$ m here)

Using these expressions the contact radius is calculated and shown in Table I for various radii,R,and surface energies,γ_s

R/ξ_g	2.0	1.5	1.0	0.5
R/nm	60	45	30	15
γ_s/J/m²				
6.0	10.3	8.47	6.46	4.07
4.8	9.52	7.86	6.00	3.78
3.6	8.65	7.14	5.45	3.43
2.4	7.56	6.24	4.76	3.00
1.2	6.00	4.95	3.78	2.38

TABLE I. Contact radius **a** (in nm) as a function of particle radius **R** (expressed in extinction distance or nm) and surface energy γ_s (J/m²). The various surface energy values correspond to γ_{eff}=10,8,6,4 and 2 J/m².

5. ELECTRON DIFFRACTION FROM SMALL PARTICLES

The stress fringes in the Fe-Ni particles in Fig.1 are due to amplitude contrast which is very sensitive to small stresses or rather a small variation in displacement. In order to gain more information from contacting small particles we have simulated contrast from adhering particles with radius 0.5,1.0,1.5 and 2.0 ξ_g and with different deviations, w=ξ_gs, from the Bragg-condition (w=-1,0,1,2).
A knowledge of the displacement which is a prerequisite for calculating contrast, was obtained by integrating equation (1). The intensity at each point was then calculated from the two-beam Howie-Whelan equations (Ref.14) using a matrix multiplication method (15). The contrast was normalized against the intensity from a foil of thickness equalling the particle radius. The **g**-vector is assumed to be perpendicular to the plane of contact and a low order reflection with an extinction distance of 30 nm was used. The results of

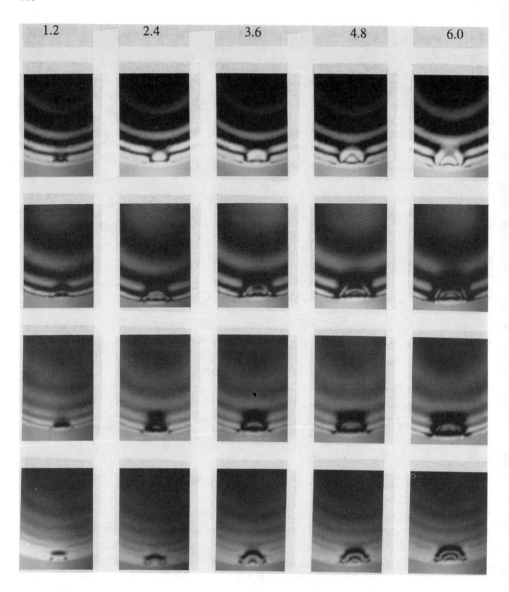

Fig 2a. Contrast from adhesion strain fields. R=2*ξ_g (=60 nm).
The surface energy is 1.2,2.4,3.6,4.8 and 6.0 J/m^2
resp. From top to bottom w=-1,0,1,2.

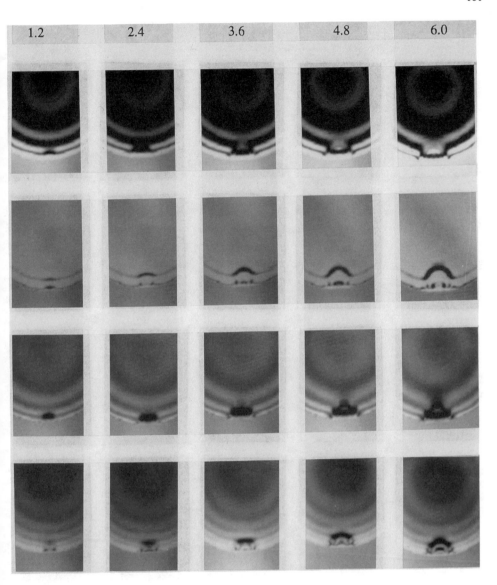

Fig 2b. Contrast from adhesion strain fields. R=1.5*ξ_g (=45 nm).
The surface energy is 1.2,2.4,3.6,4.8 and 6.0 J/m²
resp. From top to bottom w=-1,0,1,2.

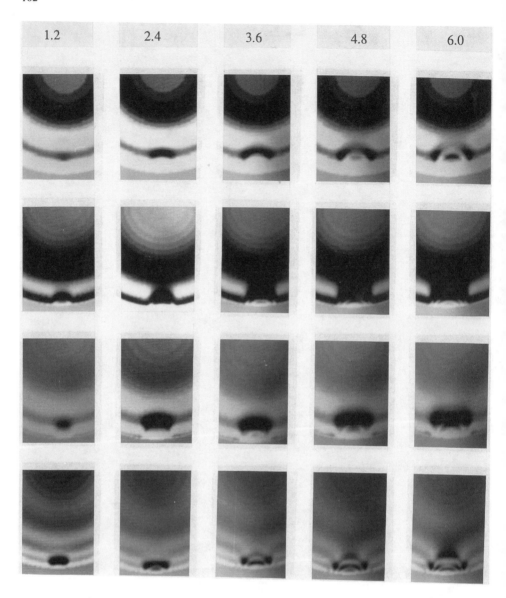

Fig 2c. Contrast from adhesion strain fields. $R=1*\xi_g$ (=30 nm).
The surface energy is 1.2,2.4,3.6,4.8 and 6.0 J/m^2
resp. From top to bottom w=-1,0,1,2.

1.2   2.4   3.6   4.8   6.0

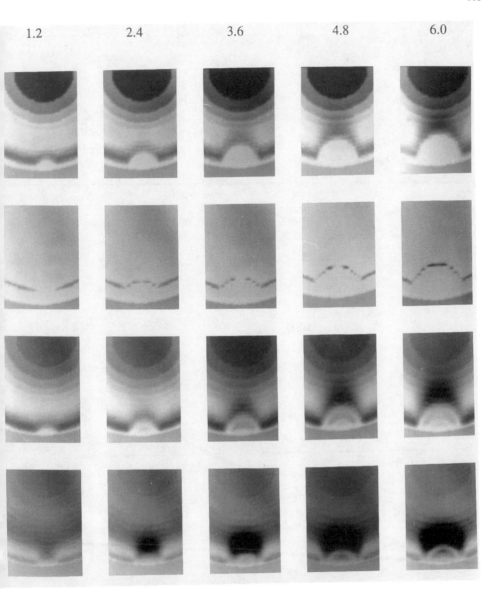

Fig 2d. Contrast from adhesion strain fields. R=0.5*ξ_g (=15 nm).
The surface energy is 1.2,2.4,3.6,4.8 and 6.0 J/m^2
resp.From top to bottom w=-1,0,1,2.

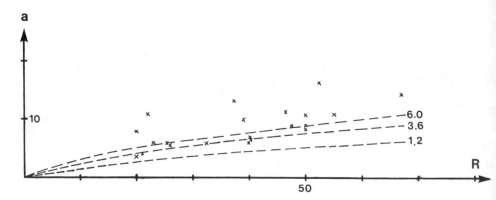

Fig.3. Measured values of contact radius **a** (nm) and particle radius **R** (nm). The theoretical relation between **a** and **R** for surface energies of 1.2, 3.6 and 6.0 J/m² is shown.

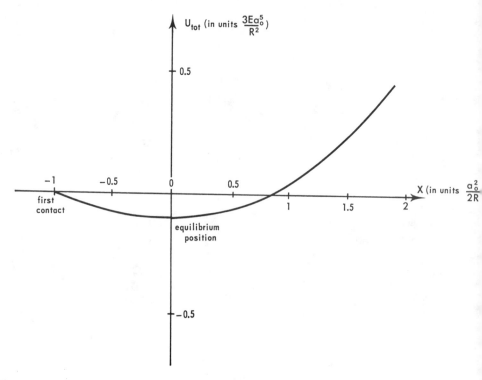

Fig.4. Potential energy, U_{tot}, of two contacting particles as a function of a coordinate $x=a^2/R$ along the centre to centre line. x=0 at equilibrium.

these calculations are shown in Figs. 2 a-d.

Many of the contrast effects in Fig. 1 are visible in the simulations and they are thus of great help in obtaining maximum information from particle contacts. Furthermore, it is difficult to make a controlled diffraction experiment with such small particles and obtain a precise information on the diffraction vector and the deviation from the Bragg condition. It is, however, in principle possible to take a series of diffraction patterns from connecting particles, but it is a tedious and not a very precise method as the particles could move slightly between exposures. A careful mapping is thus of great value to obtain information about the diffraction variables without actually measuring them one by one.

The contact radius **a** and the particle radius **R** were measured for a large number of contacts and the results are plotted in Fig.3. These results are quite scattered and lead in many cases to a larger surface energy value than expected. The results obviously show that the particles are not exactly spherical and that the local radius of the particles at contact, R_1, often is larger than the particle radius **R**. Particles could also have rolled into such a low energy position after contact (2).

6. VIBRATIONS AT CONTACT

The JKR-theory predicts the static equilibrium when two particles come together. A more careful consideration of this contact reveals that it must be accompanied by vibrations (16). The potential energy of two particles as a function of the centre to centre distance is shown in Fig.4. The curve is almost parabolic near the equilibrium position and this therefore leads to harmonic vibrations of angular frequency ω for two contacting spheres.

$$\omega = \sqrt{\frac{9Ea}{8\pi\rho R^3}} \qquad (4)$$

The frequency for nickel particles with radius 30 nm is **f**=5.70 GHz. This is close to the eigenfrequency of the spheres and coupled vibrations could easily appear in the spheres as a consequence of contact. These vibrations could then in fact be tied to twinning and martensitic transformation, which has been observed in such systems (2-4,17). Similar phenomena could also be expected in other geometries and would thus be of importance in friction.

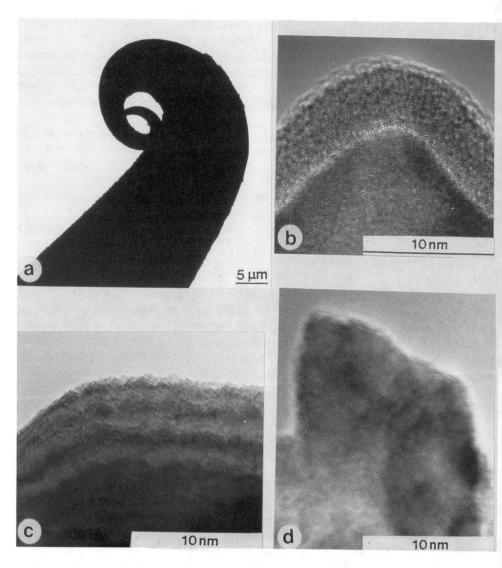

Fig.5. Transmission electron micrographs of scanning tunneling microscope tips.
a) Bent tungsten tip after use in the STM. b) Heavily oxidized tungsten tip which would show bad performance and most likely bend in the STM. c) Oxide free tip of tungsten made by electropolishing. Atomic steps are visible at the surface. This tip would show excellent performance in the STM. d) Platinum-iridium tip cut with a pair of tongs. Atomic resolution of lattice planes can be detected.

7. PLASTIC DEFORMATION

No defects besides twins are observed in small particles. The relative occurrence of these twins is related to the twin boundary energy of the metal. At many contacts (2-4) twins are observed to cross a particle, starting at the rim of particle contact. Most likely these twins are associated with and have formed due to the adhesion stress fields. However, the following relation must approximately be obeyed for twinning across a particle to occur.

$$\pi a^2 \gamma_{eff} > \pi R^2 \gamma_t \quad (5)$$

where γ_t = twin boundary energy

This relation simply states that the energy to form a twin must be taken from the excess surface energy released at contact.

8. SCANNING TUNNELING MICROSCOPY AND DEFORMATION OF STM - TIPS

A closely related geometry to particle - particle contact is found in scanning tunneling microscopy with the fine tip moving in the proximity of an uneven surface. Sometimes the tip is interacting with the surface with possible elastic and plastic deformation including transfer of material. This has obvious parallels in friction. We have made a holder to investigate STM-tips in the transmission electron microscope before and after a run in the STM. It is observed that tungsten tips sometimes give quite erroneous results (13). Comparing the same tip before and after use in the STM it was often found that the STM caused the tips to bend (Fig.5a). This bending depends on an unwanted oxide film (Fig.5b) which prevents the STM tip to perform properly. It is accordingly pressed heavier into the solid surface to obtain a tunneling current. This causes sooner or later the oxide to break and the STM is then reported to function in spite the bent tip. An improved polishing of the tip (12) allowed us to make oxide free tips (Fig.5c) with very good performance. A platinum-iridium tip cut with a pair of tongs and with no surface coating is shown in Fig.5d. These tips showed excellent performance with atomic resolution. Transmission electron microscopy of STM-tips is also very useful to study material transfer during operation, which can easily be detected as a change in geometry and/or elemental composition of the tip.

108

9. CONCLUSION

Transmission electron microscopy offers a unique possibility to investigate asperity contact in different geometries and in detail study the adhesion stress fields with elastic and plastic deformation as well as transfer of material. This makes it a very powerful tool in the study of friction. Electron microscopy can be used for surface and subsurface deformation studies before and after a friction experiment. Furthermore, it could even be used for in situ studies of friction in the electron microscope as a result of interaction between a needle and a surface.

10.REFERENCES

1) Easterling,K.E. and Thölén,A.R.Acta Met.,1972,**20**,1001
2) Hansson,I. and Thölén,A.,Phil.Mag.A.,1978,**37**,No.4,535
3) Thölén,A.R.,Acta.Met.,1979,**27**,1765
4) Thölén,A.R.,Microscopic aspects of adhesion and lubrication,Elsevier Scient.Publ.Company,1982,263
5) Johnson,K.L.,Kendall,K. and Roberts,A.D.,Proc.Roy.Soc.,1971,**A324**,301
6) Maugis,D. and Barquins,M.,J.Physique-Lettres,1981,**42**,L95.
7) Maugis,D. and Barquins,M.,J.Phys.D:Appl.Phys.,1978,**11**,1989.
8) Pollock,H.M.,Shufflebottom,P. and Skinner,J.,J.Phys.D:Appl.Phys,1977,**10**,127.
9) Pollock,H.M.,J.Phys.D.:Appl.Phys.,1978,**11**,39.
10) Ross,J.D.J.,Pollock,H.M. and Guo,Q.,Powder Technology,1991,**65**,21.
11) Bentzon,M.D.,Linderoth,S.,Pedersen,A.S. and Madsen,M.B.Sintering '91-Vancouver (in print)
12) Fasth,J.E.,Loberg,B. and Nordén,H.,J.Sci.Instrum.,1967,**44**,1044
13) Garnaes,J.,Kragh,F.,Mørch,K.A. and Thölén,A.R.,J.Vac.Sci.Technol.,1990,**A8(1)**,441)
14) Hirsch,P.B.,Howie,A.,Nicholson,R.B.,Pashley,D.W. and Whelan,M.J.,1977.Electron Microscopy of Thin Crystals,2nd ed.,R.E.Krieger
15) Thölén,A.R.,Phil.Mag.,1970,**22**,175
16) Thölén,A.R.,phys.stat.sol.(a),1980,**60**,151
17) Thölén,A.R.,Phil.Mag.A.,1986,**53**,No2,259.

Discussion *following the lecture by A R Thölén on "Particle-particle interaction in granular material".*

G McCLELLAND. I find the discussion on vibrations in the particles intriguing. Can you induce waves between contacting particles by microwave radiation, ultrasound etc?

A R THÖLÉN. That is an interesting point you raise. We have not entered any further into this question, but I would like to mention that the magnitude of vibration frequencies between contacting spheres is about the same as for standing waves in a single sphere.

F D OGLETREE. (a) In the images of the small particles, does TEM heat the particles enough to anneal them or to affect their images? (b) Regarding the images of STM tips: do you actually get contrast from a single atom at the end of the tip, or do you have to have several planes or stacks of atoms?

A R THÖLÉN. (a) Heating of particles has been much discussed, but for much (about ten times) smaller particles than those I observe. Some very nice results from e.g. Sweden and Japan show these small particles moving vividly in the TEM and even transforming in front of your eyes. The question is what is going on here, but the temperature, which is unknown, plays an important role. No such effect of temperature is observed in my larger particles (20-100 nm), but sometimes one could see whole particle chains or branching networks vibrate.

(b) Single atoms can be seen with the TEM. I do not say we do it here. We see lattice planes or columns of atoms, although tungsten atoms diffract quite heavily which could render them visible. The point resolution of our microscope is 2.3 Å.

ANON. (a) You showed beautiful necklaces of particles. Why do you get necklaces rather than a more compact shape, and why do you get the same size of particles?

(b) Why only two neighbours (for the cobalt particles)?

A R THÖLÉN. (a) Ferromagnetic particles yield necklaces, but for other particles a branching structure is obtained. The size distribution of particles is log-normal and this is thought to be due to growth by coalescence of small particles, which leads to such a distribution. (b) The particles are small ferromagnets and align accordingly. Large scale models of ferromagnetic particles show the same behaviour.

I L SINGER. When you take into account the fact that the particles probably have thin oxide layers, first of all how would that affect your interpretation of stress-strain fields in metal-metal contacts, and conversely, what can you tell us about the effect of these thin oxide layers that we have to deal with, from your data?

A R THÖLÉN. Thank you for that remark. I forgot to speak about the oxides; this is a problem, as all small particles are normally covered by oxide. But in the way we produce particles and let them interact, they sometimes meet with clean surfaces. By making them e.g. with particle extraction from a material with precipitates they will meet with clean surfaces below a liquid surface. When evaporating in a gas you have so many particles, and so much free surface that acts as a pump, that there are not enough oxygen atoms for all surfaces. When two oxidized particles meet, we have not observed the strain phenomenon in the TEM. This is most likely due to the lower surface energy values for oxides and the interaction is thus not strong enough to show up in the TEM.

MODELS OF FRICTION BASED ON CONTACT AND FRACTURE MECHANICS

Arvin.R.Savkoor

Faculty of Mechanical Engineering & Marine Technology
Transportation Division, Vehicle Systems
Delft University of Technology
2628 CD DELFT, the NETHERLANDS

ABSTRACT

Theoretical models of dry friction of solids are presented and discussed. The emphasis is placed on the basic aspects related to adhesion at the interface and the relative motion between surfaces of solids. First the standard approach of contact mechanics will be described and the Cattaneo-Mindlin theory [1] based on Coulomb friction is examined. The conventional approach recognizes static friction as a constraint on the relative tangential displacements and tacitly acknowledges adhesion, but it falls short of exploring its full implications. Instead of invoking friction laws, the more recent and basic approach treats the contact interface as a bonded joint where adhesion acts to constrain relative displacements in any direction. Considering the geometry outside the contact as an external crack, the methods of fracture mechanics are introduced and applied to study the initiation and growth of the crack that leads to the separation of solids. The effectiveness of this approach has already been proven by the JKRS theory [2], which describes the influence of adhesion between solids loaded by purely normal forces. Shear tractions arising due to friction at the interface of dissimilar materials, again loaded by normal forces, influence both the contact area and adhesion but this has a minor effect on the JKRS equation. When a tangential force is applied, depending upon its magnitude and the given situation, the contact interface responds in one of the many different ways- by peeling, by slipping, by maintaining the status quo and under certain special conditions by buckling. Tangential forces smaller than the peeling limit force cause stable normal separation which is controlled by the stress intensity factors of mode I, and the shear modes II and III. Next, the paper considers the contact interactions when the tangential force exceeds the peeling limit and continues to increase until shear fracture is initiated and slipping becomes inevitable. This point marks the limit force of static friction. The fracture characteristics of the shear mode are essentially different from those of the normal mode associated with peeling. Whereas peeling is ideally a reversible process, the process of slipping by shear mode is not. These physical aspects consider the essentially irreversible nature of the slipping process and the wear associated with it. This discussion serves as a basis of a model of an ideal process of frictional slipping. It is characterized by a two parameter model for describing the fracture strength of interfacial films in the initial virgin and in the damaged states. The model rules are set up to define the boundary conditions of shear fracture and the analysis is carried out to describe the transition from static to sliding friction under conditions of partial slip. Some thoughts are recorded to indicate how these rules may be extended to develop models for describing kinetic friction.

1   INTRODUCTION

Understanding and controlling friction, both static and kinetic, is of considerable practical importance both in engineering practice and in many other non-engineering activities of everyday life. In engineering applications, quantitative estimates of frictional forces and any other physical effects caused by friction are required for designing products and devices with the aid of simple and reliable rules. In most practical applications the designer can safely ignore forces due to natural adhesion although he is fully aware of the technology of engineering adhesives. The engineer cares little, how or why friction originates but his interest lies in describing the forces using reliable models. Putting it bluntly, adhesion without engineering is nonsense and friction a nuisance! The view that adhesion is trivial is fallacious; it will

I. L. Singer and H. M. Pollock (eds.), Fundamentals of Friction: Macroscopic and Microscopic Processes, 111–133.
© 1992 *Kluwer Academic Publishers. Printed in the Netherlands.*

become clear from the following discussion where we start from the conventional wisdom concerning friction and try to establish a factor that is common to all friction.

The usual pragmatic view is simply to postulate friction as a basic physical force of nature which arises as a reaction force when bodies in contact move or tend to move relative to one another. The two essential notional ingredients in the definition of friction are those concerning contact and relative motion of points in the interface. The words "tendency" to move are deliberately defined imprecisely to convey the notion that while no relative motion occurs, the interface can sense somehow, the remote applied force and respond to it by generating an equal and opposite reaction force of static friction. Since friction is produced without any relative motion the question arises whether or not relative motion is really essential for understanding the basic cause of friction. All one needs, is probably the contact of solids, an external tangential force and of course, Newton's laws to understand the basic nature of friction - in the static case as a pure reaction force or a constraint on displacement. Another point to be noted is that static and kinetic (sliding) friction are two quite distinct and equally important notions. The distinction becomes even more pronounced from the energetic standpoint - static friction is (an almost ideally) conservative process (if solids are perfectly elastic) whereas kinetic friction is always associated with energy dissipation. This raises the question why should considerations of energy dissipation be so important in modelling friction.

Friction is associated with a number of other complex phenomena such as material transfer and wear, heat dissipation, vibrations and noise, triboelectricity; all these are labelled as "effects" which may influence friction but are non-essential in the context of the basic cause. Bowden and Tabor [3] have conclusively shown that the influence of the normal load and surface roughness on friction can be explained in many instances through their influence on the area of (real) contact. The basic laws of friction known since the works of Amontonsand Coulomb are accepted, although not all to the same extent.

2 TRANSMISSION OF TANGENTIAL FORCES BY FRICTION
The standard approach based on Coulomb friction

2.1 GLOBAL AND LOCAL FORMS OF THE FRICTION LAW

If two rigid bodies are subjected to normal and tangential forces and furthermore are prevented from rotating, the relative motion between the bodies may be defined by two possible states, static and sliding. The actual state depends, according to the global version of Coulomb friction, upon the ratio of the tangential (T) and normal (P) force components of the resultant reaction force acting on each body. If $T < \mu P$ the state is static (no sliding) and sliding occurs if $T \geq \mu P$. The coefficient of friction μ is considered as a characteristic constant of the materials (solids and interface). How does the interface transmit tangential forces by friction of two elastic bodies? This problem was addressed independently by Cattaneo and Mindlin. Before describing their results it is important to note that one of laws of friction states that friction is independent of the nominal area of contact which suggests that the basic relations given above apply locally to any element of the contact interface. In terms of the local traction components namely, the normal pressure p and the shear traction τ, it is postulated that no relative motion can occur at a given point if $\tau < \mu p$ and if local slipping (relative motion) takes place the shear traction assumes its limiting value $\tau = \mu p$. This is the local version of the Coulomb law. The distribution of normal contact pressure has a significant influence on the state of relative motion. Generally, the normal pressure in the contact of elastic bodies is distributed non-uniformly and it is influenced strongly by the geometrical shape of the bodies. Similarly, the distribution of shear traction is generally non-uniform depending on the geometry and the state of relative motion. The aforementioned facts suggest clearly that for given normal and tangential forces such that $T < \mu P$, while there is no slip at some points, slip can occur at other locations inside the contact area.

2.2 TANGENTIAL CONTACT PROBLEM OF CATTANEO-MINDLIN

The works of Cattaneo [1a] and Mindlin [1b] have an important place in contact mechanics and tribology. These illustrate the nature of static friction as a manifestation of adhesion. The starting point of the analysis is the point contact configuration (undeformed state) of Hertz, namely two perfectly elastic and smooth spheres are pressed by a purely normal force P. The problem considered is what happens when a tangential force T is applied subsequently, at some point remote from the area of contact. Clearly, because inertial effects are not of interest, the useful range of T based on Coulomb friction, is $0 < T < \mu$ P.

Assuming that normal deformations are not affected by shear tractions due to T within the declared range, the normal contact problem is simplified and remains Hertzian. The radius a, of the contact circle and the distribution of normal pressure p, are given by the Hertzian point contact equations: (Consult the classic work of Johnson [4]) ; see also Appendix, page 596.)

$$a = \left(\frac{PR}{K} \right)^{1/3} \quad and \quad p(r) = \frac{3P}{2\pi a^2} \, (a^2 - r^2)^{1/2} \tag{1}$$

The constants R and K, are respectively, the effective geometrical and elastic parameters of the two solids (subscripts 1 and 2):
$1/R = 1/R_1 + 1/R_2$ and $K = (4/3) E^*$, where $1/E^* = (1 - v_1^2)/E_1 + (1 - v_2^2)/E_2$;

2.3 ANALYSIS ASSUMING NO SLIP (Static friction/ complete adhesion):

If the static friction in the contact interface can hold the interfaces together and prevent slip at any point inside the contact interface, such points undergo a uniform tangential displacement δ with respect to points remote from the contact area. This uniform translation in the manner of a "rigid body" is called by the term tangential "shift". It is given by

$$\delta = \frac{T\lambda}{8a} \quad where, \quad \lambda = \frac{(2-v_1)}{G_1} + \frac{(2-v_2)}{G_2} \tag{2}$$

G_1 and G_2 being the shear moduli of the materials of the two solids. The tangential force **T** corresponding to this shift, introduces shear tractions at the interface in the same direction as **T** (or δ), without causing any distortion of shape or size of the contact circle. The distribution of shear traction τ within the contact [1a,1b] is given by:

$$\tau(r) = \frac{T(a^2 - r^2)^{-1/2}}{2\pi a} \tag{3}$$

It is interesting to note that the shear traction is axially symmetric, but only in magnitude. The traction vectors are directed everywhere parallel to the applied force. Another important point to note is that for any value of T, the shear traction becomes unbounded at the edge of the contact. This is precisely the location where the Hertzian normal pressure drops to zero. Cattaneo and Mindlin argued that points close to the boundary of the contact area, according to Coulomb law cannot support the shear tractions and must therefore slip spontaneously, however small be the tangential force. The assumed boundary condition of static friction which implies tangential adhesion is overruled by the boundary condition based on (local form) of Coulomb friction.

114

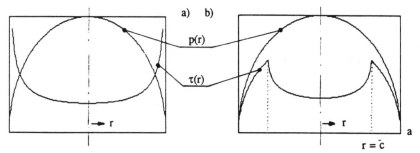

Fig.1: The distribution of normal pressure and shear traction (Hertzian configuration):
a) No-slip anywhere b) Partial Slip for T<μP in the annular region c≤r≤a (Coulomb friction).

2.4   ANALYSIS OF SLIP BASED ON COULOMB FRICTION

In view of the axial symmetry of the distribution (magnitude) of shear it is reasonable to expect that as tangential force T increases slipping will spread radially inwards from the boundary towards the centre of contact. For some finite tangential force less than that to cause total sliding, slipping is confined to an annular region concentric with the contact circle. The radius c of the inner circular region where slip is prevented by static friction is given by:

$$c = a \left(1 - \frac{T}{\mu P} \right)^{1/3} \tag{4}$$

The shear traction in the annular region is directly proportional to the Hertzian normal pressure with μ as the constant of proportionality. The distribution of traction inside the (locked) region of static friction has to satisfy the boundary condition of uniform displacement (shift) of the bulk bodies. The solution for traction is:

$$\text{In the slip region } c \leq r \leq a: \quad \tau(r) = \frac{3\mu P}{2\pi a^3} (a^2 - r^2)^{1/2} \tag{5}$$

$$\text{In the no-slip region } r \leq c: \quad \tau(r) = \frac{3\mu P}{2\pi a^3} \left\{ (a^2 - r^2)^{1/2} - (c^2 - r^2)^{1/2} \right\} \tag{6}$$

The expressions given above may be viewed as approximate solutions because it turns out that the vectors of calculated slip corresponding to the shear traction for points in the region of slip are not directed exactly in the opposite sense. The directional mismatching is small enough to be ignored.
The solution for the tangential shift δ is given by:

$$\delta = \frac{3(2-\nu)\mu P}{8Ga} \left\{ 1 - (1 - \frac{T}{\mu P})^{2/3} \right\} \tag{7}$$

Since the normal and shear traction at any point determine whether slipping occurs or not, the local version of Coulomb law functions as the criterion to predict the local failure of the adhesion. This is the standard procedure of contact mechanics for the specification of boundary conditions on the shear tractions leading to slip. Slipping also circumvents the issue of what to do about the singularity in shear traction predicted by the linear theory with boundary conditions of no-slip.

2.5   CONTACT AND FRICTION OF A FLAT PUNCH (Boussinesq)

The contact problems of Hertz and Cattaneo-Mindlin are extremely important for understanding and controlling a large number of industrial tribological situations. The problems illustrate the essential features of how the local version of Coulomb friction works in contact problems involving non-uniform distributions of pressure and traction. Extensive experimental work has been done amongst others by Johnson [5] to validate the Cattaneo-Mindlin theory and with that to establish the credibility of the local version of Coulomb law as a viable model for engineering applications. Also for basic studies, the Hertzian configuration of contact considered in this analysis is a likely candidate-model of rounded tips of asperities of surfaces (Greenwood permitting! [6]).

The Hertzian configuration is fairly simple by current standards of complexity that can be handled by the powerful methods of computational contact mechanics developed by Kalker [7] but, it is not the easiest problem to consider in combination with other formidable difficulties posed by the physical interactions at the interface. The non-conforming Hertz geometry of solids in the undeformed state is characterized by an advancing type of contact. The problem becomes non-linear due to variable boundary conditions. The contact area is not fixed but depends on normal force. A source of even greater discomfort in the analysis and interpretation is encountered in its interaction with the problem of Cattaneo-Mindlin. Because the shear and normal tractions in the interface have contrasting distributions, these give rise to the division of the contact in slip and non-slip regions which themselves vary with the applied force. The magnitude of slip is not distributed uniformly within the region of slip. While such complex aspects confirm the robustness of the Coulomb formulation, for basic studies these are not very helpful.

Are there any simpler configurations where the deformation of bulk solids lends itself to simpler interpretation of interfacial behaviour? The determination of how bulk solids distribute and redistribute remote applied forces when surface elements slip is concerned more with the mechanics of the bulk rather than the surface. In this context, a configuration described below merits some consideration. The configuration may have some use also for designing basic experiments.

The configuration in question is a rigid punch in the shape of a flat-faced circular cylinder is axially pressed against the surface of a elastic solid. This is one amongst many problems of boundary loading of a half-space, considered in Boussinesq's significant work. Clearly, the main advantage of this configuration is simplicity because, the contact geometry is fixed and symmetrical. The radius b of the contact circle can be prescribed, independently of the applied normal force P. The contact pressure p and the indentation α are given by:

$$\text{normal pressure:} \, p(r) = \frac{P}{2\pi b}(b^2 - r^2)^{-1/2} \quad \text{and indentation:} \, \alpha = \frac{(1-v^2)}{2bE}P \qquad (8)$$

In the tangential direction, the no-slip solution is identical to the corresponding one of Cattaneo-Mindlin with the trivial change in equation (3), b replacing a.

$$\tau(r) = \frac{T}{2\pi b}(b^2 - r^2)^{-1/2} \qquad (9)$$

It is immediately clear upon inspection of the solutions of normal pressure and traction that the two are proportional everywhere inside the contact, irrespective of the magnitude of the applied tangential force. The behaviour of solids assuming Coulomb friction is extremely simple and clear. For $T < \mu N$, there is no relative motion in all points, and when $T = \mu N$ the entire surface slides as a rigid surface should, probably in the way Coulomb depicted originally. The local and global versions of Coulomb friction produce identical effects. There are also other reasons why this fixed contact configuration of bodies can be useful in basic studies. One of these will be described in the section on peeling caused by tangential forces.

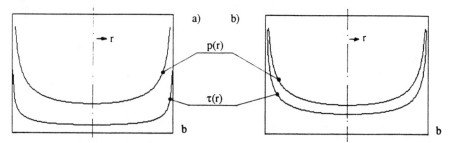

Fig. 2: The distribution of normal pressure and shear traction (Boussinesq configuration and Coulomb law): a) slip nowhere (T<µP, τ<µp) b) slip everywhere (T=µP, τ=µp sliding).

Unfortunately, in problems of tribology the true situation almost always, is far more complex than what appears initially. The simple formulae of normal and tangential tractions given above are not applicable generally; the solutions apply when the materials of the solid are similar (and a few other special material pairs). The situation with elastically dissimilar solids, as will be seen in a separate section, is not so simple. In the next part of this work, the more recent and physically motivated approach to analyze contact problems will be described.

3   FRACTURE MECHANICS APPROACH TO ADHESION AND FRICTION

3.1   FRACTURE MECHANICS BACKGROUND IN BRIEF

From the basic principles of physics it is clear that when two solid bodies are in contact their surfaces should adhere. A finite force must be applied to separate the bodies- to pull them apart in the direction of the normal to the plane of contact or to slide one body relative to the other, along the tangent to that plane. The contact interface acts like a bonded joint where adhesion constrains relative displacement in any direction. The geometry of surfaces in and around the contact may be visualized as an external crack. If an external force is applied the joint will resist separation of the solids but, the bodies deform during the process. At some stage, the joint may fracture owing to its finite ultimate strength. The characteristic features of the process that are of interest are: 1) the compliance prior to failure and 2) The force needed to initiate and propagate fracture. The first part can be treated by contact mechanics and the second by fracture mechanics of the joint. The two parts interact closely, the first poses the problem and solves. The solution is used in the second part to interpret what happens and this interpretation generating boundary conditions is fed back to the first, affecting its outcome.

The key concept involved in the theory is Griffith's idea of introducing an additional surface energy term as a complementary part of the free energy of the body. The addition is needed to compensate for the energetic contributions of the less easily prescribed surface forces acting on the body. Similarly, the key quantity which expresses the load that is sensed by the material in the vicinity of the crack tip is the "energy release rate " G (unfortunately the same symbol as that representing the shear modulus).

It was later realized, that this G actually represents the rate at which potential energy is transferred between the bulk and the surface and it clearly takes place in a local region surrounding the tip of the crack. The conclusion is that it is the local energy balance that matters. These thoughts led Irwin and Orowan to develop the energy balance formulation using the local stress field quantities which in turn, are related to G. Thus, G is the generalized force acting on the crack; it can be determined in many different ways from the local conditions, or from the somewhat more distant conditions of the elastic fields in regions enclosing the crack tip (the J and M integrals, discussed in the references). If the stress solutions are known, the most direct approach is to characterize the local stress field in terms of the stress intensity factors. The factors represent the coefficients of singular (and therefore dominating) components of stresses in the tip region indicating the strength of the singularity and are denoted by the symbol K. The K coefficients characterize both the geometry of the crack and the applied external force applied elsewhere to the solid. In the generalized form stress components in the tip region σ_{ij} at a radial distance ρ from the tip may be expressed as follows:

$$\sigma_{ij} = \frac{K}{\sqrt{2\pi\rho}} \ f_{ij}(\theta) \ + \ \text{finite stress components} \tag{10}$$

The finite terms are clearly insignificant with respect to the singular terms which dominate , as ρ tends to zero. In many simple problems the inverse-square-root type of singularity as given above applies and this leads to the typical shape of the surface near the tip. The last factor in the K term, represents the characteristic angular distribution at the tip. The main quantity of interest is the magnitude of K. The K factors are significant because of their relationship to G. The role of the K factors may be compared with that of the stresses as local measures of the applied load. Like the components of stress there are components of stress intensity factors in normal and shear.

The three stress intensity factors, corresponding to the components of traction are denoted by K_I (normal or mode I), K_{II} (In-plane shear or sliding mode II) and K_{III} (the anti-plane shear mode). The suggestion of latent crack motion in the terminology is somewhat confusing because the modes refer to possible modes of crack movements, if and when the crack begins to grow. Such movements are again referred to (justly) by modes I, II and III of fracture. In contrast, even when a crack is stationary, it may support finite values of K factors.

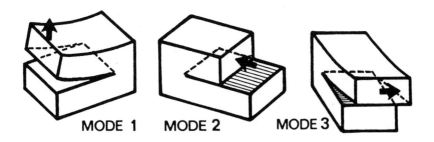

MODE 1   MODE 2   MODE 3

Fig. 3: The three basic modes of fracture or crack-movement (Mode I: normal separation, Mode II: In-plane sliding (shear), Mode III: Anti-plane sliding (shear).

In a simple tension test where only the mode I intensity factor is present, crack is initiated when the factor reaches a certain limit denoted by K_{IC}. The role of K_{IC} is similar to that of ultimate stress or the strength in tension. For more general geometrical configurations and loading all the three intensity factors may be present simultaneously, and it is their combined effect that controls the initiation of fracture. The nature of the elastic field near the crack may be described approximately by the plane strain or plane stress conditions depending upon the geometry and the loading. The necessary condition is that the energy release rate G, which combines the contributions of the three modes must attain some critical limit G_c for fracturing. Following Irwin [8] and Rice [9], G_c may be calculated for plane strain from the relation:

$$G = \frac{(1-v^2)}{E} \{K_I^2 + K_{II}^2\} + \frac{(1+v)}{E} K_{III}^2 \tag{11}$$

The relative contributions of the different K factors may influence the actual mode of fracture (crack movements). However, excluding the pure mode I fracture, the physical interpretation of mixed (combined) mode fracture is still fuzzy. More generally, the singular fields are insufficient to characterize fracture of dissipative material. The single intensity parameter is replaced by at least two parameters representing finite tractions and a finite size of the fracture region. The additional complications arising in these applications are best left to the specialists (See the references in [9] and the lecture notes of Brian Lawn [10]).

3.2   THE JKRS THEORY BASED ON FRACTURE MECHANICS:

The proper framework for the quantitative analysis of the contact mechanics of adhesion of bodies under normal forces was initiated by Johnson [11a], by considering adhesion as a constraint on relative *normal* displacements of contact points at the contact interface. The view that contact interface behaves as an adhesive joint and that its strength cannot be represented simply in terms of local stresses, lies at the root of the JKRS theory. It turns out that the new alternative interpretation of criteria of strength based on the fracture mechanics approach yields a more realistic description of the physical process.

The JKRS theory uses the Hertzian point contact configuration as the starting point. A rigid sphere (radius R) is pressed against an elastic solid (half-space) with a purely normal load P_0, which impresses a circular contact area of radius a_0. Due to the influence of adhesion (specific energy γ), the area increases to radius a_1. A (pseudo-) normal force (without adhesion), P_1 may be defined corresponding to a_1 according to the Hertz equation. This larger area of contact actually supports tensile tractions near the boundary. The equilibrium value for the size of the contact (a_1) is found using the global energy balance in the spirit of Griffith. The two most important results of JKRS theory (equations given below) concerning the effect of adhesion are: 1) The size of the contact area for a given normal force is larger than Hertzian and 2) A finite tensile normal-force - the pull-off force P_{p0} that is needed to separate the two bodies, that is when equilibrium becomes unstable (This topic has been reviewed by Pollock [11b]).

$$a_1^3 = \frac{3R(1-v^2)}{4E} \{P_0 + 3\gamma\pi R + (6\gamma\pi R P_0 + 9\gamma^2\pi^2 R^2)^{1/2}\} \tag{12}$$

The pull-off force P_{p0} is:

$$P_{p0} = -1.5 \; \gamma\pi R \tag{13}$$

The JKRS problem is an excellent example to demonstrate the power and simplicity of the stress intensity approach because the solutions of contact tractions (Boussinesq superimposed upon Hertz) are known beforehand, but for the undetermined value of the equilibrium contact radius a_1. It is then a simple matter to calculate the intensity factor K_I of the normal tensile traction (Boussinesq) and equate it to a constant say, K_{IE}, representing the strength of the interface in equilibrium. However, the actual physical meaning of the constant K_{IE} cannot be brought out without further interpretation that the corresponding energy release rate G_C is equated to the energy or work of adhesion. On the other hand, the K_C factors, decoupled from any strict physical meaning may still be useful as some measures of comparison if empirical evidence to their constancy is favourable.

3.3   CONVENTIONAL VERSUS NEW CRITERIA- ADHESION and FRICTION
What to discard and what to replace it with?

There is no dispute concerning the tangential constraint imposed in the conventional approach of contact mechanics to define effects of static friction in the sense of adhesion. However, a more explicit and consistent interpretation of adhesion is to consider it as a general constraint, which includes the JKRS case where adhesion resists normal separation of surfaces. Treating the effect of adhesion as bonding, its strength may be defined and determined using fracture mechanics. This amounts to allowing tensile and shear tractions - and even singular (integrable) fields of finite energy content associated with such tractions. The joint model also implies that the role of normal pressure in the process of friction, if any, is minor or indirect. One of the major indirect role is that it influences the real contact area of initially non-conforming surfaces. Undeformed surfaces which initially do not conform because their surfaces do not match geometrically either due to the macroscopic shapes of solids or due to microscopic asperities can be forced into conformity, at least partially by external normal loads. Another, but less clearly understood role is that the normal contact pressure can influence the strength properties of the interface by some mechanism of compaction of boundary layers adsorbed on surfaces of solids. In the first approximation, once the contact area is established the influence of normal pressure may be ignored.
Clearly, the conventional usage of the coefficient of friction as a basic parameter of the interface has no place in any basic model of friction; it should be seen merely as a non-dimensional representation of the tangential force of friction.

3.4   INFLUENCE OF FRICTION ON ADHESION OF DISSIMILAR MATERIALS
UNDER PURELY NORMAL LOADING

In the general case of contact of elastically dissimilar materials, shear tractions arise due to friction even when a purely normal force is applied. Friction influences the contact area and normal adhesion between such solids and introduces modifications, albeit minor, to the JKRS equation. When two solids are pressed into contact by purely normal resultant forces, surface points are displaced both in the normal and tangential directions. If the tangential displacements of the surfaces of two bodies are unequal, the relative displacements will give rise to slip upon contact. However, such slipping may be prevented by static friction, and some condition has to be specified to take care of the relative displacements in the contact area. The Hertzian analysis specifies the compatibility of normal displacements of surfaces but ignores the tangential components. Hence, the analysis is strictly correct for frictionless contact where relative displacements do not matter and for solids with certain special combination of materials because relative displacements are zero. In any case, any analysis that is physically consistent should consider that both adhesion and static friction will prevail simultaneously.

In the general case of elastically dissimilar bodies where relative displacements cannot be ignored, friction has to be taken into account when studying normal adhesion. The advancing nature of the Hertzian contact brings with it a special problem in the specification of boundary conditions. The points on opposite surfaces may displace freely and are traction free before making contact but, once in contact, the relative displacement is checked by static friction. (This problem does not arise in the contact of a flat-faced punch). The difficult problem of incremental boundary conditions has been transformed by Spence [12], using an elegant reasoning based on similarity. Spence extended the Hertzian analysis to include the effect of shear tractions due to friction but, he did not consider the influence of normal adhesion in the contact. Another, problem mainly of a mathematical nature, is that due to the interactive influence of the normal and the shear tractions.

Quite opposite is true of the nature of another difficulty that is typical of contacts between dissimilar materials. In treating the problem of adhesion of such bodies, the boundary tractions calculated on the basis of prescribed constraints, are not only singular but also display a peculiar oscillatory behaviour. Unlike the simple inverse square-root singularities of linear elastic fracture mechanics, these oscillating singularities do not lend themselves to a simple and unambiguous interpretation. The solutions are also not strictly correct because these also imply interpenetration of solids surfaces just outside the region of contact. The problem illustrates that the choice of physically appropriate boundary conditions for contact and crack problems is not completely free. Alternative boundary conditions have been proposed in the literature to overcome this difficulty but, there is generally no unique interpretation of the physical problem.

In the context of the present ASI meeting, it is to be hoped that research efforts of atomic and molecular dynamicists and experimenters may help in clearing up such matters concerning which boundary conditions are physically appropriate for the macroscopic elements of the interface. In the present case, the ambiguity arising from the oscillatory behaviour suggests that the stress intensity approach is not suitable for investigating the behaviour of dissimilar materials. However, it is probably reasonable to assume that any errors in the stress field in the small region at the crack tip will not seriously affect the correct calculation of the energy release rate based on the global energy balance of the system [13]. Ignoring the peculiar but localized tractions, it is expedient to follow the global energy balance approach of Griffith. Making use of the results of Spence the author [14] generalized the JKRS solution to include the effect on normal adhesion of shear tractions in the contact of dissimilar materials. The problem of two dissimilar materials is equivalent to that of a rigid punch indenting an elastic half space of appropriately modified elastic constants. The modified problem considered is that of a rigid spherical punch of radius R, in contact with an elastic half-space with shear modulus G and Poisson ratio v. When a purely normal load P_0 is applied, because of friction the resulting contact circle of radius a_0 and the distribution of normal pressure are different from those given by Hertzian equations. If the presence of normal adhesion is also considered, one may expect that the area of contact will be larger than what Spence finds. The new radius a_1 of the enlarged contact is:

$$a_1^3 = \frac{3R(1-2v)X(v)}{8G\ln(3-4v)} \{P_0 + 3\gamma\pi R X(v) + [6\gamma\pi R X(v)P_0 + 9\gamma^2\pi^2 R^2 X(v)]^{1/2}\} \qquad (14)$$

where $X(v)$ is a function of $\ln(3-4v)$ and is expressed as an infinite series. The influence of shear tractions on adhesion shows up as the variation of the contact area with v. This is the modified JKRS equation which takes both friction and adhesion into account, but the modification is only a minor one. The pull-off force P_{p0} is:

$$P_{p0} = -1.5\,\gamma\pi R\, X(v) \qquad (15)$$

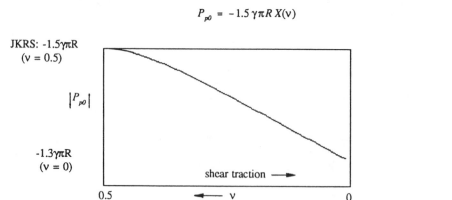

Fig. 4: The attenuation of the (normal) pull-off force due to shear tractions resulting from friction at the contact of dissimilar solids (resultant force purely normal).

The factor $X(v)$ represents the attenuation of the pull-off force due to contact friction. For dissimilar materials with $v < 0.5$, it is generally smaller than unity and it approaches unity when $v = 1/2$, that is when there is no dissimilarity or when the materials are incompressible. The study of the normal adhesion of dissimilar solids illustrates i) the interaction between adhesion and friction; it has a relatively minor effect on the JKRS equation, ii) the difficulties of formulating physically sound boundary conditions.

4 THE PROCESS OF STATIC FRICTION (Hertzian configuration)

4.1 THE THEORY OF PEELING BY TANGENTIAL FORCES

The effect of a tangential force on the contact of solids with Hertzian configuration was the subject of the analysis by the author and Briggs [15]. The starting point for this study is the JKRS solution of solids loaded by a purely normal force P_0. Owing to adhesion, the area of contact at equilibrium is larger than Hertzian. With the same (constant) normal force acting, the application of a (constant) tangential force T (say, along the x-axis), makes a net energetic contribution which depletes the total energy of the loaded system. The system therefore seeks a new state of equilibrium at the expense of the contact area which controls the potential energy of the surface. The situation is similar to the case of purely normal loading where the Hertzian contact changes to the JKRS contact in order to approach the equilibrium state. The S & B analysis considers the additional energy terms due to tangential forces and recalculates energy balance equations. It is shown that the application of a small tangential force causes the contact area to decrease by interfacial peeling. With increasing tangential force, the peeling occurs in a stable manner if the tangential force T is less than the peeling limit force T_0.

The details of the analysis are given in [15,16]. Using the same notations as those used to discuss JKRS theory, the main result is the equilibrium solution of the peeling equation as noted below.

$$a_1^3 = \frac{3R(1-v^2)}{4E} \{P_0 + 3\gamma\pi R + (6\gamma\pi R P_0 + 9\gamma^2\pi^2 R^2 - \frac{3}{16}K\lambda T^2)^{1/2}\} \qquad (16)$$

Qualitatively, the important aspect that comes to light is the interaction between adhesion and friction. The reduction of contact area due to a tangential force is similar to the 'peeling' produced by a purely normal tensile force. The mode of fracture is by normal separation in both cases but, the stress intensity factors responsible are differently composed.

According to the equation for tangential peeling, real solutions require that T is smaller than a limit force T_0 given by:

$$T_0 = \frac{4}{\sqrt{K\lambda}} \{2\gamma\pi R P_0 + 3\gamma^2\pi^2 R^2\}^{1/2} \qquad (17)$$

At the point $T = T_0$, the system attains a state of neutral equilibrium. For higher values of T, the energy release rate exceeds the rate of energy absorption by the surface. Before considering what happens next, it is worth noting that there is no slip at the interface during peeling and therefore the process so far describes the initial phase of static friction. The tangential peeling limit depends not only on the surface energy but also on the normal force pressing the solids together. In contrast to the normal pull-off force for solids (JKRS) with magnitude of $1.5\gamma\pi R$, the order of magnitude of the peeling limit force for $P_0 \gg 1.5\gamma\pi R$, is nearly the geometric mean of the applied and the pull-off forces (with an arithmetic mean of roughly $0.5 P_0$ as the upper bound). This comparison explains why frictional force is generally much larger than the normal pull-off force under similar surface conditions. In the latter case, as pointed out long ago by Tabor [17], when the normal force pressing the bodies is gradually reduced, it causes peeling which amounts to a corresponding reduction of the size of the contact. So there is hardly any contact left to withstand the tensile force applied at pull-off. In contrast, during peeling by tangential forces the contact

122

area remains fairly large. Another distinction is that peeling in the tangential case is controlled by all the three stress intensity factors which contribute jointly to the energy release rate. This aspect is well illustrated by examining an implicit assumption made in deriving the peeling equation.

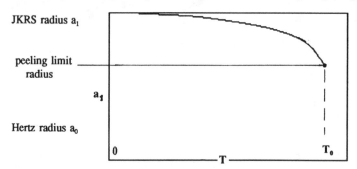

Fig. 5: The reduction of the radius of contact area (from equation 16) by peeling due to a tangential force up to the peeling limit.

Symmetry considerations: The energy release rate of peeling is calculated by assuming that while the contact changes in size it remains circular. The nature of the approximation becomes clear when it is noted that the shear tractions under the no-slip condition (Cattaneo-Mindlin) are directed parallel to the applied force. At any point on the boundary, the singular traction has the stress intensity factor K_x. It may be resolved into radial and tangential components which represent the intensity factors of the in-plane and anti-plane shear. Denoting the polar angle of the point by θ with respect to the x-direction, the stress intensity factors of the shear mode are: $K_{II} = K_x \cos\theta$ and $K_{III} = K_x \sin\theta$. The combined energy release rate of all the three modes varies with θ:

$$G(\theta) = \frac{(1-v^2)}{2\pi E} \{K_I^{\,2} + K_x^2 \cos^2(\theta)\} + \frac{(1+v)}{2\pi E} K_x^2 \sin^2(\theta) \qquad (18)$$

The result is that $G(\theta)$ is not rotationally symmetric generally except for $v = 0$. If the critical energy release rate G_C is assumed constant, the angular distribution of G will be reflected upon the shape of the contact which then becomes non-circular while retaining symmetry with respect to the axis along which the applied force acts. However, one may consider an approximate criterion where the average energy release rate given below G_{av} equals G_C:

$$G_{av} = \frac{(1-v^2)}{2\pi E} K_I^2 + \frac{(1+v)(2-v)}{4\pi E} K_x^2 \qquad (19)$$

4.2   INFLUENCE OF TANGENTIAL FORCE ON THE PULL-OFF FORCE

An interesting feature of the peeling equation is the interaction between modes I, II and III at the limit of peeling (neutral equilibrium), which determines how the interface responds to the combined action of a normal tensile force and a tangential force. This was brought to the attention of the author by Johnson [18]. The peeling limit is determined jointly by the tangential force T_0 and the normal force P_0, according to the equation given above. In the absence of a tangential force component T=0, the normal force P_0 at pull-off, denoted by P_{p0} is the JKRS value of $P_{p0} = -1.5 \ \gamma\pi R$ (For dissimilar materials $P_{p0} = -1.5 \ \gamma\pi R \ X(v)$). For a given tangential force T_0, the tensile normal force for pull-off denoted by $P_0 = P_p$ will be smaller than P_{p0}. At neutral equilibrium, the peeling equation may be rearranged:

$$\frac{P_p}{P_{p0}} = \frac{3}{64} K\lambda \left(\frac{T_0}{P_{p0}}\right)^2 - 1 \tag{20}$$

Clearly, the peeling limit is reached when the applied tangential force T, approaches $T_0 = \{8 / \sqrt{(3K\lambda)}\}$ P_{p0}; the pull-off force is then reduced to zero which means that beyond this limit the contact cannot support any tensile force at all.

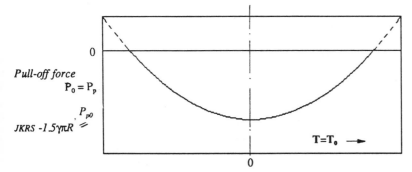

Fig. 6: Interdependence of pull-off and tangential forces at the peeling limit.

5   TRANSITION FROM PEELING TO THE STATIC LIMIT

Returning to the Hertzian configuration and to the point where the previous discussion was interrupted, the question is what happens beyond the peeling limit. If **T** exceeds **T₀**, the peeling process becomes unstable and consequently, the contact area diminishes rapidly **but**, what the analysis cannot answer is how long can peeling continue. Clearly, it is unimaginable that peeling can continue indefinitely while the bodies still support the compressive normal load. Intuitively, it would appear that peeling shall terminate at the latest when the contact area approaches its Hertzian value corresponding to the normal load. The two reasons given below support this statement. Since the area of contact at the limit of peeling is still larger than Hertzian, it will at some point become Hertzian as peeling continues. With the normal load held constant, a contact area smaller than the Hertzian is kinematically inadmissible, because otherwise it would lead to interpenetration of surfaces! This is one argument which clearly suggests that peeling shall stop when the contact is Hertzian. Another argument which supports this statement is the fact that the stress intensity factor for mode I, vanishes when the contact is Hertzian. Although, the shear mode II contributes to the energy of normal mode I for initiating peeling, it is unimaginable that it could do so all by itself, when mode I factor becomes zero. In fracture mechanics terminology, this amounts to the statement that while $G = G_c$, is a necessary condition for the initiation of fracture, it is not always sufficient. For a given mode of fracture to occur, the stress intensity of the corresponding mode should be finite.

When **T** exceeds **T₀**, the size and shape of the size and the shape of the contact remain constant and Hertzian. Since the contact is stationary, the shear tractions inside the contact, including those at the boundary, increase linearly with the applied force T. As the tangential force increases further, a point will be reached when the shear tractions become sufficiently intense to initiate shear fracture. Denoting T_c, as the applied tangential force at the point of initiation of fracture by the shear mode, the assumption that shearing is more difficult than peeling implies:
$T_c > T > T_0$; *slipping can be initiated only if* $T \geq T_c$. *The force* T_c *is the limiting force of static friction.*

An explanation is due here, to point out that certain physical aspects of the processes which hitherto have been left out of discussion, are essential to the development of the model. One of these is the implicit assumption made in the analysis of peeling that there is no slip due to shear fracture while peeling is in progress, even though the tangential force and consequently the intensity of shear traction keep on increasing. The assumption implies that it is easier to peel by normal separation than to slip in the tangential plane. An alternative possibility that is rejected here, is that slipping may start before peeling has stopped. If indeed, slipping occurs at such an early stage, it would be instrumental in arresting further peeling by dissipating excess energy in the system. (As a trivial example, the slip theory of Cattaneo-Mindlin based on Coulomb friction, predicts slipping to start directly for any value of the applied force). However, this line of reasoning, in our view, is not very fruitful to pursue because there is a large body of experimental and practical evidence that suggests that peeling is much easier than shearing.

6   PHYSICAL ASPECTS OF PEELING AND SLIPPING

6.1   PROPERTIES OF THE INTERFACE

The question one must face in developing friction models is what actually takes place at the interface when solids peel or slip? Is there any relationship between the two processes? If both processes test mechanical strength what can be said about the nature of the process concerning mass, momentum and energy balance? Whatever be the nature of the real world interfaces in practice, some explicit assumptions are necessary to develop process model reflecting ones view of what to expect and how (why) things happen. Whether such models are ultimately realistic or not can be put to test properly by designing experiments but the framework for the design of tests can only be provided within the context of a theoretical model.

Some of the well known facts upon which a model may be based are the following. The normal tensile force required to separate bodies is insignificant for all but a few combinations of materials. It is accepted generally that surface roughness and boundary layers (films) weaken the interface. However, the frictional force under these conditions has the same order of magnitude as the normal force- it is tangible. Material transfer and wear may occur as consequences of friction but, their correlation with friction is not necessarily strong. In the context of the present discussion, it is reasonable to assume that the interface is much weaker than bulk solids and that the separation of solids by external forces, takes place in the vicinity of the interface. The question of main interest is not why the strength of the interface is so much lower than that of bulk solids but, why friction is a much larger force than the purely tensile force required for pull-off, under the same given conditions.

The model considered here is based upon the following somewhat general considerations of the processes in question. From the tribologist's viewpoint, solids exposed to the atmosphere are covered by surface films adsorbed from the gaseous environment. This fact has relevance to both adhesion and friction, mainly through two kinds of characteristic properties, the physico-chemical and the mechanical. The first kind relates to the phenomena of adsorption and adhesion while the second relates to the conceptually simpler mechanical properties of strength and stiffness.

The mathematician's plane interface is in reality a three dimensional region (layer). Consider a system consisting of two solids, where the surface of each solid is covered with its own adsorbed film in some fixed environment. When the solids make contact, the two films get squeezed between the solids to form a interfacial layer having certain characteristic properties. The properties depend upon the kind of physico-chemical interactions between the two original films and the surface forces of the solids. In order to simplify the model of the interface, the effect the stiffness of the layer on the elastic deformation of the system of two solids in contact, will be ignored. The only way the layer can influence the deformation of the system is through changes in boundary conditions when the interface fails.

One of the most important property of the interface is its strength. However, because the interfacial layer is inherently heterogeneous, its strength will depend upon the regions or planes of weakness and the nature and intensity of the applied load.

6.2   IDEAL PROCESS OF PEELING

For instance, in tension the "strength" may depend primarily on, to what extent do the two separate films squeezed under pressure behave as a single layer. The governing property here is clearly the adhesion between the films of the two solid surfaces. If adhesion is weak, normal separation of solids may occur without any form of damage either by wear or by material transfer and the process is repeatable, without any need for surface manipulation. Neither the geometry (roughness) nor the physical chemistry of the surfaces alter as a result of contact followed by normal separation. Under ideal conditions of slow and controlled contact formation and separation, the process can be almost conservative energetically. This is the ideal thermodynamical process where the JKRS theory applies exactly with the energy release rate equal to the Dupré energy of adhesion.

The process of normal loading and unloading of solids in practice, may be considered a good approximation of this ideal process (which may be viewed as a Zero order wear/material transfer process). Normal separation can occur at the true interface without damaging both the bulk bodies including their surface films. The process is repeatable. This description does not apply generally to the process of friction, with one exception when peeling occurs with small tangential forces applied to the solids in the early stage of static friction.

6,3   IDEAL PROCESS OF FRICTIONAL SLIPPING

The next question is in what way frictional slipping differs from the peeling process. A process of frictional slipping analogous to the one described above (zero order process) is one where surface films of the solids slide on one another. Although atomic models of this kind have been proposed to explain the dislocation and slip phenomena, it is difficult to understand how they fit into the general pattern of macroscopic mechanics. It is also hard to explain why the processes of adhesion, static friction are almost energy-conservative but in kinetic friction energy is dissipated while slipping. One may interpret Archard's principle [19] of the nature (elastic and plastic) of contacts and argue that frictional force can be generated during slipping without necessarily invoking any inelastic or plastic deformation of solids. This basic distinction is essential to the further development of the model. For this reason, the hypothesis explored here is that slipping is associated with dissipation involving (at the least) some minimal damage.

In the order of severity in terms of damage by wear or transfer, a plausible process is one where wear or damage is restricted to surface films only. The bulk solids are not damaged or worn at all in this process. This is the ideal process of friction (Order 1). The model described next for the process of slip considers this ideal situation where energy can be dissipated involving minimal wear. Interestingly, one may still characterize this process as almost non-destructive and repeatable because fresh adsorption from the atmospheric environment can almost instantaneously repair or replace surface films damaged during the last frictional encounter.

It is important to emphasize the subtle distinction between ideal models of peeling and slipping arising from considerations of energy. The process of slipping is irreversible whereas peeling can take place ideally in a reversible manner. Clearly, the energy release rate responsible for initiating shear fracture has no direct relationship (certainly not equatable) to the reversible energy (work) of adhesion which controls peeling (JKRS and S-B equations). The aforementioned remarks do not restrict possibilities that non-ideal peeling can occur in a similar and irreversible manner as slipping. Even in these non-ideal cases of peeling, with damage to interfacial layers (Order 1 process), the strength of the layer in resisting tensile fracture may be different in magnitude than that in shear. It is quite likely that the interfacial layer (usually heterogeneous) has different intrinsic strength in different directions.

So, even in cases where interfacial layer is damaged by peeling, the normal force required may be much smaller than the tangential force needed to shear the layer.

The essential features of the model described above are qualitatively consistent with common experience and also with experimental results of Courtney-Pratt and Eisner [20], who found that purely normal loading or unloading does not produce any significant mechanical breakdown of surface films; marked disruption of films can be caused by rubbing (frictional slipping) of surfaces.

Two more aspect of shear fracture which requires some thought are the post-initiation response of the damaged interfacial layer and the aspect of recovery in the context of repeatability of the process. These aspects will be described in the next section.

Highly complex frictional processes (of higher order) where significant amount of wear or material transfer is associated are not considered here; these are still less amenable to understanding and analysis. The same holds for peeling when appreciable material is transferred or worn. The following analysis is based on fracture involving shearing of the interfacial layer at the contact. Before extending this analysis there is one more point of interest with reference to the discussion, that is,

6.4   IS PEELING A NECESSARY PRELUDE TO SLIPPING?

The straight answer is a simple no. The peeling process described in the S-B analysis occurs because the Hertzian configuration gives rise to an advancing type of contact; The area of contact is not fixed by the geometrical shape but by the physical boundary conditions imposed separately. This is not the case for the Boussinesq configuration where a rigid flat faced circular punch is pressed against an elastic solid. An additional tangential force has then no effect on the contact area and hence peeling is out of question. With increasing tangential force there is a proportional increase in shear traction in the interface. The traction is again the Cattaneo-Mindlin traction without slip. When the stress intensity is sufficiently high, shear fracture can initiate and slipping follows.

The Boussinesq configuration:

For peeling to occur in the Boussinesq configuration, the applied normal force should be tensile. In the absence of a tangential force, the pull-off force can be quite substantial depending simply upon the cross-sectional area of the punch. The problem has been solved by Kendall [21]. Since the equilibrium at the pull-off point is unstable, peeling takes place rapidly until the solids are separated. For a punch of diameter d, the pull-off force P_{p0} (ignoring dissimilarity effects) is given by:

$$P_{p0}(Kendall) = [\pi E \gamma/(1-\nu^2)]^{1/2} d^{3/2} \qquad (21)$$

The geometry of the Boussinesq configuration is well suited for experimental work on adhesion and the implementation is fairly simple provided, care is taken to prevent tilting of the punch by eliminating any parasitic moments. The configuration considered clearly demarcates the boundary between conditions leading exclusively to peeling or to slipping. Under compressive normal loads, the interface responds only by slipping while tensile loading causes peeling because peeling is easier. The effect of a tangential force on the tensile force of pull-off is similar to that for the Hertzian configuration and it is again controlled jointly by the combined intensity factors of modes I ,II and III.

Experiments performed by the author [22], using this configuration have confirmed the predicted behaviour: For compressive load, slipping is the only process observed whereas if the tensile forces are applied, peeling occurs as expected and the peeling limit forces in the normal and tangential directions interact as given by the equation for peeling under the action of combined modes. The experiments with compressive normal loads were also useful from another point of view. These results indicated that the variation of compressive normal load by a factor of 5 had only a small influence (≈ 10 %) on the frictional force (at the point of sliding) for a given punch. In contrast, the frictional force depends strongly on the size of contact, as expected it varies with the 3/2 power of the punch diameter.

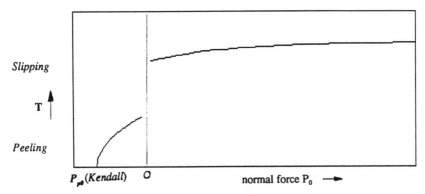

Fig. 7: Delineation of the regimes of peeling and slipping under combined tangential and normal loading for the Boussinesq configuration (flat-ended rigid cylindrical punch).

The test specimens consisted of glass cylinders (punches) and a relatively hard elastomer (modulus 4 MPa). One more advantage of the Boussinesq configuration (compared to the Hertz) is that there is practically no interaction due to surface forces outside the contact area. The action of such forces is one of the weak points of the (JKRS) contact mechanics formulation in terms of surface energy. The overall conclusion from these experiments is that the frictional force depends, if at all, only weakly on the (compressive) normal load and pressure in the contact. The small increase observed at light loads can be accounted for, by the increase in contact due to flattening of any unevenness or asperities present on the surface.

Schallamach waves:
It is believed that the hard rubber helped to overcome the experimental problems of surface instability and Schallamach waves which arise when very soft elastomers are used. No waves were visible under a low power microscope. The phenomenon of Schallamach waves [23] caused by large (non-linear) elastic strain of sub-surface material interferes with the process of friction. This aspect has been studied by Barquins et al [24], Briggs and Briscoe [25] and Roberts and Jackson [26]. The first authors observed a critical speed of translation below which Schallamach waves are not permitted. The latter authors studied the possibility of explaining the energy dissipation associated with the wave propagation purely by the mechanism of normal loading and separation of elastomer without any interfacial slip; the tangential force merely triggers the Schallamach wave mechanism. The normal separation is similar to the process of pull-off by peeling but, for visco-elastic materials the pull-off force is several orders of magnitude larger than that for an elastic solid. However, the model requires in the first place that friction (large and static) is already present in the interface, in order to induce the surface instability and in the second place the property of visco-elasticity is essential to amplify normal forces of peeling. The experience with similar kinds of motions induced by instabilities is that the applied force required to sustain the motion is greatly reduced. For example it is easier to move a carpet across the floor by creating folds rather than to slide it across. Clearly, the existence of high static friction and high intensity of shear tractions at the boundary of the contact area can promote surface instability by large elastic deformation [27] but, how this leads to the kind of translation envisaged is not clear. A detailed analysis of the problem of surface waves of viscoelastic solids proves to be very difficult. Without such an analysis, the question of how appropriate such a model is to explain the *'cause'* of friction rather than its *'effect'* is best left open.

7 ANALYSIS OF THE MECHANICS OF SLIPPING

The tangential force T_C, required for initiating slipping is the limit force of static friction. The shear stress intensity factors K_{II} and K_{III} of the shear tractions associated Cattaneo-Mindlin solution for no-slip, approach the critical strength factors K_{IIC} and K_{IIIC}. An appropriate criterion for the initiation of slipping by the shear mode of fracture may be expressed as some function of the strength factors K_{IIC} and K_{IIIC}. The precise nature of the function can only be determined empirically. In the following analysis a simple model is adapted, which ignores the distinction between the effects of modes II and III. Since the tractions are directed in the x- direction (just as T_C), it is convenient to express the stress intensity in terms of K_x and the critical shear strength of the interface in the same direction, by a factor K_{XC}, such that K_{XC} can be considered as some function of K_{IIC} and K_{IIIC}. Assuming K_{XC} as a constant property of the interface, the limiting force of static friction is given by:

$$T_C = 2\sqrt{2}\,\pi\,a_o^{\,3/2}\,K_{XC} \qquad (22)$$

The model can be completed as mentioned in the previous section, by adding additional information necessary to cover the two aspects of damage of the interfacial layer.

The pre- and post fracture response of a point in the interface which fails in shear is modelled on the following three hypotheses.

(1) When the surfaces of solids come into contact initially, the interfacial layer formed is strong and its resistance to initiation of fracture is determined by the critical shear strength factor K_{XC}.

(2) Once fracture initiation is complete and slipping starts, the damaged interfacial layer can still resist slipping but this resistance is weak. This resistance may be considered as the residual strength of the fractured interface layer to progressive failure. Since the layer has been fractured already, it is appropriate to express the residual strength in terms of a shear stress (rather than a intensity factor). The residual strength remains effective during slipping irrespective of the extent of slip. It is assumed that molecules of the gaseous atmosphere outside the contact are unable to reach the surface points so long as such points remain inside the contact area.

(3) An element of interfacial layer that is damaged in the course of slipping can regain its fracture strength when this element emerges outside the contact and becomes exposed once more to the healing influence of fresh adsorption from the environment. Under normal atmospheric conditions the process of "adsorptional healing" takes place very rapidly.

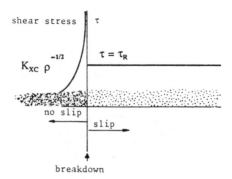

Fig. 8: The two-parameter point-breakdown model of fracture, fracture initiation strength $K_{XC}=f(K_{IIC}, K_{IIIC})$ and residual strength (Stress: τ_R).

The mechanism of slip growth for $T > T_c$ by the sliding mode of fracture is analyzed with the aid of a simple fracture model. The resistance to fracturing or breakdown of the interfacial layer is characterizes by two parameters. These are: the shear strength-intensity factor K_{XC} for the initiation of fracture and the residual shear strength τ_R, of the fractured layer resisting the subsequent propagation of a shear crack by slipping. This simple model is shown in fig. 8. It depicts a discontinuous change in fracture strength at the point of breakdown.

An extension of the fracture model containing three parameters to characterize the fracture zone in terms of a finite size and finite tractions is reported elsewhere [16]. The same reference also describes the details of the analysis of slipping based on the two models of fracture. For simplicity, the influence of material dissimilarity has been ignored. It is usual to assume in the analysis that slip starts at the boundary of the contact circle and that it is confined by an annular region which extends until the boundary of an inner circular region. There is no slip inside the inner circle which is the region of static friction. Following the reasonable assumption made by Cattaneo and Mindlin in their theory of slip, the locked region is again assumed to be a circle of radius c which is concentric with the circle of contact. It is assumed that slip occurs inside an annular ring having an inner radius r = c. The boundary conditions are imposed upon the surface of a body which is treated as an elastic half-space.

If the tangential force exceeds the static friction limit T_c, the stress intensity factor K_x cannot increase further because it saturates at the level prescribed by the critical strength factor K_{XC}. Increasing force causes slipping in an annular region $c<r<a_0$, such that fracture is initiated at the boundary r = c. The result is that an increase of tangential force leads to the growth of the slip region at the expense of the region of static friction. The boundary conditions prescribed for surface tractions and displacements are:
Displacements in $r \leq c$: uniform in the x-direction = δ (shift), and in the y direction = 0.
Tractions: in $c \leq r \leq a_0 : \tau = \tau_R$; for $r > a_0 : \tau = 0$; and at r = c : $K_x = K_{XC}$.

The solution of this boundary value problem [16] for the shear traction τ, becomes:

$$\tau(r) = \tau_R \qquad \text{prescribed in } c \leq r \leq a_0, \qquad (23) \text{ A}$$

$$\tau(r) = \sqrt{2c}\, K_{XC}(c^2 - r^2)^{-1/2} + \frac{2\tau_R}{\pi} \arccos\left\{\frac{\sqrt{c^2-r^2}}{\sqrt{a_0^2-r^2}}\right\} \quad \text{in } 0 \leq r \leq c \qquad (23) \text{ B}$$

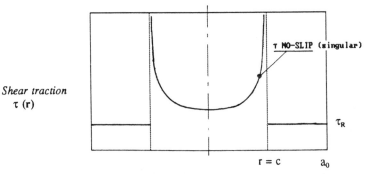

Shear traction $\tau(r)$

τ NO-SLIP (singular)

τ_R

r = c a_0

Fig. 9: The distribution of shear traction inside the contact region for partial slip, during the transition from static limit to sliding.

The shear traction inside the circle of static friction consists of two terms. The first provides the specified critical singularity for initiating the shear mode of fracture. The second term contributes to the frictional force but its effect is insignificant in the local region (point) of fracture. The distribution of shear traction sketched in fig. 9, is the alternative to the Cattaneo-Mindlin slip solution shown in fig.1b (from equations 5 and 6).

At the point where the entire contact is about to slip, c = 0, the interfacial layer is damaged over the entire contact and the force of sliding or kinetic friction T_s is simply: $T_s = \pi \, a_0^2 \, \tau_R$. For convenience, the slip boundary c, the tangential force T and the shift δ are expressed in dimensionless forms:

$$\kappa = c/a_0 , \quad T^* = T/T_s , \quad \text{and} \quad \delta^* = \{\delta/a_0\} \, \{2/(\lambda \, \tau_R)\}$$

The resulting equations for T^* and δ^* are:

$$T^* = \left(\frac{T_c}{T_s}\right) \kappa^{3/2} + \left(\frac{2}{\pi}\right) \{\arccos(\kappa) + \kappa \, (1 - \kappa^2)^{1/2}\} \tag{24}$$

$$\delta^* = \frac{\pi}{4} \left(\frac{T_c}{T_s}\right) \kappa^{1/2} + (1 - \kappa^2)^{1/2} \tag{25}$$

The tangential force T and the displacement shift δ are both functions of c. The compliance characteristics $T^* - \delta^*$ can be determined by eliminating κ from the two equations. A typical compliance curve is shown figure 10. The left plot shows curves obtained for different values of (T_c/T_s). If the process of peeling is ignored, the initial part of the curve is a straight line corresponding to equation (2) for the Hertzian contact area. If peeling is considered, the initial part will be non-linear due to the decrease in the contact area as the tangential force becomes larger. The initial effect of peeling is sketched in Fig. 10 (right, broken line).

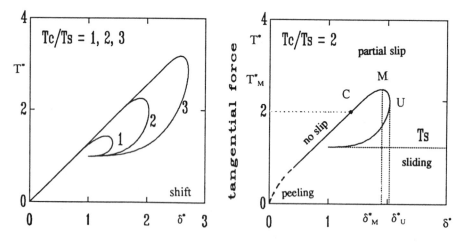

Fig. 10: **The compliance curve during transition from static to sliding friction** (the small non-linear effect of peeling is shown by dotted lines in the figure right).

As the tangential force increases slipping is initiated at point C when $T = T_C$. The curve then begins to bend and continues to rise (positive stiffness) until it reaches a point M corresponding to the maximum of $T^* = T^*_M$ at some specific value of shift δ^*_M. The frictional force cannot increase further, hence quasistatic slipping is possible only for $T^* \leq T^*_M$. The latter part of curve, beyond the point of maximum of friction is not stable and cannot be measured if the tangential force is prescribed during an experiment. However if the displacement can be prescribed (using stiff and feed-back controlled equipment) the curve where the frictional force decreases with increasing shift may be measured. Although the interface is weakened, it continues to function as a joint and it continues to accommodate relative displacement between bulk bodies until the point U (vertical tangent) where $\delta^* = \delta^*_U$, it is stretched to its limit. Larger value of δ^* must lead to sliding of the whole interface. Interestingly, the compliance curve bends backwards, so if δ^*, can be decreased carefully with reduced force the tail end of the curve can be traced. Surprisingly, although the slip region penetrates further, this part of the curve displays a positive stiffness for small deviations from equilibrium. Finally when the slipping is complete the solids are free to slide.

For prescribed shift, the friction appears as a double valued function of the shift but, considering the irreversibility of the slipping process the actual force is uniquely determined by the history of the loading process. In view of the physical postulates introduced, once an element of interfacial layer is damaged it can recover its original state only when it reappears outside the contact region into the gaseous environment. Up to the point of maximum friction, the shape of the compliance (tangential force-displacement) curve is somewhat similar to that according to the Cattaneo-Mindlin theory of slip. Beyond this point the behaviour is quite different; under ordinary circumstances where the shift displacement cannot be controlled one would expect a discontinuous transition in friction. The frictional force drops down suddenly from its maximum value to its sliding value inducing a dynamic response from the system. This sudden drop in the frictional force from quasi-static to kinetic friction is a phenomenon that is commonly associated with dry friction of solids. It is well known that this frictional behaviour is one of the primary sources of the discontinuous motion of sliding bodies popularly known in the engineering practice as the "stick-slip" phenomenon.

9 THE PROCESS OF KINETIC FRICTION

A model of the process of kinetic friction may be developed by considering sliding as a series of sequential processes of static friction. However, the detailed physical model becomes complex and speculative. Assuming that the large scale models can be applied to model surface asperities, one may depict a process where two asperities of opposite surfaces come into contact, adhere instantly and deform in that process. The micro-interface again behaves like a miniature adhesive joint, and the same methods of contact and fracture mechanics may be applied to predict the compliance and the strength of the micro-layer. Since kinetic friction involves cyclic contacts of asperities, the model needs to consider what happens between the point when the interface layer of a given asperity is damaged by slipping and the point when the same asperity meets to make contact with a different asperity of the opposite surface. The third hypothesis described earlier, offers an explanation of the mechanism whereby surface films of asperities are recycled between two successive encounters. If adsorption can take place rapidly the damaged surface film of an asperity may be repaired by "adsorptional healing". Therefore, when a new engagement cycle starts, the interfacial layer would have probably recovered its original fracture strength (K_{XC}).

A second aspect of the kinetic friction, not considered here, is that contact between a pair of convex asperities of opposite surfaces occurs not at the summits but at some points on their flanks. The analysis of this problem is more complex [16] but the physical models developed here still apply.

10   CONCLUSIONS

(1) The standard approach to analyze contact mechanical situations is based on the conventional concepts of Coulomb friction. It works reasonably well in many practical situations where the variation of material properties, surface conditions and applied loads is well within the empirically tested limits. However, for other conditions the predictability is largely lost because the approach shirks going into the physical aspects of problems. It remains essentially an empirical tool because it totally ignores normal adhesion and uses a single constant to condense the entire complexity of the interfacial tribology.

(2) The relatively new approach is more basic in the sense that the friction is considered as a mainly mechanical process rather than as an intrinsic property. The property of adhesion plays a central role in the theory, so the first major challenge is to explain why the pull-off force is so small compared with the frictional force under similar tribological conditions. Starting from the JKRS theory based on the fracture mechanics concepts, the S & B theory gives a plausible explanation of this paradox. The fracture mechanics approach provides the working hypothesis for defining strength of the interface in the normal and in the tangential directions. The friction process can be studied within the broadened framework of contact and fracture mechanics. The model is used in the analysis of peeling and slipping at the interface.

(3) The application of a tangential force can lead to either peeling or slipping of the contact but, the physical mechanisms in these two processes are quite different. The process of peeling occurs when the geometrical configurations or conditions of loading give rise to advancing contacts. When both the normal and the tangential forces influence peeling, the peeling limit is controlled jointly by the three modes of fracture. The different mechanisms of peeling and slipping may explain the distinction between reversible and irreversible nature of the two processes. Below the static limit, the process is conservative whereas slipping involves damage at least to the interfacial layer. In an ideal friction process the damage by slipping may be restricted to only the adsorbed surface-films on the solids. A two-parameter shear-fracture model has been proposed to describe the film strength. The model is used in the analysis of slipping to describe the transition from the static limit of friction to fully developed sliding.

4) An approach is indicated to extend the model to study the mechanics of sliding friction. It is based on rules which assign strength properties in the "original", "damage" and "healed" states of surface films and the concept of recycling of surface films. It is envisaged that when the surface film of an element is damaged by shearing, its strength (defined by a stress) is reduced. There is no opportunity to recover its fracture strength (intensity) while it remains in contact but, once it leaves the interface it is exposed to the environment and it regains its strength (to fracture initiation) by "adsorptional healing".

11   REFERENCES

1A   Cattaneo, C. (1938) 'Sul Contatto di due corpi elastici', Rend. Accad. Naz. dei Lincei, 27, ser. 6, 342.

1B   Mindlin, R.D., (1949) 'Compliance of elastic bodies in contact', Trans. ASME ser. E, J. Appl. Mech., 16, 259.

2   Johnson, K.L., Kendall, K and Roberts, A.D. (1971) 'Surface energy and contact of elastic solids', Proc. R. Soc. London, A324, 301.

3   Bowden, F.P. and Tabor, D. (1951) Vol I, (1964) Vol. II, Friction and Lubrication of Solids, Oxford Univ. Press, London.

4   Johnson, K.L., (1985) 'Contact Mechanics', Cambridge Univ. Press, Cambridge.

5   Johnson, K.L., (1955) 'Surface interaction between elastically loaded bodies under tangential forces', Proc. Roy. Soc., A230, 531.

6   Greenwood, J.A., (1991) 'Surface Roughness', Fundamentals of Friction, NATO ASI in Braunlage (Germany), 29 July -9 Aug. 1991.

7   Kalker, J.J., (1990) 'Three-Dimensional Elastic Bodies in Rolling Contact', Solid Mechanics and its Applications, Kluwer Academic Press, Dordrecht.

133

8   Irwin, G.R., (1960) 'Fracture mechanics' in Structural Mechanics, ed. Goodier,J.N., and Hoff, N.J, Pergamon Press, Oxford, 557.

9   Rice, J.R., (1968) 'Mathematical analysis in the mechanics of fracture', in Fracture, ed. H. Liebowitz, Vol. 2, 196.

10   Lawn, B.R., (1991) friction processes in brittle fracture', Fundamentals of Friction, NATO ASI in Braunlage (Germany), 29 July -9 Aug. 1991; this book.

11a Johnson, K.L., (1958) 'A note on the adhesion of elastic solids', Brit. J. Appl. Phys., 9, 199.

11b Pollock, H.M., (1991) 'Surface forces and adhesion', Fundamentals of Friction, NATO ASI in Braunlage (Germany), 29 July -9 Aug. 1991; this book.

12   Spence, D.A., (1975) 'Self similar solutions to adhesive contact problems with incremental loading', Proc. Roy. Soc. Lond., A305, 55.

13   Dundurs, D. and Comninou, M., (1979) 'The interface crack in retrospect and prospect' Proc. 1st USA-USSR symp. on fracture of composite materials held in Riga, Slithoff and Nordhoff, Alphen a.d. Rijn, 93.

14   Savkoor, A.R., (1981) 'The mechanics and physics of adhesion of elastic solids' in Microscopic Aspects of Adhesion and Lubrication', ed. J.M. Georges, Elsevier Sci. Publ. Co. Amsterdam, 279.

15   Savkoor, A.R. and Briggs, G.A.D., (1977) 'The effect of tangential force on the contact of elastic solids in adhesion', Proc. Roy. Soc. Lond., A 356, 103.

16   Savkoor, A.R., (1987) 'Dry Adhesive Friction of Elastomers', Doctoral Thesis, Delft Univ. of Tech., Delft.

17   Tabor, D., (1975) 'Interaction between surfaces: adhesion and friction', in Surface Physics of Materials 2, ed. Blakely, J.M., Academic Press.

18   Johnson, K.L., (1985) Private communication, notes September 1985.

19   Archard, J.F., (1953) 'Contact and rubbing of flat surfaces', J. Appl.Phys., Vol. 24, 981.

20   Courtney-Pratt, J.S. and Eisner, E., '(1957) 'The effect of a tangential force on the contact of metallic bodies', Proc. Roy. Soc. Lond., A238, 529.

21   Kendall, K., (1971) J. Phys. D. : Appl. Phys. 4, 1186.

22   Savkoor, A.R., (1990) 'Analysis of experiments on adhesion and friction of smooth rubber', in International Conference on Frontiers of Tribology, Inst. of Physics conference (Chairman Roberts, A.D.), held in Strattford upon Avon U.K., 15-17 April 1991, (Lecture to be published).

23   Schallamach, A., (1971) 'How does rubber slide?', Wear, 17, 301.

24   Barquins, M., Courtel, R. and Maugis, D., (1976) 'On the domain of existence of interfaceons', Letter to the editor, Wear, 38, 193.

25   Briggs, G.A.D. and Briscoe, B., (1978) 'How rubber grips and slips - Schallamach waves and the friction of elastomers' Philosophical Magazine A, vol. 38, No. 4, 387.

26   Roberts, A.D. and Jackson, S.A., (1975) 'Sliding friction of rubber', Nature, vol. 257, Sept. 11, 118.

27   Best, B., Meijers, P. and Savkoor, A.R. (1981) 'The formation of Schallamach waves', Wear, 65, 385.

FRACTURE, DEFORMATION
AND INTERFACE
SHEAR

..... The tables that list purported values of the friction coefficient of "steel on steel" or "copper on copper" and the like, are all false, because they ignore the factors which really determine μ.

R P FEYNMAN, R B LEIGHTON, and M SANDS, "The Feynman lectures on physics" vol 1, Addison-Wesley, 1963, p 12-5, quoted by H CZICHOS, in "Tribology in the 80's", NASA Conference Publication 2300 (1984) p 71.

..... If the formation and breaking of bonds is the main process consuming energy, how does it come about that rolling friction is so much smaller than sliding friction?

F E SIMON, in "A discussion on friction", Proc. Roy. Soc. **A212**, 439 (1952).

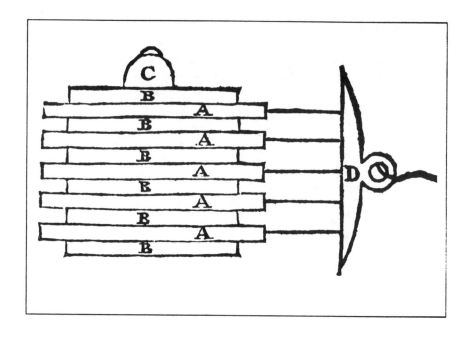

*Amontons' representation of friction between multiple surfaces (Amontons, 1699)
[From D. Dowson, <u>History of Tribology</u> (Longman, London, 1979) p. 155, with
permission.]*

FRICTION PROCESSES IN BRITTLE FRACTURE

BRIAN LAWN
Materials Science and Engineering Laboratory
National Institute of Standards and Technology
Gaithersburg, MD 20899, U.S.A.

ABSTRACT. In this paper we consider the interrelations between friction and fracture in highly brittle materials. First we examine frictional effects in the mechanics of crack formation at elastic and elastic-plastic contacts on brittle surfaces. Then we consider how fundamental intersurface forces manifest themselves as "internal friction" at crack interfaces in "model" solids like mica and glass, with special reference to environmental chemistry. Finally, we examine the controlling role of frictional processes in the strength and toughness of modern ceramic systems.

1. Introduction

Solids with ionic-covalent bonding, "ceramics", are limited in structural applications by one factor, *brittleness*. Such materials, by virtue of their innate inability to sustain energy-absorbing plastic deformation, have a singularly low resistance to the initiation and propagation of cracks. Modern fracture mechanics seeks methodologies for quantifying this brittleness, and, more importantly, of overcoming it [1].

A major player in the fracture properties of brittle ceramics is *friction*. Friction can be *deleterious*, e.g. by augmenting tensile stresses in surface contacts, with adverse effects on strength, abrasion and wear; or *beneficial*, as in toughness, by "shielding" the crack tip and dissipating a portion of the work of applied loading. Its manifestations can be obvious, as in contact configurations, or subtle, as in the crack-tip energy dissipation that attends non-equilibrium fracture in environmentally-assisted crack growth.

We shall illustrate some of the ways friction can play a role in brittle fracture with a selection of examples, drawing from traditional fracture mechanics and modern-day ceramics science.

2. Friction in Contact Fracture

Indentation stress fields usefully simulate real particle contacts in strength degradation, wear and abrasion [2], as depicted in the scheme of fig. 1. In this section we examine some of the indentation crack patterns that form in brittle surfaces, focussing on homogeneous solids like glass and mica. Reference is made to review articles for greater detail [1-4].

In the context of brittle fracture one may usefully classify contacts as "blunt" or

137

I. L. Singer and H. M. Pollock (eds.), Fundamentals of Friction: Macroscopic and Microscopic Processes, 137–165.
© 1992 *Kluwer Academic Publishers. Printed in the Netherlands.*

138

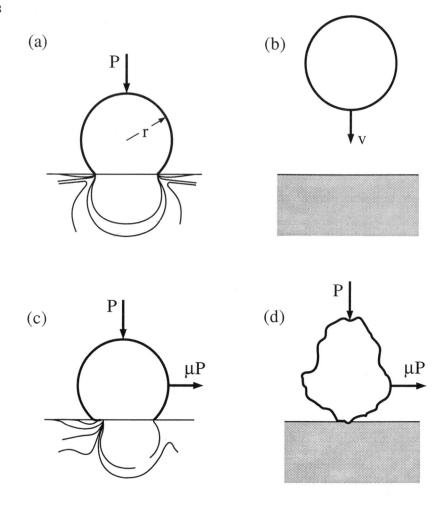

Figure 1. Model for simulating surface damage from particle contact. (a) Static blunt particle (Hertz), sphere radius r and normal load P; (b) impacting sphere, normal velocity v; (c) sliding sphere, with tangential force μP superposed onto normal load P; (d) irregular, sharp sliding particle. Equi-stress contours of tension indicated in (a) and (c).

"sharp". The archetypical *blunt* contact is a sphere on a flat surface in simple Hertzian geometry. The deformation is entirely *elastic* and the ensuing crack at critical loading is the well-documented *Hertzian cone*. Fig. 2 summarises the evolution of this crack type during the indentation loading cycle. *Sharp* contacts are typified by fixed-profile pyramidal indenters, such as the Vickers and Knoop. The deformation is *elastic-plastic*, and the cracks have the *radial* and *lateral* geometry depicted in fig. 3.

2.1   BLUNT CONTACTS AND MODIFIED HERTZIAN CONE CRACKS

The geometrical path and energetics of Hertzian fracture are predetermined by the elastic stress field between two curved surfaces [5]: the crack path by the conical trajectory of maximum tensile stress; the energetics by the subsurface stress falloff along this trajectory.

Initiation. Contact with a sphere of radius r on a flat surface generates tensile stresses on the surface immediately outside the contact circle. These stresses fall off dramatically along the conical path of the prospective crack, fig. 2b. At some point during normal loading P a shallow ring crack develops from a pre-existing surface flaw. With further increase in P the ring extends stably downward until, at a critical load P_C, the ring becomes unstable, and propagates abruptly into the full cone. A detailed fracture mechanics analysis [5] of this instability condition yields

$$P_C = ArR_0 \tag{1}$$

where $R_0 = 2\gamma$ defines an intrinsic resistance to crack propagation, with γ an appropriate surface energy. The dimensionless coefficient A is the "Auerbach constant", after the discoverer of the empirical relation $P_C \propto r$ one hundred years ago [6]. Whereas satisfaction of Eq. 1 is contingent on the pre-existence of suitable starter microcracks, P_C is (by virtue of the stable growth of the ensuing ring crack) independent of the *initial size* of these microcracks, a prediction borne out by experiments on glass surfaces with controlled, abrasion-induced flaws [7].

Propagation. After propagating downward through the ever-diminishing stress field, the crack arrests in its truncated-cone geometry at a depth $c \approx 3a$ (fig. 2b). The ever-widening circular crack front implies the simple stress-intensity factor relation for penny-like cracks [1]

$$
\begin{aligned}
K_P &= \chi P/c^{3/2} \\
&= (E'R_0)^{1/2} = T_0
\end{aligned}
\tag{2}
$$

where χ is a crack-geometry term and $E' = E/(1 - \nu^2)$, E Young's modulus and ν Poisson's ratio. Eq. 2 is a statement of Griffith equilibrium: the quantity K defines a mechanical driving force on the crack; T_0 defines the intrinsic resistance to crack propagation, or "toughness". The subscript P is to indicate that the stress intensity is operative at maximum load (cf. Eq. 6 below). Thus $c \propto P^{2/3}$, which is the Roesler relation [8] for equilibrium cone cracks. Again, the result is independent of starting flaw size, as expected for a far-field solution.

Friction. The best-studied case of a contact with friction is that of a sliding sphere with complete interfacial slipping, i.e. lateral stress at each element within the contact area equal to friction coefficient μ times normal stress at that element [9]. The effect of

140

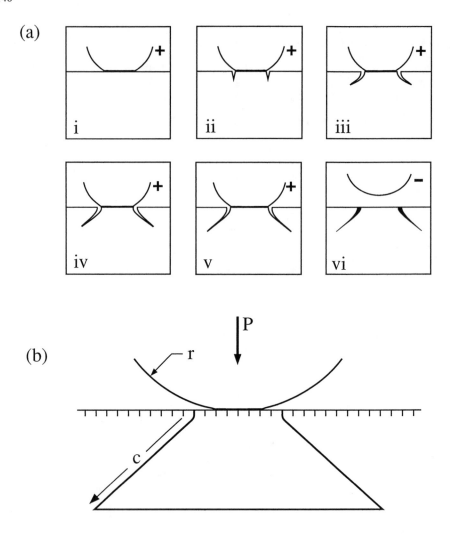

Figure 2. Hertzian cone crack system. (a) Evolution during complete load (+) and unload (-) cycle: (i) preexisting surface flaws experience tensile stresses outside contact circle; (ii) a favourably located flaw runs around contact circle to form a surface ring crack; (iii) at increasing load, the ring crack extends incrementally downward in a rapidly diminishing tensile field; (iv) at critical load the ring becomes unstable and flares downward into the full (truncated) Hertzian cone; (v) at continued loading the arrested cone grows further downward in stable fashion; (vi) at unloading, the cone closes. (b) Essential parameters of system.

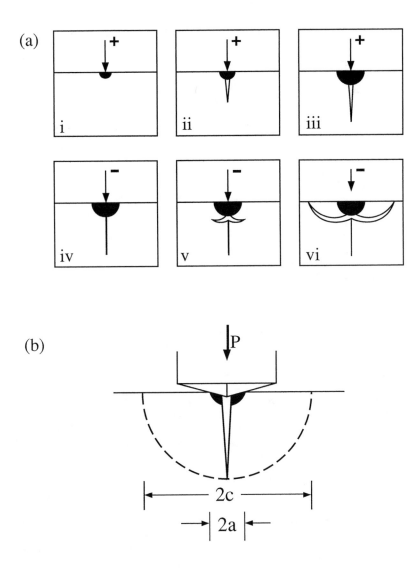

Figure 3. Evolution of radial-lateral crack system during complete load (+) and unload (-) cycle: (i) The sharp point induces plastic deformation; (ii) at critical load shear faults within the deformation zone pop-in to form subsurface radial cracks on median planes; (iii) at increased loading, the radial crack expands further downward; (iv) on initial unloading, the radial cracks expand further, from the action of residual wedge-opening stresses in the near-surface region; (v) at further unloading, the surface radial crack continues to expand, and lateral cracks begin to grow just beneath the free surface; (vi) at full unload, radial and lateral cracks achieve their final, equilibrium penny-like geometry in the residual stress field. (b) Essential parameters of system.

increasing μ is to enhance the tensile field markedly at the trailing edge of the contact, and conversely to suppress it at front, fig. 1c. The critical conditions for crack initiation therefore depend sensitively on the frictional tractions, i.e. $A = A(\mu)$ in Eq. 1 [10,11].

Because of the loss of axial symmetry in the sliding contact stress field the cone cracks are asymmetric and only partially developed [10]. Figure 4 shows cone crack tracks resulting from a steel sphere sliding on glass [12]. Since most of the applied load is supported by the material below the cone frustrum, the resultant geometry may be closely approximated by tilting the cone along the resultant applied load axis, fig. 5 [12]. The magnitude and angular displacement of the load vector in fig. 5 are

$$P = P_N(1 + \mu^2)^{1/2} \tag{3a}$$
$$\beta = \arctan \mu \tag{3b}$$

with P_N the normal component. The crack size is determined from Eqs. 2 and 3 as

$$c = (\chi P_N/T_0)^{2/3}(1 + \mu^2)^{1/3} \tag{4}$$

which is relatively *insensitive* to μ. On the other hand, as is apparent in fig. 4, the *density* of cone cracks increases strongly with μ.

2.2 SHARP INDENTERS AND RADIAL-LATERAL CRACKS

The case of elastic-plastic contact fields in normal loading beneath a sharp pyramidal indenter is more complex. A finite load cannot be supported by an infinitesimal point, so the specimen deforms inelastically at the near-contact. The indenter thereby leaves a permanent impression, the size of which is an (inverse) measure of the hardness.

Initiation. The inelastic deformation strongly influences the ensuing crack pattern by modifying the stress field, generating the radial-lateral crack geometry depicted in fig. 3. The stress intensity about the indenter point approaches the theoretical limit of the structure, causing the material to "fail" along discrete "shear faults" [13,14] in the highly compressive contact near field in order to accommodate the penetrating indenter [15,16]. An example of shear faults at a Vickers indentation in glass is given in fig. 6a. The shear faults, analogous to dislocation slip surfaces in metals but not constrained to crystallographic planes, act as their own sources for subsequent radial and lateral crack "pop-in" - there is no need for pre-existing flaws.

Working models of radial crack initiation [15] treat the faults as shear cracks with restraining interfacial friction, fig. 6b. This interfacial friction persists on unloading the indenter, accounting for the residual impression in brittle materials. The residual impression in turn manifests itself as an outward pressure on the elastic matrix encasing the deformation zone, as represented in "expanding cavity" models of the elastic-plastic field. Radial cracks extend from the ends of the faults at indentation corners, into the outer hoop-tensile field. For fixed-profile indenters (e.g. Vickers) the intensity of the stress field is determined by the size-invariant contact pressure, or "hardness" H. The threshold condition for pop-in is determined as the load at which the shear fault, which scales with the impression diagonal, reaches an unstable size [17,18]:

$$P_C = \Theta T_0(T_0/H)^3 = \Theta E'^2 R_0^2/H^3 \tag{5}$$

with $\Theta = \Theta(E/H)$ a dimensionless coefficient. Hence for sharp indenters the critical load is a strong measure of toughness *and* hardness. The fact that the shear faults act as

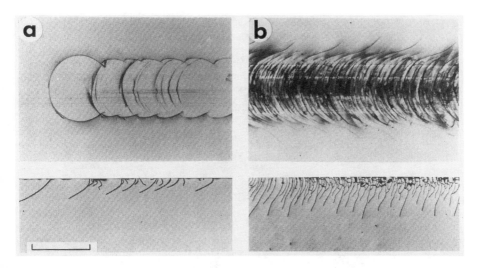

Figure 4. Surface and section views of partial-cone-crack damage on soda-lime glass from sliding steel sphere, r = 3.17 mm, P_N = 20 N: (a) μ = 0.1, (b) μ = 0.5. Index marker 500 μm. From [12].

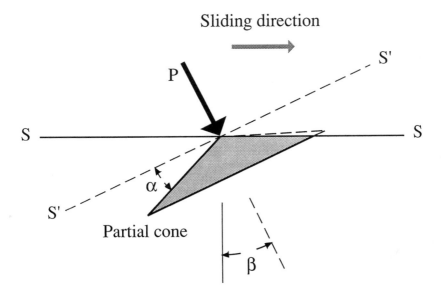

Figure 5. Cone crack geometry for sliding elastic contact of sphere on flat surface. Superposed frictional contact onto normal contact rotates load axis, effectively tilting surface SS (and hence cone axis) through S'S'. Shown here for β = 26.5° (μ = 0.5) and α = 22° (glass).

144

(a)

(b)

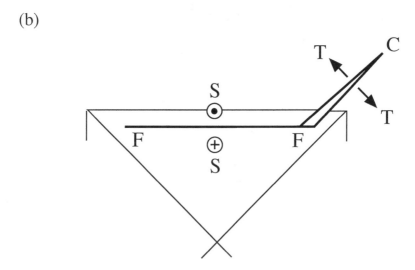

Figure 6. (a) Vickers indentation in soda-lime glass [16]. Surface etched to reveal traces of shear faults, which converge on curved surfaces beneath impression. Width of field 40 μm. (b) Model of radial crack initiation FC from shear fault FF. S denotes frictional shear stresses on fault interface, T tensile stresses in outer residual field.

controlling flaws means that, in contrast to blunt indenters, P_C is the same for pre-damaged and pristine surfaces.

Propagation. On pop-in, the radial and lateral cracks arrest at $c \approx 3a$. Again, both cracks expand on an essentially circular front, so the stress-intensity factor relation for equilibrium "penny-like" cracks is satisfied,

$$K_R = \chi P/c^{3/2}$$
$$= (E'R_0)^{1/2} = T_0, \tag{6}$$

in analogy to Eq. 2 for blunt indenters. Now, however, the maximum stress intensity is operative in the *unloaded* state - K_R in Eq. 6 relates to the *residual* field in the elastic-plastic contact [19]. The generic relation $c \propto P^{2/3}$ once more holds; and, as in Eq. 4, we expect the modifying effect of contact friction to be weak.

Chemistry. The residual contact field implicit in Eq. 6 can have a profound influence on the crack response in reactive environments. An example is given in Fig. 7. Micrograph (a) shows a radial crack system in glass immediately after completion of contact in inert environment. The residual field is apparent from the persistent birefringence in polarised light. Micrograph (b) shows the same crack pattern after prolonged exposure to laboratory atmosphere. Moisture from the atmosphere diffuses along the crack interface and interacts chemically with the walls in the near-tip region [1,20,21], causing the crack to propagate in a rate-dependent manner. Ultimately, the system reaches a new equilibrium state, determined by the reduced interface energy (Sect. 2.2 below), and the crack comes to rest.

Even more dramatic is the combined effect of residual stress and moisture on radial-crack pop-in. Exposure to water vapour lowers the threshold load for initiation in Eq. 5 relative to inert environments by over two orders of magnitude [15]. Moreover, at intermediate loads the pop-in is "delayed", i.e. occurs spontaneously after completion of the contact cycle.

3. Chemistry in fracture and adhesion

The susceptibility of fracture in brittle solids to environmental interaction raises fundamental questions as to how active species, particularly water, penetrate and interact with crack interfaces. In this section we consider the nature of chemical interactions with intrusive water at fracture interfaces in mica. Mica is an ideal material for fracture studies because of its atomically smooth cleavage, which allows one to study healed as well as virgin interfaces. We emphasise, however, that the behaviour described below is representative of brittle solids in general [1].

3.1 EQUILIBRIUM CRACKS AND ADHESION ENERGIES

The interface adhesion energy (Dupre work of adhesion) for mica interfaces at different relative humidities have been measured in fracture experiments [22-24], using an automated version (fig. 8) of earlier wedge-opening cleavage test arrangements [25,26]. From measurements of crack size c and specimen dimensions one evaluates the mechanical-energy-release rate G (= K^2/E) for double-cantilever specimens [1,22]

$$G = 3Eh^2d^3/4c^4 \tag{7}$$

 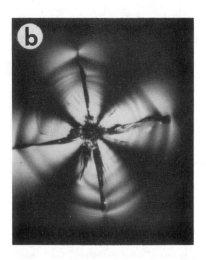

Figure 7. Vickers indentation in soda-lime glass: (a) immediately after completion of contact cycle; (b) same, 1 hr later after exposure to moisture. Note development of both radial and lateral (faint subsurface fringes). Courtesy D.B. Marshall.

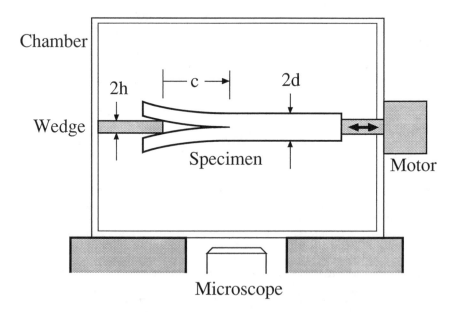

Figure 8. Wedge-opening geometry used to measure adhesion energies and crack velocity curves in mica [22]. Assembly is enclosed in environmental chamber, allowing inert (dry nitrogen gas) and interactive (humid gas) atmospheres. Chamber sits on microscope stage to allow *in situ* viewing (including VCR recording) of crack. Crack motion is effected by translating specimen relative to clamped wedge via stepper motor.

The interface energy is then determined at *Griffith equilibrium*, $G = 2\gamma$. Note $dG/dc < 0$ in Eq. 7, so the equilibrium is stable; i.e. the crack does not go spontaneously to failure as it does in the uniform tensile stress state characteristic of standard strength tests, but advances steadily at constant length c with continued wedge insertion.

Interface energies measured in this way are plotted as a function of relative humidity in fig. 9. Data are included for healed as well as for virgin interfaces. The healed interfaces are obtained by two procedures: the first by simply reversing the wedge motion in an incompletely cleaved specimen, allowing the interface to retain its original coherence; the second by rotating and recontacting fully separated cleavage halves. The primary source of adhesion in the virgin mica is a Coulombic interaction between potassium ions and negative charge in the mica sublayer [27]. Capillary condensation screens these solid-solid electrostatic interactions [28], in the manner of fig. 10 [23,24]. At larger humidities the Kelvin radius of the meniscus, hence the degree of screening, increases until, at saturation, the adhesion is dominated by the Laplace pressure term (equal to twice the surface energy of water - in bulk water even this term is lost). We see that the adhesion is diminished at the healed-coherent interfaces, presumably because of occlusion of water molecules. It is diminished even more at the healed-incoherent (rotated) interfaces, suggesting that loss of lattice registry negates the greater part of the Coulombic interaction [23,24,26].

Some comparative adhesion energy data from pulloff experiments on recontacted mica sheets in the crossed-cylinder configuration of an Israelachvili surface forces apparatus are included in fig. 9. The energies in these experiments were evaluated assuming rigid spheres held together exclusively by capillary action [29]. Maugis [30] has argued that essential complementarity in the macroscopic fracture mechanics exists between brittle-crack and contact-adhesion configurations. In the present case such complementarity applies specifically to the healed-incoherent interfaces, since the opposing mica sheets in the crossed-cylinder geometry are generally recontacted in arbitrary misorientation. That the contact-adhesion data tend to lie below their fracture counterparts in fig. 9, especially at low RH, is attributable in part to failure to account for elastic deformation in the spheres [31,32]. As a final comment concerning the relative merits of brittle-fracture and contact-adhesion methodologies, we note that the former has greater versatility in its capacity to determine the relative influence of lattice registry (healed-coherent vs healed-incoherent interfaces) and occluded water (healed vs virgin interfaces).

Recent experiments on *dis*similar interfaces, e.g. mica-silica, show that the adhesion in dry atmospheres is greatly enhanced by bulk charge transfer across the interface [24,33].

3.2   KINETIC CRACKS AND VELOCITY CURVES

As indicated in Sect. 1.2, brittle cracks exhibit kinetic characteristics in their growth mechanics in reactive atmospheres, with particular sensitivity to water. The *rate* of growth generally increases with increasing displacement of the system from equilibrium. These kinetics are traditionally represented on velocity v-G curves [1,34]. A family of such curves for virgin cracks in mica at different relative humidities is shown in fig. 11. Note that the velocity at each humidity goes rapidly to zero at some "threshold", defined by the equilibrium state $G = 2\gamma$ for that environment (fig. 9).

The micromechanics of rate-dependent crack growth are attributed to diffusion of intrusive species along the crack interface. The atomic structure at the interface near the crack tip provides energy barriers to the diffusion, resulting in energy dissipation by internal friction. Figure 12, computed from linear elasticity solutions for near-tip displacement fields [35], demonstrates how these barriers become less constraining to the penetration of water in mica as G, hence the crack opening, increases. These structure-

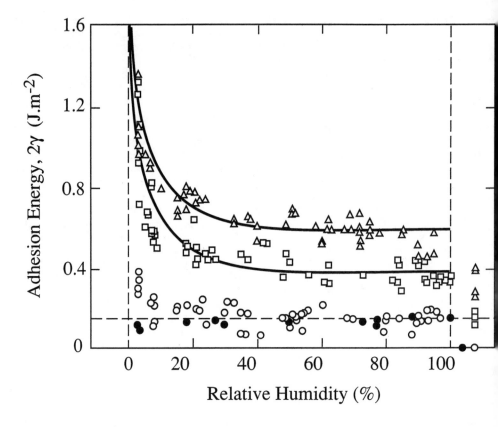

Figure 9. Interface energy of mica as function of relative humidity from fracture experiments. Data for virgin (triangles), healed (squares) and healed-misoriented (circles) interfaces. Points to right of RH 100% correspond to tests in bulk water. Closed symbols are data from Israelachvili-type crossed-cylinder apparatus. Horizontal dashed line denotes capillary contribution, $2\gamma = 144$ mJ·m^{-2} (twice surface tension of water). Solid curves are predictions from theoretical model assuming dielectric screening of solid-solid Coulombic interactions by capillary condensation (fig. 10) [24].

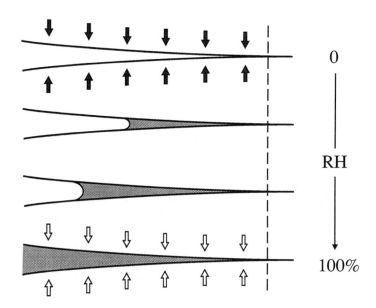

Figure 10.   Reduction of interface attraction by capillary condensation. As relative humidity increases, capillary progressively fills crack, replacing strong solid-solid interactions (closed arrows) with weak solid-liquid-solid interactions (open arrows).

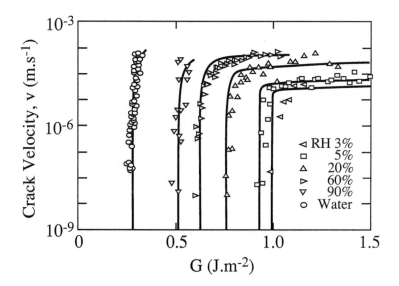

Figure 11.   Crack velocity as function of mechanical-energy-release rate for virgin interfaces in mica at specified RH [22].

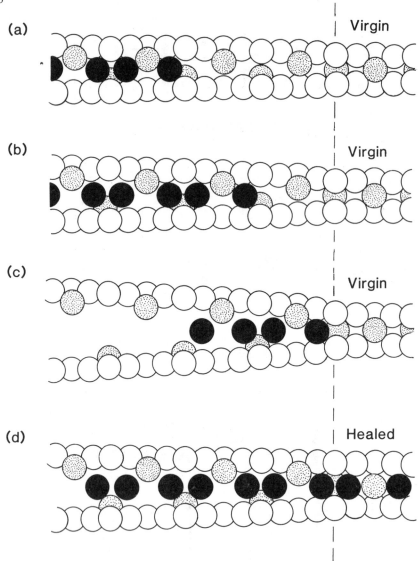

Figure 12. Crack profiles for mica cleavage interfaces [35]. Loading at (a) G = 100 mJ·m^{-2} , (b) G = 200 mJ·m^{-2}, G = 800 mJ·m^{-2}, at virgin interface. Unloading at (d) G = 100 mJ·m^{-2}, showing occlusion of intrusive water at healed interface. Elastic sphere representation, with atom sizes from ionic radii and centres from calculated linear elasticity displacement fields. Oxygen - open spheres; potassium - shaded spheres; water - solid spheres.

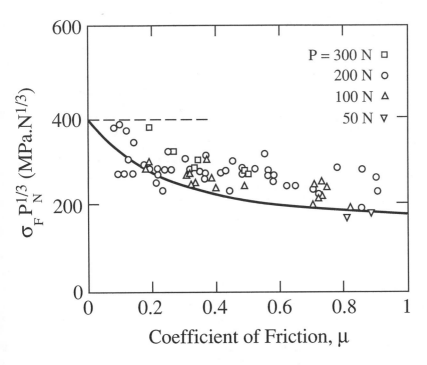

Figure 13. Plot of $\sigma_F P_N^{1/3}$ against μ for glass surfaces subjected to sliding contact with steel sphere at specified loads [12]. Dashed line denotes reference strength level at normal load.

related diffusion barriers are highly sensitive to the crack opening, accounting for the steep velocity increases in fig. 11. At limiting values of G, depending on the humidity, the water vapour can no longer maintain pace with the accelerating tip, and the velocity saturates at the "plateaus" [1,34].

The understanding of velocity curves of the type shown in Fig. 11 is of great practical interest to ceramic engineers who seek methodologies for predicting "lifetimes" of brittle components. In ceramics, flaws as small as 1 μm in inherently unstable stress geometries, i.e. where dK/dc or dG/dc > 0 [1,36], can propagate slowly but decisively to failure at sustained stresses well below the nominal laboratory "inert" strength [37].

4. Friction in strength and toughness of ceramics

Finally, we investigate some practical implications of friction in the strength and toughness properties of modern ceramics. How can we make use of our understanding of contact mechanics and crack-interface energetics in developing brittle materials for structural applications? As we shall show, friction is a critical factor in designing reliable high-toughness materials with strong flaw tolerance.

4.1   STRENGTH DEGRADATION

The strength of brittle surfaces can be severely compromised by the inception of crack-like flaws from contacts with extraneous bodies, of the kind depicted in fig. 1 [38]. Most offensive are the penetrating cone and radial cracks considered in Sect. 1. Friction, by virtue of its enhancement of the contact tensile stresses, is a potentially important player in strength degradation.

Just *how* important friction is depends on whether or not the contact exceeds the threshold for crack initiation. We have already seen in Sect. 1 that the size of well-developed cone or radial cracks is insensitive to μ. We illustrate this insensitivity in fig. 13 with results of strength tests on glass surfaces damaged by sliding spheres under conditions similar to those in fig. 4 [12]. The data are results of strength tests as a function of sliding friction μ, for specified normal loads P_N. The solid curve is a prediction obtained by combining Eq. 4 with the conventional strength equation for failure [1],

$$\sigma_F = T_0/\psi c^{1/2} \tag{8}$$

where $\psi = \psi(\alpha,\beta)$ is a crack geometry factor (fig. 5):

$$\sigma_F P_N^{1/3} = f(\mu)_\alpha T_0^{4/3} \tag{9}$$

with $f(\mu)$ a dimensionless function [12]. We see that σ_F diminishes with μ, but slowly. Hence to the structural engineer who seeks to use ceramic components in mechanical environments where contact-induced cracking is accepted as inevitable (e.g. dust-impacted turbine blades), friction is not a critical design factor.

On the other hand, in ultra-high strength applications where the incidence of a single contact crack is unacceptable, as in freshly-drawn optical glass fibres and polycrystalline ceramics with refined microstructures [1], friction is indeed a critical issue. On such pristine surfaces the spurious contact with a single sharp grit particle as small as 1 μm (fig. 1d) and loads as small as 0.01N (1 g) may be sufficient to reduce the strength of glass by two orders of magnitude [38]. We have already indicated how interfacial friction might play a strong role in enhancing crack pop-in: with blunt indenters, by dramatically diminishing the critical load for cone-crack pop-in (Sect. 1.1); with sharp indenters, via enhanced radial- and lateral-crack initiation from the chemical interaction of moisture at shear-fault interfaces (Sect. 1.2).

4.2   TOUGHNESS

The intrinsic intersurface forces that account for the adhesion energy $G = R_0 = 2\gamma$ in ideally brittle ceramics provide insufficient toughness for most structural applications. The toughness can be raised to an acceptable level only by incorporating highly dissipative elements into the microstructure, thereby introducing an additional component of internal friction [1]. These dissipative elements operate in a "shielding zone" around the crack tip, but do not interact directly with the fundamental bond-separation mode of crack extension. Shielding arises because the dissipative elements effectively oppose the crack opening from the externally applied loads. The most effective form of shielding occurs in metals, via an extensive crack-tip plastic zone. However, there is no analogous dislocation activity at crack tips in ceramics [1,39] (it is this very absence of dislocation activity that gives ceramics their innate brittleness), so other forms of energy dissipation must be sought.

For common polycrystalline ceramics the most effective shielding elements take the form of frictional crack-interface "bridges", e.g. interlocking grain facets, whiskers, fibres.

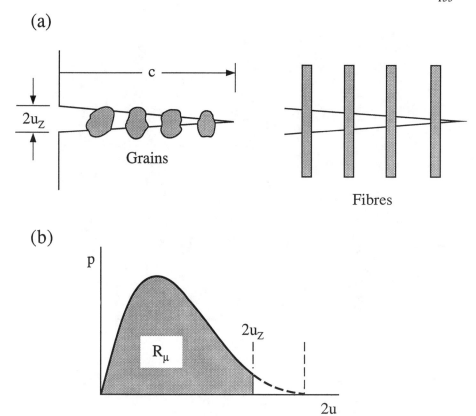

Figure 14. Examples of bridging in ceramic materials. (a) Interlocking grains and embedded fibres. Bridging elements resist crack opening, and thereby "shield" the crack tip from the externally applied loading. (b) Stress-separation function p(u). Area under curve denotes energy absorbed by bridges as they pull out of matrix.

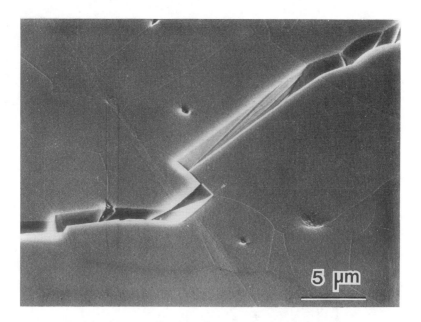

Figure 15. Grain bridging in polycrystalline ceramic. Scanning electron micrograph of crack in alumina, showing facet-facet sliding at interface well behind crack tip [45].

Grain-facet and embedded-fibre bridges are depicted in fig. 14a. The principal requirement for formation of such bridges is that there exist weak grain or interphase boundaries for deflecting the crack [40]. To be effective, the frictional shear stresses τ at these boundaries should be sufficient to dissipate large amounts of energy as the bridging elements pull out. One way of enhancing τ is to build in internal compressive stresses σ_R at the debonding interfaces, giving rise to Coulomb frictional tractions [41]

$$\tau = \mu\sigma_R. \tag{10}$$

The bridging constraint is quantified by a constituent stress-separation function p(u), i.e. restraining stress p exerted by the interfacial elements on the crack walls as a function of crack-opening displacement 2u, as in fig. 14b [1,42,43]. Generally, p is proportional to τ and dependent on characteristic microstructural dimensions like grain size or fibre diameter.

The condition for equilibrium crack extension for a material with bridging becomes

$$G = R_0 + R_\mu = R. \tag{11}$$

The shielding term R_μ is expressible as the area under the stress-separation curve in fig. 14b

$$R_\mu = \int_0^{2u_z} p(u)du \tag{12}$$

with $2u_z$ the crack opening at the edge of the bridging zone [1,42]. Together with the relation $u_z(c)$, $R_\mu(u_z)$ defines the so-called "R-curve", $R(c)$. The R-curve rises with increasing c from its base level R_0 at c = 0, up to the steady-state level at which the bridging elements ultimately disengage from the crack walls.

4.2.1 *Polycrystalline ceramics*. Grain-facet bridging in polycrystalline ceramics can lead to substantial R-curve behaviour. Typically, for alumina of grain size 25 μm, R increases from a base value 20 J·m^{-2} to a respectable 70 J·m^{-2} over a crack extension 2 mm and crack-opening displacement $2u_z = 1$ μm [1].

The mechanics of grain bridging have been studied using *in situ* microscopy of crack growth in alumina during loading to failure [44-47]. These observations reveal an extremely interesting failure evolution. The indentation cracks do not remain stationary up to the critical instability stress in the manner of classical Griffith flaws; instead, they grow steadily, grain by grain, for several hundred μm prior to this instability. The flaws are *stabilised* by the rising R-curve.

In addition, the *in situ* observations reveal vital information on the nature of the underlying bridging mechanism [44-47]. Figure 15 shows a micrograph of a bridging site in alumina. Opposing grains remain in intimate sliding contact at adjoining facets across the interface, even though the crack walls have separated through ~10% of the grain diameter, some 1 mm behind the crack tip. High-friction contacts occur principally at those facets under compressive thermal expansion anisotropy stresses in the alumina. The evidence for high frictional stresses is even more vividly demonstrated in cyclic loading, fig. 16 [48]. The cumulation of debris reflects a susceptibility to cyclic *fatigue*, as a progressive degradation of the contacts. The fundamental nature of the friction at the sliding facets has barely been questioned, and leaves us with important material and tribological issues for further study.

The flaw stabilisation from the R-curve manifests itself in a powerful manner in the strength characteristics. The failure condition no longer depends on the *initial* flaw size, but rather on the *extended* flaw size at instability (the "tangency point" on the R-curve [1,36]). Hence materials with pronounced R-curves tend to be *flaw tolerant*. To enhance this flaw tolerance we need to increase the area under the p(u) curve in fig. 14b. One way of doing this is to enhance the friction stress in Eq. 10, via σ_R, by incorporating a second phase with large differential expansion coefficients relative to the matrix [49]. Aluminum titanate is a prime candidate in alumina matrices: its c-axis coefficient is -3x10^{-6} °C^{-1} compared to 10x10^{-6} °C^{-1} for alumina. Figure 17 shows how incorporation of 20% aluminum titanate into a 6 μm alumina matrix produces a striking improvement in flaw insensitivity. Such flaw insensitivity is of intense interest to ceramicists: to the engineer, for the potential to design to a well-defined stress level, with greatly reduced susceptibility to subsequent in-service damage and with a built-in stable crack growth system to warn of imminent failure; and to the processor, for tailoring materials with specifiable properties for specific applications.

4.2.2 *Fibre-reinforced ceramics*. As with polycrystalline ceramics, the toughness of ceramic-matrix composites reinforced with fibres is controlled largely by interfacial friction properties. Accordingly, much effort is being devoted by the ceramics community to the measurement of friction in "pull-out" or "push-in" tests. In the *push-in* test, friction stresses are evaluated by measuring the hysteresis in the load-displacement function through one cycle [50].

A recent variant on the *pull-out* test has provided a vital clue as to the nature of friction in some fibre-composite systems [51,52]. A fibre protruding from both faces of a matrix section is pulled axially until it debonds and slips steadily through the section.

156

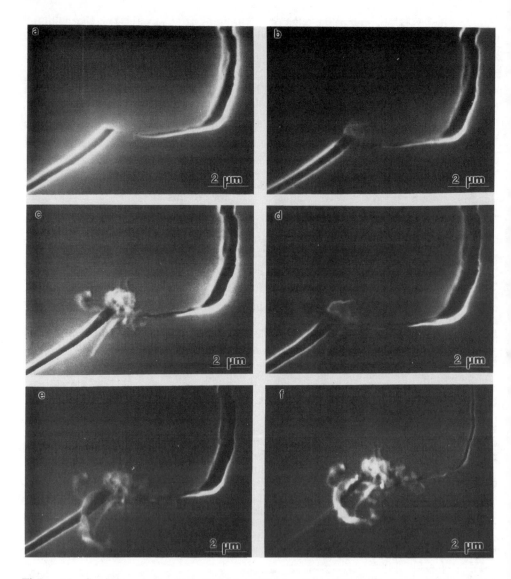

Figure 16. Scanning electron micrograph of frictional grain facet in alumina at sequential stages of cycling [48]. At maximum load, after: after (a) 0, (b) 7000, (c) 20000, (d) 27000, (e) 45000 cycles. After complete unload, (f). Note cumulation of frictional debris at facet.

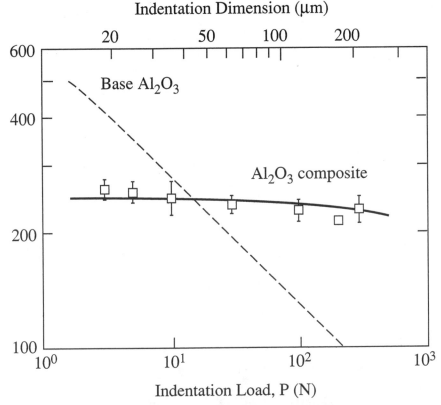

Figure 17. Strength of alumina matrix composite (grain size 6 μm) with aluminium titanate phase, as function of indentation load and flaw size (indentation diagonal). Dashed curve denotes behaviour for base alumina matrix. Note strongly enhanced flaw tolerance. After [49].

Before pull-through is complete the specimen is inverted and the fibre pulled back in the opposite direction. Figure 18 shows the two load-displacement half-cycles for a silicon carbide fibre in a borosilicate matrix. The steep rise in the curve at left reflects the initial debonding and fibre unseating as a shear crack spreads along the interface. At full debonding through the section the entire fibre, after some stick-slip, begins to pull out steadily. The relatively flat curve at right in fig. 18 indicates a steady reverse friction until, at the original fibre location, the system exhibits an abrupt reseating drop. This last feature, first reported by Jero and Kerans [51] in push-in tests, immediately suggests that there is a strong surface roughness component in the friction characteristic.

A detailed asperity model accounting for this surface roughness has been advanced [52]. Essential details are shown in fig. 19. A random distribution of spherical elastic asperities is presumed to exist on both fibre and matrix surfaces, so that interfacial contact is made through a distribution of Hertzian junctions. Microscopic examination of pulled-

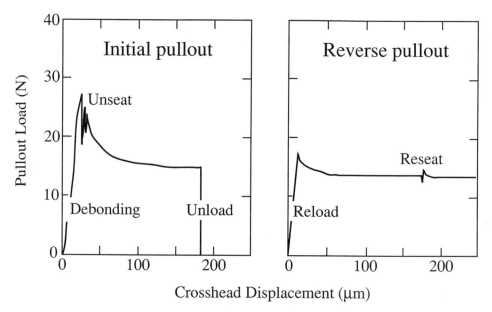

Figure 18. Measured force-displacement data for pullout of SiC fibre from borosilicate glass matrix [52]. *Left curve*, first pullout. *Right curve*, reverse pullout. Note initial steep peak at extreme left, and load drop at extreme right.

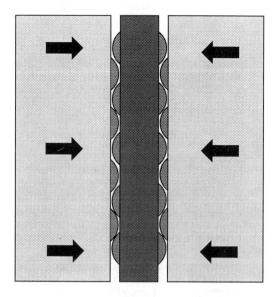

Figure 19. Asperity model of fibre pullout [52]. Asperities are treated as Hertzian contacts, under compressive loading from thermal mismatch stresses.

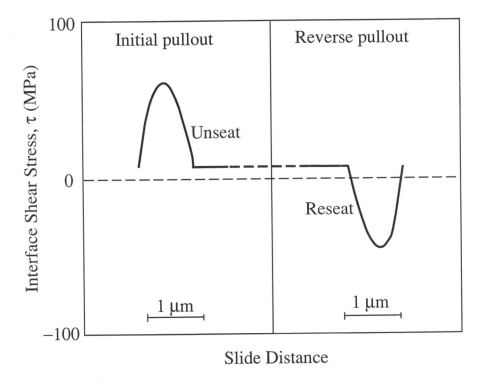

Figure 20. Calculated interfacial shear stress as function of sliding distance for model in fig. 19. Maximum at left indicates correlated elastic forces opposing unseating of asperities from their original interlock configuration. Corresponding minimum at right indicates reseating on reversed pullout. Cf. fig. 18. From [52].

out fibre surfaces confirms the existence of such asperities. In the original state the junctions of one surface seat in the valleys of the other, clamped by thermal mismatch stresses. The calculated shear resistance stress τ, using measured asperity dimensions and elastic parameters characteristic of the SiC/glass fiber-matrix system described above, is plotted as a function of fibre displacement in fig. 20. This plot reflects the broader features of the experimental pullout-displacement function in fig. 18. The initial resistance is large because of increasing resistance from both elastic deformation and the associated friction, accounting for the load peak at left. Once the asperities reach the first summits of their opposing counterparts the local normal components of the randomised Hertzian contacts average out to zero over the interface, leaving only the friction components, which always act in concert to oppose subsequent motion. The net friction thereafter maintains its steady-state value as sliding proceeds. This value is the same on reversing the pullout (albeit with opposite sign) until, ultimately, the asperities reseat and the system experiences the load drop.

5.   Conclusions

(i) Contact fracture is influenced by friction at the loading interface, propagation less so than initiation. Indenters may be classified as "blunt" (elastic) or "sharp" (elastic-plastic). The latter are characterised by strong residual stress fields.

(ii) Brittle fracture is chemistry-enhanced. Fracture and adhesion are governed by intersurface forces, which may be profoundly weakened by intervening fluid species (especially water). Interface energies determine equilibrium states, interfacial diffusion barriers kinetic states.

(iii) Friction is an important element in designing with structural ceramics. Strength is degraded by friction, less so in the domain of propagation than of initiation. Toughness is enhanced by friction at interfaces between microstructural elements in ceramic systems.

Acknowledgements

The author acknowledges many discussions and valuable contributions on various facets of this work with S.J. Bennison, L.M. Braun, E.P. Butler, E.R. Fuller, R.G. Horn, P.D. Jero, R.J. Kerans, S. Lathabai, J. Rödel, D.T. Smith, K-T. Wan. Funding was provided by the U.S. Office of Naval Research.

References

1.   B.R. Lawn (1992). "Fracture of Brittle Solids". Cambridge University Press, Cambridge.

2.   B.R. Lawn and T.R. Wilshaw (1975) "Indentation Fracture: Principles and Applications", *J. Mater. Sci.* **10** 1049-81.

3.   A.G. Evans and T.R. Wilshaw (1976) "Quasi-Static Solid Particle Damage in Brittle Solids", *Acta Metall.* **24** 939-56.

4.   B.R. Lawn and S.M. Wiederhorn (1983), "Contact Fracture in Brittle Materials", in Contact Mechanics and Wear of Rail/Wheel Systems", pp. 133-47, ed. J. Kalousek, R.V. Dukkipati and G.M. Gladwell. University of Waterloo Press, Vancouver.

5.   F.C. Frank and B.R. Lawn (1967) "On the Theory of Hertzian Fracture", *Proc. Roy. Soc. Lond.* **A299** 291-306.

6.   F. Auerbach (1891) "Measurement of Hardness", *Ann. Phys. Chem.* **43** 61-100.

7.   F.B. Langitan and B.R. Lawn (1969) "Hertzian Fracture Experiments on Abraded Glass Surfaces as Definitive Evidence of Auerbach's Law", *J. Appl. Phys.* **40** 4009-17.

8.   F.C. Roesler (1956) "Brittle Fractures Near Equilibrium", *Proc. Phys. Soc. Lond.* **B69** 981-92.

9.   G.E. Hamilton and L.E. Goodman (1966) "The Stress Field Created by a Sliding Contact", *J. Appl. Mech.* **33** 371-76.

10.   B.R. Lawn (1967) "Partial Cone Crack Formation in a Brittle Material Loaded with a Sliding Spherical Indenter", *Proc. Roy. Soc. Lond.* **A299** 307-16.

11.   D.R. Gilroy and W. Hirst (1969) "Brittle Fracture of Glass Under Normal and Sliding Loads", *Brit. J. Appl. Phys.* **2** 1784-87.

12. B.R. Lawn, S.M. Wiederhorn and D.E. Roberts (1984), "Effect of Sliding Friction Forces on the Strength of Brittle Materials", *J. Mater. Sci.* **19** 2561-69.

13. J.T. Hagan and M.V. Swain (1976) "Indentation Plasticity and the Ensuing Fracture of Glass", *J. Phys. D: Appl. Phys.* **9** 2201-14.

14. J. Hagan (1980) "Shear Deformation Under Pyramidal Indentations in Soda-Lime Glass, *J. Mater. Sci.* **15** 1417-24.

15. B.R. Lawn, T.P. Dabbs and C.J. Fairbanks (1983) "Kinetics of Shear-Activated Crack Initiation in Soda-Lime Glass", *J. Mater. Sci.* **18** 2785-97.

16. T.P. Dabbs, C.J. Fairbanks and B.R. Lawn (1984) "Subthreshold Indentation Flaws in the Study of Fatigue Properties of Ultrahigh Strength Glass", in Methods for Assessing the Strength and Reliability of Brittle Materials, pp. 142-52, ed. S.W. Freiman and C.M. Hudson. A.S.T.M. Special Technical Publication **844**, Philadelphia.

17. B.R. Lawn and A.G. Evans (1977) "A Model for Crack Initiation in Elastic/Plastic Indentation Fields", *J. Mater. Sci.* **12** 2195-99.

18. K. Puttick (1980) "The Correlation of Fracture Transitions", *J. Phys. D: Appl. Phys.* **13** 2249-62.

19. B.R. Lawn, A.G. Evans and D.B. Marshall (1980), "Elastic/Plastic Indentation Damage in Ceramics: The Median/Radial Crack System", *J. Amer. Ceram. Soc.* **63** 574-81.

20. S.M. Wiederhorn (1967) "Influence of Water Vapor on Crack Propagation in Soda-Lime Glass", *J. Amer. Ceram. Soc.* **50** 407-14.

21. S.M. Wiederhorn and L.H. Bolz (1970) "Stress Corrosion and Static Fatigue of Glass", *J. Amer. Ceram. Soc.* **53** 543-48.

22. K-T. Wan, N. Aimard, S. Lathabai, R.G. Horn and B.R. Lawn (1990) "Interfacial Energy States of Moisture-Exposed Cracks in Mica", *J.Mater. Res.* **5** 172-82.

23. K-T. Wan and B.R. Lawn (1990) "Surface Forces in Mica in the Presence of Capillary Condensation", *Acta Metall.* **38** 2073-83.

24. K-T. Wan, D.T. Smith and B.R. Lawn (1992) "Contact and Adhesion Energies of Mica-Mica, Silica-Silica and Mica-Silica Interfaces in Dry and Moist Atmospheres", *J. Amer. Ceram. Soc.*, in press.

25. J.W. Obreimoff (1930) "The Splitting Strength of Mica", *Proc. Roy. Soc. Lond.* **A127** 290-97.

26. A.I. Bailey and S.M. Kay (1967) "A Direct Measurement of the Influence of Vapour, of Liquid and of Oriented Monolayers on the Interfacial Energy of Mica", *Proc. Roy. Soc. Lond.* **A301** 3421-27.

27. S.W. Bailey (1984) "Review of Cation Ordering in Micas", *Clays and Clay Minerals* **32** 81-92.

28. G.L. Gaines and D. Tabor (1956) "Surface Adhesion and Elastic Properties of Mica", *Nature* **178** 1304-05.

29. H. Christenson (1988) "Adhesion Between Surfaces in Undersaturated Vapours - A Reexamination of the Influence of Meniscus Curvature and Surface Forces", *J. Colloid Interf. Sci.* **121** 170-78.

30. D. Maugis (1985) "Subcritical Crack Growth, Surface Energy, Fracture Toughness, Stick-Slip and Embrittlement", *J. Mater. Sci.* **20** 3041-73.

31. A. Fogden and L.R. White (1990) "Contact Elasticity in the Presence of Capillary Condensation: I. The Nonadhesive Hertz Problem", *J. Colloid Interf. Sci.* **138** 414-30.

32. D. Maugis (1991) "Adhesion of Spheres: the JKR-DMT Transition Using a Dugdale Model", *J. Colloid Interf. Sci.* **150**, 243-269.

162

33. D.T. Smith and R.G. Horn, "The Effect of Charge Transfer on the Adhesion Between Dissimilar Materials", in preparation.

34. K-T. Wan, S. Lathabai and B.R. Lawn (1990) "Crack Velocity Functions and Thresholds in Brittle Solids", *J. Europ. Ceram. Soc.* **6** 259-68.

35. B.R. Lawn, D.H. Roach and R.M. Thomson (1987) "Thresholds and Reversibility in Brittle Cracks: An Atomistic Surface Force Model", *J. Mater. Sci.* **22** 4036-50.

36. Y-W. Mai and B.R. Lawn (1986) "Crack Stability and Toughness Characteristics in Brittle Materials", *Ann. Rev. Mater. Sci.* **16** 415-39.

37. J.E. Ritter (1974) "Engineering Design and Fatigue Failure of Brittle Materials", in Fracture Mechanics of Ceramics, Vol. 4, pp. 667-86, ed. R.C. Bradt, D.P.H. Hasselman and F.F. Lange. Plenum Press, New York.

38. B.R. Lawn, D.B. Marshall, P. Chantikul and G.R. Anstis (1980) "Indentation Fracture: Applications in the Assessment of Strength of Ceramics", *J. Aust. Ceram. Soc.* **16** 4-9.

39. B.R. Lawn, B.J. Hockey and S.M. Wiederhorn (1980) "Atomically Sharp Cracks in Brittle Solids: An Electron Microscopy Study", *J. Mater. Sci.* **15** 1207-23.

40. J. Cook and J.E. Gordon (1964) "A Mechanism for the Control of Crack Propagation in all Brittle Systems", *Proc. Roy. Soc. Lond.* **A282** 508-20.

41. S.J. Bennison and B.R. Lawn (1989) "Role of Interfacial Grain-Bridging Sliding Friction in the Crack-Resistance and Strength Properties of Nontransforming Ceramics", *Acta Metall.* **37** 2659-71.

42. A.G. Evans (1990) "Perspectives on the Development of High-Toughness Ceramics", *J. Amer. Ceram. Soc.* **73** 187-206.

43. Y-W. Mai and B.R. Lawn (1987) "Crack-Interface Grain Bridging as a Fracture Resistance Mechanism in Ceramics: II. Theoretical Fracture Mechanics Model", *J. Amer. Ceram. Soc.* **70** 289-94.

44. F. Deuerler, R. Knehans and R. Steinbrech (1986) "Testing Methods and Crack-Resistance Behaviour of Al_2O_3", *J. de Physique* **C1** 617-20.

45. P.L. Swanson, C.J. Fairbanks, B.R. Lawn, Y-W. Mai and B.J. Hockey (1987) "Crack-Interface Grain Bridging as a Fracture Resistance Mechanism in Ceramics: I. Experimental Study on Alumina", *J. Amer. Ceram. Soc.* **70** 279-89.

46. J. Rödel, J. Kelly and B.R. Lawn (1990) "In Situ Measurements of Bridged Cracks Interfaces in the SEM", *J. Amer. Ceram. Soc.* **73** 3313-18.

47. G. Vekinis, M.F. Ashby and P.W.R. Beaumont (1990) "R-Curve Behaviour of Al_2O_3 Ceramics", *Acta Metall.* **38** 1151-62.

48. S. Lathabai, J. Rödel and B.R. Lawn (1991) "Cyclic Fatigue From Frictional Degradation at Bridging Grains in Alumina", *J. Amer. Ceram. Soc.* **74** 3340-48.

49. S.J. Bennison, J. Rödel, S. Lathabai and B.R. Lawn (1991) "Microstructure, Toughness Curves and Mechanical Properties of Alumina Ceramics", in Toughening Mechanisms in Quasi-Brittle Materials, pp. 209-33, ed. S.P. Shah. Kluwer Academic Publishers, Dordrecht, the Netherlands.

50. D.B. Marshall (1984) "An Indentation Method for Measuring Matrix-Fiber Frictional Stresses in Ceramic Composites", *J. Amer. Ceram. Soc.* **67** C259-60.

51. P.D. Jero and R.J. Kerans (1991) "Effect of Interface Roughness on the Frictional Shear Stress Measured Using a Pushout Test", *J. Amer. Ceram. Soc.*, in press.

52. E.R. Fuller, E.P. Butler and W.C. Carter (1991) "Determination of Fiber-Matrix Interfacial Properties of Importance to Ceramic Composite Toughening", in Toughening Mechanisms in Quasi-Brittle Materials, pp. 385-403, ed. S.P. Shah. Kluwer Academic Publishers, Dordrecht, the Netherlands.

Discussion *following the lecture by B R Lawn on "Friction processes in brittle fracture*

H M POLLOCK. You have been explaining the role of local, microscopic friction on brittle fracture - could you turn the argument round, and calculate the role of fracture processes in determining the macroscopic friction that the experimenter measures? (Maybe through some kind of energy argument.)

B R LAWN. It is well documented that one can obtain inordinately high friction coefficients at pristine surfaces, especially those contacting surfaces not yet exposed to contaminating environments. I believe such conditions prevail at our sliding grain facets, accounting for the estimated pull-out coefficients of the order or unity.

Despire these high coefficients, the associated pull-out resistance stresses at the crack interface are $\sim$ 500 MPa, as against the fundamental cohesive stresses $\sim$ 50 GPa.

K L JOHNSON. I think one of the things we could say is that if you are getting sliding coefficients of the order of unity, then the work done in sliding will be orders of magnitude greater than the work done in propagating those cracks.

B R LAWN. Moreover, the pull-out *distance* is $\approx$ 1μm, much greater than the range of cohesive atomic stresses, $\approx$ 1nm. Hence one gets large energy expenditure (area under stress-separation curve) from the pull-out, and accordingly high toughness.

B J BRISCOE. In some earlier sliding sphere experiments on elastomers we attempted to analyse the energy losses in various sources of sliding damage, including cracks. Can you perform energy analyses of that type on your cone crack data?

B R LAWN. Few attempts have been made to estimate the percentage of work absorbed by cone cracks in sliding damage on highly brittle solids. Some work on glass by Wilshaw in the 1970s suggested that only a small fraction (typically < 5%) is absorbed by cone cracks. Despite this low value, fracture remains a primary source of strength degradation and wear in brittle materials.

P J BLAU. Would you comment on the role of friction in fibre pull-out in reinforced ceramic matrices in special relation to fibre surface coatings?

B R LAWN. This is an area that is receiving a lot of attention in the ceramics community, and yet remains poorly understood at the fundamental level. Fibre pull-out and push-in tests are being used to measure the interface friction stresses. It is recognised that the nature of the interface is crucial to these stresses, and thus to the performance of the system. The friction stresses are necessary to dissipate energy, but they must not be *too* high, for otherwise interface debonding will not occur and cracks will run straight through the fibre without energy-absorbing pull-out. The proper design of the fibre-matrix interface is a matter of subtle compromise.

Coatings are one way of controlling the interface friction stresses. Roughness is another important factor, as discussed in Sect. 3.2.2.

D TABOR. Work on reinforced cements and concrete points to an important role of interface shearing in pull-out of the fibre reinforcement. The shearing leads to compaction of the porous matrix, and thereby significant energy absorption. Are your ceramic composites dense?

B R LAWN. Our ceramics are virtually fully dense, so compaction is not a factor. On the other hand, Mai and others have shown that Poisson effects can be important in the energetics (confirmed by comparing pull-out vs push-in responses) for ceramic-matrix composites.

In the context of this discussion, it may be acknowledged that the roots of the bridging concept used in my paper originate from precedents in the concrete literature.

T E FISCHER. I can add to the point concerning the effects of environment on friction and fracture. There is much evidence, from the work of ourselves and others, that dry environments increase friction and suppress crack propagation. These elements will be discussed in my talk later on.

ANON. In relation to the effects of friction on strength and wear, I should point out that you have only considered sliding in a single pass. It is well known that degradation is most severe in conditions where repeated and reciprocating sliding occur.

ANON. What is the role of dislocation generation and motion in brittle fracture?

B R LAWN. There is now compelling evidence from transmission electron microscopy studies of crack tips in highly brittle solids, notably by Hockey, to indicate that such crack tips are almost totally free of any kind of dislocation activity. It needs to be appreciated that the state of stress at crack tips is strongly hydrostatic-tensile. That is the fundamental reason why ceramics are so brittle. Even in "softer", ionic ceramics (e.g. MgO) where limited dislocation activity has been observed, generation occurs at sources adjacent to, but not *at* the actual tip. In those cases the dislocations simply *shield* rather than *blunt* the tip.

M AKKURT. What is the effect of sliding-induced fracture on the wear of brittle materials?

B R LAWN. It can be profound. Cracking produces the most severe forms of wear. This is especially so in polycrystalline materials with high internal stresses, which can enhance grain boundary microfracture. The internal stresses are the same as those responsible for enhancing crack-interface bridging (Sect 3.2 of paper), so increased toughness may be counterbalanced by decreased wear resistance.

ANON. What is the effect of temperature on fracture?

B R LAWN. Increased temperature can greatly enhance crack velocities, by accelerating the rate of fluctuations over the local energy barriers (Sect 2.2). On the other hand, temperature is unlikely to affect the thresholds on the v-G curves much, since surface (or interface) energies (Sect 2.1) are typically not highly temperature-sensitive.

In relation to strength and toughness, the effects of temperature can be much stronger, especially above 1000° C. At these high temperatures ceramics undergo a brittle to ductile transition, and creep can become a dominant factor.

As is well known, one can also get strong temperature effects in frictional sliding. In certain instances the temperature may be sufficient to cause local surface melting, even in materials like sapphire with melting points above 2000° C.

I L SINGER. How can we model brittle systems with protective coatings?

B R LAWN. Little has been done in this area. Evans, Hutchinson and others have used fracture mechanics to analyse the failure of surface coatings by surface fragmentation and delamination. Hard coatings could be used to enhance the wear resistance on ceramic substrates with high internal stresses to reduce the incidence of local microfractures.

FRICTION OF ORGANIC POLYMERS

B.J. Briscoe
Imperial College
Department of Chemical Engineering,
London SW7 2BY, UK

ABSTRACT. A short review of a fundamental basis for interpreting the friction of organic polymers is described. The approach offered is based upon a two-term non-interacting model with adhesion and ploughing components. The somewhat anomalous position of Coulombic and Schallamach friction processes is considered.

1. Introduction

Very many reviews, and a few books, have been published in recent years on the subject of polymer tribology and many of these works contain specific reference to the friction of organic polymers. There are also Conference proceedings as well as articles in the archival literature. A selection, somewhat arbitrary and based upon a personal preference, are included in the Reference section of this paper [1]. Few, if any, of these publications deviate from the theme of the general applicability of the two- term non-interacting model of friction [2]. In some cases, it is implicitly assumed and in others taken for granted. The approach, which is discussed elsewhere in this volume, has many advocators and some detractors. In the context of polymer friction, it probably has more of the former than the latter. The position is however complicated a little by the two models of friction; Schallamach Waves and Coulombic Friction which do not fit easily into the two-term model.

This primer will introduce the two-term model and its two components; the adhesion component and the ploughing component. Each will be described separately with the aid of selected examples. Short sections will then deal with Schallamach Waves and Coulombic Friction. In a short concluding discussion the merits and weaknesses of these and other models will be considered.

2. The Two-Term Model

This model has been variously described but a pictorial view is shown in Figure 1[3]. The basic premise is that the frictional work may be dissipated in two distinct regions; an interface zone and a subsurface zone. As an aside, it should be stressed that any credible model of friction should account for the work done; friction is merely a special mechanism(s) for dissipating work. The former is commonly called the adhesion component (Figures 1(c) and 1(d)) and the latter the deformation or ploughing component (Figure 1(b). The processes occurring in one are considered not to influence those occurring in the other.

167

I. L. Singer and H. M. Pollock (eds.), Fundamentals of Friction: Macroscopic and Microscopic Processes, 167–182.

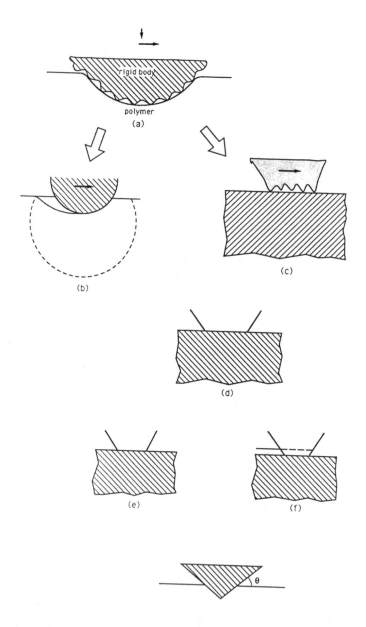

Figure 1. A pictorial description of the two-term non-interacting model. (a) generalised contact, (b) ploughing component, (c) multiple asperity contact adhesion component, (d) single asperity contact, (e) true interface sliding, (f) interface zone shear with transfer. Also shown is the parameter θ for a conical indentor; strain is ca. $\tan \theta$.

The pictorial sketch also indicates why two types of contact configuration are normally adopted for experimental studies. The study of the deformation component, Figure 1(b), will involve the sliding or rolling of a rigid body over the polymer. The action of rolling suppresses the adhesion component although efficient lubrication will produce the same effect. The adhesion component is sensed when the deformable polymer is slid over a smooth clean rigid substrate; Figure 1(c). Various subgroups of the adhesion component are shown in Figure 1(d), (e) and (f).

The distinction between the two components is arbitrary but the general approach is to consider that there is some critical asperity attack angle, θ, above which the deformation component contribution becomes significant (Figure 1).

The idea of adding two separate contributions to account for friction has a long history, as does the construction of experiments to probe one and minimise the contribution of the other. The idea was originally practiced with metals but with polymers it has a particular significance. The dissipation of frictional energy in a contact is governed, in general, by two characteristics of the system. One is the magnitude of the region over which the work is dissipated and the other is the intrinsic energy dissipation characteristic of the polymer; the counterface is rigid and does not dissipate work although, of course, it may dissipate heat. The important point is that there is a primary energy dissipation zone, or velocity accommodation locus (VAL), where mechanical work is done [4]. The heat produced is then radiated out of the zone. To pursue the consequences of this idea it is necessary to consider the volumes of material generally thought to be associated with these components. The volumes will be a product of a zone thickness and a zone area. More is said later on this topic but for the deformation component the appropriate area is of the order of the apparent contact area. The corresponding zone thickness is of the order of the contact length; Figure 1(a). For the adhesion component, the area is about the value of the real contact area. The corresponding zone thickness appears generally, for organic polymers, to range between 10 and 100nm, although the reasons for the existence of this characteristic size range are not clearly established.

The nominal rates of strain in the two regions will thus scale with the ratio of the zone thicknesses. For deformation components a few reciprocal seconds would be common. In the adhesion component, in contrast, mega (reciprocal seconds) are predicted. Similarly, there will be large differences in specific energy density input and also the consequent temperature increases. Further, there may be a significant hydrostatic stress component in the interface zones.

The point of all this is that the energy dissipation potential will be very different in the two zones. In fact, the deformations produced in the ploughing components are normally readily reproduced with some certainty in simple mechanical deformation experiments. This is not the case for the rheology of the interface adhesion component. The mechanical and particularly the energy dissipation characteristics of polymers are extremely sensitive to strain rate and temperature, amongst other external variables, and a judgement of the conditions prevailing in the zones is necessary in order to implement a credible model or correlation.

The next two sections say more about the two components.

3. Ploughing Components

Organic polymers will show the complete spectrum of mechanical response encountered in material science, even within one material, with relatively modest changes in temperature and timescale. Hence, rather than choose a polymer it is more economic to choose a nominal response. Clearly, ductile polymers behave like ductile metals; brittle polymers like some ceramics and so on. Here, viscoelastic, plastic, viscoplastic and brittle response will be described. The viscoelastic response was well studied in the context of rubbers many years ago and deserves a special comment (see reference 1(i)).

3.1 VISCOELASTIC RESPONSE IN RUBBERS [5]

In general, the friction of elastomers in this regime is explained with the aid of the relationship

$$F = \phi \, \alpha \tag{1}$$

for the frictional work done per unit sliding or rolling distance. The parameter, ϕ, describes the deformation work done, say per unit sliding (or rolling) distance, and α the fraction dissipated. It is conventional to put α as $\pi \tan \delta$ where $\tan \delta$ is the loss angle at the appropriate temperature and deformation rate. Several expressions are available to compute ϕ as a function of material characteristics and the geometry of the imposed deformation.

Prediction and experiment have been found to be in good accord. Figure 2 shows an example for PTFE adapted from Ludema and Tabor [6].

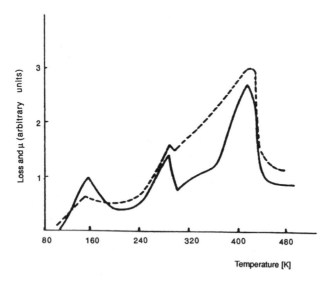

Figure 2. An illustration of the interrelationship between ploughing friction (actually rolling friction) and the viscoelastic loss parameter $\tan \delta$; the ratio of the real to the imaginary modulus. Shown is the rolling friction and the quantity $E^{-1/3} \tan \delta$ (termed loss) as a function of temperature. E, $\tan \delta$ and the friction were measured at comparable deformation frequencies and temperatures. Adapted from Ludema and Tabor [6].

3.2 PLASTIC PLOUGHING

Several first order analyses are available. For a cone of semi empirical angle β, the coefficient of friction is given by:

$$\mu = \frac{2}{\pi} \tan \theta \qquad (2)$$

where $\theta = (2\pi - 2\beta)$; Figure 1(g). An indication of the quality of the argument embodied in equation (2) is shown in Figure 3 [7].

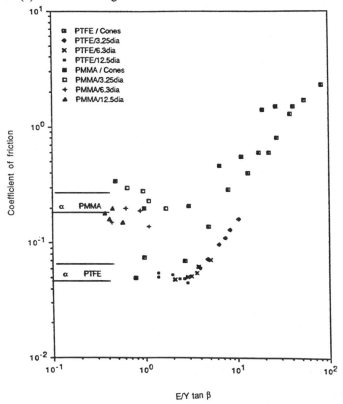

Figure 3. Friction coefficient, for various indentors, against E/Y tan θ or E/Y (r/R) for a poly (tetrafluoroethylene) and a poly (methylmethacrylate) at 20°C; unlubricated. For rigid conical indentors θ is given in Figure 1. Data for various spherical indentors are shown (radii, R, are indicated) and the quantity r is the contact width. tan θ and r/R are an indication of the imposed strain. E and Y are estimated Young's moduli and yield stresses. α is the pressure coefficient defined in equations (5) and (10).

The plot shows the general validity of the two-term model. When the magnitude of the ordinate exceeds 2 the slopes are near to 2/π.

3.3 BRITTLE, VISCOELASTIC-PLASTIC CONTRIBUTIONS

In the case where additional frictional work is dissipated by material rupture this may be included in the work done as an additional contribution. Tabor adopted this argument for the scratching of elastomers [8]. The brittle contribution is indicated in Figure 4 for a gamma-damaged PTFE [9]. The brittle or tearing contribution was estimated from the stick-slip characteristics of the frictional motion. In cases where an appreciable viscoelastic recovery is associated with plastic flow (viscoelastic-plastic friction) the viscoelastic recovery may be argued as being capable of supporting an additional fraction of the normal load by load local support at the trailing face. Hence equation (2) may be modified to give

$$\mu - \frac{x}{\pi} \tan \theta \tag{3}$$

for the case of a conical indentor. x may range from zero to two and reflects the extent of the permanent plastic flow.

3.4 PLOUGHING COMPONENTS

The geometric characteristic of the indentor and the normal load define the work transmitted to the polymer sub-surface. The work dissipated is largely as would be anticipated from a knowledge of the rheology of the polymer. In addition, this rheology is accessible by conventional means. Although the models and analyses could doubtlessly be improved, predictions and experiment are in reasonable accord.

4. Adhesion Components [10]

A popular model supposes that the interface frictional work may be described by

$$F = \tau A \tag{4}$$

where A is the real contact area and τ is a parameter termed the interface shear stress. This is the adhesion model of friction and it is supposed that the work is transmitted to the surface layer by the action of adhesive forces operating at the areas of real contact. The form of equation (4) resembles that of equation (2); the real area replaces what is essentially the apparent area and τ replaces the dissipation parameter α in equation (1) or the x parameter in equation (3).

Using the model implied in equation (4) presents two major difficulties; the computation of A and the acquisition of the magnitude and variation of the quantity τ. The estimation of A is dealt with elsewhere in this volume (see for example J. Greenwood, A. Savkoor).

5. The Interface Shear Stress, τ

Initially [11], the quantity, τ, was likened to the bulk shear stress of the adjacent solid regions. The idea was that the interface zone contained many asperity contacts which were sheared, reformed, and sheared again in a perpetual way. In many cases, cohesive rupture, as opposed to interface rupture, of the interface was noted. The rupture is, of

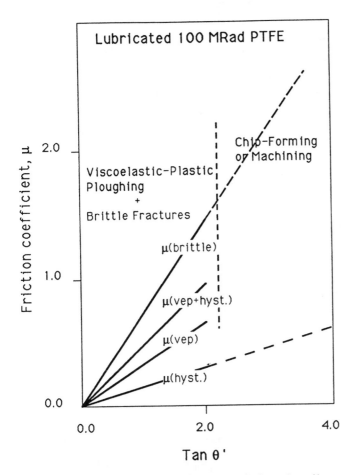

Figure 4. Coefficient of friction against tan θ for a heavily gamma damaged poly (tetrafluoro ethylene); 100 MRad. Open symbols unlubricated; closed lubricated. μ (brittle); estimated brittle work contribution estimated from stick slip response: μ(hyst) and μ(vep) are the estimated viscoelastic losses and viscoelastic ploughing contributions respectively. For the lubricated contacts the friction is composed of viscoelastic and plastic contributions. In the unlubricated case there is a sizeable contribution from brittle cracking. For "sharp" cones chip forming occurs and the geometry of the deformation is now prescribed by an internal shear plane. From reference 9.

course, nominal Mode II fracture. For some polymers cohesive rupture leads to transfer film formation and a wear process described as transfer wear. In some cases no evidence of transfer is found and we suppose that sliding is actually accommodated at the interface. In any event work is dissipated by either interface zone shear or by true interfacial sliding within a narrow zone.

It turns out that the quantity τ may be measured with some certainty using thin polymeric films and smooth non-dissipative substrates [13,14,15]. As a matter of experiment the following have been established

$$\tau = \tau_0 + \alpha\, P \tag{5}$$

$$\tau = t_0{}' \exp(Q/RT) \tag{6}$$

$$\tau = \tau_0{}'' \ln\left(\frac{V}{h}\right) \tag{7}$$

$$\tau_0 = \tau_0{}'' \exp\left(\frac{V}{D}\right) \tag{8}$$

with, in each case, other variables being fixed,

where

        P : mean contact pressure W/A
        T : temperature
        V : sliding velocity
        D : characteristic contact diameter or length
        h : dissipation zone thickness

The ratios (V/h) and (V/D) correspond to the mean strain rates (ε) and contact times (t_c) respectively. The parameters, α, Q, $\tau_0{}'$, $\tau_0{}''$ and $\tau_0{}'''$ are material characteristics. Within the limited experimental evidence available,

$$\tau = \tau_0^{*} \exp\left(\frac{Q}{RT}\right) . \ln\left(\frac{V}{h\theta}\right) + \alpha_0 \exp\left(\frac{V}{D\phi}\right) P \tag{9}$$

$\tau_0{}^{*}$ is another parameter called the intrinsic shear stress and α_0 is the value of α at long contact times [15]. The parameters θ and ϕ are characteristic frequencies which are probably functions of pressure but certainly temperature. It will be clear that there are certain inconsistencies in equation (9) in various limits of say V, h and temperature. These may be resolved by additional terms although no data exist to confirm the adoption of these forms. Data have only been obtained over rather specific ranges. The following is a summary.

6. The Influence of Contact Pressure

The relationship

$$\tau = \tau_o + \alpha\,P \tag{5}$$

for all other variables fixed is a good approximation. The expression converts to

$$\mu = \frac{\tau_o}{P} + \alpha \tag{10}$$

which is a relationship with an interesting history. At high loads, and sometimes high contact pressure, $\mu \rightarrow \alpha$, the pressure coefficient. Generally, the values of μ for a model contact of the sort shown in Figure 1(d) range from very high values and approach α at high loads. Figure 5 is an example [16]. The values of α, found by experiment, range from a general lower limit of ca. 0.05 for PTFE and poly(ethylene) to values in excess of 0.5 for glassy or highly elastic polymers. Figure 6 shows selected data for τ as a function of contact pressure; the values of α resemble those found for bulk shear yield stresses as a function of hydrostatic stress [17].

7. Variation with Temperature [18]

Equation (6) is generally found to hold quite well with activation energies of either zero or about 18 kJ mol^{-1}. The former as seen with glassy polymers below their T_g. The T_g is also a function of contact pressure. Semi crystalline polymers and glassy polymers above their T_g have finite activation energies. The activation energies, phase transformations and T_g's observed are to some extent consistent with what would be expected based upon the thermomechanical properties of the polymer under conditions which might be reasonably supposed to exist in the contacts.

8. Sliding Velocity and Contact Time [15]

The nominal shear rates may exceed $10^6\,\mathrm{s}^{-1}$ with contact times of less than $10^{-2}\mathrm{s}$ even in model experiments. These conditions are not encountered in conventional mechanical studies. Adiabatic heating and retardation in compression may occur. The effect of velocity seems, where data is available, to follow equation (7). Possible retardation effects have been detected for some polymers. Since the contact time is often small and the influence of pressure on τ is not likely to be immediate, it is supposed that a fictive pressure P_f operates such that

$$P_f = P \exp\left(\frac{t_c^*}{t_c}\right) \tag{11}$$

This term is included in equations (8) and (9).

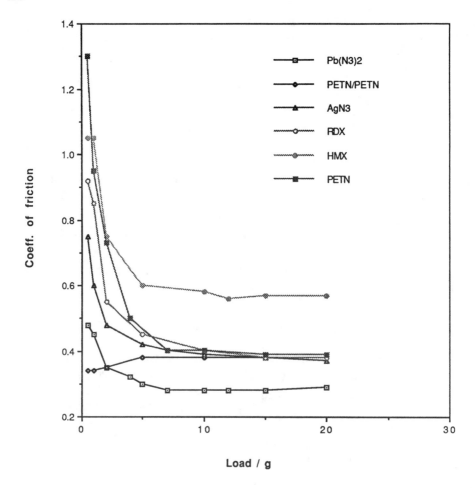

Figure 5. A typical example of friction coefficient against normal applied load for a series of explosives where single crystals are slid over smooth clean glass (and in one case another similar crystal) at 20°C and 0.2 mms⁻¹.

PETN on PETN; and against glass counterfaces;

HMX,   RDX, PETN ,   AgN$_3$,   Pb(N$_3$)$_2$. The data in most cases follow equation (10). Adapted from reference 16.

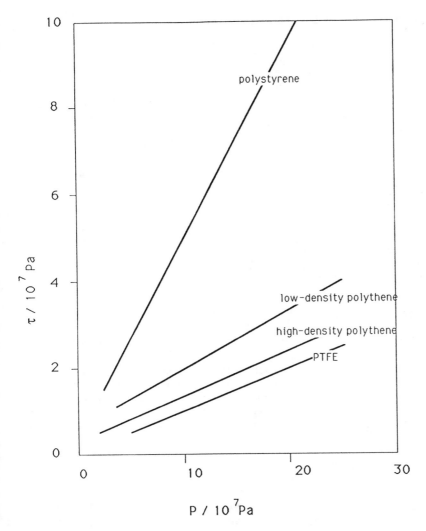

Figure 6. Interface shear stress, τ, as a function of the mean contact pressure, P, for a range of organic polymers as thin films at 20°C; high density poly(ethylene), low density poly(ethylene), (poly(styrene), PTFE. From reference 14 and references therein. To a good approximation, the data follow equation (5).

9. Interface Friction - Summary

The model requires a knowledge of the real area of contact and the characteristics of the interface shear stress, τ. Where it has been applied its predictive capability has been adequate. For example, it explains well the load and temperature dependence of friction for many polymeric systems.

10. Schallamach Waves and Coulombic Friction

These processes are real and do easily fit into the previous dual component approach. Schallamach Waves are a friction dissipation mechanism which involves adhesion but significant subsurface deformation. Coulombic Friction certainly evolves from geometric interactions of asperities but probably corresponds to an interface shear sliding process. Each of the two contains elements of both the ploughing and the adhesion models. The existence of these processes underlines the intrinsic simplifications introduced in the two component non-interacting model. However, to a first order the Schallamach and Coulomb processes may be considered as a subgroup of the adhesion model or component.

A pictorial view of Schallamach Wave Friction   and Coulombic Friction is given in Figure 7. A few comments on both are appropriate.

Coulomb Friction derives its name from a belief that it was Coulomb's favourite means of explaining friction. Basically, asperities had to move up and down over each other and hence dissipated friction through the loss of potential energy. This is an old model which is now discredited (see reference 19). The model now enjoys a new vogue in the context of atomic or molecular asperities. In practice, there are areas where Coulombic notions are useful; in particular in powder or fibre frictions. The latter has recently been addressed by the present author with others; essentially we have advocated that an extension of the adhesion model is appropriate [20].

The Schallamach friction process was introduced into the friction literature exactly twenty years ago by its namesake [20]. Relative motion is accommodated not by sliding at the interface but by the formation and propagation of a macroscopic dislocation. The dislocation is described as theSchallamachWave. In the following ten years exhaustive studies were made [22]. Various conclusions were drawn but two are significant. First, it now seems that such macroscopic waves can only be present when the substrates can accommodate reversible strains of the order of unity; they are restricted to rubbers. Second, the frictional work dissipated is well accounted for by the work done in the reversible peeling of the contact as the wave passes; hence adhesion is a fundamental requirement for the operation of the mechanism.

11. Concluding Remarks

The interpretation of polymeric friction fits easily into the two-term non-interacting model. The Coulombic and Schallamach Friction processes are arguably special cases of the adhesion component. However, their inclusion under this "adhesion component" heading does undermine the simplicity of the two models embodied in Figure 1. The Coulombic component may be interpreted as an adhesion component where the frictional forces influence the contact area; a junction growth or contraction model (see D. Tabor and A. Savkoor; this volume) may be invoked. Thus the contact area is a function of some additional variable which accounts for the gross geometry of the contact configuration. It is obvious that in such a system stick-slip motion will be the rule.

The excitement initially produced by the discovery of Schallamach waves had various roots. One was that the wave models were able to accurately predict friction for nominal sliding; there was an acceptable precedent to explain friction albeit in terms of an adhesion experiment of a simpler sort. The idea of an interfacial dislocation was so attractive - was this the way that all interfacial sliding processes were accommodated? Actually, it is likely that dislocations, or something like them, and their motion are important in the interfacial rheological response as manifest in τ. Naturally, if this were to be the case it would have

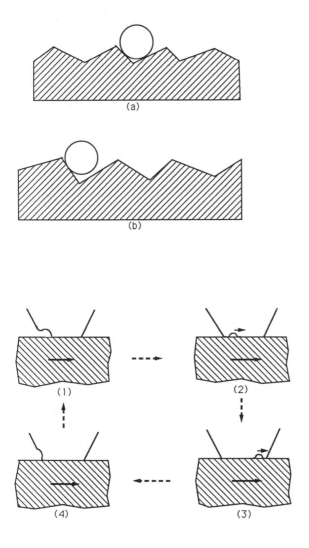

Figure 7. Schematic diagrams to represent Coulombic Friction (I) and Schallamach Friction (II).

(I) In Coulombic Friction motion is achieved by a discontinuous "jumping" between geometric engagements. Adapted from reference 20.

(II) A cross-section of a contact which is forming an interfacial ruck (dislocation) at (1) which propagates through the contact zone in (2) and (3). A repeated propagation of waves accommodates relative motion in a discontinuous way by the dissipation of peeling energy as the interface peels and heals during wave propagation. See references under 22 and reference 1(n).

important implications in the development of models to predict the absolute magnitude of τ and its functionality.

We may reflect, in conclusion, upon our ability to make *a priori* predictions of the friction of polymers and what is required to achieve such a goal. Essentially what is required is an extremely comprehensive description of the rheology of these materials over an extensive range of stresses, strain rates, temperatures, hydrostatic stresses and so on. The nature of the sliding process is such that, in many cases, these conditions <u>cannot</u> be reproduced in the appropriate combination by other means such as simple tensile testing. Hence, detailed predictions cannot be made by correlation with other types of deformation; there are cases where this may be achieved with some confidence such as in the rolling friction of elastomers. Now, we can finally note that the *a priori* prediction of friction of <u>a</u> polymer will require a comprehensive *a priori* prediction of the rheological response of the same. Needless to say this has yet to be done.

12. References

1. General References

(a) "Recent Advances in Polymer Friction and Wear", ed. L.H. Lee, Gordon & Breach 1973; almost twenty years old but still contains definitive work.

(b) "Polymer Wear and its Control", ed. L.H. Lee, ACS Symposium #287 1985; a rerun of (a).

(c) J.K. Lancaster, "Friction and Wear of Polymers"; a Material Handbook, ed. A.D. Jenkins, North Holland Pub. Co. 1972; the first substantial review.

(d) B.J. Briscoe and D. Tabor, "Friction and Wear of Polymers", Chapter 1 *Polymer Surfaces*, ed. D.T. Clark & J. Feast, Wiley 1978; our first attempt at a systematic general review.

(e) G.M. Bartenev and V.V. Laurentev, "Polymer Friction and Wear", trans. D.B. Payne, Chemistry Publisher, Leningrad 1972 - a Soviet view.

(f) Y. Yamaguchi, "Tribology of Plastic Materials", Elsevier (1990) - an interesting Japanese perspective.

(g) "Friction and Wear of Polymer Composites", ed. K. Freidrich, Elsevier 1986 - in many respects one of the most useful recent texts.

(h) "The Wear of the Non-Metallic Materials", ed. D. Dowson, M. Godet and C.M. Taylor, Mechanical Engineering Publications Ltd., London 1978.

(i) The Friction and Lubrication of Elastomers", D. Moore, Pergamon Press 1972 - a substantial compilation.

(j) "Wear of Materials" 1975, and biannual to present ASME New York - good selection of research papers.

(k) "Physicochemical Aspects of Polymer Surfaces", ed. K. Mittal, Plenum Press, N.Y. 1983 - several comprehensive reviews.

(l) R.P. Stein, "Friction & Wear" in *Failure of Plastics*, ed. W. Brostow and R.D. Coreliussen - a reflective review from an author who produced one of the earlier reviews of polymer friction.

(m) Y.A. Bely, A.I. Sviridyonok, M.I. Petrokovets and V.G. Savkin, "Friction and Wear in Polymer Based Materials", Pergamon Press 1982 - a product of one of the major Byelorussian Schools.

(n)   "Friction and Traction", ed. D. Dowson, C.M. Taylor, M. Godet and D. Berthe, Westbury Press 1981 - several seminal reviews.

Specific References

2.   F.P. Bowden and D. Tabor, "The Friction and Lubrication of Solids", Clarendon Press, Oxford 1986.
3.   From various sources but see reference 1(d).
4.   Y. Berthier, M. Brendle and M. Godet, "Boundary Conditions; Adhesion in Friction" in *Interface Dynamics*, ed. D. Dowson, C.M. Taylor, M. Godet and D. Berthe, Elsevier 1988, p.11.
5.   J.A. Green wood, H. Minshall and D. Tabor, *Proc. Roy. Soc.*, **A259** (1961), 480 but also see reference 1(d) and (g) and D.G. Flom and A.M. Bueche, *J. Appl. Phys.* **30** (1959), 1725.
6.   K.D. Ludema and D. Tabor, *Wear*, **9**, 1966, 329.
7.   P.D. Evans, "Hardness and Abrasion of Polymers", PhD Thesis, University of London (Imperial College), 1987 and B.J. Briscoe and D. Tabor, *Soviet Journal of Friction and Wear* (in press).
8.   J.A. Greenwood and D. Tabor, *Proc. Phys. Soc.*, **71**, 1958, 989 and see current author in reference 1(g).
9.   B.J. Briscoe, P.D. Evans and J.K. Lancaster, *J. Phys. D: Appl. Phys*, **20**, 1987, 346.
10.   For a general outline based upon the work of Professor Tabor's Group see B.J. Briscoe, *Phil. Mag*, **A43**, 1981, 511.
11.   For example the early Cambridge work described in reference 2.
12.   Described in some detail in various references under 1 - various authors in 1(a) are good examples.
13.   Mica as a substrate is described in B.J. Briscoe and D.C.B. Evans, *Proc. Roy. Soc*, **A380**, 1982, 349 and references therein - see also elsewhere in this volume.
14.   Glasses and other substrates see for example B.J. Briscoe and D. Tabor, *J. Adhesion*, **9**, 1978, 145.
15.   See B.J. Briscoe and A.C. Smith, Reviews on the Deformation Behaviour of Materials III, No. 3, 1980, 151 and *J. Phys. D. Appl. Phys*, **15**, 1982, 579 and reference 13.
16.   Adapted from A.J.K. Amuzu, B.J. Briscoe and M.M. Chaudhri, *J. Phys. D: Appl. Phys,* **9**, 1976, 133.
17.   Taken from B.J. Briscoe, B. Scruton and F.R. Willis, *Proc. Roy. Soc*, **A333**, 1973, 99.
18.   As an example see B.J. Briscoe and A.C. Smith, *J. Appl. Polymer Sci*, **28**, 1983, 3827.
19.   D. Dowson, History of Tribology, Longmans, 1981.
20.   M.J. Adams, B.J. Briscoe and T.K. Wee, *J. Phys. D: Appl. Phys.*, **23**, 990, 406.
21.   A. Schallamach, *Wear*, **17**, 1971, 301.
22.   As examples see A.D. Roberts and A.R. Thomas, *Wear*, **33**, (1975), 45, G.A.D. Briggs and B.J. Briscoe, *Phil. Mag.* **A38**, 1978, 387, M. Barquins, *Wear*, **91**, 1983, 103, G.A.D. Briggs and B.J. Briscoe, *Wear*, **35**, 1975, 357.

Discussion *following the lecture by B J Briscoe on "Friction of organic polymers".*

K L JOHNSON. What is the role and/or significance of surface energies in friction processes? I am thinking here of the occurrence or not of transfer films and wear.

B J BRISCOE. In some cases surface forces, as described by a surface free energy, are a critical factor in the friction process; this is the central idea in the adhesion model. The adhesive forces must be sufficient to allow strain energy to be transferred to the regions adjacent to the contact. The strain energy is then dissipated. If the adhesion (interfacial work of adhesion) is zero, no work will be done or dissipated; the friction is zero. This is the essence of lubrication practice. I think that it is attractive to believe that the thermodynamic work of adhesion should scale with the friction work. There are several such correlations in the literature; they are most certainly fortuitous in many cases. The evidence is that the adhesion should be enough but that the dissipation level is a bulk, albeit in a very thin layer, characteristic.

Similarly, the resistance to true interfacial sliding, or interface zone shear with transfer, is primarily a bulk property. With polymers, transfer is usually accompanied by some isothermal interface morphological re-ordering or thermal softening.

FRICTION OF GRANULAR NON-METALS

M. J. ADAMS
Unilever Research
Port Sunlight Laboratory
Wirral
UK

ABSTRACT. A comparison is made between the frictional properties of single particles and those of gross bodies. It is shown how relatively simple models can be developed to adequately predict the friction of powders against smooth walls when they flow as a plug. The bulk deformation of powders occurs through internal shear planes which are many particle diameters in thickness. The internal coefficient of friction associated with these planes is rather more difficult to model and involves Coulombic interlocking. However, computer simulation has proved a viable technique. This is exemplified for estimating the fracture strength of particulate solids which is greatly influenced by the extent of interparticle friction arising from the stress field in the vicinity of the propagating crack tip.

1. Introduction

Coarse powders, such as sand, in the dry state are pourable or, as often termed, 'free-flowing'. However, such powders can also be formed into a stable heap with a characteristic angle of inclination known as the 'angle of repose'. This arises because shear stresses are developed in the heap and at the supporting surface which act to resist the body forces imposed by gravity. The shear stresses are due to the friction forces developed at individual particle contacts which is the topic of this paper.

Savage (1) termed the various regimes of powder behaviour as grain-inertia (rapid flow), transitional and quasi-static (slow flow). Inertia effects dominate in the rapid flow regime and energy dissipation arises during collisions which is often lumped as a coefficient of restitution. This paper will be concerned with the much denser packing associated with quasi-static flows where stresses are transmitted through the assembly of particle-particle contacts. Under these circumstances, the friction at contact points crucially influences the behaviour of the assembly.

The deformation or 'flow' of dense phase powders proceeds by the displacement of blocks or dead zones in which there is no significant relative motion between the particles. The regions between the blocks

I. L. Singer and H. M. Pollock (eds.), Fundamentals of Friction: Macroscopic and Microscopic Processes, 183–207.
© 1992 *Kluwer Academic Publishers. Printed in the Netherlands.*

are known as failure zones in which localised shear occurs over a thickness of about 20 particle diameters (2). A simple example is the flow down a rough-walled container. If the wall is sufficiently rough, the particles adjacent to the wall are prevented from sliding. In order to accommodate the flow, failure zones are developed in the regions close to the walls and the bulk of the powder away from the walls moves as a plug (3). This velocity vector field is identical to that which is induced during the flow of a Bingham fluid under the action of a pressure gradient. However, the shear zone for powders does not resemble that for such a fluid. For example, the shear stress is relatively insensitive to gross strain rate and is a function of the hydrostatic pressure. A more realistic analogy is rigid-plastic deformation. For smooth walls, homogeneous sliding (plug flow) occurs and the frictional energy dissipation is confined to the wall interface (the adhesion component of friction). At high wall roughnesses, only bulk deformation occurs (the deformation component of friction) which is known as redundant work in plasticity terminology.

Currently, the analytical solutions generally employed to describe the internal deformation behaviour of powders are based on continuum mechanics (eg ref 4). The failure zones are treated as planes and the critical shear stress to initiate 'flow' is calculated from the product of the normal stress acting on the planes and the coefficient of friction between the planes which is termed the 'internal coefficient of friction'. This is simply a special form of a yield function known as the Mohr-Coulomb yield criterion. The continuum mechanics approach has been widely practised and has proved remarkably useful for some purposes.

Predictive analytical models of bulk deformations based on scaling individual particle properties have proved to be difficult to develop although some progress has been made. Such models are based on Coulombic interlocking which will be described in a subsequent section. The ramifications will be exemplified later in the context of estimating the fracture energy of a powder compact. Currently, computational techniques offer the most viable routes. One of these is discrete element analysis and the type of data generated will also be illustrated.

The continuum mechanics approach is equally applicable under circumstances where a powder slides at a wall and undergoes plug or compaction flows. Typically, plug flow occurs along a smooth bore pipe and compaction flows are found in tableting and roll milling. Under compaction conditions, a simplification is made in continuum models whereby a monotonically changing density gradient in the flow direction is assumed. Effectively, the principal stresses are related by a Poisson's ratio.

The tractional stresses at a wall play a crucial role in the safe and efficient design and operation of equipment. Powder flows may be usefully classified as traction induced, eg screw extrusion or roll milling, or traction retarded, eg ram extrusion or tableting. A major limitation of continuum models is the assumption that the wall coefficient of friction is independent of pressure like a coherent rough body. However, often single particles behave as point contacts for which the coefficient of friction is a function of the applied load. A similar functionality is then conferred on the particle array at the boundary wall and this can lead to gross errors in the continuum mechanics solutions. In order to overcome this limitation and still preserve the simplicity of the continuum approach, semi-particulate models have been adopted. For these models, the bulk of the powder is treated as a continuum to estimate the normal stress distribution acting on the boundary wall particles which are modelled as an array of point contacts. The underlying principles of the friction of point contacts and how they can be introduced into semi-particulate descriptions of plug or compaction flows will be discussed in the proceeding sections.

2. The Friction at Point Contacts

2.1 THEORETICAL DESCRIPTION

Single particles often display frictional characteristics that are not encountered with gross bodies forming extended multiple-asperity contacts. Their acute radii of curvature often ensures that point contacts are formed because the surface topographical features may be comparable in size to the gross body dimensions. Thus even rough particles can form contacts involving one or few asperities. Consequently, the contact between a particle and smooth wall often can be represented by a sphere and a flat. This is loosely termed a point contact although at any applied load w there will be a finite contact area A. For elastically deforming particles this is calculable from the Hertz equation as follows (5)

$$A = \pi \, (3wr/4E*)^{2/3} \tag{1}$$

where the mean radius of curvature r at the contact is given by

$$r = \left(\frac{1}{r_1} + \frac{1}{r_2} \right)^{-1}$$

with the subscripts 1 and 2 representing the radii of the contacting asperities which may be equal to particle radius r_p if the particles are sufficiently smooth. For the particle-wall case, the reciprocal of the wall radius is zero of course. The composite elastic modulus E* is given by the following expression

$$E* = \left(\frac{1-v_1^2}{E_1} + \frac{1-v_2^2}{E_2} \right)^{-1}$$

where E is the Young's modulus and v is Poisson's ratio.

If the yield stress Y of the particles is exceeded then the contact area is given simply by

$$A = w/Y \tag{2}$$

An estimate of the contact area is important because the frictional force f of many particles can be described in terms of the adhesion model (6), thus

$$f = \tau A \tag{3}$$

where τ is the interfacial shear strength. This appears to be quite generally true except for some particles under ultra-clean conditions, when the development of traction cracks is possible (7). Generally however, powders which could display such effects are not subject to rigorous cleaning and contaminant organic layers will be adsorbed on their surfaces. Consequently, the interfacial rheology of even inorganic powders tend to be characteristic of organic systems with the following type of pressure dependence (8)

$$\tau = \tau_o + \alpha p \tag{4}$$

where τ_0 is the intrinsic interfacial shear stress and p (= w/A) is the mean contact pressure.

Combining equations 1-3 leads to the following expression for the frictional force

$$f = \pi \tau_o \, (3r/4E*)^{2/3} \, w^{2/3} + \alpha w \tag{5}$$

It is observed experimentally that the relationship between the frictional force and the normal load takes the following general form

$$f = k_p w^n \tag{6}$$

where n is known as the load index. For Hertzian contacts, equation 5 applies and it can be seen that the load index will be in the range 2/3 to 1 depending on the relative values of the load coefficients. On the basis of equation 5 it is possible to derive the following expression for the load index (6)

$$n = (2 + 3\xi \overline{w}^{1/3}) / (3 + 3\xi \overline{w}^{1/3}) \tag{7}$$

where

$$\overline{w} = \exp \ [(lnw_1 + lnw_2)/2]$$

and w_1 to w_2 represents the load range considered and

$$\xi = \alpha/\pi\tau_o(3r/4E*)^{2/3}$$

Equation 7 shows that the load index depends on the interfacial rheology, elastic modulus, radius and normal load of the contact. The corresponding expression for k_p is as follows:

$$k_p = \left(\frac{\xi}{\overline{w}^{1/3}} + \alpha\right) \overline{w}^{1-n} \qquad (8)$$

For cases where the yield stress of the particles is exceeded or where multiple asperity contact occurs (see Section 2.4), then the real area of contact is directly proportional to the normal load. This gives rise to a load index of unity. However, for Hertzian contacts equation 5 applies and it can be seen that the load index will have values in the range 2/3 to 1 depending on the relative values of the load coefficients.

2.2 AUTOADHESION AND CAPILLARY FORCES

The treatment in the previous section has been validated by measuring the friction of a number of polymers formed into smooth spheres of different radii (9). An alternative procedure is to examine the micro-contacts between fine monofilaments in an orthogonal configuration; such contacts are equivalent to those formed between a sphere and a flat plane. Measurements of this type exemplify another unusual feature of point contacts, that is the effective force of attraction at a contact interface arising from molecular interactions such as van der Waals can be of similar order to the applied load. An example of the effect of autoadhesion for polyethyleneterephthalate (PET) fibres is shown in Figure 1. It can be seen that in this case sliding was achieved under a negative applied load. The intercept on the negative load axis corresponds to the measured separation force between the fibres w_a; the curve in the figure was calculated using equation 5. The magnitude of this quantity for elastic point contacts can be estimated from the surface free energy γ of the solids in contact from the following expression derived in the JKR model (11)

$$w_a = 3\pi r\gamma \qquad (9)$$

The JKR model is based on an energy balance between the molecular attractive forces and the elastic stresses that act against the formation of a stable junction. The other major cause of adhesion between particles is capillary bridges arising from moisture condensation; the forces involved are considerably greater than those arising from autoadhesion (12).

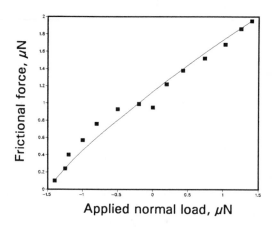

Figure 1. The frictional force as a function of applied normal load for polyethyleneterephthalate fibres in an orthogonal configuration; the theoretical curve was calculated from equation 5.

Autoadhesion or capillary forces acting between particles play a major role in retarding their flow from containers but this topic is outside the scope of this review. There is no detailed work on the effects of such forces on wall friction although they will clearly have a major effect on the wall pressure distribution and hence the friction in kilns which is described in Section 3.2. However, there is an enormous literature on measuring the effect of particle cohesion on the internal friction of powders which produce data similar in form to that shown in Figure 1; this is briefly referred to in Section 2.5. In such work, it is usual to define a cohesive strength as the shear stress at zero applied load and this quantity is incorporated in the Mohr-Coulomb yield criterion. This is a complex topic since the cohesive strength is a function of the porosity and hence loading history of the particle assembly.

2.3 BOUNDARY LUBRICATION

It has already been pointed out that powders are often contaminated by thin layers of contaminant films that reduce the coefficient of friction. This is a process of boundary lubrication which corresponds to a reduction in the interfacial shear stress. Point contacts seem particularly sensitive to the thickness of the absorbed layer. Figure 2 shows data for PET fibres treated with solutions of stearic acid having different concentrations (10). A significant reduction in the friction is found for the most dilute solution but the friction then increases with increasing concentration. The lower curve was calculated from equation 5 using interfacial rheological data for stearic acid. These results suggest that any multi-molecular layers result in an accumulation of the material in the penumbral region of the contact during sliding that effectively introduces a ploughing component of friction.

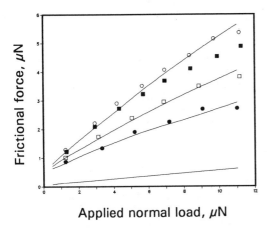

Applied normal load, μN

Figure 2. The frictional force as a function of applied normal load for polyethyleneterephthalate fibres treated in 0% (o), 0.05% (●), 0.5% (□) and 1% (■) w/w stearic acid solution in n-heptane. The lower curve was calculated using equation 5 with the measured values of the interfacial rheological parameters for stearic acid.

2.4 COMPARISON WITH MULTIPLE ASPERITY CONTACTS

According to the classical laws of friction, the frictional force should be directly proportional to load. Archard (13) describes under what conditions this would apply for elastically deforming particles. One of the limits is where the number of asperity contacts is independent of load. This corresponds to point contacts which have already been considered where the contact area at each asperity is given by the Hertz equation. Such behaviour is rarely observed. A second limit is multiple asperity contact where the number of asperity contacts increases with load. Under these conditions, the contact area is generally proportional to the load as demanded by the laws of friction. There is an intermediate case where the asperities are of constant radius and Archard showed that the contact area was then proportional to $w^{8/9}$; this has been confirmed for spherical compacts formed from maize powders (14). Glass ballotini is a simple example where point contacts are formed while mustard seeds represent an example of multiple asperity contacts (15). Point contacts are most likely to be formed when there is a wide distribution of asperity heights and the mean value is relatively large. Multiple asperity contacts will correspond to a wider distribution of relatively small asperities.

2.5 COULOMBIC FRICTION

When two pieces of abrasive paper are rubbed together, the friction is extremely high. This arises because the abrasive particles interlock with each other. The original workers who studied friction considered that mechanical locking was the primary origin of the resistance to sliding. In particular, this mechanism is attributed to Coulomb who studied the friction of wood for application to the launching of ships (see ref 16). For the extended multiple asperities formed by most bodies, their complex fixed spatial distributions prevent any interlocking of the asperities which in any case have an extremely small angle of inclination. Powders have more autonomy and are able to interlock with themselves or wall asperities provided they are sufficiently large.

Hair fibres exhibit Coulombic friction because of cuticle cells on their surfaces which have a geometric cross-section similar to the tiles on a roof. This leads to the friction being greater in sliding against the cuticle cells compared with sliding in their direction. The effect can be modelled quantitatively by assuming that the adhesion mechanism operates at contact points but taking account of the vectorial directions of the various forces (17). This can be illustrated for powders by considering three particles as shown in Figure 3.

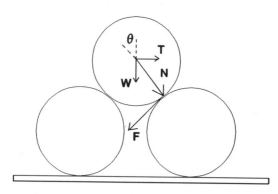

Figure 3.   Schematic representation of three particles in a Coulombic contact.

The upper particle is subjected to a normal load W and the force required to slide over the lower particles is T. The frictional force acting tangentially at the contacts is F with a normal force N. A force balance gives

$$F = T \cos \theta - W \sin \theta \qquad (10)$$

$$N = T \sin \theta + W \cos \theta \qquad (11)$$

For simplicity, letting the load index be unity so that a particle coefficient of friction μ_p is defined as F/N, then the Coulombic coefficient of friction μ_c (= T/W) is given to a good approximation by

$$\mu_c = \mu_p + \tan \theta \qquad (12)$$

Clearly the angle θ, and hence μ_c, increases with the decreasing packing density of the system. If the lower two particles represent wall asperities then, when their radii are much less than that of the upper particle, μ_c will be approximately equal to μ_p. However, as their radii approach that of the upper particle (which is known as the fully-rough condition) then the friction will be much greater. In three-dimensions the friction will be less because the upper particle has the freedom to slide between two particles which effectively reduces the angle θ.

The above ideas have formed the basis for developing particulate models of the failure zones and hence predictions of overall stress-strain behaviour. A major attraction is that two fundamental properties of particle systems can be readily interpreted. Firstly, the idea of dilatancy which was originally identified by Reynolds (18). He showed that densely packed sands expand when subject to deformation which clearly arises from particles having to move over one another in order for the failure zones to develop. Secondly, the internal coefficient of friction is always greater than that of the constituent particles which is consistent with Coulombic interlocking. Nevertheless, there has been considerable controversy in the soil mechanics literature following Rowe's (19) efforts at particulate modelling. A useful summary of the arguments is given by Horne (20). Despite the controversy, the potential for developing simple closed-form solutions continues to motivate work in this area (eg refs 21 and 22). No attempt has been made in such work to extend the models from free flowing powders to incorporate a finite cohesive shear strength. However, a number of approaches for modelling the tensile strength of cohesive particle assemblies are described in Section 4.

2.6 STATISTICAL ANALYSIS

For nominally smooth countersurfaces it has been shown that the frictional force at point contacts is a function of the mutual radius of curvature of the contact. Since most surfaces will be associated with a distribution of asperity heights then it is to be expected that this will be reflected in the frictional force. Normally, the resolution of equipment is insufficient to detect such microscopic events. However, the point contacts formed between orthogonal fibres offer a very sensitive means of determining the extent of this effect. Fibre sliding is generally stick-slip in nature and each peak in the stick phase represents a single contact point. By measuring a large number of such events it has been shown that well prescribed global distributions of frictional forces are obtained (23). For relatively smooth fibres, normal distributions are observed while those with larger surface topographical features correspond to skewed distribution such as a gamma; Figure 4 shows some typical data for cotton fibres.

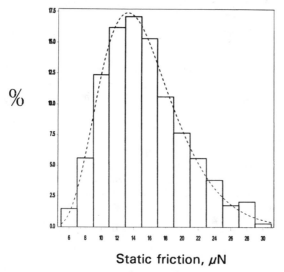

Static friction, μN

Figure 4. Histogram of frictional forces for cotton fibres fitted to a gamma distribution.

The stochastic nature of the wall friction of powder beds will be considered in Section 3.3. Such systems are even more complex because the normal forces acting on the particles will also be fluctuating with time.

3. The Wall Friction of Powders

3.1 SMOOTH PLANAR COUNTERSURFACES

In Section 2.4, a comparison was made between an Archardian body and a rough single particle. There are actually more detailed analyses available for curved elastic rough surfaces (see ref 24). Similarly, the contact between nominally flat surfaces has been modelled. In the case where the surface topography of one plane can be represented by a regular wave and the other is a flat plane, it has been shown that the total contact area increases as the load to the power 2/3. That is, each contact deforms according to the Hertz expression and the number of contacts is independent of load. However, for a randomly rough surface, the contact area is directly proportional to load. This arises because the mean contact area associated with each asperity remains constant and the number of contacts increases in direct proportion to the load. The constant mean asperity contact area is a consequence of new asperities coming into contact with increasing load which balances the corresponding increase in contact area of asperities already in contact. This is the normal type of behaviour between gross bodies and accounts for the friction being generally directly proportional to the normal load according to Amontons' law.

It might reasonably be expected that there would be these two corresponding limits for the plug flow of powders against smooth flat countersurfaces. The second case would correspond to powders with highly dispersed particle sizes. However, only the equivalent of the first case has been studied in any detail, that is when there is a narrow distribution of particle sizes. Such particulate systems can be modelled as a uniform array of spheres as shown in Figure 5.

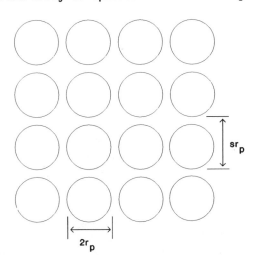

Figure 5. Schematic diagram of an open-packed cubic array of particles showing the packing parameter s.

The spacing between the spheres depends on the parameter s which will be equal to 2 when the spheres are in contact. For elastically deforming smooth particles, the frictional force for each particle is given by equation 5. If each of these contributions is summed on the basis that the normal load per unit area $\overline{W}$ is shared equally between the particles, then the frictional force per unit area $\overline{F}$ of the array is given by (6)

$$\overline{F} = \frac{\pi \tau_o}{(4E*s/3r_p)^{2/3}} \overline{W}^{2/3} + \alpha \overline{W} \tag{13}$$

where $\overline{F}$ and $\overline{W}$ are formally equal to the wall shear and normal stresses respectively.

Equation 13 shows that the wall friction of powders comprising mono-sized particles reflects that of the individual particles in that the coefficient of friction μ (= $\bar{F}/\bar{W}$) may decrease with increasing normal load. Examples are shown in Figure 6 for polyethylene beads and glass ballotini in a smooth plane walled hopper (15).

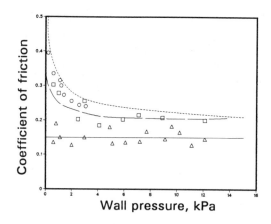

Wall pressure, kPa

Figure 6. The measured coefficient of friction as a function of wall normal stress in a smooth plane wall hopper containing polyethylene beads (o), glass ballotini (□) and mustard seeds (Δ). The theoretical curves for the polyethylene beads (short dashes) and the glass ballotini (long dashes) were calculated using equation 13.

These data points were measured at different depths in the hopper with the higher normal stresses corresponding to greater depths. As pointed out earlier, the mustard seeds have a load index of unity so that the coefficient of friction of both the individual particles and the powder are equal and independent of load.

Equation 13 shows that as the particle spacing increases (wall dilation), the friction decreases. This can be envisaged as an increase in the contact roughness of the powder. Equation 13 also shows that the wall friction is independent of particle size. This is because the contact area of each particle will increase with the particle size but this is exactly counterbalanced by the reduction in the number density of the particles.

3.2 SMOOTH CURVED COUNTERSURFACE

A simple technique for measuring the wall friction of powders involves the use of a rotating horizontal cylindrical kiln (25) which is shown schematically in Figure 7. The powder bed moves as a plug at relatively low bed depths. Initially, as the kiln rotates, the plug of powder will rotate with the cylinder because of the wall friction. Eventually the body forces acting on the powder will overcome the wall friction and the plug will slip to a lower position. The static frictional force may be calculated from the maximum angle of inclination of the powder bed during the stick-slip motion.

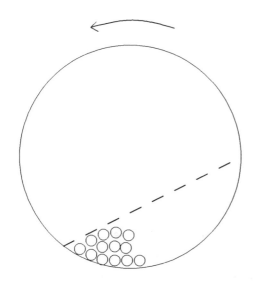

Figure 7. Schematic diagram of the cross-section of a kiln containing a power bed.

The kiln configuration is rather like the Archardian body with mono-sized asperities since the number of contacts increases with the load of particles in the cylinder. The frictional force per unit length $\tilde{F}$ can be calculated in an analogous way to that described in the previous section for the planar countersurface. This leads to the following expression (25)

$$\tilde{F} = \frac{\pi\tau_o}{(4E*s/3)^{2/3}} \left(\frac{12b}{\rho_b g}\right)^{1/9} \tilde{W}^{7/9} + \alpha\tilde{W} \tag{14}$$

where $\tilde{W}$ is the normal load per unit length calculated on the basis of equitable load sharing. The radius of the cylinder is b and the bulk density of the powder is ρ_b.

Equation 14 was derived on the basis of mono-sized particles having a load index between 2/3 and 1 (eqn. 5). The increasing number of contacts with increasing powder load has increased the load index to the range 7/9 to 1; the analogous Archardian body has a load index of 8/9. This equation has been verified using sand and glass ballotini (25); the data are shown in Figure 8. Single glass ballotini have a load index of 0.78 and the measured load index in a kiln was 0.86 compared with a calculated value of 0.84. Single grains of sand have a unit load index and, consequently, the same result is found for measurements in a kiln.

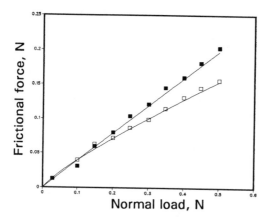

Figure 8: Frictional force as a function of normal load for glass ballotini (□) and sand (■) measured using the kiln technique.

3.3 THE INFLUENCE OF NON-UNIFORM PRESSURE DISTRIBUTIONS

An assumption in the two previous sub-sections was that the normal load acting on the wall particles was uniformly distributed. In practice, it has been found that the pressure distributions are highly non-uniform (25); Figure 9 shows data obtained for sand in a horizontal kiln.

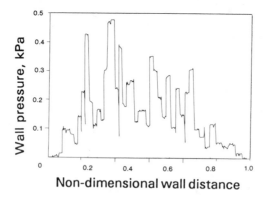

Figure 9. The measured wall pressure distribution for sand in a horizontal kiln.

In cases where the load index of the individual particles is less than one, this leads to a variation in the measured frictional force due to the corresponding fluctuation in the pressure distribution as the powder is disturbed after each measurement. This can be readily understood qualitatively. If half of the wall particles share all the load, say, then the remaining particles will be effectively redundant. That is, the effective mean spacing will be about a particle diameter greater than the actual value. This can be regarded as a load dilation where the packing parameter s in equation 14 increases. Each wall particle will bear a larger proportion of the load according to the Hertz equation but the contact area will only increase as the load to the 2/3 power rather than proportionately which is the case with the number of wall particles.

An analytical solution for the pressure distribution will be as uncertain as that for describing the surface topography of a rough body. However, the effect can be exemplified for a bed comprising of two particles. If one of the particles shares a fraction ϕ of the total load W and if the load index is n, the frictional force of the bed will be given by

$$F = k_p \left[\phi^n + (1-\phi)^n \right] W^n \tag{15}$$

For the limit of non-uniform loading ($\phi = 0$), the value of k for the bed will equal k_p whereas for uniform loading ($\phi = 1/2$) it will be equal to $2^{1-n} k_p$. Hence this simple example shows quantitatively how bed non-uniformity reduces the wall friction.

A practical illustration of the above effect was observed for the friction of coal measured in kilns (26); the coal particles studied in this work had a load index of 2/3 against smooth walls. It was found with rough walled kilns, for which the load index was unity, that water did not lubricate the coal. However, for smooth walls, when the load index was less than unity, the wall friction was reduced by the presence of water. This was ascribed to the action of viscous damping by the water which resulted in greater non-uniformities in the pressure distribution than for the dry cases. Experimental evidence supported the contention that greater relaxation was possible in the dry state. The friction was found to increase after some rotation of the cylinder and the friction increased with rotational velocity. In the presence of water, the friction was insensitive to such mechanical perturbations.

3.4 COMPACTION FLOWS

The previous sub-sections have been concerned with the wall stresses developed in horizontal cylinders. For vertical cylinders or tall containers of other geometries, the friction at the vertical walls also has a pronounced effect on the pressure distribution. The frictional force is developed at the wall boundary of any horizontal section because the mass of particles on the upper surface of the section produces a radial component of stress. That is, the powder behaves like a continuous body with a non-zero Poisson's ratio. The effect of the wall friction is that a proportion of the load acting on the slice is carried by the walls of the container. Consequently, the wall pressure distribution is considerably smaller than would be expected on the basis of a hydrostatic calculation.

Similar arguments apply to cases where a vertical load is applied to the top of a powder in confined uniaxial compaction; this configuration is shown in Figure 10. A sensitive parameter for assessing the influence of the wall friction on the pressure is the stress transmission ratio R which is the ratio of the transmitted to the applied stress. The earliest derivation of an expression for this parameter is due to Shaxby and Evans (27).

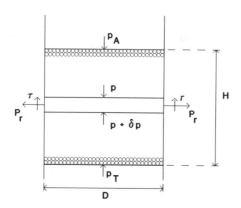

Figure 10.   The stresses acting on a powder in confined uniaxial compaction.

They applied a continuum mechanics analysis with a constant coefficient of friction μ at the wall which yielded the following result

$$R = \exp\ (-4\beta\mu H/D) \qquad (16)$$

where H is the current height of the powder and D is the diameter of the cylinder. The ratio of the radial to vertical stress was taken to be a constant equal to β.   This equation predicts that the transmission increases during the compaction process.
    Spencer et al (28) developed a semi-particulate approach and derived an alternative expression for R as follows

$$R = \exp\ (-4\lambda\mu_p H_o/D) \qquad (17)$$

where H_o is the initial height of the powder and μ_p is the coefficient of friction of the individual particles. In this case the ratio of radial to the vertical force acting on a particle was taken to be a constant and equal to λ. Unlike the previous expression this equation predicts that the stress transmission ratio is a constant with respect to the extent of compaction.
    The Spencer et al equation is consistent with some experimental data (29) but for other powders the behaviour is much more complex; Figure 11 shows results for some proprietary granules. At low applied stresses the transmission ratio increases until a plateau level is achieved.   In some cases there is a maximum in the data in the plateau region. These agglomerates behave as point contacts with a load index less than unity.

Consequently, their coefficients of friction will be large at low compaction pressures and will decrease to some asymptotic value at high pressures. This pressure sensitivity of μ_p is consistent with the observed behaviour. If the modelling procedure adopted for the kilns in Section 3.2 is employed in the Spencer et al semi-particulate model, then an expression for R is obtained which can be approximated to the following convenient form for comparison with equations 16 and 17

$$R = \exp\left[\frac{-4k\lambda^n H_o}{\sigma_A^{1-n}(s_o r)^{2(1-n)} D}\right] \qquad (18)$$

where s_o is the initial value of the packing parameter s. This expression displays the dependency on applied stress σ_A described above and reduces to equation 17 when the load index is unity. It also shows that if the particle radius r is reduced by fragmentation then there will be a reduction in the value of R as shown in Figure 11.

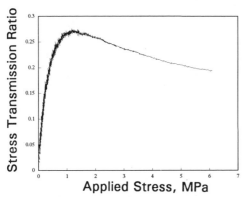

Figure 11. The stress transmission ratio in confined uniaxial compaction as a function of applied stress for some proprietary granules.

4. The Contribution of Interparticle Friction in the Tensile Failure of Particle Assemblies

4.1 SEMI-PARTICULATE ANALYSIS

The modelling of the wall friction in plug flow was described in Section 3.1. In the terminology of fracture mechanics, this can be described on a gross scale as brittle mode II (in-plane shear) failure. The term 'brittle' is used in the sense that the failure process and hence the dissipation of energy is confined to the failure plane. For cohesive powders, a similar analysis has been carried out for mode I (tensile) failure by Kendall et al (30). They considered assemblies in which autoadhesive forces operated between the particles. On the basis of the JKR model of autoadhesive junctions (11), they estimated the total energy required for tensile failure by summing the separation energies across a failure plane. This is essentially a semi-particulate model because, outside the failure plane, the assembly was treated as a rigid continuum. That is, a purely brittle failure was assumed.

The Kendall et al model has been examined by measuring the fracture energy of glass ballotini agglomerates (31). It was found that the measured values were about a factor of 10^3 greater than that calculated from the model. The origin of this discrepancy is that materials rarely display pure brittle fracture.

Fracture involves the extension of pre-existing flaws. It has been shown that the stress field in the vicinity of a crack or a flaw in body is singular at the crack tip for elastic solids (32). Real materials can not accommodate singular stresses and this leads to some stress limitation process which may involve atomic rearrangement in highly brittle materials such as some ceramics, for example, or to plastic distortion in polymers. The region ahead of the crack tip where this occurs is known as the process zone. The fracture energy is the sum of that required to rupture junctions in the failure plane and that dissipated in the process zone. For many materials, the process zone work is many order of magnitude greater than the rupture energy in the failure plane.

Particulate systems have process zone sizes which correspond to many particle diameters (33). For autoadhesive assemblies, the energy losses will be associated with frictional work developed by microslip between the particles. The problem is therefore analogous to that of estimating the wall friction of powders when the walls are very rough. The ploughing component of friction or shear zone is equivalent to the process zone in a fracture mechanics description.

4.2 NUMERICAL SIMULATION

As mentioned in Section 1, the most effective technique for understanding deformation processes at the particle scale involves computer simulation. One of the currently available techniques is discrete element analysis. This has the advantage that realistic interaction laws between the particles can be incorporated. Considerable progress has been made in this respect particularly through the efforts of Thornton (34) and Walton (36). More specifically, it is possible to include Hertzian deformation, autoadhesion and interfacial friction.

The traditional method for measuring the strength of particulate assemblies is the Brazilian test. This involves the diametric compression of cylindrical specimens between parallel platens. The failure stress is calculated from the force to fracture on the basis of Hertzian deformation.

Figure 12 shows a discrete element simulation for the diametric compression of a two dimensional particle assembly with JKR and interfacial friction interaction laws (35). The figure corresponds to the peak compression force and shows plastic zones which have initiated from the platens. Subsequently, the plastic zones grow in size. Even when the compression force has approached zero, no macroscopic cracks are evident (Figure 13).

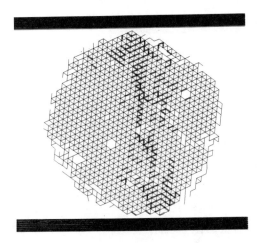

Figure 12. Computer simulation of a Brazilian test corresponding to the peak compressional force. The thin lines join the centres of particles that are in contact. The thick lines show junction ruptures which are generally preceded by interparticle sliding.

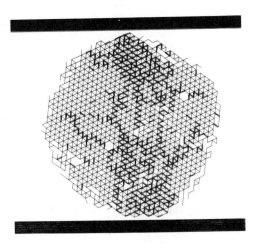

Figure 13. Simulation data corresponding to Figure 12 at a strain when the compressive force is approaching zero.

Recognisable crack growth is not observable until relatively high strains and this is initiated from the centre of the assembly (Figure 14). This behaviour corresponds to typical elasto-plastic fracture where the major proportion of energy dissipation is associated with bulk deformation.

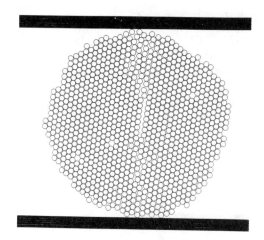

Figure 14. Simulation data corresponding to much higher strains than that in Figure 13; this figure shows the actual particle geometries.

5. Conclusions

The friction of powders at smooth walls is somewhat analogous to the multiple asperity contact of a gross body with the asperities representing the particles. However, unlike such contacts, the number of particle contacts remains approximately constant with increasing applied normal load. In principle, it is then straightforward to estimate the wall friction of a powder against a smooth wall. The major complication is that the stresses are not transmitted uniformly in powders and the pressure distribution at a wall is impossible to estimate.

Another feature of powders that distinguishes them from continuous solids is their bulk deformation characteristics which resemble plastically deforming solids. Such solids can be modelled on the basis of a simple yield criterion such as the Tresca. For powders, the shear stress at slip planes depends on the normal stresses acting on those planes. The resistance is described by an internal coefficient of friction. This is incorporated into the Mohr-Coulomb yield criterion which is formally equal to the Tresca criterion when the internal coefficient of friction is zero. The slip planes are many particle diameters in thickness. Although analytical models have been developed, more informative approaches are based on computer simulation.

References

1. Savage, S.B. (1979) 'Gravity flow of cohesionless granular materials in chutes and channels', J. Fluid Mech., 92, 53-96.

Stephens, D.J. and Bridgwater, J. (1978) 'The mixing and segregation of cohesionless particulate materials Part 1. Failure zone formation', Powder Technol., 21, 17-28.

Nedderman, R.M. and Laohakul, C. (1980) 'The thickness of the shear zone of flowing granular materials', Powder Technol., 25, 91-100.

Spencer, A.J.M. (1982) 'Deformation of ideal granular materials', in H.G. Hopkins and M.J. Sewell (eds.), Mechanics of Solids: The Rodney Hill 60th Anniversary Volume, Pergamon Press, Oxford, pp. 607-653.

Timoshenko, S. (1934) Theory of Elasticity, McGraw-Hill, New York.

Adams, M.J., Briscoe, B.J. and Pope, L. (1987) 'A contact mechanics approach to the prediction of the wall friction of powders', in B.J. Briscoe and M.J. Adams (eds.), Tribology in Particulate Technology, Adam Hilger, Bristol, pp.8-22.

Swain, M.V. (1979) 'Microfracture about scratches in brittle solids', Proc. R. Soc. Lond., A366, 575-597.

Briscoe, B.J. and Smith, A.C. (1980) 'The shear properties of thin organic films', Rev. Deformation Behaviour Materials, III, 151-191.

Briscoe, B.J. and Adams, M.J. (1991) 'The friction of powders', Powder Technol., To be published.

10. Adams, M.J., Briscoe, B.J. and Kremnitzer, S.L. (1983) 'A survey of the adhesion, friction and lubrication of polyethylene terephthalate monofilaments', in K.L. Mittal (ed.), Physicochemical Aspects of Polymer Surfaces, Vol 1, Plenum Press, New York, pp. 425-450.

11. Johnson, K.L., Kendall, K. and Roberts, A.D. (1971) 'Surface energy and the contact of elastic solids', Proc. R. Soc. Lond., A324, 301-313.

12. Adams, M.J. and Edmondson, B. (1987) 'Forces between particles in continuous and discrete liquid media', in B.J. Briscoe and M.J. Adams (eds.), Tribology in Particulate Technology, Adam Hilger, Bristol, 154-172.

13. Archard, J.F. (1957) 'Elastic deformation and the laws of friction', Proc. R. Soc. Lond., A243, 190-205.

14. Briscoe, B.J., Fernando, M.S.D. and Smith, A.C. (1987) 'The role of interface friction in the compaction of maize', in B.J. Briscoe and M.J. Adams (eds.), Tribology in Particulate Technology, Adam Hilger, Bristol, pp. 220-233.

15. Tuzun, U., Adams, M.J. and Briscoe, B.J. (1988) 'An interface dilation model for the prediction of wall friction in a particle bed', Chem. Engng. Sci., 43, 1083-1098.

16. Dowson, D. (1979), History of Tribology, Longman, London.

17. Adams, M.J., Briscoe, B.J. and Wee, T.K. (1990) 'The differential friction effect of keratin fibres', J. Phys.D: Appl. Phys., 23, 406-414.

18. Reynolds O. (1885) 'On the dilatancy of media composed of rigid particles in contact', Phil. May., 8, 22-53.

19. Rowe, P.W. (1962) 'The stress-dilatancy relation for static equilibrium of an assembly of particles in contact', Proc. R. Soc. Lond., A269, 500-527.

20. Horne, M.R. (1965) 'The behaviour of an assembly of rotund, rigid, cohesionles particles', Proc. R. Soc. Lond., A286, 62-97.

21. Nemat-Nasser, S. (1980) 'On the behaviour of granular materials in simple shear Solid Found., 20, 59-73.

22. Ueng, T. and Lee, C. (1990) 'Deformation of sand under shear- particula approach', J. Geotech. Eng., 116, 1625-1640.

23. Briscoe, B.J., Winkler, A. and Adams, M.J. (1985) 'A statistical analysis of th frictional forces generated between monofilaments during intermittent sliding' J. Phys. D: Appl. Phys., 18, 2143 -2167.

24. Johnson;, K.L. (1985) 'Contact Mechanics', Cambridge UP, Cambridge.

25. Briscoe,B.J., Pope, L. and Adams, M.J. (1985) 'The influence of particle surfac topography on the interfacial friction of powders', in I. Chem. E. Sym. Ser. N 91, I. Chem. E., Rugby, pp. 197-211.

26. Adams, M.J., Briscoe, B.J., Motamedi, F. and Streat, M. (1991) 'The Frictiona characteristics of coal particles', J. Phys. D: Appl. Phys., To be published.

27. Shaxby, J.H. and Evans, J.C. (1923) 'On the properties of powders: the variatio of pressure with depth in columns', Trans. Faraday Soc., 19, 60-72.

28. Spencer, R.S., Gilmore, G.D. and Wiley, R.M. (1953) 'Behaviour of granulate powders under pressure', J. Appl. Phys., 21, 527-531.

29. Isherwood, D.P. (1987) 'Die wall friction effects in the compaction of polyme granules', in B.J. Briscoe and M.J. Adams (eds.), Tribology in Particulat Technology, Adam Hilger, Bristol, pp 234-248.

30. Kendall, N., McN Alford, N. and Birchall, J.D. (1986) 'The strength of gree bodies', Inst. Ceram. Proc. Special Ceramics No 8, 255-265.

31. Abdel-Ghani M., Petrie, M., Seville, J.P.K., Clift, R. and Adams, M.J. (1991 'Mechanical properties of cohesive particulate solids', Powder Technol., 65 113-124.

32. Iglis, C.E. (1913) 'Stresses in a plate due to the presence of cracks and shar corners', Trans. Inst. Naval Architects, LV, 219-230.

33. Adams, M.J., Williams, D. and Williams, J.G. (1989) 'The use of linear elasti fracture mechanics for particulate solids', J. Mater. Sci., 24, 1772-1776.

34. Thornton, C. (1987) 'Computer-simulated experiments on particulate materials' i B.J. Briscoe and M.J. Adams (eds.), Tribology in Particulate Technology Adam Hilger, Bristol, pp. 292-302.

35. Walton, O.R., Braun, R.L., Mallon, R.G. and Cervellie, D.K. (1988) 'Particle dynamics calculations of gravity flow of inelastic frictional spheres' in M Satake and J.T. Jenkins (eds.), Micromechanics of Granular Materials, Elsevier Amsterdam, pp. 153-162.

36. Thornton, C., Yin, K.K. and Adams, M.J. (1991). To be published.

Discussion *following the lecture by M J Adams on "Friction of granular non-metals"*

J M ISRAELACHVILI. You made the comment that some adhesive interactions cannot be described by employing JKR theory but arise from capillary forces, implying that they are intrinsically different. I am not sure that this is correct. The origin of the surface energy is not specified in JKR theory, and indeed if there are capillary forces, and provided that the radius of the meniscus is small compared with the radius of the particles, then JKR theory should apply. Consequently, the distinction between the two interactions is artificial.

M J ADAMS. I would agree with your comment for touching spheres where the capillary forces can be treated in the same way as other forces, such as van der Waals, in the contact zone. This point was made nicely in your paper with Len Fisher (Colloids and Surfaces, **3**, 303, 1981). However, in practice, larger liquid bridges and a finite separation distance between the particles (due to packing constraints) is often of interest. Under these circumstances, when the capillary forces are significantly smaller and relatively insensitive to any elastic deformation of the particles due to the liquid bridge forces, a capillary analysis based on rigid particles is adequate (cf ref 12).

M O ROBBINS. Some numerical simulations (P A Thompson and G S Grest (1991), Phys Rev Lett 67, 1751) suggest that interparticle friction is not very important in the dynamics of particle flow, whereas you have indicated that this quantity if of critical importance.

M J ADAMS. Current numerical simulation techniques may be described either as rigid or soft particle models. The former was developed by Campbell (e.g. C S Campbell and C E Brennen, J. App. Mech. **52**, 172, 1985) and is based on instantaneous collisions with energy dissipated through a coefficient of restitution. A surface friction coefficient is not specified with this technique, only stick or slip contacts. Details of the second technique (discrete particle modelling) are given in the paper. Simulations of steady shearing flow show that the stresses depend strongly on the degree of elasticity and void volume, but are less dependent on the interparticle friction coefficient (V R Walton and R L Braun, Acta Mechanica, **63**, 73, 1986). This is in marked contrast to the wall stresses in plug flows where the particle friction is the dominating factor.

M O ROBBINS. You made the suggestion that the wall friction of coal measured using a kiln increased after initial rotation of the kiln because of the development of a more uniform state. Does this imply a deviation from plug flow, since the particles will be undergoing some spatial rearrangement?

M J ADAMS. Strictly, this would be the case, although in practice microscopic displacement between the particles would probably be sufficient to produce a more homogeneous stress distribution.

U LANDMAN. Could you explain how the JKR interactions are incorporated into the numerical simulation of particulate agglomerates?

M J ADAMS. Thornton and Yin (Powder Technology, **65**, 167, 1991) have developed the method based on co-linear JKR impact. In addition, they have extended the method to describe tangential autoadhesive impacts based on the work of both Savkoor and Briggs (Proc. Roy Soc. Lond. **A356**, 103, 1977) and Mindlin and Deresiewicz (J. Appl. Mech. Trans. ASME **20**, 327, 1953). This approach is required in soft-particle numerical simulation since it is necessary to compute the evolution of the forces; the adhesion model of friction simply provides a failure criterion.

H M POLLOCK. R J Meyer (J. Phys. D.: Appl. Phys. **22**, 1825, 1989) has applied JKRS theory to the friction of particles and claims that when the particles have flat faces, the effective friction coefficient is no longer independent of elastic constant. Thus, for example, use of a plasticiser, causing a decrease in modulus of a polymer, should produce a large rise in friction. Would you care to comment?

M J ADAMS. In this paper, Meyer cites the work of K Kendall (Nature **A319**, 203, 1986) who argues that the frictional force of a spherical particle of diameter D in a JKR contact would be given by μW_{eff}, where μ is the coefficient of friction and W_{eff} is the effective normal load obtained from JKR theory as follows:

$$W_{eff} = W + \frac{3}{2}\pi D\Gamma + [\pi DW\Gamma + (\frac{3}{2}\pi D\Gamma)^2]^{\frac{1}{2}}$$

where W is the applied normal load and Γ is the Dupré work of adhesion. Kendall employed this approach to explain the non-linear load dependence of the friction of point contacts. However, there are a number of problems. For example, point contacts still display a load index of less than unity when immersed in a perfectly wetting liquid which completely attenuates the autoadhesion (M J Adams, B J Briscoe and S L Kremnitzer in "Microscopic Aspects of Adhesion and Lubrication", J M Georges (ed), Elsevier, Amsterdam, 1982, pp 405-419). According to the Kendall treatment, the load index would be unity, since Γ would be equal to zero under these circumstances. Primarily, this is a normal load scaling argument and μ is a lumped parameter which provides no information on the mechanism of friction in terms of the energy dissipation processes. Actually, the treatment based on the adhesion model of friction which I described shows that, for point contacts, the friction will be a function of the elastic modulus (see equation 5 in the paper).

A plasticiser has two main effects on the friction of a JKR contact under an applied normal load. It will decrease the elastic modulus, leading to an increase in the real area of contact and hence increase the friction. In addition, it will decrease the interfacial shear strength and hence decrease the friction. In general, the former predominates and consequently plasticisers cause an increase in the frictional force (ref 10). For grossly plasticised contacts, it is possible that viscoelastic deformation hysteresis losses could occur which will further augment the friction. A plasticiser will also reduce the friction because of its wetting action.

For smooth flat particles, the area of contact would be independent of the elastic modulus provided the flat faces were in planar contact. Consequently, the friction would be decreased by a plasticiser, via its action on the interfacial shear strength, provided that viscoelastic losses were not introduced. Myer, who applied Kendall's arguments, confused the issue by treating flat particles as truncated spheres. In this case, plasticisers would increase the friction in an analogous way to point contacts.

DEFORMATION AND FLOW OF METALS IN SLIDING FRICTION

T H C CHILDS
University of Leeds
Leeds
United Kingdom

ABSTRACT. Bowden's and Tabor's original views [1,2] of adhesive and ploughing friction at plastic contact areas between metals are briefly reviewed, before considering in more detail the nature of surface flow (wave, wedge or chip forming) at plastic contacts of rough surfaces in the presence of adhesion. These flows are mapped on to a plane in which surface slope and contamination are the axes. For surfaces of small slope, in the absence of strong adhesion, the contact regions can be elastic. Conditions for elasticity are added to the friction mechanism map to highlight the critical range of friction coefficients from 0.3 to 0.4. For lower values, sliding flows will be elastic or plastic waves; for higher values localisations of surface plastic flow, wedge formation or transfer are likely to develop.

1. Introduction

The remit of this chapter is to review the surface deformations and the magnitudes of the friction coefficients that occur when materials which can flow plastically are slid over one another. These depend on the geometry of the sliding interface and the magnitude of the interfacial shear traction; and also on the elastic and plastic properties of the materials. In general, as illustrated for an artificially regular interface in figure 1, the interface at the real areas of contact is inclined to the sliding direction. Both the normal stress p and the tangential stress s on the interface contribute to the friction force F and normal load W: the friction coefficient μ is given by

$$\mu = F / W = (p\sin\theta + s\cos\theta)/(p\cos\theta - s\sin\theta) \tag{1}$$

If s, p or θ vary over the contact, (1) is modified by integrating over the area of contact. Two special cases are particularly simple to consider: (i) $\theta = 0$, so

$$\mu_{adh} = s/p \tag{2}$$

and (ii) s = 0, so

$$\mu_p = \tan\theta \tag{3}$$

I. L. Singer and H. M. Pollock (eds.), *Fundamentals of Friction: Macroscopic and Microscopic Processes*, 209–225.
© 1992 *Kluwer Academic Publishers. Printed in the Netherlands.*

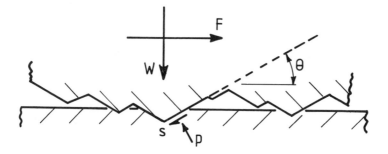

Figure 1. A sliding contact between two surfaces

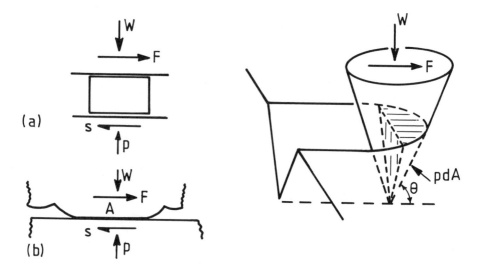

Figure 2. Combined loading and shear of (a) a block, (b) a friction junction
Figure 3. A cone scratching a flat surface

Equations 2 and 3 are the well-known expressions for the adhesive and deformation (or ploughing) components of friction [1].

At the time that equation 2 was developed it was assumed that real areas of contact between metals would be plastically stressed. The interaction between s and p was used to explain the high sliding friction coefficients, 1.0 or larger, that were observed between clean ductile metals. The key point was noted that the area of a plastically loaded junction would grow if a friction force were added to it. By analogy with the combined loading and shear of a block between two parallel platens, figure 2a, for which p and s can be related by the Tresca yield criterion equation 4a, p and s acting on an adhesive friction junction, figure 2b, may be assumed to be related by an equation of the form of 4b, where α and β are assumed to be constants, and k is the shear yield stress of the plastically loaded material.

$$p^2 + 4s^2 = 4k^2 \tag{4a}$$

$$p^2 + \alpha s^2 = \beta k^2 \tag{4b}$$

Equation 4b may be used to relate the area A of the junction when acted on by F and W together to its area A_0 when acted on by W alone:

$$(A/A_0)^2 = 1 + \alpha \, (F/W)^2 \tag{5}$$

The contact pressure when the area is A can be substituted in (2), and after some rearrangement an expression for the friction coefficient allowing for the junction growth is

$$\mu = \frac{s}{k} \left\{ \beta - \alpha \left(\frac{s}{k} \right)^2 \right\}^{-\frac{1}{2}} \tag{6}$$

If α and β are constants, then equation 4b suggests that they will be equal, because when p=0 and for an extremely clean interface, s may be expected to equal k. If $\alpha = \beta$, and when s=k, the friction coefficient according to equation 6 becomes infinite. There is a further expectation from equation 4b that $\beta = 25$, for in the absence of sliding the contact has the geometry of a hardness indent for which it is known that p=5k [2]. In that case, equation 6 indicates that if either α/β or s/k is reduced from 1.0 to 0.95, the predicted friction coefficient falls from infinity to less that 1.0: the prediction of infinity is critically dependent on the assumptions of the model (equation 4b) as well as on the cleanness of the interface.

In the absence of adhesion, the prediction of equation 3 for ploughing friction varies slightly with the details of the assumed shape of the contact. Figure 3 shows a cone scratching a metal flat. If the contact pressure is assumed uniform over the interface and the interface is assumed to be just the area of intersection between the indenter and the flat (i.e. extra contact due to deformation of the flat ahead of the indenter is ignored), integration over the interface of the components of pdA resolved in the directions of F and W leads to equation 7a for the friction coefficient. A similar calculation for a sphere ploughing a flat leads to equation 7b, where $\tan \theta$ is the slope of the sphere at the edge of the contact.

$$\mu = [2/\pi]\tan\theta \tag{7a}$$
$$\mu = [4/(3\pi)]\tan\theta \tag{7b}$$

These differ from equation 2 by a simple numerical factor.

Equations 2 to 7 present the simplest description of friction with any claim to reality. Yet as far as ploughing deformation is concerned they give no information about the flow pattern of material round the indenter, and a consideration of deformation combined with adhesive friction is avoided. In the case of adhesive friction, the theory takes s/k (which may be thought of as a surface contamination factor) as given. It has nothing to say about the nature of interfacial rheology that controls s. Finally by no means all metal sliding contacts are plastic: hard polished surfaces and many run-in surfaces suffer elastic or elasto-plastic deformation. Further review of these topics is developed in the remainder of this paper.

212

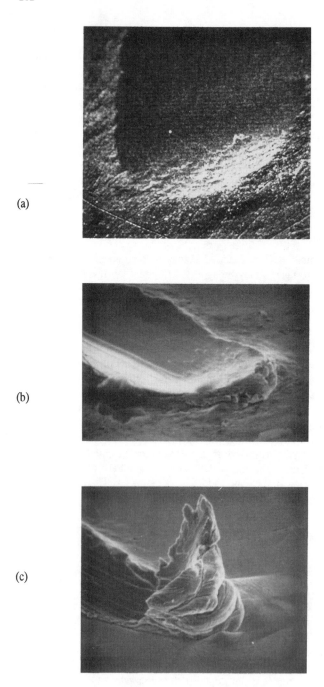

Figure 4. A silver surface scratched by diamond cones, $\theta =$ (a) 5°
(b) 25° and (c) 60°. (E*/Y) = 650

2. Plastic Flow In Deformation With Adhesion

Figure 4 shows a range of surface flows observed when cones of different slopes are slid over a flat. Metal may flow round the cone as a wave, be trapped ahead of the cone as a wedge or prow, or flow up the face of the cone as a chip. The conditions of slope and adhesion which lead to these differences may be studied for plane strain flows of ploughing by a wedge shaped asperity, for rigid plastic, non-workhardening metals, by means of slip-line field theory. The results may then be compared with experiments on three-dimensional flows with real metals. Slip-line field theory may also be used to analyse junction growth more rigorously than in the discussion of the previous section.

2.1. ESSENTIALS OF SLIP-LINE FIELD PLASTICITY ANALYSIS

The maximum shear stress in a plane strain flow of a non-hardening plastic material is everywhere k. Slip-line field theory provides a means of calculating the trajectories of the maximum shear stress (the slip lines) and the variations of hydrostatic pressure throughout the plastic region.

The hydrostatic pressure variations are the starting point. Figure 5a shows slip lines α and β which at a point 0 are inclined at ϕ to the cartesian axes x,y. Analysis of stress at 0 gives

$$\sigma_{xx} = -p-k\ sin2\phi$$
$$\sigma_{yy} = -p+k\ cos2\phi \qquad (8)$$
$$\tau_{xy} = k\ cos2\phi$$

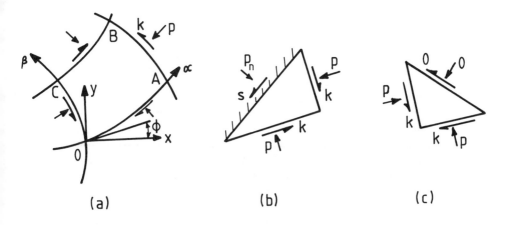

Figure 5. (a) α and β slip lines inclined at ϕ to cartesian axes, x, y at 0 and diagrams concerning (b) rough and (c) free surface boundary conditions

Substitution of equations 8 into the stress equilibrium equations

$$\partial\sigma_{xx} / \partial x + \partial\tau_{xy} / \partial y = 0; \quad \partial\sigma_{yy} / \partial y + \partial\tau_{xy} / \partial x = 0 \tag{9}$$

and considering the case when $\phi = 0$, gives

$$\begin{aligned} d(p + 2k\phi) &= 0 \text{ in the direction of an } \alpha \text{ line} \\ d(p - 2k\phi) &= 0 \text{ in the direction of a } \beta \text{ line.} \end{aligned} \tag{10}$$

Thus if p is known anywhere in the field, it may be calculated anywhere else, provided the trajectories of the slip lines are known. Equations 10 themselves provide rules for obtaining the trajectories. The variation of ϕ along an α or β line must be such that the calculated pressure change is zero in traversing, in figure 5a for example, from O to A, A to B, B to C and back to O. Two common boundary conditions to determine the extent of the field and to give a starting point for pressure calculations are the rough surface and free surface conditions. Figure 5b shows slip lines intersecting, at an angle ζ a rough surface on which the shear stress is s. Force equilibrium parallel to the surface gives

$$\zeta = 0.5\cos^{-1}(s/k) \tag{11}$$

At a free surface (figure 5c), the condition that $s = 0$ requires the slip lines to intersect the surface at an angle of $\pi/4$, while zero normal stress requires $p = k$.

A more detailed account of the construction of slip-line fields, of the calculation of velocities in the fields and of force and velocity boundary conditions may be found in standard texts [3,4].

2.2. PLASTIC MODELS OF PLOUGHING

Figure 6 shows possible slip line field models of wave and chip forming flows when a rigid wedge slides over a plastic flat [5-7]. It is required by their geometry that

$$\begin{aligned} \theta &\leq \zeta, \text{ for the wave flow} \\ \theta &\geq \pi/2 - \zeta, \text{ for the chip flow,} \end{aligned} \tag{12}$$

where ζ is defined in equation 11. The consequent range of validity of these fields is mapped on to the s/k, θ plane in figure 7a and the friction coefficients calculated from them are shown in figure 7b. For the wave flow, for example, friction coefficient is calculated from equation 1: force equilibrium normal to the wedge surface gives

$$p_n/k = p_2/k + \sin 2\zeta \tag{13}$$

Equation 10 and the free surface boundary condition give

$$p_2/k = 1 + 2\phi_1 \tag{14}$$

The value of ϕ_1 is determined by the volume conservation requirement that the tip of the wedge lies at the level of the surface:

$$\sin\theta = \sqrt{2}\sin\zeta\,\sin\left(\frac{\pi}{4}+\zeta-\phi_1-\theta\right) \tag{15}$$

Figure 7 shows that for a range of θ about 45°, and of increasing width the greater is s/k, there is no wave or chip forming flow and it may be surmised that these are the conditions that lead to prow or wedge formation. However, Petryk [8] has shown that the fields of

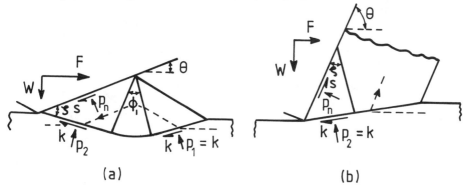

(a)   (b)

Figure 6. (a) Wave and (b) chip forming flows over a wedge tool

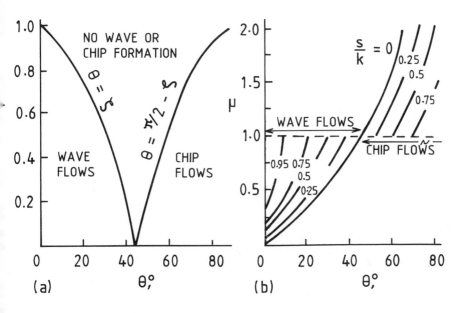

(a)   (b)

Figure 7. (a) Flow mechanism map and (b) friction coefficients for the fields of figure 6.

216

figure 6 are not the only possible ones: surface plasticity is highly non-unique. His further fields are shown in figure 8: more complex wave flows (I), combined wave and wedge flows (II), wedge flows (III), combined wave and cutting flows (IV) and cutting flows with and without stagnant zones (V and VI). Their ranges of validity are shown in figure 9a and the more complex resulting friction variations are shown in figure 9b. For $\mu \leq 0.4$ and $\mu \geq 1.0$ the situation is close to that shown in figure 7b, but at intermediate values of μ friction is not uniquely defined by s/k and θ. Its magnitude becomes less determinate the larger is s/k, as is indicated by the areas enclosed by the dashed, dotted and dash-dotted lines for s/k = 0.5, 0.75 and 1.0 respectively.

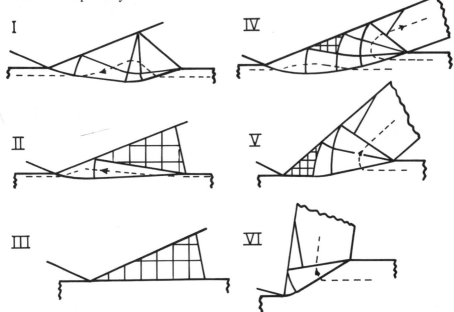

Figure 8. Further wave, wedge and chip forming flows [8]

2.3. EXPERIMENTAL STUDIES OF PLOUGHING

It is difficult if not impossible to measure s/k: what one can measure is the variation of μ with θ. In experiments with wedges it is found that, for small values of θ ($\theta < 25°$), μ is greater than $\tan\theta$; data can frequently be fitted to curves of constant s/k [9]. For large values of θ, approximately equal to 90°, used in metal machining tests in which s/k has been measured in the range 0.75 to 1.0, the ratio of the cutting to the thrust force, equivalent to μ in the current terminology, is usually less than $\tan\theta$, in the range 1 to 2. These observations are in accord with figures 7b or 9b.

Most experiments on the sliding of hard cones, pyramids and spheres over softer flats have concentrated on recording flow transitions rather than friction coefficients. Transitions from wave to prow flows have been observed in the range of θ from 15° to 30°, and transitions to chip forming flows over the much wider range of 20° to 100° [10], as could be predicted

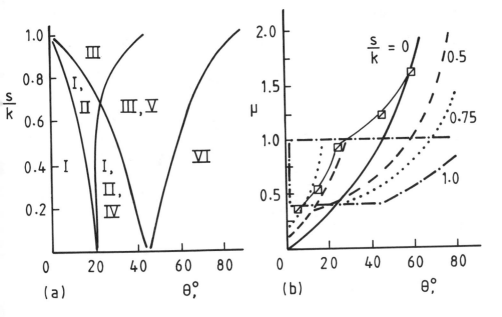

Figure 9. (a) Flow mechanism map   (b) friction coefficients for the fields of figure 8
⊠ Experimental data for the flows of figure 4 [11]

from figure 9a. However friction coefficients in the chip forming range have been found to be greater than expected from plane strain theory. The experimental points in figure 9b refer to the flows shown in figure 4 [11]. The freedom of material from the flat to escape round the edge of the cone, which is denied in plain strain conditions, and which would alter the directions of the maximum shear stresses in the flat relative to the load and friction directions, may account for the higher than estimated friction coefficients for the 45° and 60° cones.

The onset of wedge and chip forming flows are restricted to μ greater than 0.4. For smaller values, wave flows always occur. This value may be compared with the critical value of 0.25 to 0.3 for distinguishing between sub-surface and surface plastic flows in nearly elastic conditions to be considered in section 3.

2.4. JUNCTION GROWTH

Slip line field models of junction growth (section 1) have also been developed (6, 12, 13). They have shown equation 4b only approximately to describe the combined stress conditions in the junction. Little error derives in the calculation of junction growth by equation 5 but the sizes of α and β conspire to maintain μ from equation 6 less than or equal to 1.0. It would seem that friction coefficients greater than 1.0 measured between clean ductile metals require work hardening of the contacts and the formation of hard prows to generate an element of ploughing friction, as first observed by Cocks and Antler [14,15] and as already considered in the previous section.

3. THE TRANSITION TO ELASTIC CONDITIONS

The contact between metals is not always plastic. The transition of a Hertzian contact from elasticity to plasticity under increasing load has been considered in terms of the parameter $(E^*/Y)(a/R)$ by K.L. Johnson in the appx. to this book. Consideration has been extended to the contact of rough surfaces by J.A. Greenwood in the previous chapter, by introducing the plasticity index $(E^*/Y)(\sigma_s \kappa_s)^{0.5}$, where σ_s and κ_s are the surface peak height standard deviation and mean curvature respectively. As $(E^*/Y)(a/R)$ or the plasticity index increase from 2.5 to approximately 30 the contact changes from elastic, to elasto-plastic, to totally plastic. The question arises: how does sliding affect the transition from elastic to plastic deformation?

3.1. SURFACE SHEAR STRENGTH CHARACTERISTICS
An initial question is how best to describe the surface shear traction in elastic conditions? In plasticity analyses, as in section 2, the surface shear strength is usually taken to be independent of contact pressure:

$$s/k = m = \text{constant} \tag{16a}$$

In elastic analyses it is usually assumed that

$$s/p = \mu = \text{constant} \tag{16b}$$

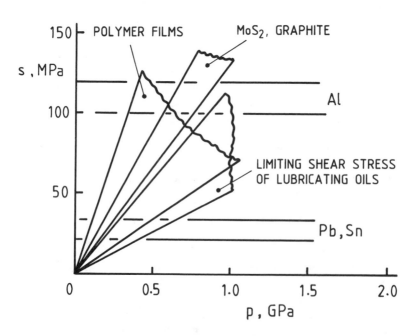

Figure 10. Dependence of shear strength on pressure for polymer [17], layer-lattice [16] and highly sheared oil [18] films

The appropriate form probably depends on the nature of the interfacial shear film. There is experimental evidence that interfacial shear strength of thin films is proportional to contact pressure for graphite and molybdenum disulphide [16], for polymer films [17], and for lubricant fluids at high shear rates [18], while for soft metal films, shear strength is independent of pressure, figure 10.

3.2. MODIFICATION OF HERTZIAN CONTACT STRESS

The addition of the surface traction 16a or b to an elastically loaded Hertzian contact modifies the maximum shear stress distribution round the contact in two ways. The region of largest maximum shear stress, which in the absence of traction lies at a depth from 0.5a to 0.8a below the centre of the contact (depending on the contact shape), is displaced towards the surface. For small values of m or μ ($m \leq 0.5$, $\mu \leq 0.25$) the size of this maximum shear stress is hardly changed from $0.3p_0$ where p_0 is the maximum Hertzian contact stress. At the same time a second region of large maximum shear stress develops at the surface. This is shown in Figure 11, in units of μp_0, for a Hertzian line contact. The two surface shear distributions, 16a and b, have been chosen to have the same mean value, figure 11a; the magnitudes of the corresponding surface maximum shear stresses, figure 11b, have been calculated from information in [19]. The constant surface traction distribution gives rise to an infinite maximum shear τ_{max} at $x/a = \pm 1$; it will therefore cause yielding at any value of p_0. However this is extremely localised. It may be important for the wear of contacts but will be ignored here in the context of friction. It will be assumed that for both traction distributions, at the surface

$$\tau_{max} = \mu p_0 \tag{17}$$

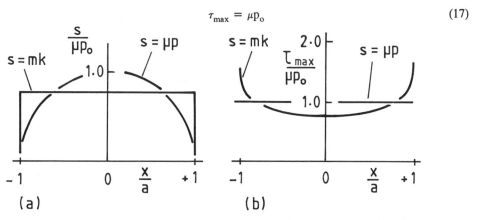

Figure 11. (a) Surface tractions and (b) the corresponding maximum surface shear stresses, for sliding line contacts under Hertzian loads

3.3. SLIDING CONTACT STRESS STATE MAPPING

The maximum shear stress of equation 17 modifies the values of the plasticity index at which transitions from elastic to plastic stressing occur. Johnson (6) has provided a map showing how the stress state depends on μ and p_o/k in a single Hertzian contact (where p_o is the maximum Hertz stress as before). It may be shown that p_o/k is approximately equal to $(E*/Y)(a/R)$, and we have already discussed the equivalence of $(E*/Y)(a/R)$ and plasticity index, so his map may be directly used with a change of scale to produce figure 12a. The $(E*/Y)(a/R)$ scale has been made logarithmic. To the left of line AA, deformation is always elastic. Repeated sliding in conditions just to the right of AA induces residual stresses that displace the elastic boundary to BB (shakedown limit). Between the lines BB and BC sub-surface plastic strain occurs but it is constrained to an elastic order of magnitude. To the right of BC plastic strain is still constrained elastically but it occurs at the surface. The position of the boundary DD above which flow is fully plastic is not known with any certainty, except that at $\mu=0$ it passes through $(E*/Y)(a/R)=30$; it has been drawn arbitrarily parallel to BC.

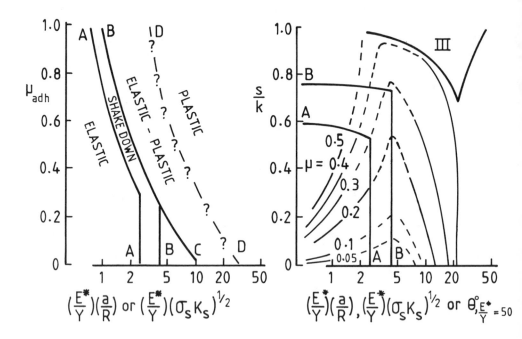

Figure 12. (a) Regimes of elastic, elastic-plastic and plastic contact stress, mapped on a base of adhesive friction coefficient and plasticity index; (b) a transposition of the elastic boundary on to an s/k, plasticity index or surface slope map with the addition of contours of constant friction coefficients, up to $\mu = 0.5$

3.4. Friction Mechanism Mapping

Within the elastic region, the deformation or ploughing component of friction is zero, apart from any elastic hysteresis. A relationship may be developed between the effective value of s/k and μ, in order to transpose the elastic boundary of the map of figure 12a on to the scale of figures 7a or 9a, in order in turn to develop a unified map for combined adhesive and deformation friction, taking elastic deformation into consideration. Equation 18 is a series of manipulations: the first is a rewriting of s/k to introduce the average contact stress $\bar{p}$; the second recognises that $s/\bar{p}$ is the adhesive friction coefficient and that $\bar{p}/k$ is cp_o/k (where c is 2/3 for a point contact or $\pi/4$ for a line contact); finally p_o/k has been replaced by $(E*/Y)(a/R)$ as previously introduced in section 3.3.

$$s/k \equiv \left(s\ /\bar{p}\right)\!\left(\bar{p}\ /k\right) \equiv c\mu(p_o\ /k) \equiv c\mu(E^*\ /\ Y)(a\ /R) \tag{18}$$

Equation 18 has been used to transform the boundaries AA and BB from figure 12a on to a map of s/k versus $(E*/Y)(a/R)$. Figure 12b is the result. The lines AA and BB from figure 12a are drawn for a line contact, $c = \pi/4$. The region III marked on the map is the region of prow formation from figure 9a, for the particular case of $(E*/Y)=50$. The reason for this choice of $(E*/Y)$ is introduced later. Thus figure 12b is a combined elastic and plastic friction mechanism map, with elastic adhesive friction in the region round the origin bounded by AA, prow or wedge formation in the region III, and elastic-plastic to plastic wave flows between AA/BB and III. At higher values of slope parameter, abrasive machining flows commence.

Contours of constant friction coefficient have also been added to figure 12b. Within the elastic region, μ is an adhesive friction coefficient and rearrangement of equation 18 gives

$$\mu = (s/k)(1/c)\ [(E*/Y)(a/R)]^{-1} \tag{19}$$

Contours are drawn for μ from 0.05 to 0.5.

In the plastic region μ depends on combined adhesive and deformation components and as already considered in section 2.2 depends on s/k and mean surface slope, either a/R, $\tan\theta$ or $(\sigma_s\ \kappa_s)^{0.5}$ as appropriate. In order to map μ values from figure 9 on to the scale of figure 12b, a value must be selected for $(E*/Y)$. For the special case of sliding steel on steel $(E*/Y)$ may be close to 50. For that value, the scale of $(E*/Y)(a/R)$ of figure 12b is then numerically approximately the same as that of surface slope in degrees (1 radian is 57°), for the range of slopes shown. Lines of $\mu = 0.2$, 0.3 and 0.4 in the plastic state have been added to figure 12b for this case. Finally the elastic and plastic friction coefficient contours have been extended into the elastic-plastic region on the basis that they should meet one another smoothly: the lines are broken to indicate that their detailed positions are speculative in that region.

3.5. COMMENTRY ON THE FRICTION MECHANISM MAP

The map (figure 12b) may be discussed in terms of three friction coefficient magnitudes: $\mu < 0.3$, $0.3 < \mu < 0.4$ and $\mu > 0.4$. For $\mu < 0.3$, surface deformation will either be elastic, elastic-plastic with sub-surface constrained plastic deformation or a plastic wave flow. In the transition range of μ between 0.3 and 0.4 there is an increasing danger, as s/k and surface slope increase, that plastic flow may concentrate on the surface and verge on wedge or prow formation. If μ should become greater than 0.4, there is a possibility that this is because of extreme surface smoothness in an elastic contact state, but if the elastic limit should be exceeded there is a rapid transition to destructive plastic wedge or machining flows. It must therefore be prudent to regard $\mu = 0.3$ as an upper value for long-lived components, quite apart from considerations of energy losses in sliding.

The map may also be used to consider minimum attainable friction coefficients in the absence of fluid film lubrication. For example, to achieve $\mu < 0.05$ for steel sliding on steel certainly needs a combined surface roughness slope less than between 5° and 10°, which is not difficult to achieve, but it also requires s/k to be less than 0.1. In section 3.1 the nature of surface shear strength was briefly discussed. Figure 10 shows that most classes of material surface contaminants seem to have a shear strength that increases with normal stress, in this case the contact stress. A sliding surface which has run-in to the region AA-BB of figure 12b will have a mean real contact stress of about 2k. There is thus a lower limit to s/k which is twice the gradient of the s/p data in figure 10. That is 0.1 for the non-metal data in that figure. For the metal data only Pb or Sn films on the hardest substrates appear to offer low enough shear strengths for $\mu < 0.05$ (this example is offered as illustrative: it is not a recommendation for a practical solution to low friction coatings). It is a matter of continuing research to understand the rheology of thin surface layers, with the objective of constructing new molecules with low shear strength yet load carrying and surface protective capacity.

4. CONCLUSION

The paper has briefly reviewed the classical ideas of adhesive and ploughing friction of plastic contacts, and has discussed in some detail slip-line field models of combined ploughing and adhesive friction which lead to expectations of wave, wedge and chip forming flows depending on the contamination and roughness of surface slopes. It has then considered the conditions of contamination and roughness that lead to elastic contact and has concluded that the friction coefficient range 0.3 to 0.4 is critical in determining whether surfaces are benignly orseverelyaffected by sliding.

REFERENCES

1. Bowden F.P. and Tabor D. (1950) The Friction and Lubrication of Solids Vol.I Ch.5, Clarendon Press, Oxford.
2. Bowden F.P. and Tabor D. (1964) The Friction and Lubrication of Solids Vol.II Ch.16, Clarendon Press, Oxford.
3. Hill R. (1950) The Mathematical Theory of Plasticity Chs.6-8, Clarendon Press, Oxford.
4. Johnson W. and Mellor P.B. (1983) Engineering Plasticity Ch.12, Ellis Horwood, Chichester.
5. Challen J.M. and Oxley P.L.B. (1979) 'An explanation of the different regimes of friction and wear using asperity deformation models', Wear 53, 229-243.
6. Johnson K.L. (1985) 'Contact Mechanics Ch.7, Cambridge University Press, Cambridge.

7. Lee E.H. and Shaffer B.W. (1951) 'Theory of plasticity applied to a problem of machining, Trans. ASME, J.Appl.Mech. 18, 405-413.

8. Petryk H. (1987) 'Slip line field solutions of sliding contact', Proc.Conf.Friction, Lubrication and Wear - fifty years on, pp987-994, Mechanical Engineering Publications,I.Mech.E., London.

9. Challen J.M. et al (1984) 'Plastic deformation of a metal surface in sliding contact with a hard wedge', Proc.R.Soc. Lond., A394, 161-181.

10. Childs T.H.C. (1988) 'The mapping of metallic sliding wear', Proc.I.Mech.E. 202 PtC, 379-395.

11. Childs T.H.C. (1970) 'The sliding of rigid cones over metals in high adhesion conditions', Int.J.Mech.Sci. 12, 393-403.

12. Johnson K.L. (1968) 'Deformation of a plastic wedge by a rigid flat die under the action of a tangential force', J.Mech.Phys.Solids 16, 395-402.

13. Collins I.F. (1980) 'Geometrically self-similar deformations of a plastic wedge under combined shear and compression by a rough flat die', Int.J.Mech.Sci 22, 735-742.

14. Cocks M. (1966) 'Shearing of junctions between metal surfaces', Wear 9, 320-328.

15. Antler M. (1964) 'Processes of metal transfer and wear', Wear 7, 181-203.

16. Briscoe B.J. and Smith A.C. (1982) 'The interfacial shear strength of molybdenum disulphide and graphite films', Trans. ASLE 25, 349-354.

17. Briscoe B.J. in Friedrich K. (ed.) (1986), Friction and Wear of Polymer Based Composites, Ch.2, Elsevier, Amsterdam.

18. Evans C.R. and Johnson K.L. (1986) 'The rheological properties of elastohydrodynamic lubricants', Proc.I.Mech.E. 200 Ptc, 303-312.

19. as reference 6, but chapters 2 and 7.

Discussion *following the lecture by T H C Childs on "Deformation and flow of metals in sliding friction"*

M J ADAMS: What would be the effect on the slip line fields you described if the Tresca were replaced by a Coulombic wall boundary condition?

T H C CHILDS: It can be done. One has to translate the local friction coefficient into an s/k factor, and you then find, for example, that the slip lines meet the tool at varying angles along the rake face. The point about many fully plastic solutions is that the predicted pressure distribution is almost uniform over the surface, implying that s is constant. However, if you have a thin layer of some particular lubricant, then the geometry could vary according to the local values of s/k, so that quite large changes in local pressure could occur.

I L SINGER: Your models for friction are purely mechanical and assume ideal plasticity. I have two questions. First of all, to what extent does N Suh's model of friction differ from yours, either in structure or in predictions? The second question relates somewhat to a model of D Maugis, who predicted "purely k-dependent" frictional instabilities without assuming any mechanical behaviour; you show a regime of "mechanically-determined" instabilities $(0.4 < \mu \ (?) < 0.6$ for ideally plastic materials. Is there some way in which one can understand the experiments as matching one or other of these models?

T H C CHILDS: Suh described three components of friction: adhesive friction, ploughing friction and a component that occurred when wear particles are trapped at the interface. I believe that the last two components are different aspects of the same process. You can either have plastic flow on its own if the asperities are steep enough, or you can get plastic flow if you have a sharp wear particle or any other piece of debris trapped in the interface. To speculate, maybe the ploughing aspect of his model could represent the wave end of the spectrum of processes, and when grit causes major damage to the surface, maybe you are towards the prow- and chip-forming end of the spectrum. In reply to your other question: I didn't in fact use the word instability. What I meant is that if you have a particular system, then one or another of these mechanisms such as wave formation or chip formation will occur, so I don't think we are talking about the same thing.

J A GREENWOOD: When trying to decide when to consider what the elasticians take as $s = \mu p$ or the plasticians take as $s = \varsigma k$, your chart of how shear strength varies with pressure missed out one interesting surface film; you didn't include iron oxide. Do you think you could comment on that?

T H C CHILDS: I don't know of anyone who has attempted to do experiments with iron oxide; I think it would be very valuable and interesting. One has moved in recent years away from the position where shear strength was regarded as effectively constant for most materials. Now the literature has gone to the other extreme, and a lot of people seem to imply that metals are about the only type of material for which this true. So for something like iron oxide, is the bonding likely to influence the pressure coefficient of shear? That is a very interesting question. Maybe you should have asked Dr Briscoe.

B J BRISCOE: Many years ago we did study a range of oxides and the trend with the oxides is not dissimilar to those that we see with the metals. There is a problem in the way those films are prepared. They often have organic contaminants present in them and this tends to increase the pressure coefficient. The other point that is worth making is that many years ago, Bridgeman in his experiments with high pressure anvils squashed many materials including oxides and obtained similar pressure coefficients. For oxides he got pressure coefficients that were of the order of about 0.03, compared with shall we say 0.3 for organic materials and something like 10^{-3} for metals.

T H C CHILDS: Yes, I certainly think that it will be well worthwhile doing more experiments and computer simulations in this area.

J A GREENWOOD: I have never understood what Bridgeman did other than to put two anvils together, apply a very big load, slide them and measure a force. I have never seen why this wasn't just an ordinary friction experiment and why what he was measuring wasn't just a coefficient of friction. How on earth out of his load and his force did he deduce yield strength or shear stress, seeing as we don't have the slightest evidence as to what his area of contact was?

E RABINOWICZ: Iron oxide as discussed in this last exchange of questions does not exist. Iron has three oxides; there is Fe_2O_3, which is a horrible material from the point of view of friction and wear, then there is FeO which is OK, and there is Fe_3O_4 which is a very fine solid lubricant. Also I would mention that it is very important to determine the moisture coefficient in such an experiment, and moreover that cobalt gives the same type of oxide, Co_3O_4, used widely as a lubricant in magnetic recording apparatus.

I L SINGER: What prevents someone from performing equivalent three-dimensional calculations for $\mu(\theta, \phi)$, by analogy with two-dimensional $\mu(\theta)$ calculations? Do we need a different method of calculation, or a better way (TV) of representing the data, such as the type of presentation used by U Landman for his MD simulations?

T H C CHILDS: Yes, slip line field analysis can only be used in 2-D cases. There isn't an equivalent analytical 3-D analysis, though of course one has finite element modelling[1].

---

[1](Editors' footnote): see also D Lebouvier, P Gilormini and E Felder, Thin Solid Films 172, 227 (1989) for a description of a 3-D incremental kinematic method that takes account of frictional as well as plastic work.

MAPPING OF FRICTION

Extended discussion contribution by K L JOHNSON, *Department of Engineering, University of Cambridge, England.*

I have been asked to pull together certain points that have been discussed during the past three days and which are relevant to the idea of the mapping of various aspects of friction. I shall refer to the presentations by J A Greenwood on roughness, A R Savkoor on sliding contact, T H C Childs on friction modelling, and myself on contact mechanics.

As a parameter for measuring or assessing the elastic or plastic deformation of a single contact, both Childs and I used the non-dimensional parameter $E^* a/(YR)$, where the symbols in turn represent the combined elastic modulus, yield stress in tension or compression, radius of contact, and radius, of the contacting sphere or single asperity. Y may be replaced by twice the yield stress in simple shear, k. Greenwood defined the plasticity index, which of course likewise gives a measure of whether an array of asperities deforms plastically or elastically. He defined it in terms of the rms profile height of the summits and the hardness, H. I refer to the simple model that involves a constant asperity radius of curvature R. If we have a varying radius of curvature, then we can replace $1/R$ by the mean summit curvature κ_s (despite the very serious difficulties about regarding a rough surface as an assembly of asperities ranging in size to below a micron, that Greenwood has described to us, most of use feel that there must be a minimum size of asperities that don't survive the first light brush of two sliding contacts. Thus I shall assume today that there is such a thing as κ_s). Thus the Greenwood plasticity index becomes $(E^*/H)\sqrt{(\sigma_s \kappa_s)}$. Greenwood shows also that the mean asperity contact radius, $\bar{a}$, is $\sqrt{(\sigma_s/\kappa_s)}$, i.e. independent of load. Thus, if we replace a in the original non-dimensional parameter by $\bar{a}$, we obtain $(E^*/Y)\sqrt{(\sigma_s \kappa_s)}$ which I will denote by Ψ; this is just three times Greenwood's plasticity index, since $H = 3Y$. So we see that we may use the same parameter for multiple contacts on rough surfaces as for the contact of one asperity.

On the map shown as figure 12b (p. 220) of the chapter by T H C Childs, we may thus identify the x-variable with the plasticity index for multiple contacts between rough surfaces. As ordinate he has the surface contamination factor $f = s/k$. For a perfectly clean surface, this ratio is equal to unity, and it falls as the surface becomes more contaminated. As he discussed, the map is divided into regimes according to the type of deformation that applies. Let us try to decide which type of model will be appropriate for particular materials having a particular type of surface finish. For example, where on the map should we place mild steel finished with a grinding wheel? In table 1, I have tabulated a few ranges of values of Ψ for typical metals, polymers and ceramics (later I will mention why elastomers have so far been omitted): I had to twist Dr Greenwood's arm a bit to

I. L. Singer and H. M. Pollock (eds.), Fundamentals of Friction: Macroscopic and Microscopic Processes, 227–234.
© 1992 *Kluwer Academic Publishers. Printed in the Netherlands.*

get rough values of σ_s and κ_s for the various types of surface.

	E^*/k	Ψ		
		SMOOTH (ground and polished)	MEDIUM (ground)	ROUGH (grit-blasted)
$(\sigma_s \kappa_s)^{\frac{1}{2}}$		0.0055	0.071	0.22
Metals	10^2-10^3	0.5-5	7-70	22-220
Polymers	20-150	0.11-0.27	1.4-3.5	4.5-11
Ceramics	50-100	0.28-0.55	3.6-7.2	11-22
Elastomers	?	-	-	-

Table 1: values of Ψ.

We can now superimpose these ranges of values on Childs' map, as shown in figure 1:

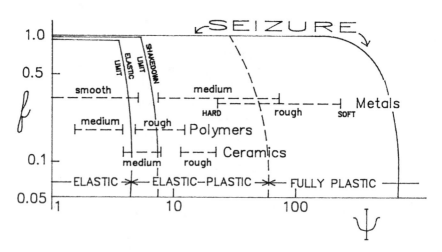

Figure 1: f as a function of Ψ

We see that for metals, for example, medium roughness brings us just into the fully plastic zone if the metal is soft, and if the surface is well polished, even softer metals are in the elastic regime.

Let us now consider the effects of strong adhesion and the type of sliding behaviour covered in A R Savkoor's chapter, for which we think of friction in terms of mode II fracture. To recapitulate one or two of his points: taking a single-asperity event, if we press a sphere into contact with a flat and then apply a tangential force T, the sphere will distort and you will have a distribution of frictional traction over the contact circle of radius a, reaching infinity at the edges if there is no slipping, i.e. with strong adhesion (gripping). The mode II stress intensity factor K_{II} is given by $T/(4\pi a^3)^{\frac{1}{2}}$, and when this reaches the critical value K_{IIc}, we would expect the fracture to occur, at a force $T_c = K_{IIc}(4\pi a^3)^{\frac{1}{2}}$. The crack propagation from the edge of the contact will be highly unstable. Using this model, a theory of friction can be worked out (see chapter by A R Savkoor, also [D1] and [D2]), but for the moment I shall not discuss that. Instead, let us ask in what circumstances the model is valid, and whether it has any physical reality. One of the restrictions on elastic fracture mechanics is that the process zone at the crack tip should be small compared with the general dimensions of the body. With plastically-deforming metals, the size of the process zone is determined by k, so that for a very soft material you have ductile shearing instead of brittle fracture. For linear elastic fracture mechanics to be valid, plastic flow must be restricted to a zone that is much less than a. This condition will give us another non-dimensional parameter: the mean stress $T_c/(\pi a^2)$ must be less than k, so that the parameter Φ, defined as $K_{IIc}/(k\sqrt{a})$, must be less than $\sqrt{\pi}/2$, i.e. approximately unity.

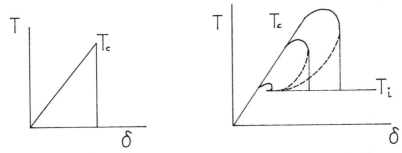

Figure 2: Stick-slip/steady sliding transition (from [D2])

Setting aside the question of Φ for the moment, we will see that there is another drawback to the mode II friction models discussed by Savkoor. If you have a mode I type of fracture, the surfaces get out of the way as the crack propagates and you have no interference between the two surfaces once the crack has opened. However, if we think of slipping as a mode II fracture, then we still have pressure forcing the two surfaces together. Even though the crack has penetrated a certain distance, within the peripheral annulus there is still contact and presumably some resistance to slipping[2]. In the very sensible model presented by Savkoor [D2], with support from experiments on rubber carried out by Barquins, the contact is supposed to develop strong adhesion if left in place without sliding, and K_{IIc} is quite large. Once slip starts, the tangential stress is then much lower, corresponding

---

[2] See reference to the suggestion of Gittus, discussed by D Tabor on page 15.

230

to the quantity s. Following the approach described by Savkoor we now consider the variation of T with deflection δ, as shown in figure 2. For strong adhesion, T rises linearly with δ until it reaches T_c since δ is equal to $T/(8E^*a)$, and then drops abruptly to zero. If slip is allowed, then at a certain point the mode II fracture starts at the edges of the contact and propagates inwards: an equilibrium calculation predicts the odd-shaped curve shown in the figure. The behaviour will differ according to the loading condition (fixed load or controlled displacements).

For fixed T conditions, when the maximum on the curve is reached, the system becomes unstable and you have a sudden transition to sliding. With an individual asperity in a multi-asperity contact, the hinterland of the body will provide fixed δ conditions so that it is possible to continue round the curve to the point where the tangent becomes vertical, before sliding occurs, at a load $T_i = \pi a^2 s$. If you were to postulate that once it came to rest again, that particular contact reforms a strong adhesive junction, then the model will allow you to re-trace a similar curve and slide again, so that stick-slip motion will result.

What is the condition for this type of stick-slip? The answer is that T_c must be greater than T_i. This condition implies that $\Gamma \gtrsim 1$, where Γ is another non-dimensional parameter defined as $K_{IIc}/(s\sqrt{a})$. For rough surfaces we should replace a by $\bar{a} = (\sigma_s/\kappa_s)^{\frac{1}{2}}$, so that Γ becomes $(\kappa_s/\sigma_s)^{\frac{1}{4}}K_{IIc}/s$, and again, we expect either stick-slip or steady sliding motion according to whether Γ is greater or less than 1.

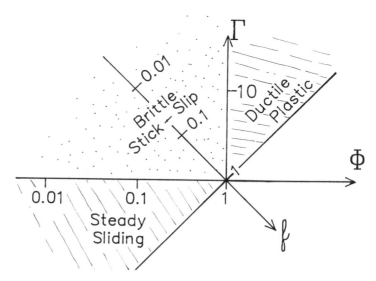

Figure 3: modes of sliding

In order to see how certain materials fit into this scheme, I have assembled the three non-dimensional parameters Φ, Γ and f, in a map. These are not independent since $\Phi/\Gamma = s\sqrt{a}/(k\sqrt{a}) = f$. In figure 3, we have Γ as ordinate and Φ as abscissa, both with logarithmic scales, so that the axis of f is a line at $45°$: the only relevant part of the map is that for which f is less than 1. Three regimes appear. Region I corresponds to ductile failure, very much as Tabor described many years ago. With strong adhesion, shearing will take place within the body of the material. In region II, stick-slip is predicted, and elastic contact with steady sliding will be seen in region III.

To conclude this highly speculative argument, I have collected some data from the materials selection charts of M F Ashby, as shown in table 2. There are no data for K_{IIc} so I have used values of K_{Ic}.

	K_{Ic}/k	SMOOTH (ground & polished)	MEDIUM (ground)	ROUGH (grit-blasted)
			Φ	
$\bar{a}(\mu m)$		27	7.1	22
Metals	5-50	1-10	1.9-19	1.1-11
Polymers	0.5-5	0.1-1	0.19-1.9	0.11-1.1
Ceramics	0.035-0.11	0.0067-0.021	0.013-0.041	0.0075-0.023
Elastomers	?	-	-	-

Table 2: values of Φ

If we insert these values onto figure 3 we see that as expected, metals will deform in a ductile manner, with no possibility of any kind of mode II fracture mechanics mechanism. However, with polymers and ceramics, log Φ can be negative: depending upon the cleanliness or otherwise of the surfaces, if the material properties and the roughness bring you into the upper left-hand rectangle in the figure, there is the possibility of discontinuous behaviour. If Γ, Φ and f are all less than 1, we expect steady sliding.

FURTHER DISCUSSION

H M POLLOCK: My question is concerned with the approximation involved in setting K_{IIc} equal to K_{Ic} for the sake of argument. K_{Ic} varies as the square root of the work of adhesion, while K_{IIc} involves forcing one sheet of atoms over the other. To say that they are of comparable magnitude sounds something like saying that the activation energy for diffusion of atoms along a surface is of the same order of magnitude as that required to desorb atoms from a surface: I should have thought that you would expect K_{IIc} to be rather less than K_{Ic}. However, even if you reduce K_{IIc} to, say, the square root of one fifth of the book value of K_{Ic}, maybe this makes no significant difference?

K L JOHNSON: There are others who I am sure would be better able to answer than I, with regard to comparing the effects of atomic forces when you are trying either to separate rows of atoms or to shear them, but my feeling is that the corresponding critical stress intensity factors are of a similar order of magnitude. The area of research that I am familiar with, and which relates to these questions, is that of interface cracks. What happens when you have an interface between two materials which is subject to both K_I and K_{II} loading? On the whole it is found that K_{II} is higher than K_I, as an experimental fact. There are attempts of course to combine their effects through squaring and adding, to obtain a total work of adhesion w, for example: $2E^* w = K_I^2 (1 + \tan^2 \theta)$, where $\tan \theta = K_{II}/K_I$, but this is a much-debated area of ongoing research [D3]. I think that such differences as there are, are rather small, given that it is hard even to say whether we are indeed likely to find this type of behaviour with given materials.

M J ADAMS: Yes, it is generally very difficult to propagate mode II cracks, and attempts to do this usually result in local mode I failure. This might explain some of the data for polymers. There is also another effect which influences the stress intensity factor, as is evident from fracture mechanics of ceramics initiated with fatigue notches. The resistance to fracture has been found to increase with crack growth due to frictional forces acting in a zone behind the advancing crack front. This has implications with respect to Johnson's point about the process zone size which is related to the ratio of the stress intensity factor and the yield stress. The effective increase in the mode II fracture resistance for sliding contacts due to the formation of failure surfaces that are not stress-free could provide the basis for the applicability of the adhesion model of friction. This is essentially a lower bound plasticity solution rather than a fracture criterion: that is to say, yielding in the interfacial zone is more favourable than mode II crack propagation in many sliding contacts.

A R SAVKOOR: In connection with the point made by Pollock, what I was suggesting in my talk was that stress intensity factors can be used in several ways. In one way you have this connection with strain energy release rate which may be a reversible process involving work of adhesion. On another occasion you could use it as a parameter for measuring or sensing the load on the crack tip. What will happen, as Adams said, is that if you get a plastic zone or a yielding region, and the energy that is consumed there is sufficiently large, you get a K_{II} which can be very high compared to K_I. So what is actually of physical significance as a limiting value of K_I or K_{II} would totally depend on what type of breakdown process is happening in a particular case. For cohesive failure, you possibly will have K_{II} greater than K_I. For adhesive failure there is at present no way to estimate their ratio, and the question is still quite open in my view.

K L JOHNSON: What those figures that I showed make fairly clear is that if one is dealing with metals then one is going to get the ductile type of failure, and in that sense K_{IIc} is an irrelevance. Thus, suggestions about describing sliding or incipient sliding in fracture mechanics terms are ruled out as far as metals are concerned; that is what the map shows, I think. The question is whether there are these possible models and mechanisms in the case of materials such as ceramics which have very high yield strength. Of course, one material which we know shows some of the fracture mechanics characteristics, is rubber, as studied by Savkoor. Therefore you might have said, why have elastomers been left off the map, given that they provide an obvious example where this type of stick-slip action is observed to take place? The problem about including rubber is that you have to include strain rate effects, and thus, another parameter. Since I have already used up the three dimensions I couldn't do anything about that today.

H M POLLOCK: Has any progress been made, since D Maugis last discussed the question[3], with the calculation of friction coefficients in cases where the interface exerts no tangential traction, but where frictional resistance depends on bulk anelastic properties?

K L JOHNSON: If an interface between two sliding solids exerts no tangential traction, whether there is any frictional resistance to sliding depends on the elastic and inelastic properties of the two solids. Consider a cylindrical or spherical slider on a flat surface:

(i) Two perfectly elastic solids will deform without hysteresis and hence without resistance.

(ii) If the slider is rigid and the flat surface deformable and inelastic there will be a resistance arising from asymmetry in the pressure distribution at the interface, i.e. the slider will be pushing against a higher resistance at the front of the contact.

(iii) If the flat surface is rigid then there can be no resistance to motion irrespective of inelasticity in the slider (which is not being continuously deformed).

All this was said by Greenwood and Tabor in 1958 when they were comparing the rolling and sliding resistance of hysteretic solids.

---

[3]His question [D4] was as follows: "When a slider passes over an element of volume beneath the surface, this element is subjected to a cycle of stresses during which energy is lost. But how does the slider feel these losses, and have a frictional resistance? Of course, through shear stresses within the area of contact; these surface stresses modify the volume stress tensor, hence the energy loss; the problem thus appears self-consistent. Can these stresses be computed by the deformation of the surface (frictional force not in the plane of sliding); can an interface offer resistance to a shear stress if adhesion forces are zero: the question is still open!".

I L SINGER: Would you care to comment on the distinctions between your "mechanical" model of friction and the "energy based" model of Heilmann and Rigney, e.g. Wear 72, 195 (1981)?

K L JOHNSON: As I understand it, Rigney's energy-based model of friction relies on deducing the friction from the plastic work done in producing the permanent deformation which is observed in a wear track. This approach is unsatisfactory because it relies on the final deformed state after the slider has passed. Beneath a slider the shear stress alternates from forward to backward and what one sees in the wake is the resultant of the deformation in theses two opposing directions. With an elastomeric material the net permanent deformation may be zero even though considerable energy has been dissipated due to the cyclic deformation beneath the indentor.

I L SINGER: How does Suh's mechanical model differ from the adhesion/ploughing model presented by Childs?

K L JOHNSON: Suh's description of friction is not materially different from the adhesion/ploughing model presented by Childs. In his description Suh follows the changes in friction in repeated sliding and attributes these different stages to different mechanisms. Stage 1 is principally ploughing a la Childs, Stage 2 shows an increase due to an increase in adhesion, and Stages 3 and 4 describe the influence on friction of wear particles generated in Stages 1 and 2. So that we possibly need to add an abrasion mechanism to that of Childs if the wear debris is harder than the original solids.

Discussion references

D1. Thornton, (1991) C J. Phys. D: Appl. Phys. **24**, 1942-6.

D2. Savkoor, A R, "Dry adhesive friction of elastomers" (doctoral thesis), Delft Technical University, The Netherlands (1987: see also his chapter in this book).

D3. Hutchinson, J W, in "Thin films: stresses and mechanical properties (M R S Fall Meeting Symp. Proc. 130)", ed by J C Bravman et al, Materials Research Society (Pittsburgh, USA) 1989, p 397; Scripta Metallurgica, to be published.

D4. Maugis, D, in "Microscopic aspects of adhesion and lubrication" (J-M Georges, ed), Elsevier (1982), p 221.

LUBRICATION BY
SOLIDS AND
TRIBOCHEMICAL FILMS

..... There are three basic problems concerning the behaviour of all types of surface films ..... First, how are the films attached to the substrate ; second, what are the strength and shear properties of the films; third, how do they break down?

D TABOR, in "Tribology - fifty years on" (I. Mech.E., ed), Mechanical Engineering Publications, London 1987, p 157.

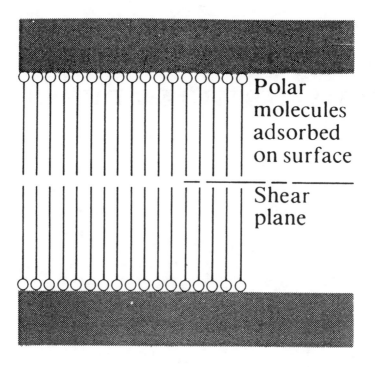

Polar
molecules
adsorbed
on surface

Shear
plane

Schematic representation of Sir William Bate Hardy's concept of boundary lubricating films. [From D. Dowson, <u>History of Tribology</u> (Longman, London, 1979) p. 354, with permission.]

SOLID LUBRICATION PROCESSES

I. L. SINGER
Code 6170
U.S. Naval Research Laboratory
Washington DC 20375   U.S.A.

ABSTRACT.   Solid lubricant behavior and lubrication by thin solid films are reviewed.
Crystalline as well as non-crystalline solids are shown to give low friction, as are surface
films as thin as a monolayer. The Bowden-Tabor adhesion model of friction is shown
to be particularly useful for measuring the shear strength of thin solid films under elastic
Hertzian contact conditions. However, the meaning of the "shear strength" as a materials
property is called into question because of the effects of atmosphere on the friction
coefficient. Mechanisms of "shear" are then examined from a microscopic point of view.
Microscopic studies of the rheological behavior of solid lubricants suggest that interfacial
films play an important role in accommodating sliding motion.   Several recent
investigations of interfacial films generated during dry sliding against coatings and
surface treatments are presented.   Surface analytical studies of interfacial films from
diamond-like carbon and MoS_2 coatings and Ti^+-implanted steel provide evidence that
tribochemical reactions take place between counterfaces, surfaces and the atmosphere.
A model that accounts for the tribochemical films generated during sliding contact is
described, and a thermochemical basis for the reactions governing film formation is
presented.   These studies suggest that the lubricity of coatings is determined by the
interfacial films generated during sliding, not by the bulk properties of the coatings
themselves.

1. Introduction

In the earlier chapter of T. Childs, we saw how the friction coefficient regulates the
stresses transmitted between two surfaces and thereby controls deformation (and wear)
behavior.   In continuum mechanics, the friction coefficient is often treated as a
contamination factor, a parameter that relates the shear strength of the interface to that
of the (weaker) solid substrate. In fact, the friction coefficient in dry sliding contact is
often determined by properties of materials that reside at the interface. Guided by bulk
properties of known materials, tribologists have placed both organic (see previous
Chapter by Briscoe) and inorganic solids at interfaces to lubricate substrates. However,

237

I. L. Singer and H. M. Pollock (eds.), Fundamentals of Friction: Macroscopic and Microscopic Processes, 237–261.

the friction and wear behavior of these materials is not always that expected of the bulk solids. Why? Because it is often the *interfacial films* formed during sliding contact, not the solid originally placed there, that controls friction properties. The purpose of this chapter is to link macroscopic behavior of solid lubricants with microscopic aspects of friction and wear processes.

In this chapter, we review and discuss both mechanical and chemical aspects of the friction behavior of thin solid films. Section 2 reviews material properties (structure, composition, etc.) of known solid lubricants, looking for one or more *properties* that would explain low friction. Failing to find a common property, we then seek a *mechanism* for low friction behavior and test the mechanism against a variety of parameters. Section 3 demonstrates that a general relationship exists between the friction coefficient, the contact pressure and a "shear strength" parameter. But we also discover that this shear strength parameter is not necessarily an intrinsic property of the material, e.g., it can be controlled by the atmosphere. Pursuing a more detailed picture of the material aspects of this parameter, Section 4 focuses on microscopic behavior of low friction interfaces. First, rheological behavior of a sliding interface is described; in particular, three modes that allow sliding motion to be accommodated wholly within the interface are presented. Then, three microanalytical studies are presented as evidence that interfacial films can have significantly different chemistry and structure than the original solid lubricant material. Finally, a wear model that accounts for the chemical and structural aspects of interfacial film formation is introduced. One of the conclusions inferred from these investigations is that the lubricity of coatings is determined by the interfacial films generated during sliding, not by the bulk shear properties of the coatings themselves.

2. Solid lubricating materials

Liquid lubricants allow heavily-loaded counterfaces to roll or slide over each other with minimum tangential resistance. Any solid material which performs the same function can be considered a solid lubricant. A solid lubricant, therefore, should ensure that the two counterfaces separate in the vicinity of the lubricant (and not inside one of the counterfaces!) and do so with the lowest friction coefficient possible. For example, friction coefficients less than 0.3 would minimize surface damage, based on contact mechanics. [See, for example, chapter by T. Childs in this book.]

The most common materials used as solid lubricants have lamellar (layered) structures based on hexagonal lattices, graphite and MoS_2 being the best known examples. In graphite, planes of σ (covalently) bonded carbon atoms give high in-plane strength, whereas the planes themselves are coupled by weaker π (van der Waals) bonds, which allow easy interplanar shear. A similar bonding picture holds for the dichalcogenide MoS_2. Covalent bonds join sulfur and molybdenum atoms in planar arrays of hexagonal S-Mo-S "sandwiches." The sandwiches are joined at the sulfur atoms by van der Waals forces, allowing for low interplanar shear strength. Other dichalcogenides of transition metal sulfides, selenides and tellurides also behave as solid lubricants. However, not all

TABLE 1. MATERIALS USED AS SOLID LUBRICANTS	
LAMELLAR...............................	MoS_2, H_3BO_3, graphite, $(CF_x)_n$, CdI_2
OXIDES....................................	CdO, TiO_{2-x}, $Co(ReO_4)_2$
HALIDES..................................	CaF_2, BaF_2, LiF_2, PbI_2, $CuCl_2$
CHALCOGENIDES......................	PbS, Sb_2S_3, CdSe, HgS, As_2S_3
SOFT METALS...........................	Pb, Ag, Au, In, Sn
GLASSES..................................	B_2O_3, $PbO \cdot SiO_2$
DIAMONDLIKE CARBON (DLC)...	i-C, a-C:H, a-C:Si
ORGANICS...............................	stearic acid, soaps, waxes
POLYMERS...............................	PTFE (e.g. Teflon), polyimide

amellar structures give low friction; for example mica gives a relatively high friction coefficient ($\mu \geq 1$) [1] [also, see chapter by Israelachvili]. Nor is it necessary to have a lamellar structure to give low friction.

A wide variety of non-lamellar and even non-crystalline materials can be used as solid lubricants (see Table 1). One extreme case is the rather recently discovered amorphous, diamond-like carbon (DLC) coating[1]. Although known primarily as a hard coating, it provides some of the lowest friction coefficients ever measured ($\mu \leq 0.01$) [2,3]. Solid lubrication can also be achieved by a variety of treatments that rely on thermal and chemical conversion processes or energetic beams (ion, electron and laser beams) to alter surfaces. These materials and surface treatments and their technological application are described in a variety of books and articles [4,5,6,7]

The ability of very thin surface films to influence friction has been known for many decades. In 1950, Bowden and Tabor [8] showed that air-formed oxide films, only 1-2 nm thick and often contaminated by a thin carbonaceous layer, protected metal substrates from almost certain seizure. Monolayer films of fatty acids, about 3 nm thick, afford similar protection and even lower friction coefficient ($\mu \approx 0.1$) [9] [See chapter by Timsit]. Submonolayer films on solid surfaces can also reduce friction coefficient considerably. Buckley [10] and colleagues at NASA (Lewis) were perhaps the first to routinely apply modern surface analytical methods to identify the thin surface films that control friction. For example, Miyoshi and Buckley [11,12] demonstrated that heat treatments can convert surfaces exhibiting high friction to low friction and identified

---

[1] DLC coatings are sometimes called i-C coatings, because they are produced with the assistance of ions, or a-C:H or a-C:Si, to indicate that the coatings are amorphous and contain H or Si, respectively.

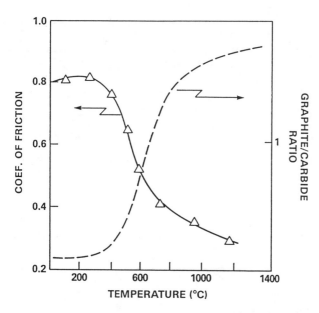

Fig. 1. *The friction coefficien[t] and chemical state of carbon fo[r] SiC in vacuum as a function o[f] temperature. Chemica[l] assignments were obtained from the C(1s) photoelectro[n] spectrum. [From Miyoshi an[d] Buckley, Ref. 12]*

a variety of surface chemical changes (e.g., surface segregation, bond dissociation, etc.) responsible for the behavior. One classic study was on SiC in vacuum. As depicted i[n] Fig. 1, the friction coefficient dropped from about $\mu=0.8$ to $\mu=0.2$ as the temperatur[e] rose from 400 °C to 800 °C [12]. In situ XPS analysis revealed that concurrently graphitic carbon replaced carbidic carbon as the top-most layer, as a result of th[e] evaporation of Si. This result illustrates the efficacy of molecularly-thin lubricatin[g] films, as well as the value of the surface analytical approaches in friction investigation [10]. More generally, these surface analytical studies provide convincing evidence tha[t] solid lubrication can be more than a "bulk" material property. However, before w[e] address these microscopic approaches to understanding friction, a review will be give[n] of the traditional, macroscopic investigations of friction behavior.

3. Mechanism of lubrication by thin solid films

The Bowden-Tabor model of metallic friction provides a good starting point fo[r] understanding how thin solid films can reduce friction [13]. As discussed elsewher[e] in this book, the frictional force can be expressed as

$$F = A \cdot S + F_p \tag{1}$$

where A is the area of contact, S is the shear strength of the interface, and F_p is th[e] plowing term. To reduce friction, the shear strength, the area of contact and the plowin[g] contribution must be minimized. Fig. 2 illustrates how this can be achieved: a har[d]

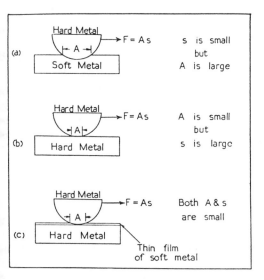

Fig. 2. *A schematic of the contact between two metal surfaces, illustrating how a thin soft film reduces friction. [From Bowden and Tabor, Part 1, p. 112]*

substrate reduces both the area of contact and the penetration (the latter minimizes plowing) while a soft film reduces the shear strength [14].

The friction coefficient, μ, is obtained by dividing both sides of Eq. (1) by the normal load, L. For coatings sufficiently thin that the plowing term can be ignored, the friction coefficient is given by

$$\mu = F/L = A \cdot S/L = S/P \tag{2}$$

where P is the mean pressure of the contact. This formula states that the friction coefficient depends independently on two terms: the shear strength and the pressure. In the following subsections, this model is tested against observed effects of coating parameters (thickness and substrate) and running conditions (load, temperature, velocity and atmosphere).

3.1. EFFECT OF FILM THICKNESS

Dependence of the friction coefficient on the thickness of solid lubricating films is represented schematically in Fig. 3 [15,16,17]. The friction coefficient is a minimum for an optimum film thickness (about 1 μm in the above example) but increases for thinner and thicker films. In the thin film limit, asperities of the substrate break through the film and the metal-metal contacts contribute to the friction coefficient. In the thick film limit, the friction coefficient is increased by plowing of the film. The general behavior can be described by the formula [18,19]:

$$\mu = \beta \, (S_s/P_s) + (1-\beta) \, S_f/P_{f,s} \tag{7}$$

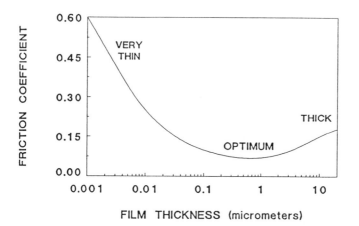

Fig. 3. *The effect of film thickness on the friction of metallic films.*

where the subscripts s and f refer to the substrate and the film. The first term gives the area fraction β contributing to substrate interactions; the second term, the contribution of the film/substrate combined. In the latter term, the pressure can range from a value equal to the hardness of the substrate alone, for thin films, to the hardness of the film itself, for thick films. More detailed models of the variations in friction coefficient with thickness for thick films can be found in the literature [20,21].

3.2. EFFECT OF SUBSTRATE

The substrate influences the friction coefficient by regulating the area of contact (hence the pressure) through its elastic or plastic properties; this is discussed in the next paragraph. The substrate can also play an important role in the endurance of a solid lubricating film. If the film cannot bond to the substrate, its endurance will be low. The endurance of a MoS_2 coating on a Ti alloy substrate, for example, can be 3-4 orders of magnitude lower than the same coating on steel [22] because of poor interfacial bonding [23]. Other examples will be given later.

3.3. EFFECT OF LOAD

Although Amontons' Law (i.e., the friction coefficient is independent of load) holds for many combinations of materials [24], a load-dependent friction behavior is common for many solid lubricants such as polymers [25,26], and thin, inorganic coatings on hard substrates [27,28,29,30]. Under elastic contact conditions, the friction coefficient often decreases with increasing load, as illustrated in Fig. 4 for In (a soft metal) on steel and for MoS_2 on Cr [4]. A model that has been used to explain this load dependence is the Hertzian contact model [26,27,28,29]. The friction coefficient is derived from Eq. (2). The shear strength S of solids at high pressures has been observed to have a pressure dependence [31] approximated by

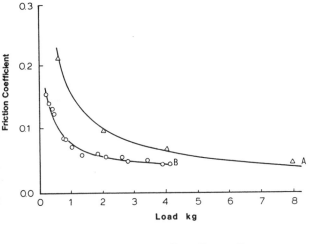

Fig. 4. *Effect of load on the friction coefficient for lubricating films of A) In on mild steel and B) MoS₂ on Cr. [From Peterson and Ramalingam, Ref. 4]*

$$S = S_o + \alpha P \qquad (3)$$

where α represents the pressure dependence of the shear strength. Putting Eq. (3) into Eq. (2) gives the following expression for the friction coefficient:

$$\mu = (S_o/P) + \alpha. \qquad (4)$$

For non-conforming surfaces (i.e. concentrated contacts), the Hertzian pressure, P_H, for elastic contacts varies as L^n, where $n=1/3$ for a circular contact and $n=1/2$ for a cylindrical (line) contact [32] [also, see Appendix by Johnson in this book]. Thus, the friction coefficient under Hertzian contact varies as

$$(\mu - \alpha) \propto S_o L^{-n}, \qquad (5)$$

Evidence for the behavior suggested by Eq. (5) is shown in Fig. 5, a log-log plot of the adjusted friction coefficient $(\mu - \alpha)$ vs load for five sets of MoS₂ friction data taken mostly from the literature. In the two cases where ball-on-flat geometries were used [33,34], an $L^{-1/3}$ dependence is obtained; in the other three cases, with cylinder-on-flat geometries [35,36,37], an $L^{-1/2}$ dependence is found. Two solid lines, with $L^{-1/3}$ and $L^{-1/2}$ dependencies, are drawn for reference. Hence, the Hertzian contact behavior represented by Eq. (5) accurately describes the load dependent friction coefficients for MoS₂.

Although Karpe [33] verified the Hertzian contact model in 1965, alternative explanations for the load dependence of friction were also presented [35,38] and, until recently [39,40], the Hertzian contact model has not been used to analyze MoS₂ coatings.

Several years ago, the full pressure-dependent friction behavior described in Eq. (4) was examined by analyzing a large number of friction measurements as a function of all

244

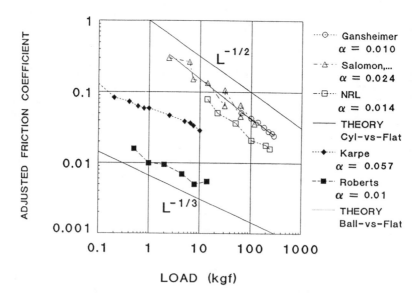

Fig. 5. *Adjusted friction coefficient (μ - α) vs load for five sets of MoS₂ friction data taken from the literature (see text for details).*

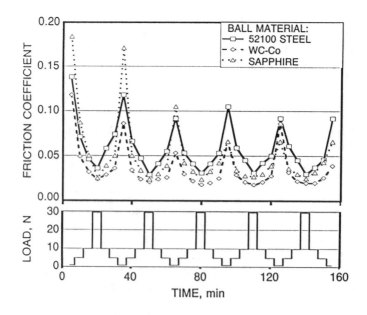

Fig. 6. *Friction coefficients (steady state) and loading schedule vs time for three ball materials against MoS₂-coated 440C steel. [From Singer, et al. Ref. 41]*

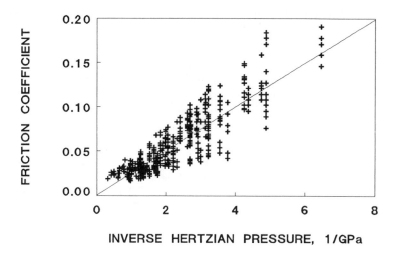

Fig. 7. *Friction coefficient vs inverse Hertzian pressure. Straight line is the least squares fit of 629 friction coefficients using Eq. 4.*

three independent variables: elastic modulus, ball radius and load [41]. According to Hertzian theory [32,42], the friction coefficient, for smooth balls and flat substrates loaded below the elastic limit (typical of bearing design), depends on the three variables as

$$\mu = S_o \, \pi \, (3R/4E)^{2/3} \, L^{-1/3} + \alpha, \tag{6}$$

where E is the composite elastic modulus of the couples, R is the radius of the ball, and L is the load. Bearing-grade balls were slid against sputtered MoS_2-coated substrates in dry air under loading conditions depicted in the lower part of Fig. 6. Steady-state friction coefficients were recorded as a function of load, starting with a low load (1 or 2 N), then incrementally raising and lowering the load between 1 and 30 N (always below the elastic limit of the couple). Friction coefficients for 1/2" balls of glass, steel and WC:Co against one of the MoS_2-coated 440C steel substrates are shown in the upper part of Fig. 6. These curves show clearly that the friction coefficient varied inversely with load. In addition, for a given load, the steady-state friction coefficient decreased as the elastic modulus increased.

Nearly 630 data points like those in Fig. 6 were subjected to power law regression analysis. Friction coefficient vs load data exhibited the $L^{-1/3}$ behavior expected for the ball-on-flat geometry. In addition, least squares fits validated the dependency of μ on R and E found in Eq. (6). Fig. 7 shows a plot of μ vs $1/P_H$ for the combined data. The best-fit straight line, also shown, gave a mean slope (S_o) of 24.8±0.5 and an intercept (α) of α=0.001±0.001 MPa. These values compare favorably with friction data for MoS_2-coated steel reported in the literature (in Fig. 5) and with the measured bulk shear strength (S_o=38 MPa) of fully dense MoS_2 [43].

Table 2. Shear strength measurements on solid lubricating films. (S = S$_o$ + α P)

FILM	SHEAR STRENGTH		REF.
	S$_o$(MPa)	α	
Sputtered MoS$_2$ (in dry air)	25	0.001	Singer et al. [41]
DLC (in dry air)	25	0.01	NRL [44]
B$_2$O$_3$ (in 50% RH air)	23	0.006	Erdemir et al. [45]
Sputtered MoS$_2$ (in vacuum)	7	0.001	Roberts [46]

Friction measurements performed under Hertzian contact conditions can be used to determine S$_o$ and α values. Table 2 lists several sets of S$_o$ and α data obtained recently by load-dependent friction measurements on thin, vapor-deposited solid lubricating coatings. It appears that MoS$_2$ [41], DLC [44] and hydrated B$_2$O$_3$ [45] coatings have comparable S$_o$ and α values in air, but MoS$_2$ has a lower S$_o$ value in vacuum [46]. These S$_o$ values are more than an order of magnitude larger than for stearic acid (on glass) [28] [also see chapters by Briscoe, Timsit and Adams in this book], but the values for α, which represents the lowest attainable friction coefficient, are roughly an order of magnitude lower. Although reliable numbers for S$_o$ and α can be obtained, it is not clear whether these numbers are the shear strength of the coatings or of the interface itself; this will become apparent as we look at the effects of running conditions (temperature, velocity and atmosphere).

3.4. EFFECT OF TEMPERATURE

Friction coefficients of many solids decrease with increasing temperature. This behavior is consistent with the increased softening of materials at high temperatures. Fig. 8 depicts the friction coefficient and tensile strength of Ag (powder) as a function of temperature [47]. A definite correlation of friction coefficient and bulk strength can be seen. Measurements performed by Peterson and Ling [48] on a variety of metals at one-half their melting points also show that the friction coefficient varies linearly with tensile strength. For some materials, however, the friction coefficient decreases or increases dramatically at certain temperatures [49]. Marked decreases in friction coefficient with increasing temperature may indicate a change from solid to liquid lubrication, as exemplified by the drop in the friction coefficient of boric oxide from 1.4

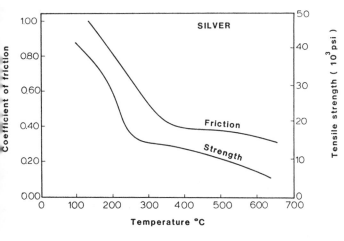

Fig. 8. *Friction coefficient and tensile strength of Ag (powder) as a function of temperature [From Peterson and Ramalingam, Ref. 47].*

to 0.1 over the temperature range 450 to 650 °C [50]. Sudden increases in friction coefficient generally indicate coating failure, attributable to oxidation (e.g. MoS$_2$ at above 500 °C) or decomposition (polymers - see previous chapter), or to poor coating adhesion.

3.5. EFFECT OF VELOCITY

In general, the friction coefficient of both metals and non-metals in sliding contact decreases with increasing speed. The decrease is attributed to frictional heating [51].

3.6. EFFECT OF ATMOSPHERE

Gases, or the absence of gases, can have a profound effect on the friction and wear behavior of solid surfaces in general [see Fischer's chapter] and solid lubricants in particular. An example of the effects of atmosphere on the friction coefficient is shown schematically in Fig. 9. for four well-known, low friction materials. Two of the materials, MoS$_2$ [22,52] and DLC [2,3] coatings, have very low friction coefficients (μ=0.01-0.02) in vacuum but much higher friction coefficients (μ=0.2) in air. The other two, graphite [53] and diamond [54], display the opposite behavior. The contrast in friction behavior is more striking when one recognizes that both MoS$_2$ and graphite have easily-sheared, lamellar structures, which intrinsically *should* give low friction. The atmospheric effects, attributed primarily to moisture (or hydrogen, in the case of diamond), demonstrate the importance of surface chemistry in friction processes. But the chemistry cannot be simple; otherwise, the three forms of carbon -- graphite, diamond and DLC -- would be expected to exhibit similar behavior.

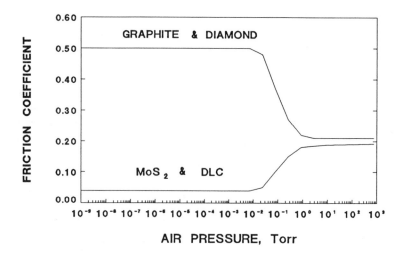

Fig. 9. *Schematic of effect of atmospheric air on the friction coefficient of four materials. [After Buckley, Ref. 10, p. 574.]*

3.7. EFFECT OF REPEATED PASSES

The endurance of a solid film, especially a thin solid film, depends on many of the above parameters such as load, atmosphere and type of substrate. Some examples are given here:

- MoS$_2$ coatings, both burnished [55] and ion-beam sputtered [56], survive 100-1000 times longer in dry air than in moist air.
- Monolayers of fatty acids have 1 to 2 orders of magnitude higher endurance on glass than on steel [57] [See, also, the chapter by Timsit]
- MoS$_2$ coatings can last twice as long on ceramics as on steel [22,58].

In the first case, the loss of endurance in moist air is due to a change in wear mode of the coatings. In the latter two cases, wear of the substrate contributes to the reduced endurance of the coatings on steel.

Moreover, there is no general trend of endurance with friction coefficient. Thin MoS$_2$ coatings, for example, show decreasing endurance as the friction coefficient decreases (i.e., as the load increases). Generally, though, durable solid lubricating coatings can have extremely low wear rates, less than a fraction of an atomic layer per pass. Hence, coatings no thicker than 1 μm may survive a million or more passes.

4. Microscopic aspects of solid lubrication

We have seen that substrates coated with *selected* solid lubricants can give low friction in the *proper* atmosphere. To understand this lubrication process, it is necessary to understand:

- how the coating accommodates to sliding, and
- how an interfacial film is formed by the rubbing action of the coating against the counterface in a gaseous atmosphere.

These issues are addressed in this section.

4.1. RHEOLOGY OF THE INTERFACE

Godet and co-workers [59,60] have discussed the many ways (both "where" and "how") that a "third body" can accommodate relative motion at the interface of two solids. Here, following more simplified considerations described by Peterson et al. [4,39,61], we illustrate in Fig. 10 three possible ways that sliding is accommodated between a counterface and a coating (film)-covered substrate:

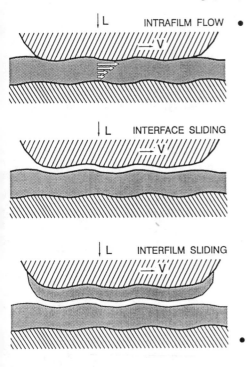

Fig. 10. *Schematic representation of three ways that sliding can be accommodated between a counterface and a film-covered substrate. [After Ives and Peterson, Ref. 61]*

- <u>Intrafilm flow</u> occurs when the film adheres strongly to both surfaces and flows like a viscous fluid (Couette flow, see chapter by Dowson) to accommodate relative displacements of the two surfaces. Although normally associated with noncrystalline materials, viscous-like shear flow can also take place with crystalline materials at relatively high temperatures. The friction behavior of Ag, shown in Fig. 8, would be such an example; the correlation of shear strength with friction coefficient clearly indicates that the Ag undergoes plastic flow to accommodate motion, and the shear strength of the film itself determines the friction coefficient. Platelets of lamellar materials can also exhibit flow by intercrystalline slip [62,63].

- <u>Interface sliding</u>, which allows the two original surfaces to slide over one another, would occur only if there were no adhesion to the rider. This has not been demonstrated conclusively in macroscopic sliding tests, but may occur under low load, single asperity contacts, accessible by methods like atomic force microscopy.

- <u>Interfilm sliding</u> occurs when the lubricant adheres to both surfaces but separates into two distinct films (at least for a moment or so) that slide across one another. In this case, neither of the two original surfaces is in contact. The shear strength is, therefore, that of the two films sliding against one another.

Real solid lubricants often undergo both flow and sliding as they accommodate to the normal and shear stresses by thinning or breaking apart. Sliding increases the complexity of interfacial rheology by exposing the two interfaces to the surrounding atmosphere; this is the subject of the next section. However, it is clear that the "S_o" term in Eq. (4) need not represent the "shear strength" of a bulk material, nor that of a "third body;" it might represent a property of a film/gas/film interface. Until this issue is resolved, a more accurate term for "S_o" might be the "velocity accommodation parameter."

4.2. INTERFACIAL FILMS

In this section, we describe the "third body" found at sliding interfaces and, working backwards, model the mechanisms by which these products are formed. First, results from recent studies of DLC and MoS_2 coatings and Ti^+-implanted steel are presented. Then, a model of transfer film formation is proposed and analyzed in terms of thermochemical processes.

4.2.1 *DLC coatings.* Carbon coatings in general, and DLC coatings in particular, exhibit a wide range of friction and wear behavior depending on atmospheric conditions. In this section, we present evidence that the atmospheric effects can be related to tribochemical reactions among the counterface, coating and the gases.

Sugimoto and Miyake [64] have demonstrated that DLC coatings (a-C:Si) rubbed against a steel ball in vacuum can give friction coefficients $\mu \leq 0.01$. In order to identify the interfacial products formed, a test was interrupted when the friction coefficient reached 0.007 and infrared (IR) microscopy was used to analyze all interface surfaces. They observed that the rubbed coating and the debris in the track had IR features characteristic of the original DLC material. However, the film that transferred to the ball showed very different features than the DLC (see Fig. 11), indicating selective transfer of an sp^3 hydrocarbon film with C=C bonds (at 1630 cm^{-1}) and no Si-H bonds. In addition, polarized IR spectra (not shown) of the transfer film revealed that the hydrocarbon molecules were oriented along the sliding direction. Thus, an adherent, oriented, aliphatic-type hydrocarbon transfer film was formed tribochemically during sliding in vacuum, and the low friction coefficient resulted from interfilm sliding between it and the parent DLC coating.

Kim et al. [65] have observed that DLC coatings run at atmospheric pressure exhibit distinct friction and wear behaviors, depending on the gases present. An IR microprobe was used to identify the debris and film structures observed on a variety of interfacial films formed by a Si_3N_4 ball rubbed against DLC-coated silicon under dry Ar, dry air

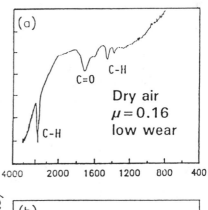

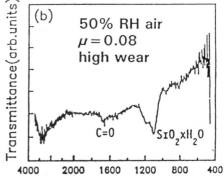

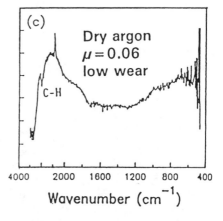

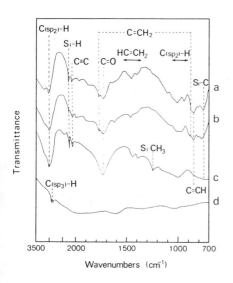

Fig. 11. *IR microprobe spectra from Si-containing DLC film: (a) unrubbed film, (b) rubbed trace, (c) wear debris, (d) film transferred to the ball surface. [From Sugimoto and Miyake Ref. 64]*

Fig. 12. *IR microprobe spectra from debris in tracks of DLC films rubbed against Si_3N_4: (a) 0% RH air; (b) 50% RH air; (c) 0% RH argon. [From Kim et al Ref. 65]*

and moist air conditions. Spectra of debris generated under these three conditions are shown in Fig. 12:

(a) In dry air, which gave relatively high friction but low wear, the transfer film and debris were a carbonyl compound formed tribochemically by oxidation of hydrocarbons in the DLC coating.

(b) In moist (50% RH) air, lower friction but high wear resulted from a transfer film composed of oxidized hydrocarbon and hydrated silica (from the ball).

(c) In dry Ar, a transfer film of the original DLC covered the ball. The lowest friction coefficient and lowest wear resulted from a DLC transfer film sliding against the DLC surface.

Thus we find that the range of tribological behavior of DLC coatings in the four different atmospheres can be attributed to tribochemical reactions, which result in distinctly different transfer films. Nonetheless, it is assumed from the low friction (and reasonably long life) that interfilm sliding was the velocity accommodation mode, but this has not been proven. Further studies on the chemical basis for the magnitude of the velocity accommodation parameters of this system are encouraged.

4.2.2. MoS₂ coatings. Sliding tests between a steel ball and an MoS_2-coated steel flat were performed in a vacuum chamber equipped with an Auger analyzer. Auger analysis indicated that some MoS_2 transferred to the steel ball and that the surfaces of the interfacial films had the identical compositions [66]. Subsequent friction and wear behavior of MoS_2 coatings on steel was investigated (in a 4-ball tester) as a function of gas (Ar or air) and ball material (steel, WC:Co and sapphire) [58,67]. In all cases, well-defined transfer films were observed in the area of contact (roughly the Hertzian diameter) of initially uncoated balls, with compacted and loose debris observed outside this area. These features can be seen in the optical micrograph, shown in Fig. 13, of a film formed on WC:Co rubbed in Ar for 10 minutes. Transfer films like those in the figure were then analyzed by transmission electron microscopy (TEM) and Auger sputter depth profiling. Specimens for TEM analysis were obtained by stripping debris from the

Fig. 13. *Optical micrographs of the contact area on an uncoated WC:Co ball rubbed for 10 minutes in Ar against MoS₂-coated steel. [From Fayeulle, et al. Ref. 67]*

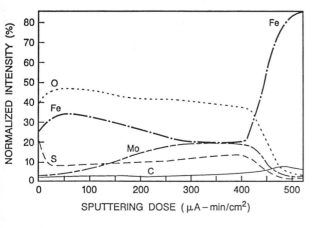

Fig. 14. *Auger sputter depth profile of a debris particle on the contact spot of a steel ball after 10 minutes sliding in dry air against MoS₂-coated steel. [From Fayeulle, et al. Ref. 67]*

contact zone.

An Auger sputter depth profile of a debris particle from the contact spot on a steel ball rubbed in air for 10 minutes is shown in Fig. 14. This debris appears to be a layered mixture of Fe oxide and (Fe,Mo) oxide containing S. The O and S profiles complement one another, suggesting that the S was being replaced by O during oxidation of MoS₂. The profiles of both Fe and O suggest the Fe and O diffused through the film. In general, Auger analysis indicated that MoS₂ transfer films and the interface between transfer films and the ball were being oxidized or converted to Fe-Mo oxides. But with Auger analysis, the best one can do is speculate on the phases being formed.

TEM, however, was able to identify the phases of the interfacial film detected in the contact zone. The phases, listed in Table 3, are clearly products of complex tribochemical reactions: four of the five phases are reaction products of MoS₂ with O and Fe or Co, and three of the four products are ternary compounds. In addition, TEM showed that the MoS₂ and the CoMoO₃ crystallites were textured: their basal planes were oriented parallel to the interface.

Table 3. Phases identified by electron diffraction of debris stripped from balls slid against MoS₂-coated steel.

RIDER	PHASE
steel	MoS_2, $FeMoO_4$, Fe_2MoO_4, MoO_3
WC:Co	MoS_2, $CoMoO_3$, MoO_3

4.2.3. Ti-implanted steels. Ti⁺ or (Ti⁺+C⁺) implantation into steel can lower the friction coefficient from 0.6 to 0.2 and substantially reduce wear [68]. Auger analysis of low friction surfaces show a thin, mixed (Fe,Ti) oxide on top of a carburized Ti-rich subsurface layer. TEM studies found that transfer films generated by sliding on implanted surfaces yielded an α-Fe_2O_3/$FeTiO_3$ phase with a thin, plate-shaped morphology [69]. By contrast, sliding on nonimplanted surfaces produced an Fe_3O_4 phase with curled or spherical morphology. An explanation for these results will be presented later.

4.3. INTERFACIAL FILM FORMATION

Oxides and oxide films on solids act as solid lubricants in dry sliding contact [70] [also see chapter by T. Fischer]. In this section, we examine a mechanism for generating oxide films and the thermochemical basis for oxide film formation.

4.3.1. *Model for interfacial film formation.*
As suggested earlier, sliding against an adherent solid lubricating coating probably begins with some intrafilm flow or interface sliding, while loose material is pushed out of the way and the coating becomes thinner. After run-in, a steady state is reached in which interfilm sliding dominates. The model, shown schematically in Fig. 15, has been proposed to describe steady state wear via transfer film and debris formation [71]. First (left), a thin layer is removed from the coating and transfers to the rider. This could be, for example, the oxidized surface of MoS_2. Next (center), the transfer layer reacts with surrounding gases (e.g. O_2) and the rider material, M (e.g. Fe or Co). Although solid-phase reactions are not normally expected to reach equilibrium at low temperatures (e.g. below 100 °C), thin transfer layers could react to completion because diffusion lengths are small (only several atomic layers thick) and defects would enhance diffusion. Finally (right), as more layers accumulate and the transfer film thickens, debris particles break and fall onto the wear track. In the following sections, we ignore the mechanical aspects of the wear process and restrict our attention to the gas and solid phase chemical reactions.

4.3.3. *Thermochemistry and Film Formation.*
According to the above model, a transfer film that forms on steel sliding against MoS_2 in air should consist of the stable reaction products of MoS_2, O_2 and Fe. The stable phases expected for Mo-S-Fe-O reactions can be found in equilibrium phase diagrams obtained experimentally [72] or by

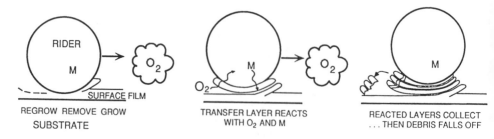

Fig. 15. *A schematic of the wear process observed for the three low friction, low wear treatments during sliding in air. [From Singer, Ref. 71]*

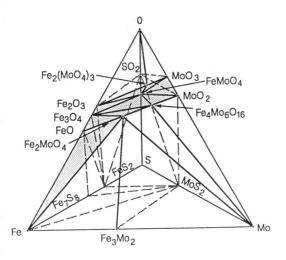

Fig. 16. *Simplified quaternary phase diagram of Mo-S-Fe-O. [From Fayeulle, et al. Ref. 67]*

thermochemical calculations based on the second law of thermodynamics [73]. Once found, the results can be presented graphically in the form of an isothermal section of a quaternary phase diagram [74].

A simplified quaternary phase diagram for the Mo-S-Fe-O system [67] is shown in Fig. 16. The diagram was constructed from four ternary diagrams: three oxide ternaries were attached to the edges of the Mo-S-Fe ternary diagram, then "folded up" to form a pyramid. The Fe-Mo-S diagram was derived from experiments at T=600 °C [75], while the three remaining ternaries were calculated according to the method suggested by Beyers [76]. The compounds that remain stable (co-exist) in one-another's presence are connected by "tie-lines" and "tie-planes." The tie-lines in the Mo-S-Fe ternary diagram, at the base of the pyramid, show that Fe co-exists with MoS_2. However, with oxygen present, many reactions are possible. MoS_2 can oxidize to MoO_2 and ultimately to the more stable oxide MoO_3, and Fe can oxidize to any of the three stable oxides [77]. These Mo and Fe oxides, in contact with one another, can also react to form the four ternary phases seen in the Fe-Mo-O ternary diagram. Ultimately, all possible quaternary compositions generated within this contact will oxidize to one of the phases in the shaded region of the diagram. The quaternary phase diagram accounts for the three oxide compounds observed by TEM (Table 3) -- MoO_3, $FeMoO_4$, Fe_2MoO_4) -- and for the non-stoichiometric mixtures found by Auger depth profiles (Fig. 14) in transfer films. Similar thermochemical arguments have been used to account for the adhesion of MoS_2 coatings to oxidized Fe surfaces and the "release" of debris from the interface after reactions have gone to completion [67].

4.3.2. *Thermochemistry, Film Formation and Low Friction.* Thermochemistry can also provide an explanation for the friction processes accompanying wear. If wear occurs by the removal of a thin surface layer, then the interface between the layer and the substrate may be an easily sheared plane. MoS_2 is a well-known example of a material with easy

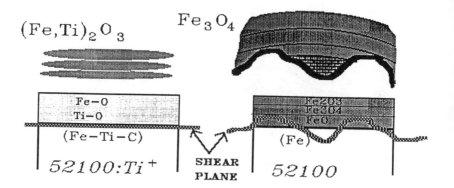

Fig. 17. *Schematic of two types of wear debris generated by differences in the shear planes of Ti+-implanted and nonimplanted 52100 steel.*

shear planes, located along the S-S bonds oriented parallel to the basal planes. Less well understood is the mechanism of easy shear of thin oxide films that leads to the low friction of TiN [78] and TiC [69,79] coatings or the Ti+-implanted steel mentioned above [69,80].

A possible explanation is that oxide film growth is diffusion limited, and the film that forms -- albeit thin -- attains a single, stable phase. Then, oxide and substrate phases that co-exist at the interface, like TiO_2 on TiC, should exhibit minimal driving force for adhesion. In Ti+-implanted steel, the TiC-rich interface might act as a diffusion barrier to hinder interdiffusion of Fe and O between the surface oxide and the substrate. The oxide and substrate phases would then co-exist and have minimal adhesion. By contrast, nonimplanted steels continue to oxidize and would form an unstable, graded oxide layer (Fe_2O_3 on Fe_3O_4 on Fe).

The low energy, oxide/substrate interface of Ti+-implanted steel is believed to be an easy shear plane, from which stable oxides are stripped, as depicted in the left panel of Fig. 15. The easy shear plane would give low friction and account for the fully-oxidized phase, labeled α-$(Fe,Ti)_2O_3$, and its plate-like morphology (see Fig. 17). By contrast, the lack of a well-defined shear plane in steel would result in a meandering shear plane, with transfer films that contain unoxidized Fe from the substrate. Partial oxidation of the underlying metal could result in a substoichiometric oxide like Fe_3O_4 and stress-induced curling of the transfer film, both features characteristic of debris from nonimplanted steel.

5. Summary and Conclusions

This chapter has presented friction behavior from a materials perspective. Macroscopic behavior and microscopic analysis and models were combined to give a tentative physical and chemical basis for the friction behavior of thin solid films. We may conclude that:

- Low friction coefficients can be achieved with a variety of materials, not just the most obvious lamellar materials. Surface analytical investigations confirm that films as thin as one monolayer can provide solid lubrication.

- The Bowden-Tabor model of friction applied to thin solid films on hard substrates can account for much of the observed behavior. In addition, the model can be made quantitative in cases where the contact is elastic and obeys Hertzian mechanics, and the film is sufficiently thin. In these cases, Amontons' Law is not obeyed.

- Microscopic examination of the interface after sliding contact on long-lived, low friction coatings suggest:
  - transfer films form during sliding contact;
  - transfer films control both friction and wear;
  - tribochemical reactions produce thermochemically stable phases at or near room temperature.

- The "shear strength" term, S_o, is a velocity accommodation parameter of the sliding interface and may not necessarily represent the "shear strength" of the original coating or of the transfer film itself.

Future issues that have been raised in this chapter include:

- How do gas-surface reactions alter the velocity accommodation parameter (and wear behavior) of sliding contacts?

- How can we predict the magnitude of the velocity accommodation parameter ("shear strength") from the properties of interfacial films?

- What are the mechanical processes that create the quasi-stable transfer layers represented in Fig. 15?

- How can we tailor the formation of "lubricating" transfer films to reduce friction and minimize wear?

Acknowledgement

The author thanks his major collaborators, R. Bolster, S. Fayeulle, P. Ehni, and J. Wegand for their contributions, and M. Peterson for ongoing advice and encouragement. This work was made possible by funds from DARPA, SDIO and ONR.

References

1. Bowden, F.P. and D. Tabor, The Friction and Lubrication of Solids Part 2, (Clarendon Press, Oxford, 1964) p. 201.

2. Enke, K., H. Dimigen, and H. Hubsch, Appl. Phys. Lett. 36 (1980) 291; Enke, K, Thin Solid Films 80 (1981) 227.

3. Miyake, S., S. Takahashi, I. Watanabe, and H. Yoshihara, ASLE Trans. 30 (1987) 121.

4. Peterson, M.B. and Ramalingam, S., in Fundamentals of Friction and Wear of Materials, edited by D.A. Rigney (A.S.M., Metals Park OH, 1981), p. 331.

5. Dorinson, A. and Ludema, K.C., Mechanics and Chemistry in Lubrication (Elsevier, Amsterdam, 1985). Chapter 19.

6. Singer, I.L. in New Materials Approaches to Tribology: Theory and Applications, edited by L. Pope, L. Fehrenbacher and W. Winer, MRS Symposium, 140 (MRS, Pgh. PA, 1989) p. 215.

7. Bhushan, B. and Gupta, B.K., Handbook of Tribology (McGraw-Hill, New York, 1991). Chapters 5 and 13.

8. Bowden, F.P. and D. Tabor, The Friction and Lubrication of Solids Part 1, (Clarendon Press, Oxford, 1950), Chapter VII.

9. Bowden and Tabor, Part 1, Chapter IX; Bowden and Tabor, Part 2, Chapters XVIII and XVIX.

10. Buckley, D.H., Surface Effects in Adhesion, Friction, Wear, and Lubrication, (Elsevier, New York, 1981). Chapters 8 and 9.

11. Miyoshi, K., and D.H. Buckley, Wear, 110 (1986) 295.

12. Miyoshi, K., D.H. Buckley and M. Srinivasan, Am. Ceram. Soc. Bull. 62 (1983) 494.

13. Bowden and Tabor, Part 1, pp. 110-121; Bowden and Tabor, Part 2, pp. 158-185.

14. Bowden and Tabor, Part 1, p. 112.

15. Bowden and Tabor, Part 1, p. 115.

16. Spalvins, T. and B. Buzek, Thin Solid Films 84 (1981) 267.

17. Shafei, T., R.O. Arnell and J. Halling, Trans. ASLE, 26 (1983) 481.

18. Bowden and Tabor, Part 1, p. 105.

19. Rabinowicz, E., Friction and Wear of Materials (Wiley, New York, 1965) Chapter 8.

20. Briscoe, B.J. and A.C. Smith, ASLE Trans., 25 (1982) 349.

21. Kato, S. K. Yamaguchi, E. Marui, and K. Tachi, J. Lube Technol., 103 (1981) 236.

22. Roberts, E.W. and W.B. Price, New Materials Approaches to Tribology: Theory and Applications, edited by L. Pope, L. Fehrenbacher and W. Winer, MRS Symposium, 140 (MRS, Pgh. PA, 1989) p. 251.

23. Stupian, G.W. and A.B. Chase, J. Vac. Sci. Technol., 14 (1977) 1146.

24. Bowden and Tabor, Part 1, p.98.

25. Bowden and Tabor, Part 2, p. 214.

26. Bowers, R.C., J. Appl. Phys. 42 (1971) 4961.

27. Bowers, R.C. and W.A. Zisman, J. Appl. Phys. 39 (1968) 5385.

28. Briscoe, B.J., B. Scruton and F.R. Willis, Proc. R. Soc. London A 333 (1973) 99; B.J. Briscoe and D.C.B. Evans Proc. R. Soc. London A 380 (1982) 389.

29. Shafei, T., R.O. Arnell, and J. Halling, Trans. ASLE, 26 (1983) 481.

30. Pope, L.E., J.K.G. Panitz, Surf. Coat. Technol. 36 (1988) 341.

31. Bridgeman, P.W., Proc. Amer. Acad. Arts Sci. 71 (1936) 387.

32. see K.L. Johnson, Proc. Instn. Mech. Engrs., 196 (1982) 363 and references therein for an overview of Hertz contact.

33. Karpe, S.A., Trans. ASLE 8 (1965) 156.

34. Roberts, E.W., Proc. Inst. Mech. E. Tribology -- Friction, Lubrication and Wear, Fifty Years On, (Inst. Mech. Eng., London, 1987) p. 503.

35. Gansheimer, J., Schmiertechnik, 11, 271 (1964) (in German); Trans. ASLE 8, 175 (1965) (discussion paper). Gansheimer, who obtained friction coefficient vs load curves similar to those presented in Fig. 1, contended that heating, produced by increased frictional stress at higher loads, drove off moisture from MoS_2, thus reducing the friction coefficient.

36. Salomon, G. and A.W.J. DeGee, Trans. ASLE 8 (1965) 176 (discussion).

37. Unpublished data from Hohman tests on sputtered MoS_2, R.N. Bolster, NRL, 1989.

38. Winer, W.O., Wear, 10 (1967) 422 and references therein.

39. Kanakia, M.D. and M.B. Peterson, "Literature Review of Solid Lubrication Mechanisms," Report No. BFLRF-213, July 1987 (AD A185010), (NTIS, Springfield VA 22161).

40. see Roberts and Price, Ref. 22; E.W. Roberts, Thin Solid Films 181 (1989) 461.

41. Singer, I.L., R.N. Bolster, J. Wegand and S. Fayeulle, B.C. Stupp, Appl. Phys. Lett., 57 (1990) 995.

42. See chapter by K.L. Johnson in this book.

43. Tyler, J.C. and P.M. Ku, "Fundamental Investigation of Molybdenum Disulfide as a Solid Lubricant," Report No. SWRI-RS-501, June 1967 (AD 818439/2ST) (NTIS, Springfield VA 22161).

44. data of J. Wegand, NRL, 1991, unpublished.

45. Erdemir, A., R.A. Erck and J. Robles, Surf. & Coating Technol., 49 (1991) 435.

46. Values calculated by author from data in Ref. 34.

47. Peterson and Ramalingam, Ref. 4, p. 342.

48. Peterson, M.B. and F.F. Ling, J. Lubrication Technol. 92 (1970) 535.

49. Peterson and Ramalingam, Ref. 4, p. 341.

50. Peterson, M.B., J.J. Florek and S.F. Murray, Trans. ASLE 2 (1960) 225.

51. Bowden and Tabor, Part 2, Chapter XXII.

52. Spalvins, T., J. Vac. Sci. Technol. A5 (1987) 212.

53. Bowden and Tabor, Part 2, p. 191.

54. Bowden and Tabor, Part 2, Chapter X.

55. DeGee, A.W.J., G. Salomon and J.H. Zaat, Trans. ASLE, 8 (1965) 156.

56. Bolster, R.N., I.L. Singer, J.C. Wegand, S. Fayeulle, and C.R. Gossett, Surface and Coatings Technol. 46 (1991) 207.

57. Cottington, R.L., E.G. Shafrin and W.A. Zisman, J. Phys. Chem. 62 (1958) 513.

58. Fayeulle, S., P.D. Ehni and I.L. Singer, Surf. & Coatings Technol. 41 (1990) 93.

59. Berthier, Y., M. Brendle and M. Godet, STLE Tribology Trans. 32 (1989) 490.

60. Berthier, Y., L. Vincent and M. Godet, Wear, 125 (1988) 25.

61. Ives, L.K. and M.B. Peterson, Fundamentals of High-Temperature Friction and Wear With Emphasis on Solid Lubrication for Heat Engines, edited by F.F. Ling (Indust. Tribology Inst., Troy NY, 1985) p. 43.

62. Holinski, R., ASLE Proceedings -- First International Conference on Solid Lubrication 1971 (ASLE, Park Ridge IL, 1971) p. 41.

63. Sliney, H.E., ASLE Trans., 21 (1978) 109.

64. Sugimoto, I. and S. Miyake, Appl. Phys. Lett., 56 (1990) 1868.

65. Kim, D.S., T.E. Fischer and B. Gallois, Surf. & Coatings Technol. 49 (1991) 537.

66. Ehni, P.D. and I.L. Singer, in New Materials Approaches to Tribology: Theory and Applications, edited by L. Pope, L. Fehrenbacher and W. Winer, MRS Symposium, 140 (MRS, Pgh. PA, 1989) pp. 245.

67. Fayeulle, S., I.L. Singer, and P.D. Ehni,in Mechanics of Coatings, Leeds-Lyon 16 Tribology Series, 17, edited by D. Dowson, C.M. Taylor, M. Godet (Elsevier, GB, 1990) p. 129.

68. Singer, I.L., Appl. Surface Science 18 (1984) 28-62 and references therein.

69. Fayeulle, S. and I.L. Singer, Materials Sci. and Engineering A115 (1989) 285.

70. Bowden and Tabor, Part 2, p. 41.

71. Singer, I.L., Surf. & Coatings Technol., 49 (1991) 474.

72. For example, as found in Phase diagrams for ceramists, edited by Roth et al. Vols. 1-5, (Am. Ceramics Soc., Columbus, Ohio, 1964 - 1981).

73. Kubaschewski, O. and C.B. Alcock, Metallurgical Thermochemistry, 5th edition (Pergamon Press, Oxford, 1979)

74. West, D.R.F., Ternary Equilibrium Diagrams, (Macmillan, New York, 1965).

75. Levin, E.M., C.R. Robbins, and H.F. McMurdie, "Phase diagrams for ceramists" Vol. 1, (Am. Ceramics Soc., Columbus, Ohio, 1964) Fig. 3970. p. 521.

76. Beyers, R., J. Appl. Phys. 56 (1984) 147; R. Beyers, R. Sinclair and M.E. Thomas, J. Vac. Sci. Technol. B2 (1984) 781.

77. Kubaschewski and Alcock, Ref. 73, pp. 380,381.

78. Singer, I.L., S. Fayeulle and P.D. Ehni, Wear, 149 (1991) 375; also in Wear of Materials 1991 ed. K.C. Ludema and R.G. Bayer (ASME, NY 1991) p. 229.

79. Fayeulle, S. and I.L. Singer, Proceedings of the 18th Leeds-Lyon Symposium of Tribology edited by D. Dowson, C.M. Taylor, M. Godet (Elsevier, GB, 1992).

80. Singer, I.L. and R.A. Jeffries, J. Vac. Sci. Technol., A1 (1983) 317.

Discussion *following the lecture by I L Singer on "Solid lubrication processes"*

J LARSEN-BASSE. I would expect that you would have iron hydroxides in the debris if steel were run in normal (with moisture) air.

I L SINGER. We have not seen iron hydroxides. But my collaborator, Serge Fayeulle, has identified a molybdenum hydroxide as one of the phases that forms when MoS_2 is run in moist air. I have not discussed hydroxide phases here, because we have not studied them thoroughly and it is more difficult to get thermodynamic data for the hydroxides than the oxides. Nonetheless, research on hydroxide phases is very important for understanding normal air environments.

J N ISRAELACHVILI. How should one apply the ternary phase diagram when you have solids in equilibrium with each other? Solids generally don't have enough time to equilibrate at low temperatures, and in the sliding contact presumably we're talking about phases which are forming in a very short time. Are you suggesting we should use high temperature phase diagrams, but that everything freezes rather quickly?

I L SINGER. That's one explanation: high temperature phase diagrams may account for phases created when sliding contact creates a "hot spot" and then everything quenches in. An alternative, described in section 4 of my paper, suggests that one doesn't have to evoke high temperatures; rather, since the wear takes place with such thin transfer layers that phase formation occurs in short times even at low temperatures.

J GREENWOOD. Can I thank you Irwin, for putting to rest my doubts of whether we can get a shear strength out of a friction force measurement. Your technique of varying independently the radius, load, and elastic modulus is what is required.

FRICTION WITH COLLOIDAL LUBRICATION

Jean-Marie GEORGES, Denis MAZUYER, Jean-Luc LOUBET and Andre TONCK

Laboratoire de Tribologie et Dynamique des Systemes U.A. C.N.R.S. 855,
ECOLE CENTRALE DE LYON
36, rue Guy de Collongues ECULLY 69130 FRANCE.

ABSTRACT The paper concerns   boundary lubrication in the regime of mild wear .
The understanding of friction processes has greatly advanced as a result of combining contact mechanics with more detailed studies of the lubricant substances present at interface and the rheology of thin films .
We analyze the different physicochemical processes induced by lubricants, the principles of the formation of friction films, and we look for the relevant mechanical properties of the friction films.
New experimental techniques are described for measuring both the static and dynamic interactions of very thin lubricant films between two very smooth surfaces as they are moved normally or laterally relative to each other. A large range of the contact pressure range is studied, from 10^4 to 3.10^9 Pa .
The lubricant analysed here is a colloidal suspension of overbased calcium carbonate stabilised in pure dodecane. Using first a surface force apparatus, we have observed the squeeze effect and the frictional resistance of the colloidal suspension using a sphere and plane geometry at nanometric scale. The compaction of an adsorbed layer composed of colloidal particles is responsible for the different values of the tangential force, and particularly for the friction instabilities. Secondly, using other tribometers that can support heavy loading, we have analyzed the high domain of pressure. The common feature of the experiments is that, after a critical pressure of $\approx 10^6$ Pa, the colloidal particles do not flow within the film at the interface, but form a compacted " mattress " between the two surfaces. Sliding now takes place between solid surface and this compacted film, via a squeezed hydrocarbon layer. The mean shear strength of the interface can then be expressed by a relation of the form: $\tau = \tau_0 + \alpha p$, where p is the mean contact pressure .

CONTENTS

263

I. L. Singer and H. M. Pollock (eds.), Fundamentals of Friction: Macroscopic and Microscopic Processes, 263–286.
© 1992 *Kluwer Academic Publishers. Printed in the Netherlands.*

A THIN FILM WITH ELASTIC CONTACT

A1 Introduction

The paper concerns the boundary lubrication in the mild wear or ultra-mild regime of metals and oxides. Boundary lubrication is an old and difficultsubjectofresearch[1](2){3}{4}, essentially because its multidisciplinarityneeds a great number of tools.

Boundary lubrication implies first, the study of the conditions at which the boundary film is subjected. Then, depending on the nature of the surfaces and of the environment, different types of films are created. Because, many lubricants and solid dispersions are colloidal, we analyze the friction with a colloidal solution.

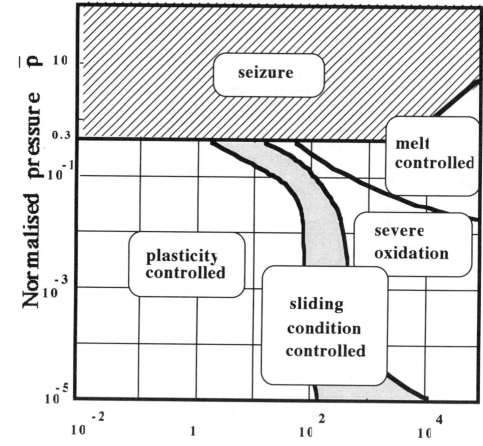

Figure 1 Contours of different regimes of frictional behaviour for the unlubricated sliding of steel surfaces. These regimes are separated by field boundaries taken from the wear- mechanism map for steel. -(after S.C. Lim, M.F. Ashby, J.H. Burnton) {6}.

Lim and Ashby{5}{6}, recently reviewed the wear mechanism and the variations of the friction coefficient μ (ratio of the tangential force T, to the normal force F) for the dry sliding of steel on steel. A wide range of sliding conditions is presented in the form of a wear-regime map and friction-regime map (Figure 1). For a given sliding geometry and contact area, μ is a function of the mean contact pressure p, acting across the sliding surfaces. Their relative velocity V, their initial temperature θ, and the thermal, mechanical and chemical properties of the materials, that meet at the surface, are related for each of the mechanisms by a relation of the form :

$$\mu = f\{p, V, \theta\} \tag{1}$$

There is no single ideal choice of variables for such a problem. Nevertheless the authors suggest as operating variables, the normalized contact pressure and the normalized sliding velocity defined by :

$$\overline{p} = \frac{F}{A} \cdot \frac{1}{H_s} \tag{2}$$

$$\overline{V} = V \cdot \frac{a_h}{\alpha} \tag{3}$$

Here A is the nominal (apparent) contact area of the sliding surfaces, H_s is the room-temperature hardness of the softer material in contact, α is the thermal diffusivity and a_h is the radius of the circular nominal contact area.

At very high loads and velocities, a layer of molten metal forms between the sliding surfaces reducing μ to very low values. At higher velocities (V > 1m/s) the initial surface conditions are replaced by one characteristic of the wear process, and μ depends on the pressure and sliding velocity and not the surface state. At slow sliding speeds (V< 1m/s) μ greatly depends on the state of the surface : friction between rough surfaces is greater than that between surfaces which are smooth. In this case the friction is due to the combined effects of three components: adhesion, abrasion by the asperities, and shearing of the interfacial film.

We would like to focus our attention in this paper on the case of slow sliding speed, where the temperature of the environment dominates. But, as reported by the literature, "temperature, like patriotism is not enough"{7}. It is important to take into account the nature of the sliding surfaces and, for instance, some catalytic effects between the surfaces and the environment.

A2 Nature of the contact : 3 length scales

Three scales play a role in the boundary regime.

At the macroscopic scale, the form, the mechanical properties of the bulk materials and the mechanical conditions impose the mean contact pressure. Here we assume that the contact is macroscopically elastic, and consequently that A in the equation (2) is given by the classical Hertz theory. Therefore, the normalized pressure value is less than 0.3:

$$\overline{p} = \frac{F}{A} \cdot \frac{1}{H_s} \leq 0.3 \tag{4}$$

At the microscopic scale, the surface roughness plays an important role. All surfaces are rough, so that the contact of surfaces is essentially the contact of the higher summits on such surfaces. The roughness of surfaces is due to two cumulative effects: asperities of the solids and solid deposits created by the friction process. Thus, the knowledge of the nature of the deformations at the level of asperities is crucial, because it determines the nature of the friction. A plasticity index Ψ was proposed to indicate the nature of deformation. Ψ is defined by the relation (5) {8}:

$$\psi = \frac{E_s}{H_s} \cdot \sqrt{\frac{\sigma}{\beta}} \tag{5}$$

where E_s and H_s are the elastic modulus and the hardness of the material, σ and β the standard deviation of asperity heights and mean radii of curvature. Moreover, Halling{9} has confirmed the relationship between the plasticity index Ψ and the dimensionless pressure $\overline{P}$ to ensure elastic contact, the ideal low wear conditions being obtained when the contact is elastic Figure 2).

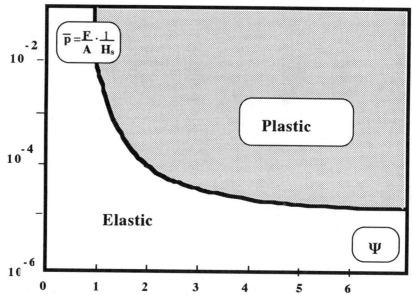

$$\overline{p} = \frac{F}{A} \cdot \frac{1}{H_s}$$

Figure 2 : Relationship between the plasticity index Y and the dimensionless pressure $\overline{p}$ to ensure elastic contact. According to J. Halling {9}.

At the nanoscopic scale, the physico-chemical nature of the solids and the interface explains the behavior of interfacial film. The mechanical behavior of the film varies with the Hill number (which is the ratio of the film thickness to the contact length), and the mechanical properties of the film itself.{10}{11}{12}{13}.

B PRINCIPLES OF BOUNDARY LUBRICATION

B1 PHYSICO-CHEMICAL PROCESSES WITH LUBRICANTS

In industrial lubricants, antiwear additives are always associated with other additives, for example, detergents, dispersants, anti-corrosives, anti-foam agents, viscosity index improvers, anti-oxydants. Some interferences have been readily observed. These interactions between additives are of three types. First, chemical reactions occur between additive molecules, in the oil phase, prior to friction. For example, zinc dithiophosphate (ZDTP) is a weak acid which reacts with overbased sulfonate detergent {19}, Second, the adsorption or chemical reaction processes of additives towards metallic surfaces are in competition, leading to different surface products. Third some additives such as detergents prevent paste agglomeration during friction, thus modifying or disturbing any film formation. These phenomena occur with certain types of ZDTP and neutral calcium sulfonates {20}. The real nature of the sliding material is also very important ; therefore it is essential that the cleaning procedure before the friction test can remove even the most strongly adsorbed layers.

Prior to motion and friction, the environment reacts with the surfaces. For instance, the additives or lubricant bases are adsorbed or react with the metallic surfaces to create new condensed products{14}{15}. The components of the metallic surface are generally hydrated oxides and adsorbed lubricant, depending on machining and cleaning procedures used. Because of its patchy nature, the chemical and structural determination of its composition is difficult even with modern analytical tools (ESCA, AUGER, SIMS...).

The possibility of compound fixation on the surface can be determined by the application of Pearsons Hard.Soft.Acid.Base. principle{16}. The HSAB principle indicates is the activity between different compounds."Hard acids bind strongly to hard bases and soft acids bind strongly to soft bases" . Most of the metals are "soft" acids , and some of them "soft" bases. Some authors, like Mori and Imaizumi{17} have showed that fresh steel surface is so active, that chemical reactions, such as benzene decomposition and sulfide formation, can occur on it even at room temperature. The reason for the high activity may be the presence of lattice defects on the nascent surface, formed by mechanical working. In this condition the metal behaves like a "soft" acid. Three groups of organic compounds has been studied with nascent steel surfaces (Table B1): (i) unreactive, (ii) relatively unreactive, (iii) highly reactive. Organic compounds adsorb through electron donation to the metal surface. Polar compounds, such as carboxylic acids, and amines are classified as "hard" bases. Carboxylic acids act as Bronsted acids as well as Lewis bases, because they have pair electrons on their oxygen atom. Benzene (π electron), 1-hexene and diethyl sulfide are "soft" bases, which exhibit a very high activity on the "soft" acid (nascent metal). Non-polar molecules, such as alkanes, do not adsorb on the "fresh" metals' surfaces.

Moreover, it is commonly assumed that polar compounds such as fatty acids, amines and alcohols adsorb on metal surfaces. According to the literature {18} , polar compounds do not adsorb on the metallic surfaces, but on the oxides.

Table 1 : Adsorption activities of organic compounds on the nascent steel surfaces
(according to S. Mori, Y. Imaizumi 1987){17}

Organic compounds	Adsorption	Activity (sec^{-1})
	Mild steel	Stainless steel
n-hexane	-----	-----
cyclo-hexane	-----	-----
1-hexene	0.14	0.28
benzene	0.54	0.39
diethyl-disulfide	0.23	0.35
methyl proprionate	0.12	0.18
proprionic acid	0.01	0.05
propyl amine	0.01	0.01

--------- no adsorption observed

B2 PRINCIPLES OF THE FORMATION OF A FRICTION FILM

In order to interpret the results of tribological tests and the mechanism of lubricant action, it is necessary to have a lubrication model. We have proposed one (Table 2) {21}.
Prior to friction, additives or the lubricant base are adsorbed or react with the metallic surface to create new condensed products. The nature of the layer depends on environmental conditions. Its thickness is

Table 2   Components of the Friction films {21}

INTERFACE MATERIALS
   ◊ Chemical reaction products (oxides, sulfides....)
   ◊ Adsorbed compounds (surfactants,polymers, resins, colloids....)
   ◊ Colloidal bulk (floc, compact, ...)

TRIBOLOGICAL CONDITIONS
   ◊ Pressures (contact, solid hardness, ...)
   ◊ Temperatures (mean, sur- ,...)
   ◊ Geometry of the solids interface (shape of the convergent entry, roughness,..)

SOLID THIN FILM
   ◊◊ Characteristics
      ◊ Size of the particle units
      ◊ Nature of the binding
      ◊ Mechanical transformations (compaction, shear, ...)
   ◊◊ Nature
      ◊ Consequences of the rheology :(continuous, patchy, lumps, rolls,...)
      ◊ wear mechanisms (scratches, delamination,...)

generally small (monomolecular scale). These products are the interface materials, which, in the favorable areas of the friction interface, produce the colloidal paste during friction under very high local pressures and high temperature. The interface under boundary lubrication can be considered as a capillary whose sides are in relative motion. In the convergent inlet, surface products are first picked up and mixed and then, during their transit in the interface, give a colloidal medium. This paste is considered as the entity primarily responsible for the formation of individual wear scratches, depending on its rheological behaviour. This paste may be chemically transformed during friction, as is demonstrated by the electrical contact resistance study, and produces adherent films. Depending on the nature of the compounds different types of films are created: (i) friction-polymer films, (ii) tribo-reaction films, (iii) compacted films. These films are sometime called " third-bodies"{25} We now analyze these films.

(i)   Friction- polymer films

In 1973, Furey{22} presents a new concept for the reduction of wear by the use of particular compounds capable of forming protective polymeric films directly on rubbing surfaces. Due to the high surface temperatures in regions of greatest contact and the possibility of the added catalytic action of certain freshly exposed surfaces, very thin protective polymeric films will form in these areas. The first step may involve concentration of the polymer former on the solid surface by adsorption. In a dynamic system, the polymer films are continually being formed and worn away. However, the system tends to have a built-in control: the formation of the polymer film will tend to reduce contact and friction and therefore the rate at which the film continues to form. As the polymer film is physically worn away, friction and surface temperatures will tend to increase, thereby causing the rate of film formation to increase. The interposed film is a deposited one and may be relatively thick. Its function is to reduce contact and adhesion between solids. According to Furey, the evidence in support of the in situ polymer film mechanism include: (a) the superiority of polymer-forming compounds over closely-related non polymer-formers in a variety of boundary lubrication tests; (b) indications of localized and adherent surface films,(c) the existence and solvent removal behaviour of pronounced beneficial carryover effects, (d) the results of mechanism studies in a dynamic sliding system with radioactively-labeled compounds, and. (e) the overall success of the approach in developing new classes of effective additives for reducing wear, scuffing, and surface damage.

Furey {22}, Meyer{23} Mazuko and al{24} have described the performance of friction-polymer type additives . A fatty acid such as palmitic acid adsorbs on the surface with its carboxyl group. The adsorbed molecules are desorbed from the surface by both mechanical and thermal disturbances owing to the friction. When a fatty acid desorbs, the lubricity decreases markedly. With hydroxycarboxylic acids, which have two polar groups hydroxyl and carboxyl in the molecule, the tribochemically polymerized molecules are supposed to be produced on the surface. Since they have several ester groups, which are adsorption sites, it would be necessary for the adsorbed polymer molecules to lose all of these adsorption sites before they lose their lubricity. The probability of losing all adsorption sites for the polymer must therefore be much lower than that for fatty acid. The authors assume that a friction-polymer is produced by the tribochemical process,which generates an intermolecular esterification of hydroxyycarboxylic acids.

(ii) Tribo-reaction films

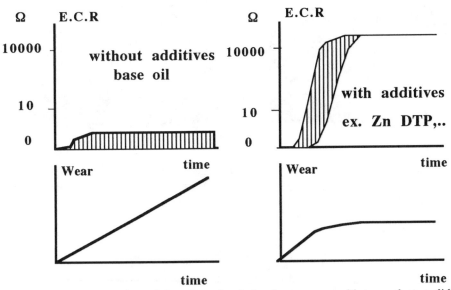

Figure 3 a: Typical evolutions of the contact electrical resistance measured between the two sliding surfaces (ECR), and of the wear for two situations: pure base oil and additive base. The second situation reveals the formation of a friction film (an electrical insulator).

Zinc dithiophosphate (ZDDP), extensively used in engine oils, is a good example of an anti-wear additive which creates a tribo-reaction-film. Its action has been studied by analysis of wear particles generated during a friction test between two ferrous substrates. Martin{26}, Belin {27} carried out friction tests using a plane on plane tribometer consisting of an AISI 52100 steel ring rubbing against a lamellar graphite cast iron plane. The specimens are immersed into a 2% weight by weight solution of di-isopropyl zinc dithiophosphate (ZDDP). The temperature is kept constant at 70°C. The mean contact pressure and the low sliding speed (20 mm/s) generates a boundary regime of lubrication in the contact (figure 3a). Wear particles are collected at the end of the test, rinsed in hexane and centrifuged. Electron microscopy is directly performed on the so prepared wear debris. A set of results presented here was obtained on wear debris from a test at 6.10^6 Pa. A typical micrograph of a foil like wear fragment is shown in figure 3b .

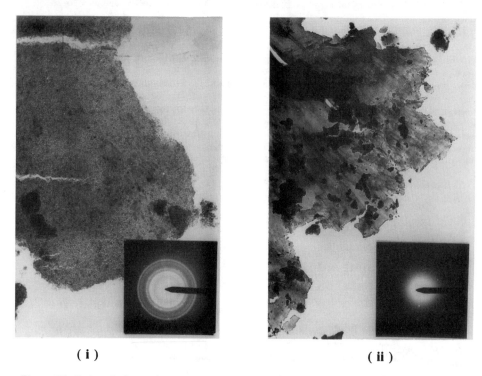

(i) (ii)

Figure 3 b:(i) A typical wear fragment obtained by the friction of the couple steel-cast iron lubricated with pure dodecane (contact pressure 0.1GPa, speed 1cm/s). The particle corresponds to an accumulation compacted of oxide crystallized Fe_2O_3 .
(ii) A typical wear fragment obtained with the same conditions as in (a),but lubricated with a solution of 10^{-2} ZDTP in dodecane.

The particle presents an inhomogeneous structure mainly composed of small crystallites, having 5 to 10 nm diameter, containing an amorphous matrix. The nature of the interfacial film is complex. It contains elements from surfaces, from the environment and from the degradation products of the additives
The exact role of sulphur (from ZDDP molecule) is still unclear; indeed sulphur rich species have already been identified {28}{29}, but their interactions with the phosphate matrix are still unknown. The role of phosphorous in the formation of the anti-wear amorphous film has been recently highlighted. As shown by EXAFS and Analytical Electron Microscopy (AEM), a phosphate glass is formed, phosphorous acting as a network former, iron and zinc being modifier cations.

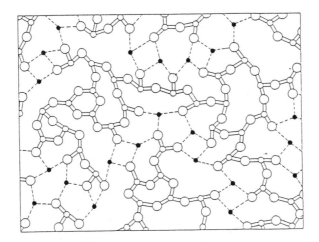

Figure 4 : Possible arrangement of atoms in an oxide glass material.   According G.N. Greaves{27}.

O	:	oxygen atoms
o	:	glass "former" cations (P,S),
.	:	"modifier" cations (Fe,Zn),
solid lines	:	covalent bonds,
dotted lines	:	ionic bonds.

According to recent developments in the structural knowledge of mineral compounds in geology, it can be stated that iron cations, in the Fe^{3+} oxidation state, act as network modifiers, the network being itself constituted by phosphate chains (poly or pyrophosphates)(Figure 4). This point suggests that the phosphate is first formed by thermal processes and that crystallized iron oxide could be "digested" in the structure by friction induced diffusion processes of Fe^{3+} cations in the network{28}. Consequently, this can explain the absence of severe abrasive wear in the steady state regime in a friction test of ferrous material lubricated with ZDDP. Recently Bell and al.{52} show that ZDTP films are composed by three successive layers: near the steel surface an inorganic layer (iron sulfide and iron oxides), then a glassy polyphosphate and finally an organic rich layer.

(iii) Compacted films

If we consider the macroscopic contact of two solids rubbing together, the real nature of the solid surfaces in contact, and of the lubricant, is very often colloidal {10} . Both the solids and the lubricant consist of assemblies of particles {2}. The size of each isolated particle is larger than common molecules, being greater than 1nm. But it is smaller than 1μm, which is small in comparison with the contact area. Mineral oils and greases contain such colloids, especially in the form of additives. For instance, a dispersion of overbased calcium salts is widely used as an anti-wear agent in diesel engines{28}{30}. Overbased calcium salts have a wide range of applications especially as anticorrosive and detergent additives in the automotive engine oils as well as in marine diesel engine. Antiwear properties of lubricants containing this type of product are also very well-researched. It is known that the antiwear mechanism of such additives is essentially dependent on the solid film formation in the solid contact {31}. Recently it was also shown that the detergency properties are controlled by the formation of a brittle burned solid oil film {32}.

B 3 WHAT ARE THE RELEVANT MECHANICAL PROPERTIES OF FRICTION FILMS?

The important mechanical properties of friction films are their bulk properties, such as elastic moduli, yield strength and viscosity; and the film-substrate interface properties, e.g. the shear strength of the interface. In order to obtain such information on tribochemical films, experiments have been conducted using a quasi static sphere plane tribometer {34}(called Tr.2, in the next chapter). In static conditions, tangential force and displacement are measured with the help of very rigid equipment. The sphere (4.75 mm radius) is fixed to the machine frame through a piezoelectric force transducer in order to measure the tangential force T at the contact. An oscillating motion of the flat plane is obtained by means of an electrodynamic shaker.. The resolution and the stiffness of the measurement transducer are 0.01 N and 6.10^9 N/m respectively. The resolution and stiffness of the excitation transducer are 10^{-3} N and 10^7 N/m, so the resolution of displacement is 10^{-4} μm. For very small tangential displacement, if the sphere is loaded with a constant normal force, the tangential displacements of the interface are almost entirely elastic. The measured tangential compliance δ / T can give an approximated value of the elastic shear modulus of the film under pressure: δ / T = 1/ k + 1/ k*, where k is the tangential static stiffness of the sphere-plane contact without film, and k* the tangential stiffness of the film. The shear modulus of the film G_f is given by the following relation {34} G_f = k* t / 0.9 A, where t is the film thickness, and A the real contact area covered by the film. Taking in account that the films cover only 30% of the contact zone, the values given in table3 are obtained.The sphere material is AISI 52100 steel, and the surface of the AISI 52100 steel plane is covered with tribochemical films formed in boundary lubrication on a plane on plane tribometer.

Four sample lubricants representative of four types of products have been used for the film formation (table 3): (A) a pure paraffin (n-dodecane), (B): zinc sec-butyl-dithiophosphate at one percent by weight in dodecane, (C): overbased calcium sulfonate at five percent by weight in dodecane, (D): a friction modifier, a complex ester (molecular weight = 4300) at one percent by weight in dodecane. These films have different composition and small thickness (120 nm for film A, 60 nm for films B, C, D). Films B and C are formed by circular spots over only a part of the surface while films A and D are apparently continuous.

Table 3 Elastic shear modulus of films measured with a quasi-static sphere - plane tribometer.

FILM	Type surface coverage		Color (optical microscope)	Thickness (nm)	Elastic in For Contact p= 0.7 GPa	Modulus GPa Pressure p=1.3GPa
A	continuous	≈ 100%	grey	120-200	3.3	4.3
B	circular spots	≈ 30%	brown, blue	50-70	10.5	12.6
C	circular spots	≈ 30%	blue	50-70	6.3	9.6
D	continuous	≈ 100%	blue	50-70	2.9	1.8

Surfaces of the AISI 52100 steel covered with four tribochemical films formed in boundary lubrication with a plane on plane tribometer.{34}

film A : a pure paraffin (n- dodecane)

film B : zinc sec butyl dithiophosphate at one percent by weight in dodecane

film C : overbased calcium sulfonate at five percent by weight in dodecane

film D: a friction modifier, a complex ester (molecular weight = 4300) at one percent by weight in dodecane.

It appears that elastic shear moduli obtained for the four films do not vary much with the normal load. Furthermore, the values obtained are rather similar for the different films : the elastic shear moduli values are between 2 and 13 GPa. These values are very much lower than for metals but similar to those of a polymer (for instance ≈ 1GPa for the polystyrene). The results show that the different friction and wear behavior of films is not directly correlated to their elastic properties. This seems to indicate that the important properties are related to the plastic behavior and or the boundary properties of the film, as we shall see later.

C   BEHAVIOR OF A COLLOIDAL FILM

We analyze, first, the mechanical properties of the overbased calcium detergent films, and secondly, their tribological properties of friction

C I   MECHANICAL PROPERTIES OF OVERBASED CALCIUM DETERGENT FILMS

The purpose of this chapter is to compare the mechanical properties of such solid films formed with a well defined lubricant. The mechanical properties of such boundary films are very dependent on the process conditions. They are essentially due to the physico-chemical transformations of the colloid. Three types of films are compared : a film obtained by the drying of the base solvent (D.F.), a film due to the trapping and compaction of the colloid between two steel surfaces (C.F.), a friction film obtained with this type of lubricant (F.F.). Adsorption, compaction and shear transformations induce a solid film. The elastic modulus and the hardness of the adsorbed and then compacted film depend on the volume fraction of the colloid in the film. Friction gives a much harder film, with a higher elastic modulus.

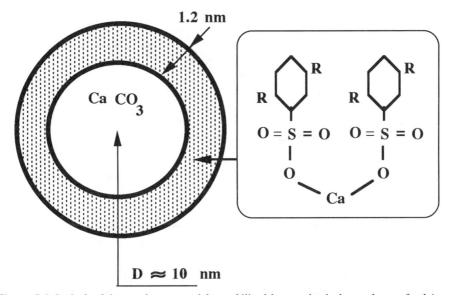

Figure 5 Spherical calcium carbonate particles stabilized by an adsorbed monolayer of calcium didodecyl benzene sulfonate surfactant,. The particles have an diameter, measured by light scattering and electronic microscopy, is found to be 9.5 ± 0.5 nm .The calcium carbonate of the core has an amorphous structure and is surrounded by a monolayer of surfactant, whose thickness is 1.2 nm .

Lubricant

The colloid system chosen consisted of spherical calcium carbonate particles suspended in n-dodecane.The n-dodecane, obtained from Merck-Schuchart (<99%), was used as received. The particles were stabilized by an adsorbed monolayer of calcium didodecyl benzene sulfonate surfactant, and prepared by Elf Co {33}. The particles have been well characterized. Their external diameter, measured by light scattering and electronic microscopy, is found to be 9.5 ± 0.5 nm {35}. The calcium carbonate of the core has an amorphous structure and is surrounded by a monolayer of surfactant, whose thickness is 1.2 nm {35}. Liquids were stored over P_2O_5 prior to use in order to scavenge any trace amounts of water.

Instruments used

Two types of instruments were used: surface forces apparatus S.F.A. and tribometers Tr. (table 4.

(i) First a S.F.A.is used. A new piece of equipment, allowing the continuous and simultaneous measurement of static and dynamic forces, displacement and current voltage between an indentor (sphere or pyramid) and a plane, is used. This equipment is already described in the literature {36}. Two types of experiments can be conducted. Nano-indentation measurement S.F.A.1: in this case the force/displacement indentation curve can be obtained, when the trigonal (90°) penetrator

Instrument	Ref.	Nature of the contact	Use
Surface Forces App.	S.F.A.1	trigonal-plane	indentation
Surface Forces App	S.F.A.2	Sphere- plane	squeeze film
Surface Forces App	S.F.A.3	Sphere- plane	sliding friction
Tribometer	Tr 1	plane-plane	sliding friction
Tribometer	Tr 2	Sphere- plane	static measurements
Tribometer	Tr 3	Sphere- plane	sliding friction
Tribometer	Tr 4	Sphere- plane	sliding friction

Table 4 Instruments used for the tribological measurements

indents the flat specimen. In the second type of experiment S.F.A.2, we used a sphere of 15.4 mm diameter and a plane made of 52100 heat-treated steel. They are polished with a diamond paste and cleaned with alcohol soxhlet. The height of asperities does not exceed 5 nm.

(ii) Four tribometers are also used. A plane on plane friction machine Tr1, already described elsewhere {31} is used to obtain tribochemical films, as described later. A sphere on plane vibrotribometer Tr2, allows to test the interface in static {37}. Friction force and electrical contact resistance can be simultaneously recorded. The tribometers Tr3, and Tr4 are described later.

Dry deposit (DD) of the colloid

In order to evaluate the mechanical properties of the very concentrated colloid solution, we are interested in deposits left by the evaporation of a drop (0.05 ml) of a dodecane solution and overbased calcium carbonateThe liquid drop spreads over a hot plane (100°C) due to the wetting phenomenon and has a diameter equal to 20 mm. After one hour of evaporation, a solid deposit remains on the plane. During the evaporation process the drop, which is originally liquid, reaches, after a certain time a gel state because the solvent (dodecane) dries off, and therefore the colloid concentration increases. In the gel state the mechanical behavior of the drop is viscoelastic {32}. The Young's modulus $E(\Phi)$ of the product increases, when the volume fraction Φ of the colloid increases according the law{38}:

$$E \, \alpha \left[\Phi - \Phi_g\right]^m \qquad (6\ )$$

where $m \approx 3$ to 6, and Φ_g is the concentration of the gel state. Indentation hardness measurements were conducted with a trigonal diamond on the film residue in the area where the thickness is less than 0.3 μm. We found a plastic hardness of $H \approx 1.7.10^8$ Pa and $E \approx 9.5.10^8$ Pa. (cf.Table 5 and figure 6).

Compaction of an adsorbed film (CF)

The compression process of an adsorbed film of colloid on steel surfaces is characterized with the surface force apparatus S.F.A.2. A dilute solution (volume fraction, $\Phi \approx 10^{-4}$) of dodecane and colloidal carbonate is introduced into the steel-plane interface. At a constant temperature (25°C), 8 hours are necessary for the colloid to cover the steel surfaces with a layer which includes an almost complete "monolayer" of calcium carbonate particles, with some incomplete layers on top of it {39} (figure 7). The total attraction between two particles is weak, the solution does not flocculate for a period of a month. The attraction between the particles and the steel surface is higher.

Table 4 Instruments used for the tribological measurements	Thickness t (nm)	Elastic modulus E X 10^7 Pa	Hardness H X 10^7 Pa	E / H	Ref.
DD	300	90 -100	16 -18	5 -6	(a)
CF	10 -70	8.10^{-3} - 1.8	10^{-4} -10^{-1}	14-63	(b)
FF	60	1600 - 2800			(c)
FF	60	1500 - 3000	135 - 150	10 - 22	(a)
CaCO$_3$		3100	135	10 -22	(d)
CaCO$_3$			150 -163	19 - 22	(e)

DD	Dry deposit	(a)	Trigonal indentation measurement
CF	Compacted film	(b)	Surface forces measurements
FF	Friction Film	(d)	calcite data [40]
		(e)	aragonite data [40]

Table 5 Mechanical properties of the carbonate films

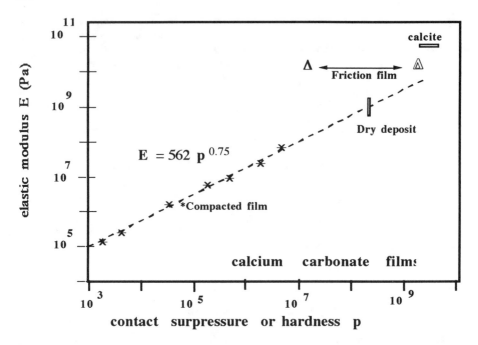

Figure 6 Evolution of the elastic modulus of different films versus the contact pressure (p). Three types of carbonate film are plotted on the same graph. We note that the compacted film (CF) and dry deposit (DD) data are on the same curve $E = 562 \; p^{0.75}$. It is not the case for the friction film (FF) data. The FF elastic modulus is given for two contact pressure values : the estimated contact pressure during the friction test, and the pressure during the hardness test . In the two cases, values are higher than that of the compacted film. This is, maybe, due to the crystallization of calcium carbonate during the friction process.

The Hamaker constant value, for steel-carbonate (4.10^{-20} J) is higher than that for carbonate-carbonate ($1.2 \; 10^{-20}$ J). During the squeeze of the layer caused by the sphere/plane indentation, the two layers first overlap each other. A gel phase is created in the interface (figure 7). Then, the colloid concentration increases in the interface. Finally, a compact colloid film is obtained. During the process, the yield pressure increases. It is limited by the deformation of steel surfaces. Indentation data allow us to follow the mechanical properties of the film {39}. In particular, the elastic modulus of the layer depends on the volume fraction of colloid according to a relation given by equation (6), but also on the compaction pressure p. We found (figure 6):

$$E_f = 562. \; p^{0.75} \; (Pa) \tag{7}$$

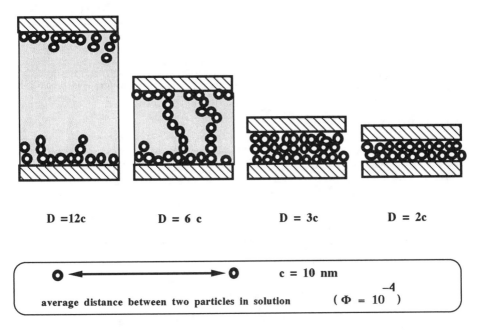

D =12c   D = 6 c   D = 3c   D = 2c

○ ◄————————————► ○   c = 10 nm

average distance between two particles in solution   (Φ = 10⁻⁴)

Figure 7 With the surface force apparatus S.F.A.2, we observe the compaction of a colloidal carbonate layer introduced into the sphere-plane interface (sphere-plane distance D). At a constant temperature (25°C), 8 hours are necessary for the colloid to cover the steel surfaces with a layer which includes an almost complete "monolayer" of calcium carbonate particles (diameter c), with some incomplete layers on top of it (D=12c). During the squeeze of the layer caused by the sphere/plane indentation, the two layers first overlap each other. A gel phase is created in the interface (D=6c). Then, the colloid concentration increases in the interface. Finally, a compact colloid film is obtained (D=3c or D=2c)).

Friction films (FF)

Tribochemical films were obtained in a boundary lubrication regime with the plane on plane tribometer Tr1. We use a solution of dodecane to which calcium carbonate ($\phi = 0.05$) has been added, and a 52100 steel plane sliding against a cast iron plane. The apparent contact pressure is 4.10^6 Pa, which corresponds, taking in account that the film covers only 30 % of the contact zone, to an estimated real pressure of $1.4 \ 10^7$ Pa. The sliding distance is 50 m and the average sliding speed is 10^{-2} m/s. The microscopical observation of the film created on the 52100 steel surface during the friction process shows that circular islands (5 μm diameter) cover almost 30 % of the apparent contact area. This colored film (refractive index 1.4) is 60 nm thick (measured with different methods {34}). The film is microindented with a trigonal penetrator S.F.A.1. Different loadings and unloadings are made during the test. The unloading curve permits the evaluation of the elastic modulus and the hardness of the layer, {5}, and we found that the hardness of the film is $H = 1.35 - 1.5.10^8$ Pa with $E \approx 1.5\text{-}3.10^{10}$ Pa. These values are in agreement with those of shear modulus G measured by the vibrotribometer Tr2 (chapter B3). Taking into account that $E \approx 2.6$ G (with a Poisson ratio equal to 0.3), we obtain $E = 2.5.10^{10}$ Pa, for an average pressure of contact $p \approx 1.3.10^9$ Pa.

It is interesting to compare the elastic properties of the different studied films (Table 2) .If the CF and DD films follow the same law $E \propto p^{0.75}$, the FF film has a different behavior. We attribute this phenomenon to the fact that in the FF film, the core of the colloid undergoes a shear transformation and that the physical state of the core is no longer amorphous (like in the CF or DD film) but is in a crystalline state. This result agrees with the electron microscopic observations of the wear debris {34} and with the data of calcite or aragonite {40}. The elastoplastic ratio E/H of CF and DD is very similar to polymer materials with Van der Waals binding taken at a temperature below the transition temperature {40}.

C III FRICTION BEHAVIOR

The friction force between two sliding solids lubricated with a very thin film usually results from three phenomena {41}: the ploughing of the surfaces by the asperities, the adhesion at the many small areas of intimate (or real) contact, and the shear of the interfacial film. If the surface roughness is very small, the shear properties of the thin film are dominant .

Bowden and Tabor {3} established the relation giving the friction coefficient μ for the situation where the surfaces in their contact region are separated by a thin solid film. The critical shear stress of the film τ, is less than that of the bulk material of the two solids in contact.

$$\mu = \frac{T}{F} = \frac{\tau}{p} \qquad (8)$$

Here T and F are the tangential and the normal forces and p is the mean pressure in the contact. Bowers and Zisman {43} showed that the shear strength of the film depends on the real applied pressure . Consequently they write:

$$\mu = \frac{\tau (p)}{p} \qquad (9)$$

It is difficult, in fact, to find in the literature, the shear strength value of the material for a given pressure. Nevertheless, some authors {44}{45} prefer writing the pressure dependence in the following form

$$\tau = \tau_0 + \alpha\, p \qquad (10)$$

If this expression appears to be quite general for thin solid films, the question is to know, if the quantity τ is a true measure of the shear strength of the film material or a measure of the shear strength of the interface between the solid and the film material. According to Briscoe and Tabor {44}, τ is a true measure of the shear strength of the film material, which is less than the critical shear stress of the bulk material.

We report, using three tribometers, the frictional behavior of a thin colloidal film. We analyze the shear behavior for a large range of pressure. We show, that for such film, τ measures the shear strength of the interface.

Three series of experiments are carried out with three different instruments, using a large range of normal loads as reported in Table 6. The colloïd system chosen is again spherical calcium carbonate particles suspended in n-dodecane.

(i) Using a surface forces apparatus S.F.A.1, S.F.A.2, we have observed first the squeeze effect of the colloidal solution between a rigid sphere and a plane. Then secondly, we have studied the effects of the sliding. The data, already published{39}{46}, are here summarized. The specimens are made with heat-treated steel 52100. Their surfaces are very highly polished, the asperity height being below 5nm. Due to the adsorption process and a slight flocculation phenomenon, the surfaces are covered by a solid layer of agglomerated colloids. So, with a solution of $\Phi = 10^{-4}$, after 10 hours of adsorption, each surface is covered with a 3.5nm thick layer.

Table 6 Tribometers and materials used in the experiments.

Tribometers	Normal load (N)	Sliding speed	Materials	E (GPa)	Lubricant $\Phi*$
Surface forces S.F.A.3	10^{-6} 10^{-3}	1nm/s	Steel 52100	210	Solution ($\Phi = 10^{-4}$) Adsorbed film
ball on flat Tr 3	10^{-2} 10	10μm/s	Sapphire	420	Dry film ($\Phi = .5$)
ball on flat Tr 4	400	1cm/s	Steel 52100	210	Solution ($\Phi = .1$)

.* Φ Volume fraction of calcium carbonate particles diluted in pure dodecane.

During the indentation process, the colloidal layer is compacted and an average value of the volume fraction Φ of colloids in the interface can be evaluated. The sliding process is conducted for two given distances D maintained constant during the sliding. For the first distance D=30nm =3c, which corresponds to a mean volume fraction of the interface Φ=0.3, the extra normal pressure is 1.4 $.10^{4}$ Pa.(absolute mean pressure p=1.14.10^{5} Pa) . In this condition, the initial mean sliding shear stress is τ = 1.4 .10^{4} Pa, but increases rapidly during the sliding. The layer flow in the interface leads to a consolidated film. For a distance D=20nm =2c, the indented layer is more compacted (Φ = 0.5). The mean contact pressure and the mean shear stress are respectively p = 2 10^{6} Pa and τ = 1.2 10^{6} Pa. We note that the tangential stiffness is not stable, indicating that the flow is not continuous in the interface.

(ii) With a second tribometer Tr3, a smooth sphere of radius R = 4.75 mm, is loaded against a smooth plane upon which a colloidal film is deposited , whose volume fraction is about 0.5. The thin film {47}, is about 0.1μm thick. In these experiments, the specimens used are made of sapphire, and have a low surface roughness (asperity height is close to 3.5 nm). The transparency of the plane permits the observation of the contact interface, with an optical microscope. Using Newton's interference rings, it is possible to measure the contact area. Then, the plane describes a reciprocating motion with an amplitude of 100μm. Optical observations of the contact zone during the sliding process give an evaluation of the contact radius.

The contact pressure, for smooth sphere and plane loaded below the elastic limit can be calculated as the mean "Hertzian" pressure,when the interfacial film has a negligible effect on the contact area A. The contact radius a_h is related to the normal load F by the relation:

$$F = \frac{a_h^3}{1.3} \frac{E_s}{R}$$
(11)

where E_s is the composite elastic modulus of the samples, R is the radius of the sphere.

We found, that the experimental data agree this law, for F > 2N. At this load, the contact pressure corresponds to the hardness of a compact colloidal film (H = 2.10^{8} Pa). When the contact pressure value is lower, the two solids are not elastically deformed, they are rigid. They deform elastically the thin film , whose elastic modulus is E_f. The contact radius a of the rigid sphere, which indents the film, is given by the elastic foundation model {8}:

$$F = \frac{\pi}{4} \cdot E_f \cdot \frac{a^4}{R \cdot t}$$
(12)

where t is the film thickness before indentation. But the elasticity of the film varies with its compaction, and then with the contact pressure according to the equation (7). Taking into account the relations (7) and (12), we found that for low load, F is proportional to a^{10}, in agreement with experimental data.

(iii) For the high pressure experiments, we used the same type of contact , with 52100 steel specimens. The roughness is now in the range of 150nm for peak to valley. The lubricant is a solution of colloids in n-dodecane (Φ =0.1). The sphere describes a circular track on the plane, with a sliding speed of the order of 1cm /s. The contact pressure is less than the elastic limit of the contact (2 GPa). The stabilized friction force is obtained after few revolutions. The optical observation, after the test, clearly shows a thin white deposit on the contact zone, the size of which corresponds exactly to the computed Hertzian area of contact.

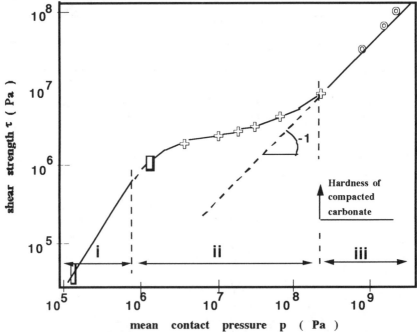

Figure 8 Pressure dependence of the shear strength of a film of colloidal carbonate at 20°C. Sliding speeds of the order of 1nm/s to 1 cm/s. The experiments are conducted with three different tribometers. (i) For the low range of the contact pressure, a surface force apparatus is used (data are represented by a rectangle). (ii) For the medium range of the pressure, a sapphire sphere is loaded against a glass plane.(iii) For the high range of pressure, high polished steel surfaces are used. If for the first situation (i), the flow of the colloidal solution can be homogeneous, for the cases (ii) and (iii), a compacted film of colloids adheres to the sphere, and the sliding takes place between the film and the plane

Figure 8 indicates the evolution of mean shear stress as a function of the computed averaged pressure for the three series of experiments. For the high pressure region,(index iii Fig.8) the pressure p is higher than the compacted film hardness. The hardness of a dry film (deposit of colloidal calcium corresponds to the hardness of the core which is amorphous calcium carbonate {7}. In this pressure region, the two solids in contact are elastically deformed, and the mean shear stress τ is almost proportional to the pressure p: $\tau = \alpha$ p. We found $\alpha = 0.04$ for p = 3.10^8 Pa, and $\alpha = 0.055$ for p = 3.10^9 Pa.

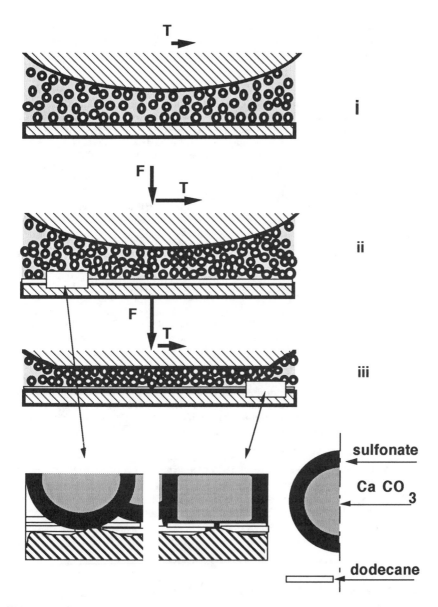

Figure 9 Schematic representations of the interfaces sphere-plane lubricated with the colloidal solution, for different pressures. Fig.9 i, the distance sphere- plane is maintained constant during the sliding, and the normal pressure (10^5 -10^6 Pa) has a low value. The colloidal solution can continuously flow in the interface. Fig.9 ii, the colloidal film is compacted, the sliding takes places

between the plane and the film. Fig.9iii, for a pressure higher than 2.10^8 Pa, the sphere-plane contact is elastically deformed as also is the colloid itself. The shell of sulfonates surrounding the colloid is destroyed, and the sliding involves dodecane molecules.

These values, which are the same as those obtained by Bridgeman {50} for paraffins, can correspond to the shear of dodecane molecules aligned on the surface. At this pressure, the dodecane is a solid (the solidification pressure at 20°C is $1.8.10^8$ Pa,{49}). The slip would occur involving CH2:CH2 interactions as described by the " crank-shaft" (or wrinkle) model proposed by Tabor {48}. In these experiments the slip takes place at the interface (made of dodecane molecules aligned on the surface),between the steel (or sapphire) plane and the compacted film adherent to the sphere. We note that at this pressure, the surrounding shell of sulfonate is probably removed as shown, recently, by Electron Energy Loss Spectroscopy{27}. The friction force does not depend here on the nature of the substrate, as already found for MoS_2 thin coatings {45}.

For the medium range of pressure (index ii Fig.8), τ is almost independent of the pressure $(\tau = \tau_0 \sim 3.10^6 \, Pa)$. We note that, in this situation, a slab of compacted film governs the pressure. The mean shear strength τ_b of the compacted material, which constitutes the film, can be evaluated as being $\tau_b = H / 6$, then $\tau_b = 3.10^7$ Pa; therefore $\tau_b / \tau_0 = 10$. Briscoe {44} mentions such a ratio value for the polymers.

We conclude that τ_0 does not correspond to the strength of the carbonate film, but of the shear stress of the interface carbonate film - plane. The carbonate film adheres to the sphere and the shear stress value corresponds to the shear stress of the interface steel plane - $CaCO_3$ core as described in Figure9. This interface is constituted by dodecane molecules and the surfactant of the colloid (didodecyl benzene sulfonate), which presents head of CH_3. A value of $\tau_0 = 3.10^6$ Pa is given in the literature for the sliding of CH_3 against CH_3 at low contact pressure{44} (Figure9).

For the low range of pressure (index i Fig.8), we have observed, by mechanical measurement, the consolidation of an adsorbed layer of colloidal particles on solid surfaces during the normal approach of the substrates. Hence we have shown that the frictional resistance is strongly dependent upon the consolidation, leading to instabilities for a critical volume fraction of colloids in the interface. This corresponds to a pressure of 2.10^6 Pa, where the sliding takes place between the plane and the compacted film.

In conclusion, the common feature of the experiments is that ,after a critical pressure evaluated at 10^6 Pa ,the colloidal film does not flow in the interface of solids in contact. But it realizes a compacted " mattress " between the two surfaces . The sliding takes place between the "mattress" and the plane surface, squeezing an hydrocarbon layer..

ACKNOWLEDGMENTS
The authors are indebted to B. Constans for helpful discussions.

REFERENCES

{1} .Hardy W.B :*Collected papers*   Cambridge Univ.Press (1936)

{2}   Beerbower A.: *Boundary Lubrication*, (Off. Dep. of Army Washington) AD 747 336 (1972)

{3}   Bowden F.P.and Tabor: D.: *The Friction and Lubrication of Solids* Part 2 (Clarendon, Oxford, 1964)

{4}   Akhmatov A.S:. *Molecular Physics of Boundary Friction,* (Israel Program for Scientific Translations Jerusalem 1966) 381.

{5}   Lim S.C.and Ashby: M.F.: Wear-Mechanism Maps Acta metall. 35, 1,   (1987) 1.

{6}   Lim S.C., Ashby M.F.and Brunton J.H.: Wear-Rate Transitions and their Relationship to Wear Mechanism. Acta metall. 35, 6, (1987) 1343.

{7}   Cameron A.: Lubricant Chemistry and Tribology Chemistry, Boundary and Extreme Pressure Lubrication. I. Mech. C250 (1987) 355.

{8}   Johnson: K.L.' *Contact mechanics*   (Cambridge University Press 1985) 105.

{9}   Halling :J: The Tribology of Surface Coatings, Particularly Ceramics Proc. Instn. Engrs. 200 C1(1986) 31.

{10} Georges J.M.: Colloidal Behaviour of Films in Boundary Lubrication, in *Microscopic aspects of Adhesion and Lubrication*, J.M. Georges Eds.Tribology series 7, Elsevier, Amsterdam, (1982) 729.

{11} Hamer J.C., Sayles S.C.and Ionnanides E.: Partic le Deformation and Counterface when Relatively Soft Particles are Squashed, Tribology transact. 3, (1989) 281.

{12} Hill R.: *"The mathematical theory of plasticity"*, (Oxford Clarendon press, 1967) 240.

{13} Kendall R.: Relevance of Contact Mechani cs to Powders, Elasticity, Friction and Agglomerate Strength, *"Tribology in particulate technology"* , Ed. B.J.Briscoe and M.J. Adams, pp. 91-99, (1987).

{14} Frewing J.J, The Heat of Adsorption of Long Chain Compounds and their Effect on Boundary Lubrication Proc. Royal Soc., A182   (1944) 236.

{15} Grew W.J.S. and Cameron A. Thermodynamics of Boundary Lubrication and Scuffing, Proc. Royal Soc., A327, 47-59 (1972)

{16} Ho T., *Hard and Soft Acids Bases Principle in Organic Chemistry*, Academic Press (1977)

{17} Mori S., Imaizumi Y., Adsorption of Model Compounds of Lubricant on Nascent Surfaces of Mild and Stainless Steels under Dynamic Conditions S.T.L.E. 31, 4, (1987) 449.

{18} Spikes H.A.: Additive-additive and Additive-surface Interactions in Lubrication, Lubrication Sci. 2,1,(1989) ,3.

Cann P.M.E., Johnston G.J. and Spikes H.A.: The formation of thick films by phosphorous based Antiwear-additives I.Mech. E. Conf. on Tribology Friction Wear and Lubrication C208/87 543.

{19} Sakurai T., Ikoda S., and Okabe H,: The Mechanism of Reaction of Sulphur Compounds with Steel Surfaces during Boundary Lubrication, ASLE . Trans. 25, (1982) 117.

{20} Coy R.C.and Jones R.B.: The thermal degradation of E.P. Performance of Zinc Dialkyl Dithiophosphate on Steel, ASLE Trans.24 (1981) 24.

{21} Georges J.M., Mathia T.,Kapsa Ph., Meille G. and Montes H.: Mechanism of Boundary Lubrication with Zinc Dithiophosphate, Wear, 53 (1979) ,9.

{22} Furey M.: The Formation of Polymeric Films Directly on Rubbing Surfaces to Reduce Wear, Wear, 26 (1973) 369.; and Models of Tribopolymerisation as an Anti-Wear Mechanism : Proceed. of the Japan Int. Tribology Conf. Nagoya (1990) 1089.

{23} Meyer K ,Weh K.and Berndt H.: Antiverschleisswirksamkeit polykondensationfâhriger Additive, Schmierungstechnik, 12, (1981) 691.

{24} Masuko M,Ohmori T.and Okabe H.:Anti-Wear Properties of Hydroxycarboxylic Acids with Alkyl Chains, Tribology Int. 21,4 (1988) 199.

{25} Godet M. : The "third-body" approach : a mechanical view of wear., Wear 100 (1984) 437.

{26} Martin J.M.,Mansot J.L.,Berbezier I.,and Belin M.:Microstructural ASpects of Lubricated Mild-Wear with Zinc Dithiophosphate, Wear, 107, pp. 355-366 (1986).
Belin M.,Martin J.M. and Mansot J.L.: Role of Iron in the Amorphisation Process in Friction-Induced Phosphate Glasses, STLE Trans. 3(1989) 410.

{27} Greaves G.N.: Local Order in Oxide Glasses, Proc. Int. Conf. Frascati, Italia (1982).

{28} Watkins R.C.: The Anti-Wear Mechanism of ZDDP, part II, Tribology Int., 15, pp. 13-15 (1982).

{29} Hsu S.H.,Klaus E.E,and Cheng H.S. : A Mechano-Chemical Descriptive Model for Wear under Mixed Lubrication Conditions , Wear, 128 (1988) 307.

{30} Marsh J.F.: Colloidal Lubricant Additives, Chemistry and Industry, 20 (1987) 470.

{31} Kapsa Ph., Martin J.M., Blanc C., Georges J.M.: Anti-Wear Mechanism of ZDDP in the Presence of CAlcium Sulfonate Detergent, ASME Trans,Tribology J. 102 (1981) 486.

{32} Georges J.M., Loubet J.L.,Tonck A., Mazuyer D.and Hoonaert P.: ON the Mechanical Properties of Overbased Calcium Detergent Films, Proceedings of Japan International Tribology Conference , Nagoya, (1990) 517.

{33} Elf Research Center Solaize 69 France.

{34} Tonck A., Kapsa Ph, Sabot J. : Mechanical Behavior of Tribochemical Films under a Cyclic Tangental Load in a Ball-Flat Contact, ASME Trans., Tribology J., 108 (1986) 117.

{35} Mansot J.L.,Martin J.M., Dexpert H., Faure D., Hoornaert P, Gallo R.: Local Structure AnaLysis in Overbased Reversed Micelles,, Physica B, 158 (1989) 237.

{36} Tonck A., Georges J.M.and Loubet J.L.: Measurements of Intermolecular Forces and Rheology of Dodecane between Alumina Surfaces , J. Colloid and Interface Sc., 126, 1 (1988) 150.

{37} Tonck A.,Sabot J.and Georges J.M. : Microdisplacements between Two Elastic Bodies Separated by a Thin Film of Polystyrene, ASME Trans., Tribology J., 106 (1984) 35.

{38} Ehrburger F.and Lahaye J.: Behaviour of Colloidal Silicas during Uniaxial Compaction, J. Phys.˙France, 50, 11 (1989) 1349.

{39} Georges J.M., LoubetJ.L. and Tonck A.: Molecular Contact Pressure in Tribology, in "New Materials Approaches to Tribology; Theory and Application", L.E. Pope, L.L.Fehrenbacher, W.O. Winer Eds. M.R.S. (MRS Pittsburg, PA.) 140 (1989) 67.

{40} Ashby M.F.and Jones D.H.: Int. Series. 34, Pergamon Press (1985).

{41} Tabor D.: Friction and Wear -Developments over the last fifty years, Inst. Mech. Eng. C 245 (1987) 157.

{42} Martin J.M., Belin M.and Mansot J.L.: Exafs of Calcium in Overbased Micelles, J. Phys. France, 47, 12, 887, (1986).

{43} Bowers R.C. and Zisman W.A.: Pressre Effects on the Friction Coefficient of Thin-Film Solid Lubricants, J. Appl. Phys. 39, 12, (1968) 5385.

{44} Briscoe B.J. and Tabor D.: Shear Properties of Thin Polymeric Films, J. Adhesion, 9 (1978) 145. and Briscoe B.J.: The Interfacial Friction of PolymerComposite Materials: General Fundamental Principles, in Friction and Wear of Composite Materials K. Freidrich Eds. (Elsevier Scientific Publishers 1985) chapt.2, 39.

{45} Singer, I.L.,Bolster R.N.,Wegand J.,Fayeulle S.and Stupp B.C.: Hertzian Stress Contribution to Low Friction Behavior of Thin MoS_2 Coatings, Appl. Phys. Lett. 57, 10 (1990) 995.

{46} Georges J.M.,Mazuyer D.,Tonck A.and Loubet J.L.: Friction of Thin Colloidal Layer,J.Phys. Cond. Mat. 2, (1990) 399.

{47} By thin, we mean films whose thickness is much less than the Hertzian contact radius, which in these experiments varied from 10 to 200 mm.

{48} Tabor: D.: The role of Surface and Intermolecular Forces in Thin Film Lubrication, Microscopic aspects of Adhesion and Lubrication, J.M. Georges Eds. Tribology series 7, Elsevier, Amsterdam, (1982) 651.

{49} Ducoulombier D., Zhou H., Boned C., Peyrelasse J., Saint-Guirons H., and Xans P: Viscosity of Liquid Hydrocarbons, J.Phys. Chem. 90 (1986) 1692.

{50} Bridgeman P.G.: Pressure Effect on hydrocarbons, Proc. Am. Acad. Arts Sci. 71, (1936), 387.

{51} Hallouis M., Mansot J.L.and Martin J.M.: Colloidal Anti-Wear Additives part II , Tribological, behavior., J. Colloid Int. To be publish.

{52} Bell J.C. Delargy K.M.and Seeney A.M. The removal of substrate material through thick Zinc Dithiophosphate Antiwear Films . 18Th Leeds- Lyon Symposium 1991 (To be published.)

Discussion *following the lecture by J-M Georges on "Friction with colloidal lubrication"*

J FERRANTE. I have a question about the nano-load/displacement technique that you used to determine the elastic modulus. One of the common situations in regard to the testing of thin films in general is the so called anvil effect, that is to say, what is the indentation depth compared with the thickness of the film. Can you detect any influence of the substrate on your data, by looking at the change in slope of the load/displacement curve during the identation?

J-M GEORGES. This question is a little outside the scope of my talk. Our case fortunately is the case of a soft layer on a hard substrate, which is much simpler than when it is the other way round. A bigger problem that we have with our soft layers is in fact the curvature of the contact.

D P LETA. It looks as if your values of pressure are extremely low. In order to change the shape from spherical (in other words, the lowest possible surface area) to any other shape, if you put in additional sulphonate, does that change the pressure at which you have fracture and then go to the correspondingly different slope? I was also curious to know if you had tried removing sulphonate and then doing the nano-tribometer experiment again, when you will have just an uncoated calcium carbonate particle?

J-M GEORGES. If you load at high pressure with calcium carbonate using only true compression, you do not remove the sulphonate. But sometimes before you remove it, you affect the core of this calcium carbonate particle. We have not done the experiment you suggest, because we never find an uncoated calcium carbonate particle *without* any transformation (crystallisation) of the core.

EFFECT OF SURFACE REACTIVITY ON TRIBOLOGICAL PROPERTIES OF A BOUNDARY LUBRICANT

R.S. TIMSIT
Alcan International Ltd.
Kingston R & D Centre
P.O. Box 8400
Kingston, Ontario
Canada K7L 5L9

ABSTRACT. The paper examines the shear-strength and tribological properties of molecular layers of stearic acid on glass, aluminum and gold surfaces. These properties on the various surfaces are correlated with the chemical reactivity of stearic acid with the solids.

1. Introduction

Indirect experimental evidence has suggested that the performance of a boundary lubricant relates in part to the adhesion strength of lubricant molecules to the sliding surfaces[1-3]. Bowden and Tabor[1] showed that fatty acid films deposited on unreactive substrates lubricate less efficaciously than films on reactive surfaces. These observations lead to the conjecture that weakly-adhering "lubricants" do not lubricate effectively because they are easily expelled from the working interface. The present paper confirms the validity of this conjecture by examining the shear strength and tribological properties of stearic acid on glass, aluminum-and gold-coated glass. Experimental evidence is presented that stearic acid adheres to glass and aluminum in air, but adheres less well to gold. This evidence is used to establish the relationship between reactivity (or adhesion strength in this case) and the tribological properties of stearic acid on these surfaces. Reactivity was inferred from spectral changes in selected lubricant infrared absorption bands, induced by chemisorption. The factors affecting stearic acid durability in a sliding interface, where the lubricant adheres strongly to both sides of the interface, are discussed. The effect on friction of lubricant adhesion to only one side of a sliding interface is also evaluated.

2. Experimental Technique

2.1. SHEAR-STRENGTH MEASUREMENTS

All measurements of shear strength and durability were performed at room temperature with a reciprocating slider-on-flat apparatus[4]. The flat surface consisted of bare borosilicate glass (surface rms roughness of 1-2 nm) or of glass coated with a vacuum-sputtered film of aluminum

I. L. Singer and H. M. Pollock (eds.), Fundamentals of Friction: Macroscopic and Microscopic Processes, 287–298.
© 1992 Kluwer Academic Publishers. Printed in the Netherlands.

or gold, with a thickness of 100 and 280 nm respectively. The glass surfaces were prepared as described previously[5] and were lubricated with a stearic acid Langmuir-Blodgett (LB) film. It is not clear that the weak bonding of the acid to Au (as will be shown later) yielded layers of similar molecular arrangements on the Al and Au surfaces. The slider consisted of a glass hemisphere with a radius ranging from 200 μm to 6 mm[5]. Friction was measured at a mechanical load ranging from 5×10^{-3} to 0.1 Kg and at speed of 60 μm s^{-1} over a track length of 1.5 mm. The shear strength (τ) of a sliding interface was evaluated from a measurement of the frictional force (F) and a knowledge of the area (A) of the sliding contact through the equation

$$F = A\tau \tag{1}$$

with A given as[6]

$$A = \pi(kRP)^{2/3} \tag{2}$$

where R and P are respectively the slider radius and the contact load. The quantity k is given as $3(1 - v^2)/2E$ where v (=0.2) and E (=7.1×10^{10} Pa) are respectively the Poisson ratio and the elasticity modulus of glass. Equation (2) ignores the elastic deformation of the metal-coating because the ratio of the elastic moduli of glass and Al is approximately unity and the contact radius is always much larger than the metal film thickness[7]. For each measurement of frictional force, shear strength was evaluated from Eq. (1) by substitution of Eq. (2) using the measured values of R and P. Similarly, the compressive stress was evaluated as P/A.

2.2. INFRARED SPECTROSCOPY

The strength of the interactions of stearic acid with glass, Al and Au was assessed qualitatively by Infrared Reflection Absorption Spectroscopy (IRAS) from changes in the spectrum of the lubricant monolayer both on adsorption at room temperature and on heating of the substrate. Spectra from LB films on Al and Au were recorded in single reflection, using radiation incident to the surface at an angle of 85° to the normal. Spectra from molecular films on glass were obtained using p-polarized light incident at an angle of 70°[8]. The work was performed with a Fourier Transform spectrometer adapted with a separate chamber in which the grazing-incidence reflection optics were located[4]. All spectra were collected with a 4 cm^{-1} resolution and corrected for background by subtracting the spectrum obtained from the substrate before deposition of the LB layer.

3. Results And Discussion

3.1. INFRARED SPECTRA

Stearic acid ($CH_3(CH_2)_{16}COOH$) attaches itself to a solid surface via the polar COOH group[1,9] as illustrated in the inset in Fig. 1(b). The material owes its lubricant property to the weak interaction of the saturated CH_2 and CH_3 groups with other solids, such as the mating sliding surface, or with other stearic acid molecules. Figs. 1(a) and (b) show the infrared (IR) absorption spectra of (a) the bulk acid at room temperature, and of (b) an LB film on Al respectively after deposition (spectrum I) and following heating at 100°C for 2 hours in dry nitrogen (spectrum II).

The spectra have been described in detail elsewhere[4]. The bands in the 2800-3000 cm[-1] region originate from vibrations of the CH$_2$ and CH$_3$ groups. Several of the bands in the 1400-1800 cm[-1] frequency range arise mainly from vibrations of the COOH polar end[10]. The bands in the 1100-1350 cm[-1] region originate from wag and twist vibrational modes of the CH$_2$ group. The difference in relative intensities of the lines in the 2800-3000 cm[-1] region in the spectra of Figs. 1(a) and (b) is due to molecular orientation effects in the LB layer[11]. The differences in the 1400-1800 cm[-1] range originate from absorption frequency shifts in the chemisorbed layer. These shifts indicate that the polar group is bonded chemically to the native aluminum oxide film on Al[12,13], probably to form a salt. The bonding strength to Al$_2$O$_3$ appears to be large because the absorption bands undergo little decrease in intensity on heating the LB film to 100°C for 2 hours, as illustrated in spectrum II (Fig. 1(b)).

The nature of the chemical bond of stearic acid with glass could not be determined as readily as with Al because IRAS spectra of thin films on dielectric materials are distorted by the optical properties of the substrate[8]. Figure 1(c) shows spectra recorded from an LB layer on glass respectively at room temperature (spectrum I) and after heating for 5 minutes at 80°C (spectrum II). The CH$_2$ and CH$_3$ bands in the 2800-3000 cm[-1] region are clearly defined in spectrum I. The similarity of the relative intensities of these bands with those observed from the bulk material suggests that the molecular assemblies on borosilicate glass are not well ordered. The origin of the bands in the 1200-1800 cm[-1] region is unclear. The peaks at 1470 and 1510 cm[-1] may arise from carboxylate ions attached to metallic components such as Na, K, Al etc... in the glass[14]. The loss of IR absorption intensity after heating at 80°C, stemming from thermal desorption from the glass, is appreciable. This loss is larger than that observed after a more severe thermal treatment of the LB layer on Al[15] (Fig. 1(b), spectrum II) and suggests that the acid chemisorbs more strongly on Al than on borosilicate glass.

The infrared spectra of Figs. 1(b) and 1(c) indicate that stearic acid reacts chemically both with the native oxide surface film on Al and with borosilicate glass to form a chemisorbed rather than a weakly physisorbed layer. The thermal desorption data suggests that the acid adheres more strongly to oxidized Al than to glass.

Figure 1(d) shows the infrared absorption spectrum of a stearic acid film on gold at room temperature. Note that the relative intensities of the stretch vibrations of the CH$_3$ and CH$_2$ groups in the 2000-3000 cm[-1] region (spectrum I) do not differ greatly from those obtained from the bulk specimen (Fig. 1(a)), but are noticeably different from those yielded by stearic acid on Al. These observations suggest that the layers on Au consist of randomly oriented molecules rather than of self-organized assemblies[13]. The absorption peak near 1700 cm[-1] corresponds to a dimer band[10,12]. An appreciable number of acid molecules are thus paired via their COOH end-group. The presence of this absorption band in spectrum I indicates that the stearic acid COOH group does not interact as strongly with gold as it does with Al or glass. The presence of the dimer band and the observation of absorption intensities approximately twice the magnitudes measured from stearic acid on Al suggests that the stearic acid film on Au consists approximately of a double-layer.

Spectrum II in Fig. 1(d) was obtained from a lubricant layer on Au after heating to a temperature slightly higher than the melting point of stearic acid (71.5°C) for approximately 5 minutes. There was no detectable evidence of residual lubricant either when the substrate was hot or immediately after it was cooled to room temperature. These observations, along with those indicating no shift or distortion of absorption bands on adsorption of the lubricant on Au, indicate that the acid desorbs easily from gold and thus that it is loosely bound to the metal. This is in strong contrast with the adhesion properties of stearic acid on glass and oxide-covered Al.

In summary, IRAS spectra indicate that stearic acid reacts chemically with the native oxide

290

surface film on Al to form a strongly-bonded film. The thermal desorption data indicates that the chemical reactivity of the acid with borosilicate glass (and hence its bonding strength to glass) is weaker than with Al. Finally, the shape of the infrared absorption bands and the thermal desorption data of stearic acid on Au, indicate that the acid does not react chemically with gold and that it adheres to that metal considerably less strongly than to both glass and Al.

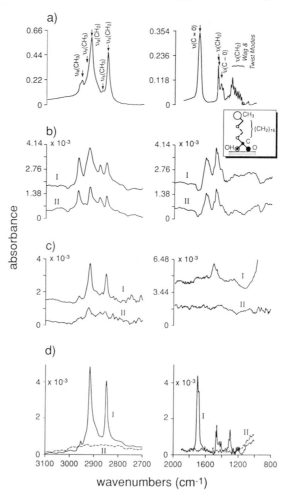

Figure 1. Infrared spectra of stearic acid as:
(a) bulk solid,
(b) an LB layer on aluminum immediately after deposition (spectrum I) and after~2 hours at 100°C (spectrum II),
(c) an LB layer on a borosilicate glass immediately after deposition (spectrum I) and after 5 minutes at 80°C (spectrum II).
(d) a molecular layer on gold immediately after deposition (spectrumI), and after 5 minutes at 72°C (spectrum II).

3.2. SHEAR-STRENGTH AND DURABILITY MEASUREMENTS

3.2.1 *Stearic Acid in glass/glass and glass/Al interfaces*. Depending on the compressive stress and the thickness of the Al coating, the kinetic friction coefficient measured at lubricated glass/glass and glass/Al interfaces ranged from 0.04 and 0.15 and was reproducible to approximately $\pm$ 0.015. Examinations by optical and electron microscopy yielded no clear evidence of mechanical damage to the slide or slider after a single traverse.

Although no wear marks could be identified on the Al surface, there were occasional indications of slight Al pickup on the slider after a single pass in the glass/stearic acid/Al system[5]. The detection by Auger spectroscopy of a relatively large concentration of carbon-containing material in the contact area of the slider after a single traverse was interpreted as residual stearic acid. The presence of acid on the slider suggests that the measured shear strength stemmed from shearing of a lubricant transfer-layer on the slider with lubricant on the slide, rather than from shearing of the lubricant/substrate interface.

The shear strengths calculated from the friction measurements are shown in Fig. 2. The shear strength of stearic acid layers on Al films increases from ~ 10 MPa to 70 MPa over the compressive stress range of ~100 to 1000 MPa. The shear strength of the acid on glass is noticeably lower than on Al. Over the contact stresses investigated, the dependence on normal stress of the shear strength of stearic acid measured in a glass/glass contact may be represented by the following least-squares-fit expression

$$\tau = -0.16 + 0.059\,\sigma + 1.13 \times 10^{-5}\,\sigma^2 \tag{3}$$

where τ and σ are respectively the shear strength and normal stress, all expressed in MPa. The standard deviations of the first term and the linear and quadratic coefficients in the right-hand-side (RHS) of Eq. (3) are respectively 1.42, 0.006 and 0.4×10^{-5}. The first term thus does not differ significantly from 0. Expression (3) is in excellent agreement with the earlier data of Briscoe et al.[3]. Similarly, the dependence on normal stress of the shear strength of stearic acid in a glass/Al contact is given by

$$\tau = 5.8 + 0.068\,\sigma \tag{4}$$

The standard deviations of the first term and the contact-stress coefficient in the RHS of Eq.(4) is 0.004 and 0.005 respectively. Within the experimental uncertainty, the rate of increase of τ with σ is thus identical in expressions (3) and (4) and the absolute shear strengths differ only by a constant amount, independent of contact stress. This suggests that the effect of the Al film on shear strength stems largely from a slider/slide interaction at light loads (σ~0), possibly due to adhesion.

It may easily be shown that the increase in the shear strength of stearic acid on Al films over the values measured on glass does not arise from ploughing[5]. We surmise that the increase stems in part from the transfer of Al from the slide to the slider. This transfer would increase friction due to slight metal/metal adhesion during sliding and would thus increase the effective shear strength. As mentioned earlier, there was unambiguous evidence of slight metal transfer to the slider[5].

The data presented above indicates that stearic acid is as efficacious a lubricant on glass as it is on Al, despite its lower binding energy to glass as indicated by the desorption data of Fig. 1(c). These observations suggest that a lubricant film need not adhere very strongly to rubbing solids

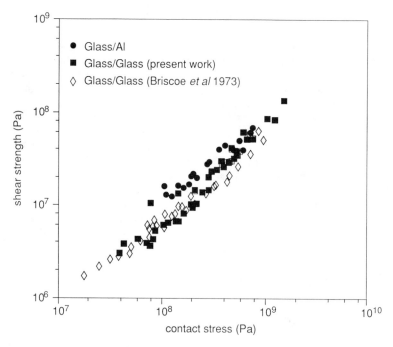

Figure 2. Dependence of shear strength of stearic acid on compressive stress, for several of the friction couples investigated. The shear strength properties of the lubricated Al/Al interface were similar to those measured from glass/Al couples, as explained in the text.

to remain in the sliding interface and function as a lubricant. The evidence for the existence of a threshold for this bonding strength, below which lubricant action is impossible, will become apparent when the frictional properties of gold/stearic acid/gold sliding contacts are considered.

The durability of stearic acid layers was measured with bare glass sliders both on bare glass and on Al-coated slides. Durability was also evaluated with Al-coated sliders (100 nm thick Al) on Al-coated slides. The variation of average kinetic friction with number of traverses, for the lubricated glass/glass interface at a contact stress of 200 MPa, is summarized in the lower curve of Fig. 3. There was no evidence of mechanical damage to either surface after 200 successive passes. The durability of the LB layers on glass appears excellent.

The typical durability of stearic acid monolayers on Al films, at a contact stress of 200 MPa, is summarized also in Fig. 3. The friction coefficient measured both with bare glass and Al-coated sliders rises rapidly with successive passes and reaches values almost characteristic of unlubricated glass/Al interfaces[5] after approximately 12 traverses. At that juncture, the mechanical damage to the Al film was similar to that generated in dry friction[5]. The metal film on the Al-coated slider was also severely disrupted. The durability of LB films in glass/Al and Al/Al interfaces is clearly lower than in glass/glass contacts, but is similar to the durability of stearic acid on steel[1]. The low durability on Al stems from metal transfer to the slider and the subsequent metal-to-metal adhesion which eventually leads to tearing of the Al surface. We surmise that metal transfer entraps the lubricant and effectively buries it.

The evidence of Fig.3 suggests that under conditions where a lubricant reacts chemically with the rubbing solids, lubricant durability is affected more by the wear rate of the sliding solids (hence by the rate of material exchange between the sliding surfaces) than by the degree of chemical reactivity of the lubricant with the solids. Thus, durability is high in glass/glass contacts because there is no significant exchange of glass between the sliding components. This result contrasts with the durability on Al where metal transfer across the sliding interface acts to deplete the lubricant in the wear track.

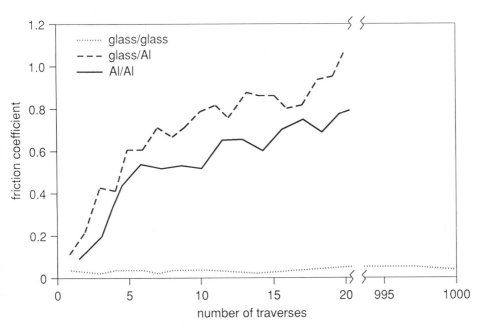

Figure 3.   Variation of kinetic friction coefficient with number of traverses, measured with glass sliders on stearic acid respectively on glass and on Al-coated glass. The durability of stearic acid LB films on glass is considerably higher than on Al.

3.2.2 *Stearic Acid in gold/gold, glass/gold and aluminum/gold interfaces.* The friction coefficients measured between glass and stearic acid-covered gold was appreciably smaller than that yielded by dry glass/Au interfaces[5]. No gold deposit could be detected by Auger spectroscopy on the glass slider surface. Because the acid adheres weakly to gold, as evidenced by the IRAS spectra in Fig. 1(d), the low friction in the lubricated interface is probably due to transfer of acid to the glass slider and to subsequent dragging of the alkyl chain over the gold.

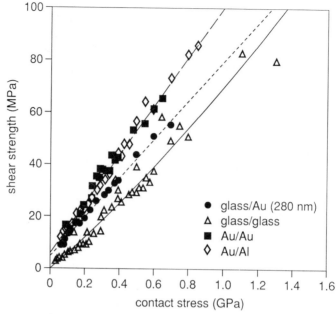

Figure 4.   Shear strength of stearic acid as measured in various interfaces.

The results of shear-strength measurements are shown in Fig. 4. A least-squares fit to the experimental data yields the following dependence of shear strength on applied stress for stearic acid in a glass/Au interface,

$$\tau = 3.7 + 0.076\,\sigma \tag{5}$$

where all the stresses are expressed in MPa. The standard deviations of the first term and the contact-stress coefficient in Eq. (5) are 0.79 and 0.002 respectively. Over most of the contact stress range investigated, the shear strength is slightly larger than that reported above for stearic acid both on Al and on glass. This observation may have an important implication. If friction in the lubricated glass/Au system stems primarily from dragging of the alkyl chain of acid molecules on gold, the observation suggests that the chain interacts more strongly with the Au surface than it does with another lubricant layer. The latter interaction occurs in lubricated glass/glass and Al/glass sliding interfaces.

One aim of the present investigation was to determine the load-bearing properties of stearic acid in a Au/Au interface, where the acid adheres weakly to both sides of the sliding contact. The capability of the acid either to sustain shear or to function in roller-bearing-like fashion in this interface would provide new insights into the action of lubricants in general.

Measurements of shear strength in lubricated Au/Au interfaces revealed that stearic acid does not sustain shear in this interface and is expelled from a Au/Au contact. An examination of the sliders used for shear-strength measurements, after the single traverse, revealed that the gold had been stripped away from one of the contacting surfaces. Figure 5(a) shows an SEM micrograph of the wear track on a lubricated gold slide following a typical traverse with a Au-coated slider. Initial contact with the slider was established on the left side of the track. The large flake in the area of initial contact is a fragment of gold removed from the slider and attached to the slide. Mechanical attachment of the flake could only have occurred after complete expulsion of the stearic acid from the slider/slide interface to allow for mechanical bonding of the contacting gold surfaces. This bonding in turn caused detachment of the gold film from the slider during the initial stage of motion.

The effective shear strength measured from lubricated Au/Au sliding interfaces is also shown in Fig. 4. Clearly, the shear strength pertains to lubricated glass/Au interfaces since the measurements were obtained from kinetic friction coefficients determined after detachment of the gold film from the slider. This interpretation is supported by the similarity in shear-strength properties of the lubricated Au/Au interface with those of the lubricated glass/Au system, also shown in Fig. 4. Note that the effective shear strength of the lubricated Au/Au interface is appreciably larger than that of the lubricated glass/Au interface over the entire range of contact stresses investigated. This larger effective shear strength may arise from additional resistance to sliding incurred by the dragging of gold debris across the wear track. Remnants of such gold film debris are clearly evident in the wear track on the slide in the micrograph of Fig. 5(a).

 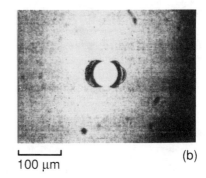

100 μm

(b)

Figure 5.   (a) SEM micrograph of wear track on lubricated gold after a traverse by a Au-coated glass slider. The large debris at the beginning of the track (left) is a fragment of gold peeled off from the slider.
(b) optical micrograph of Au-coated slider after a traverse on a lubricated Al film.

Figure 4 also shows the shear strength of stearic acid measured in Au/Al interfaces. The measurements were obtained without detectable evidence of mechanical damage to the gold on the slider or to the Al on the slide after a single traverse as illustrated in Fig. 5(b). The least-squares fit to the shear-strength data obtained from Au/Al interfaces yields

$$\tau = 4.27 + 0.096 \, \sigma \tag{6}$$

where all the stresses are expressed in MPa. The standard deviations of the first term and contact-stress coefficient in the RHS of this expression are 0.3 and 0.003 respectively. For completeness, Fig. 4 also reproduces the shear-strength data from the lubricated glass/glass interface described earlier. Note that shear strength is larger in a Au/Al or glass/Au interface, where the lubricant is attached to only one sliding surface, than in the glass/glass system where the lubricant is chemically attached to the two sides of the interface. As was explained earlier, this trend may stem in part from slight metal transfer across the sliding junction (where metal is present), which acts to increase the effective shear strength of the interface. If metal transfer is negligible, the trend would suggest that the free alkyl group of lubricant molecules attached to Al or glass (via the functional group) interacts more strongly with the opposite gold sliding surface than with another stearic acid layer.

The durability of LB layers on gold films was determined with bare glass sliders at a compressive stress of 200 MPa. Since stearic acid adheres weakly to gold and is easily transferred to the glass slider, the durability properties examined were in fact those of acid layers collected on the slider. The durability of the lubricant in Au/Al interfaces was also examined.

Figure 6 shows the results of three separate evaluations of the lubricant durability in a glass/Au interface. The thickness of the gold film on the glass slide was 280 nm. There is a steady increase in friction over 270 traverses. As the data suggest, the reproducibility of durability properties was excellent. Detailed examinations by SEM revealed appreciable gold transfer to the slider and damage in the form of scratches on the gold film after 10 to 15 traverses. The friction coefficient at that juncture was approximately the same as reported at dry glass/Au interfaces[5]. This suggests that stearic acid was depleted from the interface after this number of traverses.

Figure 6 also shows the durability of LB films in a Au/Al interface, at a contact stress of 280 MPa. The first traverse generally yielded some stiction and an average kinetic friction coefficient of~0.07. The friction coefficient then generally rose to~0.12 after a few traverses and remained approximately at that magnitude for several tens of traverses. After 80-100 traverses, the friction coefficient rose to~0.5 within a few additional traverses. This large friction increase occurred concurrently with severe mechanical damage to the Al film. The experimental evidence indicates that metal transfer led to the initial increase in friction in Al/Au interfaces[5].

The information in Fig.6 supports the conclusions derived from Fig.3 that lubricant reactivity has a smaller effect on durability than material transfer across the sliding interface, even where the reactivity is weak with one of the rubbing solids (i.e. gold). The durability of stearic acid in glass/Au and Au/Al interfaces is higher than in glass/Al contacts because the rate of metal transfer of Al to gold appears to be smaller than that of Al to glass.

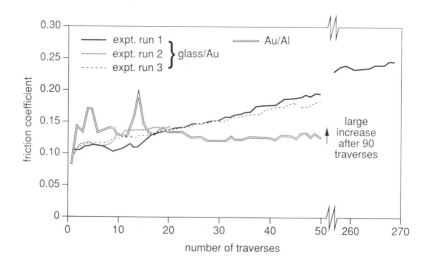

Figure 6.   Results of evaluations of durability of stearic acid at glass/Au and Au/Al interfaces.

4. Summary And Conclusions

The present paper has provided experimental evidence correlating the boundary-lubricant properties of stearic acid with the chemical reactivity (the bonding strength) of the acid with the sliding solids. The absence of chemical reactivity of stearic acid with gold and hence the weak adhesion of the lubricant to that metal inferred from the IRAS spectra, explains the inefficacy of stearic acid as a lubricant in gold/gold sliding contacts. The weakness of adhesion prevents the acid from sustaining mechanical shear in a gold/gold interface and thus from maintaining a coherent thin film in that interface. The experimental evidence demonstrates that stearic acid does not act in roller-bearing-like fashion in a Au/Au sliding interface. Lubricant function is restored when one side of the interface is replaced by a surface to which the acid adheres. These observations indicate that a "lubricant" needs to be chemically reactive[1] with only one side of a sliding interface to lubricate the interface.

The effective shear strength of stearic acid in glass/Au and Au/Al interfaces is larger than in a glass/glass contact. The larger shear strength in interfaces where metal is present may arise from the transfer of minute quantities of metal during sliding, thus increasing friction. If the effect of metal transfer on friction is negligible, the larger shear strength in glass/Au and Au/Al interfaces would suggest that the alkyl chain of the acid (bound to glass or Al) interacts more strongly with a gold surface than with a stearic acid layer chemisorbed on the mating surface, such as in a lubricated glass/glass contact.

On soft materials where lubricant adhesion is strong, such as stearic acid on Al, lubricant durability is limited by the rate of transfer of metal to the slider. This transfer may bury the lubricant, effectively removing it from the sliding interface. On hard surfaces where lubricant adhesion is also relatively strong, such as stearic acid on glass, lubricant durability is very high because material transfer across the sliding interface is minimal.

5. Acknowledgments

The authors thank C.V. Pelow and P. Hamstra of the Kingston R&D Centre for technical assistance during the course of this work.

References

1. F.P. Bowden and D. Tabor, The Friction and Lubrication of Solids, (Oxford Press, 1950) Vol. 1.
2. B.J. Briscoe, B. Scruton and F.R. Willis, Proc. Roy. Soc. (London), A333, 99 (1973).
3. B.J. Briscoe and D.C.B. Evans, Proc. R. Soc. Lond. (London), A380, 389 (1982).
4. R.S. Timsit, G. Stratford and M. Fairlie, Tribology and Mechanics of Magnetic Storage Systems, STLE Special Pub. SP-22, 4, 98 (1987).
5. R.S. Timsit and C.V. Pelow, J. Tribology, to appear (1992).
6. S. Timoshenko and J.N. Goodier, Theory of Elasticity Co., New York, 1951).
7. J.M. Leroy and B. Villechaise, Mechanics of Coatings, Proc. 16th Leeds-Lyon Symposium on Tribology, (Elsevier, Amsterdam, 1990) 195.
8. A. Udagawa, T. Matsui and S. Tanaka, Appl. Spectrosc., 40, 794 (1986).

9. A.S. Akhmatov, Molecular Physics of Boundary Friction, (Israel Program for Scientific Translations, Jerusalem, 1966).

10. I.R. Hill and I.W. Levin, J. Chem. Phys., 70, 842 (1979)

11. P.A. Chollet, J. Messier and C. Rosilio, J. Chem. Phys., 64, 402 (1976).

12. M. Fairlie, B.R. Pathak and R.S.Timsit, Proc. 1985 Conf. on Fourier and Computerized Infrared Spectroscopy, SPIE 553, 496 (1985).

13. D.L. Allara and R.G. Nuzzo, Langmuir, 1, 52 (1985).

14. J.P. Williams and D.E Campbell, Encyclopedia of Industrial Chemical Analysis, (J. Wiley & Sons, Inc., New York, 1971), 403.

15. R.S. Timsit, R. Hombek and H. Sulek, paper presented at the STLE Conference, May 1990, Denver CO.

CHEMICAL EFFECTS IN FRICTION

Traugott E. Fischer
Department of Materials Science and Engineering
Stevens Institute of Technology
Hoboken NJ 07030, USA

ABSTRACT. Tribochemistry designates the interaction of chemistry and friction. This interaction usually takes the form of a modification, most often an acceleration, of reaction rates. This paper deals with the tribochemical reactions of the sliding solids where the worn volume can be used to determine reaction rates and where the morphology and composition of the wear tracks can give information on the reaction products. We concentrate on tribochemistry in slow sliding to eliminate the obvious effects of high temperatures. We discuss the tribochemical reactions of various ceramics with ambient humidity for the information they provide on the chemistry: we find that the oxidation of covalent ceramics (Si_3N_4 and SiC) and the dissolution of its reaction products in water modifies the surface morphology and contact stresses, resulting in decreases of wear rates, and that oxide ceramics are susceptible to adsorption embrittlement which increases wear rates. The oxidative wear of steel is examined for its ability to provide insight into the mechanisms and kinetics of tribochemical reactions. In general, friction lowers the temperature at which reactions are observed by several hundred degrees.

INTRODUCTION

Since friction, in its simplest form, involves the making and breaking of adhesive bonds between the sliding bodies and is always influenced by surface films that modify this adhesion, it is almost obvious that chemical reactions of the bodies with each other, with the environment or with liquid lubricants play a major role in tribology. It is widely observed that chemical reaction rates are strongly modified, usually accelerated, by the simultaneous occurrence of friction. This phenomenon bears the name of tribochemistry and has been extensively studied in the Institute of Peter-Adolf Thiessen in Berlin for more than twenty years [1,2]. Their work was primarily focused on the exploration of the tribochemical reactions produced in ball mills. We will concentrate our attention on the reactions that occur in sliding, with or without lubrication.

Singer [3,4] has analyzed the thermochemical aspects of such reactions and described the new compounds that can be formed from the interaction of

299

L. Singer and H. M. Pollock (eds.), Fundamentals of Friction: Macroscopic and Microscopic Processes, 299–312.
1992 Kluwer Academic Publishers. Printed in the Netherlands.

the sliding bodies. We shall concern ourselves here with the kinetics of tribochemical reactions; namely, with the modification of the reaction rates by simultaneous friction [5]. We first describe various mechanisms by which friction modifies reaction kinetics, we then review the tribo chemistry of ceramics to gain an overview of the different phenomena that are encountered, finally we examine the mechanisms of tribochemical acceleration of reaction rates in the possible oxidative wear of steels.

MECHANISMS OF TRIBOCHEMISTRY

Frictional Heat

The most obvious mechanism by which friction increases the rate of chemical reactions is frictional heat which has been used since the earl iest times for the production of fire. As discussed elsewhere in this volume, one distinguishes between general increase in temperature and flash temperature [6-8,48]; the latter consists of the temperature flashes that occur at contacting asperities.

These temperatures increase linearly with sliding velocity and more slowly with load, because the latter determines the real contact area over which the friction power is dissipated. The increases in reaction rate caused by frictional heat are no different than those caused by other increases in temperature; they constitute the majority of tribochemical phenomena and will not concern us further here. When the sliding velocity and load are kept low to avoid frictional heating, one observes that other mechanisms operate by which friction increases reaction rates.

Exposure of Clean Surfaces

It is well known that reactions with clean surfaces are usually rapid and that their rates are controlled by diffusion of reagents through the layer of reaction product that is formed on the surfaces. Wear exposes fresh surfaces and accelerates the reaction; in steady state, the reaction rate equals the rate of removal of the reaction product, as we shall see in the oxidative wear of steels.

Transformation of the Material

Wear constitutes a severe deformation of material on a small scale. Strains in the order of 1000% are commonly observed in metals; in ceramics one observes plastic deformation and microfracture. These defects serve as high-energy sites with increased reactivity on the surface and as dif fusion paths that cause large increases in diffusion limited reaction rates in the subsurface region. Razavizadeh and Eyre, for example, observed relatively thick layers of oxide on the surface of friction tracks in aluminum [9].

Triboelectricity

Rubbing and fracture cause charge separations in ionic materials and set up large electrostatic potentials that can lead to discharge in the surrounding gas. With metallic bodies separated by a lubricant, electrochemical potentials are often observed. In addition, exoelectron emission is often quoted as a cause of tribochemical reactions despite the very low emission currents usually observed.

Direct Mechanical Stimulation of Reactions

Consider two atoms that are separated by large mechanical stresses. As the distance of these atoms increases, the energy splitting between the highest occupied molecular orbital (HOMO, usually bonding) and the lowest unoccupied molecular orbital (LUMO, usually antibonding) decreases. This diminishes the activation energy of the electron transfer taking part in a chemical reaction. Such phenomena occur only where atoms are separated; namely, in friction and in fracture. In solids, such a mechanism participates in the phenomenon of adsorption-induced fracture or stress corrosion cracking; it is responsible for the wear increase of oxide ceramics by water and polar hydrocarbons.

CHEMICAL EFFECTS ON CERAMIC TRIBOLOGY

The tribochemistry of ceramics takes several forms, depending on the materials, the environment and the mechanical conditions of rubbing; it can consist in modifications of surface composition and topology that decrease wear [10,11], in a purely chemical form of wear (by dissolution in the liquid environment) [12,13], in the chemisorption and boundary lubrication effectiveness of inert hydrocarbon [11,14], or in chemically induced cracking that increases wear rates [14].

Wear Reduction by Tribochemical Oxidation

The ambient humidity has a pronounced effect on the wear of silicon nitride and other ceramics [11,15] not only the amount of wear, but the wear mechanism itself is modified by humidity. Silicon nitride wears rapidly in dry argon, but if the environment contains various amounts of water vapor, the wear rate decreases by as much as two orders of magnitude. Under these conditions, the wear scar is much smoother than after sliding in dry gases and is covered with an amorphous silicon oxide which is probably strongly hydrated [10,16]. The result is a reduction of the local stresses responsible for the mechanical wear (Figures 1 and 2).

In the absence of friction, measurable oxidation rates of silicon nitride are obtained only above 1000K [17]; they are increased one thousandfold by the presence of humidity in the air [18]. During friction, massive oxidation is obtained even at room temperature. How this occurs exactly has not been determined yet. One can speculate that the reaction is accelerated because the hydroxide formed on the surface is continuously removed and a fresh surface is exposed by friction, but clear experimental

Figure 1. Wear track of silicon nitride after sliding in dry argon to avoid tribochemical effects. The arrow shows the sliding direction of the counterface and is 10 μm long. Note the relatively rough surface and the loose fine wear debris.

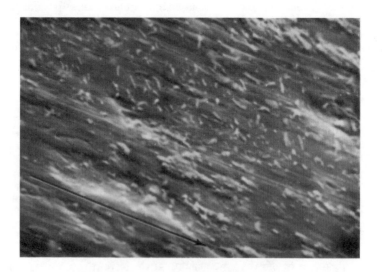

Figure 2. Wear track of silicon nitride after sliding in 98% humid argon. The arrow shows the sliding direction of the counterface and is 10 μm long. Note the smoother surface and the coherent, protective, layer of wear debris. The latter consists mainly of hydrated silicon oxide.

vidence for any one mechanism is still lacking.

When silicon nitride slides in water, the tribological reaction is a
issolution [12] of the material at the contacting asperities that produces
ltraflat surfaces and completely eliminates mechanical wear; the resultant
urfaces are so flat that hydrodynamic lubrication [13] is obtained even
n water at low sliding speeds (Figure 3).

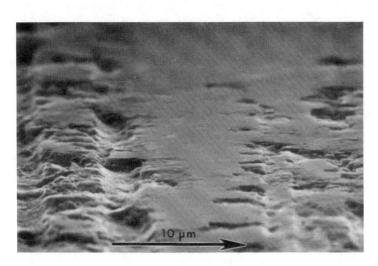

igure 3. Wear track of silicon nitride after sliding in water. The arrow
s 10 μm long. Note the flat contact surface caused by tribochemical dis-
olution of the asperities.

he formation of lubricious oxides.

At elevated temperatures, sliding in humid air reduces friction as
ell as wear in a limited range of load and sliding velocities [19]: the
urfaces are covered by a smooth layer of silicon oxide that continuously
eplenishes itself as it wears. As the severity of sliding (i.e., the
roduct of load and sliding velocity) is increased beyond a certain value,
riction and wear are high and the wear surfaces are rough. The passage
rom low-to-high friction is the result of a competition between the kine-
ics of formation of the lubricious oxide layer and its wear.

Lubricious oxides have recently been formed on silicon nitride by pre-
xidation [20]. Friction coefficients as low as μ = 0.05 have been
btained with a very smooth and flat surface that avoids ploughing. The
mooth surface is achieved by friction in water as described above [13];
ubsequent oxidation in air for a few minutes produces a low-friction
urface.

Gates, Hsu and Klaus [21] have shown that sliding in water causes a
ecrease of the friction coefficient of alumina against itself from μ = 0.6

to μ = 0.25. This lowering of friction is attributed to the formation o stable aluminum hydroxides which are modified to a layered structur (trihydroxide-bayerite) by the frictional shear stress.

Chemically induced fracture in oxide ceramics.

Wallbridge, Dowson and Roberts [22] have reported that wear of alumin sliding in water is higher than in air. With zirconia, water increases th wear rate by an order of magnitude over wear in dry nitrogen [14]. Thi increase in wear occurs by intergranular fracture; caused by chemisorptio embrittlement. According to Wiederhorn and Michalske and their coworker [23,24], this phenomenon occurs by the attack of the bonds between neigh bor217ing metal and oxygen ions at a crack tip by water; it is related t the well-known tendency of oxide ceramics to form hydroxylic surfaces [16]

Adsorption-induced fracture occurs also in the presence of hydrocarbo lubricants [14]. In the case of zirconia, paraffin causes an increase i the wear rate of about 50% over the wear rate in dry nitrogen where th friction coefficient is 0.7. When sliding occurs in paraffin with 0.5 stearic acid, which is a classic boundary lubricant, the friction coeffi cient decreases to 0.09 but wear increases by another factor of 3. Chem ical attack of the grain boundaries and intergranular fracture are th causes of this increase in wear.

Boundary lubrication by paraffins.

Polar molecules such as fatty acids, alcohols and esters adsorb ont metallic surfaces and cause boundary lubrication; nonpolar hydrocarbon (for instance, paraffins) do not act as boundary lubricants for metals In the case of all ceramics investigated or known to us, paraffins act a boundary lubricants, with friction coefficients in the neighborhood of = 0.12. We do not have a proven explanation for this difference; it prob ably has the same origin as the catalytic activity of ceramics for th cracking and isomerization of hydrocarbon; the acid sites on the surfac of ceramics, strong enough to break carbon-carbon bonds at elevated tem peratures in catalytic cracking [25], are capable of absorbing the mole cules at room temperature and provide boundary lubrication [11,14].

THE OXIDATIVE WEAR OF STEEL

The wear of a metal in dry sliding (i.e., without a lubricant) can b severe or mild, depending on the hardness of the material, the load and th sliding velocity [26-28]. Often the initial wear is severe and occurs b plastic deformation and fracture. After a certain time, small wear debri forms a layer that decreases the local stresses and the wear rate. Unde these conditions, material is removed from the sliding surface by oxidativ wear: An oxide grows on the steel surface; because of the differing crysta structures of steel and oxide, incompatibility stresses at the steel/oxid interface reduce the adhesion of the oxide so that it is removed by th action of friction, and a clean metal surface is exposed and re-oxidizes Oxidation and removal of the oxide is thus a quasiperiodic process [29-31].

In the absence of friction, the oxidation rate of steel is controlled by the diffusion of the reacting species through the layer of the reaction product. At higher temperatures the oxidation of iron follows a parabolic law [32]; with the oxide layer thickness increasing with the square root of time:

$$\xi = C t^{\frac{1}{2}} , \qquad (1)$$

where ξ is the thickness of the oxide layer, t is the average growth time, and C is the parabolic rate constant. At room temperature, the growth rate is logarithmic:

$$\xi = a \ln \left(\frac{t}{t_0} + 1 \right) , \qquad (2)$$

where both a and t_0 are constants [33]. The parabolic law corresponds to thermally activated diffusion through the oxide layer, and the logarithmic law corresponds to the electromigration of ions in the large electric field formed by the oxygen anions adsorbed on the oxide surface. These anions are formed by quantum mechanical tunneling of electrons through the oxide.

The quasiperiodic removal of the oxide scale layer by friction is a fundamental mechanism of oxidative wear: it increases the net oxidation rate, and is responsible for the material removal that constitutes wear. It is convenient to approximate this phenomenon by taking an average time τ between successive scale removals and to assume that the oxide grows from thickness zero during that time according to some kinetic law. This mechanism leads to an oxide growth rate that is linear in time:

$$X(t) = \xi(\tau) \frac{t}{\tau} , \qquad (3)$$

where t is the total sliding time, τ the average growth time, and ξ the thickness of the oxide layer. If the oxidizing area is A = L/H, where L is the load and H the hardness of the steel, the volume of oxide grown and removed at each step is

$$\Delta V = \frac{L}{H} \xi(\tau) , \qquad (4)$$

and the total amount of material removed by oxidative wear is

$$V(t) = \frac{L}{H} \xi(\tau) \frac{t}{\tau} . \qquad (5)$$

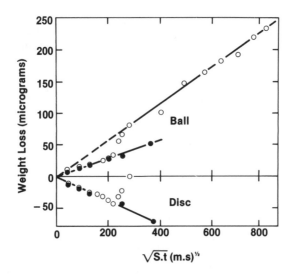

Figure 4. Oxidative wear of AISI 52100 steel. Weight loss of the ball an‹ weight gain of the disk. s is the sliding, t the time. The velocity wa‹ varied. o = constant time, varying distance, • = constant distance, vary‹ ing time. Load: 9.8 Newton.

Figure 5. Oxidative wear of 52100 steel. Scanning electron micrograph o‹ the transferred oxide layer on the disk. The bar indicates 1 μm. Notic‹ that the transfer layer consists of very thin oxide platelets.

The wear rate is determined by the kinetics of oxide growth and of scale removal. Quinn [31,32] has proposed a theory for the oxidative wear of steel at sliding velocities where frictional heat causes high temperatures. He assumed that the oxide is removed whenever it reaches a critical thickness ξ_c after having grown according to the parabolic law. In this case $\xi(\tau) = \xi_c$ and Equation 1 yields $\tau = \xi^2_c/C^2$ so that Equation 5 becomes

$$V_{Quinn}(\tau) = \frac{LC^2}{H\xi_c} \cdot t . \tag{6}$$

The surface temperature is proportional to the sliding velocity v [6-9] and determines the oxidation rate constant C of Eqs. 1 and 6 through an Arrhenius law, so that (6) becomes, with v = S/t,

$$V = \frac{LC_s^2 \ EXP \ (-2Q/Kt)}{H\xi_c v} \cdot t . \tag{7}$$

where S is the sliding distance, v the velocity, Q an activation energy, k Boltzmann's constant, and T the surface temperature of the sliding bodies. Equation 6 is a simplified version of Quinn's formula that contains terms allowing for a real contact area divided into a number of spot contacts. Much of the remaining modeling in Quinn's theory relates T with the load, the sliding velocity, and the details of the contact geometry. Notice that the wear rate depends on the velocity through the temperature that is generated by friction.

Fischer and Sexton [34] have designed their experiments to keep frictional temperatures increases insignificant ($\Delta T \approx 2K$) by using small loads and velocities. They observed that oxidative wear is linear in sliding distance and varies with velocity as

$$V = \frac{LC}{H\sqrt{\lambda}} \cdot \frac{S}{\sqrt{v}} = \frac{LC}{H\sqrt{\lambda}} \cdot \sqrt{vt} = \frac{LC}{H\sqrt{\lambda}} \cdot \sqrt{St} , \tag{8}$$

where λ is an average sliding distance (defined below). This relationship is illustrated in Figures 4 and 5. Equation 8 is obtained directly from Equation 5 if one assumes that the oxide film is removed, whatever is thickness, every time the surfaces have rubbed over the average distance λ. The weakness of Equations 7 and 8 is that they contain in an unknown parameter (ξ_c or λ). By invoking a physically realistic thickness value [$\xi(\Delta\tau) > 2$ nm] for the film at removal, one can set a lower limit to λ (>10 nm) and to C in Equation 8.

Comparison with experimental values of ΔV then leads to the conclusion that friction changes the growth law of the oxide film at room temperature from logarithmic to parabolic and causes the oxidation rate to be as large as it would be at 900 K in the absence of friction [34]. During rubbing of the surfaces, diffusion becomes more rapid than electromigration. Diffusion is greatly facilitated by structural defects such as grain boundaries, dislocations, and vacancies; it is also well known that large number of such defects are introduced by plastic deformation in the surface of materials. Therefore, structural modification of the material, especially the creation of defects, constitutes another mechanism by which friction increases chemical reactions [34]. The existence of a layer of material with highly excited defects is further demonstrated by such phenomena as triboluminescence and exoelectron emission.

CONCLUSION

This brief selection of examples has shown us that during sliding, chemical reactions can occur with rates that otherwise require much higher temperatures. These tribochemical reactions can increase or decrease the wear rate, depending on the material and the gaseous and liquid environment. Decreases in wear have been observed in silicon nitride because the reactions occurring at contacting asperities produce smooth surfaces that eliminate the concentration of contact stresses and the resultant mechanical failure mechanisms. The oxidative wear of steel is less severe than the purely mechanical wear because the interfacial stresses between substrate and oxide cause the delamination of the latter, and consequently a much shallower removal of material than caused by plastic deformation. In oxide ceramics, by contrast, reaction with humidity cannot remove material at asperities, but causes a weakening of the material bonds at crack tips. The consequence is an increase in the wear rate by intergranular fracture.

The oxidative wear of steel by presented a phenomenon: the harder ball loses material which is transferred to the softer disc in the form of oxide. In addition the wear rate presented a clear kinetic law that was amenable to modeling and that showed that the oxidation is not only accelerated, but follows a different kinetic law that pointed to increased diffusion. In the case of ceramic tribochemistry, the dependence of wear rate on sliding velocity is much more complex and we have not yet been able to identify with certainty the mechanisms by which friction modifies the chemical reactions. The challenge in providing an understanding of tribochemistry will be to design experiments that isolate individual causal relationships as a first step. Once we have gained more insight into the simple situations, it will be possible to explore simultaneous phenomena that represent technical situations more realistically.

Tribochemistry is not limited to the examples we have looked at. Diamond and carbon films are known to undergo tribochemical reactions. Diamond, for instance, cannot be used as a cutting tool for steels as it dissolves in the latter. We also are convinced that the polishing of diamond by jeweler's rouge is a simple tribochemical oxidation of carbon and reduction of iron oxide, similar to the reaction that is used in the reduction of ore. Diamond-like carbon films likewise are known to interact with the environment [35]; modifications of lubricants in sliding service are well known. Polymer molecules dissolved in the oils to decrease the temperature dependence of the viscosity are straightened and aligned or cracked by the high shear, leading to temporary or permanent loss of viscosity. The action of antiwear and "extreme pressure" additives in lubricants is based on tribochemical reactions.

We feel that a more thorough understanding of the principles of tribo-chemistry will further our understanding of the principles of friction and wear and will provide a foundation for the development of advanced lubricants and tribological materials.

ACKNOWLEDGEMENTS

This work was supported by grants from the National Science Foundation and the Dow Chemical Foundation.

REFERENCES

[1] G. Heinicke, 1984, Tribochemistry, Munich: Carl Hanser Verlag.
[2] P. A. Thiessen, K. Meyer, and G. Heinicke, eds., 1967, Grundlagen der Tribochemie, Berlin, Akademie Verlag.
[3] I. L. Singer, this volume and Surf. & Coatings Technol., 49 (1991) 474.
[4] S. Fayeulle, I. L. Singer and P. D. Ehni in Mechanics of Coatings, Leeds-Lyon 16 Tribology Series, 27, edited by D. Dowson, C. M. Taylor, M. Godet (Elsevier, 1990) p. 129.
[5] T. E. Fischer, Ann. Rev. Mater. Science, 18 (1988) 303.
[6] J. P. Archard, Tribology 7 (1974) 213.
[7] D. Tabor, Proc. R. Soc. London Ser. A251 (1959) 378.
[8] M. F. Ashby, J. Abulawi and H-S. Kong, STLE Tribol. Trans. October 1991.
[9] K. Razavizadeh and T. S. Eyre, Wear 79 (1982) 325.
[10] T. E. Fischer and H. Tomizawa, Wear, 105 (1985) 29.
[11] S. Jahanmir and T. E. Fischer, Tribology Trans., 31 (1988) 32.
[12] T. Sugita, K. Ueda and Y. Kanemura, Wear, 97 (1984) 41.
[13] H. Tomizawa and T. E. Fischer, ASLE Transactions, 30 (1987) 41.
[14] T. E. Fischer, M. P. Anderson, S. Jahanmir and R. Salher, Wear, 124 (1988) 133.
[15] H. Shimura and Y. Tsuya, Wear of Materials, 1977, K. Ludema, ed. ASME, New York (1977) 452.
[16] R. K. Iler, The Chemistry of Silica, John Wiley & Sons, New York (1979).
[17] A. J. Kiehle, L. K. Heung, P. J. Gielisse and T. J. Rockett, J. Am. Ceram. Soc., 58 (1975) 17.
[18] S. C. Singhal, J. Am. Ceram. Soc., 59 (1976) 82.
[19] H. Tomizawa and T. E. Fischer, ASLE Transactions, 29 (1986) 481.
[20] T. E. Fischer, H. Liang, W. M. Mullins, Mat. Res. Soc. Symp. Proc., 140 (1989) 339.
[21] R. S. Gates, S. M. Hsu and E. E. Klaus, Tribology Trans., 32 (1989) 357.
[22] N. Wallbridge, D. Dowson and E. W. Roberts, Wear of Materials, 1983, K. C. Ludema ed,.., ASME, New York 202.
[23] S. M. Wiederhorn, S. W. Freiman, E. R. Fuller and C. J. Simmons, J. Materials Sci., 27 (1982) 3460.
[24] T. A. Michalske and B. C. Bunker, J. Appl. Phys., 56 (1984) 2686.
[25] H. Knozinger and P. Ratnasamy, Catal. Rev.-Sci. Eng., 17 (1978) 31.
[26] J. F. Archard, J. Appl. Phys. 24 (1953) 981.

[27] J. F. Archard and W. Hirst, Proc. R. Soc. London Ser. A 236 (1956) 397-410.
[28] W. Hirst and J. K. Lancaster, J. Appl. Phys. 27 (1956) 157.
[29] T. F. J. Quinn, Tribol. Int. 16 (1983) 305.
[30] T. F. J. Quinn, Tribol. Int. 16 (1983) 257.
[31] M. D. Sexton and T. E. Fischer, Wear 96 (1984) 17-30.
[32] M. H. Davies, M. T. Simnad and C. E. Birchenall, J. Met. 3 (1941) 889.
[33] E. A. Gulbransen, Trans. Electrochem. Soc. 81 (1942) 327.
[34] T. E. Fischer and M. D. Sexton, 1984, Physical Chemistry of the Solid State; Applications to Metals and their Compounds, ed. P. Lacome, p. 97, Amsterdam: Elsevier (1984).
[35] D. S. Kim, T. E. Fischer and B. Gallois, Surf. & Coatings Technol., 49 (1991) 537.

Discussion *following the lecture by T E Fischer on "Chemical effects in friction"*

B D STROM. You've discussed the possibility of a tribological interface lowering the activation energy for chemical reactions. This is one possible explanation for tribochemical reactions taking place at low temperatures. Do you suppose that mechanical energy may provide the energy for chemical reaction directly, without existing as thermal energy and regardless of changes in activation energy?

T E FISCHER. This is a very interesting question. I have not described a direct excitation of a chemical reaction by mechanical energy in the text because we do not have any evidence for it in solids. In liquids, this is well known: the very high shear rates in a lubricated contact (as large as 10^7 sec^{-1}) causes a straightening and alignment of dissolved polymer molecules; it even causes cracking of large polymers. In solids, I believe such a phenomenon is unlikely. Even very large sliding speeds (10 m/s = 10^{10} nm/s) are slow with respect to thermal vibration speeds.

There remains the question of the definition of temperature. As a bond between atoms of the two bodies is broken, it is quite possible that these atoms vibrate with energies much larger than the kT of the flash temperature (see Uzi Landman's contribution to this volume), and that these vibrations excite chemical reactions. Now, do we wish to call this thermal energy or direct mechanical excitation?

Notice also that it is not easy to distinguish the "decrease in activation energy" due to atom separation from direct mechanical excitation. The splitting between the HOMO and LUMO levels is caused by the mechanical energy; one can describe this phenomenon in different words: the tensile stress stores energy in the bond, and this energy is available to the chemical reaction. Thus the same phenomenon can be described as decreasing the activation energy and as direct mechanical stimulation.

I find it attractive that this problem stimulates us to revisit the fundamental definitions of temperature and energy.

J T YATES. The question of whether mechanical deformation can cause chemical reactions has an analogy with current chemistry studies in which molecules are accelerated as high energy beams against a surface, and reactions seem to occur. A model being proposed for methane molecules activated by this acceleration/collision process is that when the molecule flattens against the surface, three of the hydrogen bonds bend, bringing the carbon atom closer to the surface, which promotes the breaking of one of the carbon-hydrogen bonds. The analogy here is that we put the mechanical energy into the molecule and cause it to react with the static surface.

T E FISCHER. This model is indeed an interesting analogue to tribochemistry. I tend to agree with you that this is a case of modification of the molecular orbitals and their energies, with consequent modification of the symmetry (Woodward Hoffman) rules and activation energies for reaction. This is also an analogue for the fundamental mechanisms of catalysis, where the adsorption of molecules onto a surface causes distortion of the molecular shape and shifts in the molecular orbitals.

R S POLIZZOTTI. A comment and a question regarding zirconia. Zirconias are basic oxides, unlike silica, and perhaps the hexadecane and stearic acid results are related in the following way: hexadecane has a reasonably high solubility for water and oxygen in air. The oxidative thermochemistry of hexadecane is such that, in the presence of light, there is a good probability (even at room temperature) of generating significant concentrations of hexadecanoic acid. As a result the acid/base chemistry associated with the chemical attack on the surface may be responsible for the pitting you are seeing in that system. Also, if zirconia is treated with certain mineral acids like sulfuric acid, it converts the surface into a hydrated, sulfated zirconia, which is in fact a super acid and is an extraordinarily strong cracking catalyst. It might be worthwhile pretreating zirconia surfaces with these mineral acids and also neutralizing them to see whether these chemicals have an effect. And finally a question: have you looked at pH effects of water with regard to some of the wear?

T E FISCHER. The oxidation of hexadecane is an interesting problem. The resultant formation of hexadecanoic acid is often invoked to explain the low friction that we obtain. A simple experiment provides evidence against that: this oxidation of hexadecane is independent of the sliding material and should cause boundary lubrication in all cases. We obtain boundary lubrication only with ceramics, not with metals. We haven't looked at pH effects, but we know that in Si_3N_4 they should be very important because silica is much more soluble in basic water than in acid water.

K L JOHNSON. In the oxidative wear of steels, what is the size of the wear debris, and does that lead to a situation in which there is a layer of granular material separating the two surfaces?

T E FISCHER. The wear debris have the form of very thin platelets of oxide on the disk; they are so thin that we were not able to see their thickness in the SEM. No wear debris are observed on the ball. As the experiment progresses, there is mechanical wear on the disk itself, and larger debris are formed. Our model applies only to the early sliding where the oxide worn off the ball accumulates on the disk.

D TABOR. During the second World War, I was involved in a group that was polishing glass lenses (presumably for military purposes). They discovered that if you had an acidic glass, the rate of removal of material and the surface finish depended on having a certain amount of sodium hydroxide in the water used in the polishing wheel. On the other hand, if you had a basic glass you would put a slight amount of acid in the water to increase the polishing rate and improve the surface finish.

T E FISCHER. This is a clear case of tribochemistry used before the name was coined by Heinicke and Thiessen. The semiconductor industry also finds that changing the pH of water improves the polishing of silicon wafers.

D DOWSON. Did you estimate the thickness of the water film on those very smooth surfaces?

T E FISCHER. Yes, by applying a standard calculation of hydrodynamic lubrication, given the load, the velocity, and the known viscosity of water, we find that this thickness is less than ten nanometers. Such a thin fluid film can provide hydrodynamic lubrication only for the very smooth surfaces that tribochemical wear forms on silicon nitride.

SURFACE SCIENCE AND EXTREME PRESSURE LUBRICATION - CCl_4 CHEMISTRY ON Fe(110)

John T. Yates, Jr., V.S. Smentkowski and A.L. Linsebigler
Surface Science Center
Department of Chemistry
University of Pittsburgh
Pittsburgh, PA 15260, USA.

ABSTRACT. The application of surface science methods to the investigation of model lubricant-metal surface chemistry processes is described. The chemistry of CCl_4 on an atomically clean Fe(110) surface has been investigated and the lubricant, $FeCl_2$, has been postulated to nucleate as clusters on defect sites. A method to quench this extreme pressure lubricant nucleation and hence to promote lubricant dispersion is described.

1. Introduction

Extreme pressure tribology occurs during cutting and grinding of metals when cutting fluids or grinding aids are employed. Very little is known about the molecular details of extreme pressure (EP) tribology. This is because the experimental techniques of surface science, with their ability to measure surface phenomena at the atomic level of resolution, have not been widely applied to tribology. Pioneering work by Buckley [1,2] involved EP lubricants containing either chlorine or sulphur atoms chemically bound in model lubricant molecules. It was discovered that surface chemical reactions between the lubricant molecule and the metal surface were induced by the production of nascent metal surfaces during mechanical abrasion. These nascent surfaces are really atomically clean surfaces, exposed from the bulk by the removal of the surface of the material. Auger spectroscopy was employed as a surface analytical tool in these studies, often using stainless steel as a substrate. However, using Auger spectroscopy alone, the basic surface chemical reaction steps involved could not be studied in detail. Indeed, the main papers and reviews of this field do not present firm information about chemical reaction steps or even definitely identify surface chemical species which compose the lubricant layer [3-6].

The problem of EP lubrication is complex. Under conditions where grinding or machining of metals occurs in the presence of a fluid added to increase efficiency, one

I. L. Singer and H. M. Pollock (eds.), Fundamentals of Friction: Macroscopic and Microscopic Processes, 313–322.

is dealing with a complex problem in which surface chemistry is mixed with physical effects due to the deformation of the solid substrate. These include:

- production of nascent surfaces
- production of surface defects at the nascent surface
- production of mechanical strain at the surface
- generation of high local temperatures
- decomposition of the lubricant molecule

There is strong evidence that tribological phenomena exhibit an underlying character which is dependent on molecular level issues. For example in the early work of Zisman, it was shown, using Langmuir-Blodgett films, that the lubrication of a slider on glass was related to just a monolayer of lubricant, and that the coefficient of friction was systematically dependent on the molecular weight of the lubricant molecule [7]. More recent elegant molecular tribometry work by Israelachvili [8] and by Granick [9] has detected changes in lubrication parameters which are related to monolayer-by-monolayer changes in thickness of the lubricant film involving only a few molecular layers of lubricant above an atomically smooth mica surface. Careful ultrahigh vacuum (UHV) pin (sapphire)-on-disk experiments have detected a decrease in the coefficient of friction for Cl_2 adsorption on iron [10]. And finally recent atomic force microscopy experiments have detected stick-slip phenomena at the atomic level of resolution on graphite [11]. In cases where surface chemistry occurs (as in EP lubrication at reactive surfaces) it is most certain that molecular level phenomena are of importance, because molecular events such as defect formation on the surface and chemical bond breaking in the adsorbed lubricant molecules occur.

2. Tribochemistry on Atomically Clean Iron Surfaces

2.1. Model Nascent Iron Surfaces

In a dynamic grinding experiment [12], or in abrasive pin-on-disk experiments [13], fresh surfaces are being produced continuously as metal is moved or removed from the surface. This process is schematically shown in Figure 1. Here an abrasive particle or a tool edge produces a chip by breaking metal-metal bonds. The new surfaces produced (labeled A and B) are initially atomically clean surfaces, having never been exposed to the atmosphere. These nascent surfaces contain a high density of defect sites, produced during mechanical separation [14]. The reactivity of the nascent metal surfaces with atmospheric gases (if present), or with molecules of a fluid added for lubrication purposes, will be high for many metals. Atmospheric oxygen (itself an extreme pressure lubricant for iron [12]) has a sticking coefficient near unity [15]; thus the lifetime of the clean nascent surface will be of the order of only 0.05 μ sec in air at one atmosphere total pressure.

In order to investigate the surface chemistry which occurs on such nascent surfaces, we have utilized atomically clean iron single crystals which are studied in ultrahigh

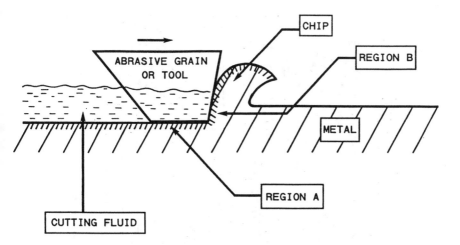

Figure 1. Schematic Formation of Metal Chip during Machining or Grinding [12].

vacuum (pressure $< 1 \times 10^{-10}$ torr). Under these conditions, hours are required for a monolayer of background gas to adsorb. Excellent control of the deposition of model EP lubricant molecules, using a calibrated beam doser [16,17], can be achieved in an apparatus shown in Figure 2. Temperature programming of the crystal containing the adsorbate has been employed to study the surface chemistry [17].

In order to investigate the surface chemistry which occurs on defect sites on the iron crystal, the surface may be made purposely defective by controlled sputtering using an Ar ion gun and controlling the relative degree of surface damage to the clean iron surface by measuring the number of Ar^+/cm^2 incident [18,19].

The use of surface science methods like those above is uncommon today in tribology experiments. They allow for control of the following important factors:

1. Surface structure of the atomically clean substrate.
2. Defect density.
3. Temperature.
4. Coverage of adsorbates.

2.2.   Recent Surface Science Studies Related to Chlorine-Based Extreme Pressure Lubrication

A brief summary is presented of the experiments designed to understand the surface chemistry of a chloroalkane, CCl_4, on an atomically clean iron surface, Fe(110). CCl_4 is a model EP lubricant [20]. Similar behavior has been seen for C_2Cl_4 [21] and Cl_2 [22,23].

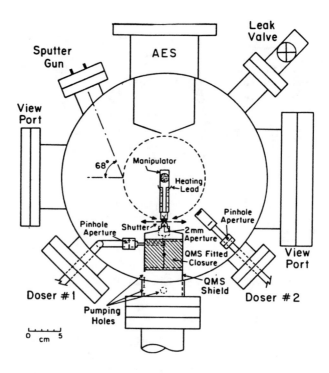

Figure 2.   Ultrahigh Vacuum System Used for Model Tribochemistry Experiments on Atomically Clean Surfaces [17]. The major measurement methods include temperature programmed desorption using the line-of-sight multiplexed mass spectrometer (QMS) and Auger spectroscopy (AES). In addition accurate adsorbate coverages may be achieved using either of two collimated beam dosers [16].

It was found that near 100 K, CCl_4 dissociatively adsorbs on the atomically clean iron surface. The adsorption of CCl_4 to coverages greater than one-half monolayer leads to a situation in which the free radical, $:CCl_2$, is produced and is detected as a desorbing species at ~ 180 K. The thermal desorption measurements for these halogenated hydrocarbon species are shown in Figure 3, where the adsorption of CCl_4 is investigated by programming the temperature upwards from 90 K. Here it may be seen that monolayer and multilayer CCl_4 species are detected in the $CCl_3{}^+$ desorption trace, and that another species not exhibiting a $CCl_3{}^+$ ion fragment, is also produced near 180 K (cross hatched peak). This species is the $:CCl_2$ radical [24]. This surprising production of a free radical on an iron surface at low temperatures is almost unprecedented [25], and is schematically illustrated in Figure 4. Because of the higher strength of the Fe-Cl bond [83 kcal/mol] compared to the C-Cl dissociation energy for $CCl_4 \rightarrow \cdot CCl_3$ [71 kcal/mol] and for $CCl_3 \rightarrow :CCl_2$ [67 kcal/mol], it is predicted on thermochemical grounds that the removal of Cl ligands may occur down to $:CCl_2$ [24].

Beyond this point, the dissociation energy of the C-Cl bond exceeds the Fe-Cl bond strength and this energetically disfavors further decomposition of :CCl_2 [24].

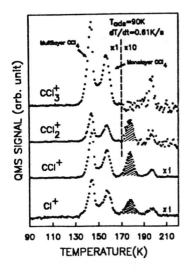

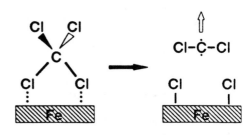

Figure 4.   Schematic Illustration of :CCl_2 Production on Fe(110).

Figure 3.   Thermal Desorption of Species Derived from CCl_4 (2.2 Monolayer) on Fe(110) [24].

An additional desorption product, $FeCl_2$, is observed by line-of-sight mass spectrometry near 560 K [26]. We believe that $FeCl_2$ is actually the EP lubricant formed under real tribological conditions. Characteristic thermal desorption spectra are shown for $FeCl_2$ as a function of initial CCl_4 coverage in Figure 5. It is noted that the desorption is detected above about 0.5 monolayer of CCl_4. The shape of the desorption peak is characteristic of desorption kinetics which are zero order in the coverage of $FeCl_2$. Zero order kinetics, indicative of $FeCl_2$ clustering, yield a congruent leading edge for the various peaks which is independent of the initial $FeCl_2$ coverage. Measurements of the activation energy for $FeCl_2$ liberation by zero order kinetics yields $\Delta H = 48.7 \pm 8.2$ kcal/mol [26]. Since the enthalpy of sublimation of bulk $FeCl_2$ is $\Delta H_s = 44.0 \pm 3.0$ kcal/mol [27-30], it is likely that the desorption of $FeCl_2$ occurs from nucleated $FeCl_2$ species which behave, in sublimation, as bulk $FeCl_2$ particles on the surface. The formation of a metal compound lubricant is somewhat analogous to lubricant soap formation by carboxylic acids on metal surfaces.

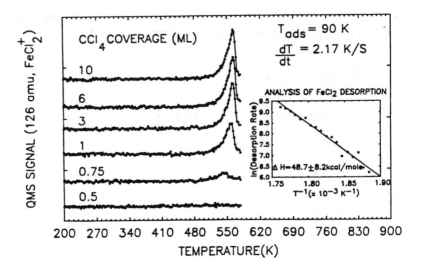

Figure 5. Temperature Programmed Desorption of $FeCl_2$ as a Function of Initial Coverage of CCl_4 [26].

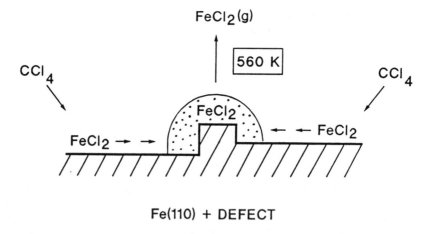

Figure 6. Schematic Diagram of $FeCl_2$ Nucleation at Defect Sites on Fe(110), followed by $FeCl_2$ Desorption at 560 K.

The nucleated $FeCl_2$ species are postulated to agglomerate at surface defects naturally present on the Fe(110) crystal [18,26]. Using Ar^+ bombardment, the surface defect concentration may be significantly enhanced, causing the desorption yield of $FeCl_2$ to increase by a factor of 9 [22]! Annealing the artificial defects away at 400 K, prior to CCl_4 adsorption, causes the enhanced yield of $FeCl_2$ to disappear [18].

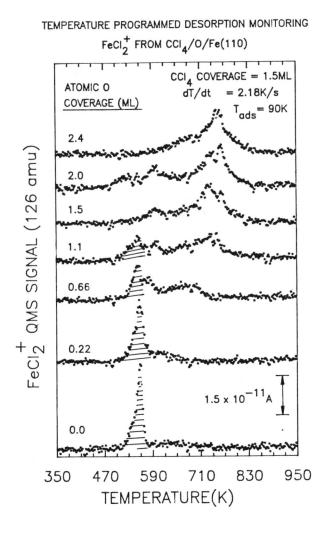

TEMPERATURE PROGRAMMED DESORPTION MONITORING

$FeCl_2^+$ FROM $CCl_4/O/Fe(110)$

Figure 7.   Quenching of $FeCl_2$ Nucleation on Defects by Chemisorbed Oxygen [19].

Auger spectroscopic measurements also indicate that heating the CCl_4-generated layer to about 500 K leads to agglomeration of $FeCl_2$ into clusters which exhibit partial screening of Cl Auger electron emission [18].   A schematic diagram of the nucleation and zero order desorption of $FeCl_2$ is given in Figure 6.

2.3.   Chemical Quenching of $FeCl_2$ Nucleation on Defect Sites

The defect sites, which are postulated to act as nucleation centers for $FeCl_2$ cluster formation, may be quenched efficiently by the chemisorption of oxygen prior to CCl_4

adsorption. This interesting phenomenon has been tested in a sequence of experiments in which the initial oxygen coverage has been systematically increased, as shown in Figure 7. A mobile O_2 precursor to dissociation [31] is postulated to deliver chemisorbed oxygen to defect sites preferentially. The 560 K-$FeCl_2$ desorption process (cross hatched peak) is attenuated and converted to an $FeCl_2$ desorption process which occurs at much higher temperatures. A plot of the attenuation is shown in Figure 8. It was also demonstrated that oxychlorides of the form $Fe_xO_yCl_z$ do not desorb from O + CCl_4 layers [19].

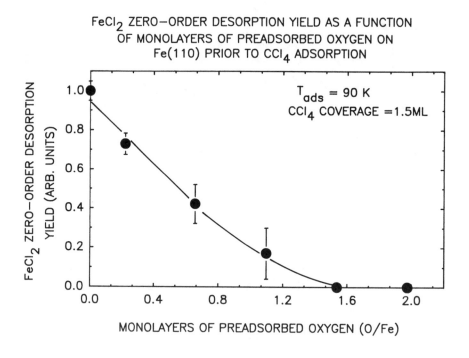

FeCl$_2$ ZERO−ORDER DESORPTION YIELD AS A FUNCTION
OF MONOLAYERS OF PREADSORBED OXYGEN ON
Fe(110) PRIOR TO CCl$_4$ ADSORPTION

Figure 8.   Yield of $FeCl_2$ in Zero-Order State for Various Oxygen Coverages [19].

A schematic model in which chemisorbed oxygen preferentially deactivates the defect sites on the iron surface is shown in Figure 9. The effect of this is to retain $FeCl_2$ dispersion and to shift the $FeCl_2$ desorption temperature upwards by about 250 K as seen in Figure 7. This causes the $FeCl_2$ EP lubricant to be retained to a higher temperature.

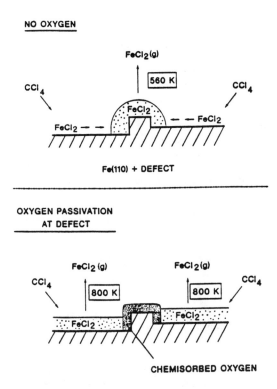

Figure 9.   Schematic Diagram of the Effect of Preferential Oxygen Chemisorption at Defect Sites on the Formation of FeCl₂ Clusters by Nucleation [19].

2.4. Connection to Tribological Issues

These studies have demonstrated that CCl_4 produces $FeCl_2$ on atomically clean Fe surfaces. $FeCl_2$ nucleation occurs readily at defect sites. In addition, chemisorbed oxygen efficiently quenches this nucleation process. We would therefore expect that oxygen and possibly other antinucleating agents might be effective in maintaining $FeCl_2$ EP lubricant films in a more highly dispersed condition, and hence in enhancing the EP lubricant efficiency of CCl_4 and other similar molecules. In addition, iron oxides are to be expected to play a role as EP lubricants themselves [12,31].

Further studies, using other tools of surface science should provide more detailed information about the molecular processes which occur in the application of model lubricant molecules to surfaces which are models for the nascent surfaces being produced in cutting and grinding technologies.

322

3. References

1. Buckley, D.H. (1974) ASLE Trans. 17, 36; 17, 206.
2. Buckley, D.H. December (1973) NASA Technical Note, TN-D-7528.
3. Dorinson, A. and Ludema, K.C. (1985) "Mechanics and Chemistry in Lubrication," Elsevier, Amsterdam Ch. 11, pg. 255.
4. Buckley, D.H. (1981) "Surface Effects in Adhesion, Friction, Wear and Lubrication," Elsevier, Amsterdam.
5. Mould, R.W., Silver, H.B. and Syrett, R.J. (1973) Wear 26, 27.
6. Ferrante, J. (1989) Surface and Interface Analysis 14, 809.
7. Levine, O. and Zisman, W.A. (1957) J. Phys. Chem. 61, 1068; 61, 1188.
8. (a) Israelachvili, N.J., McGuiggan, P.M. and Homola, A.M. (1988) Science 240, 189; (b) McGuiggan, P.M., Israelachvili, J.N., Gee, M.L. and Homola, A.M. (1989) Mat. Res. Soc. Symp. 140, 79; (c) Israelachvili, J.N., Fisher, L.R., Horn, R.G. and Christenson, H.K. (1982) Tribol. Ser. 7 (Microsc. Aspects of Adhes. Lub.), 55.
9. (a) Alsten J.V. and Granick, S. (1988) Phys. Rev. Lett. 61, 2570; (b) Alsten, J.V. and Granick, S. (1989) Mat. Res. Soc. Symp. 140, 125.
10. Pepper, S.V. (1976) J. Appl. Phys. 47, 2579.
11. Mate, C.M., McClelland, G.M., Erlandsson, R. and Chiang, S. (1987) Phys. Rev. Lett. 59, 1942.
12. Duwell, E.J., Hong, I.S. and McDonald, W.J. (1969) ASLE Trans. 12, 86; Duwell, E.J. and McDonald, W.J. (1961) Wear 4, 384.
13. (a) Quinn, T.F.J. and Winer, W.O. (1987) J. Tribol. 109, 315; (b) Hong, H. and Winer, W.O. (1989) J. Tribol. 111, 504; (c) Hong, H. and Winer, W.O. (1989) Mat. Res. Symp. 140, 301.
14. Kohn, E.M. (1965) Wear 8, 43.
15. Smentkowski, V.S. and Yates, J.T., Jr., (1990) Surf. Sci. 232, 92.
16. Bozack, M.J., Muehlhoff, L., Russell, J.N., Jr., Choyke, W.J. and Yates, J.T., Jr. (1987) J. Vac. Sci. Tech. A 5, 1.
17. Smentkowski, V.S. and Yates, J.T., Jr. (1989) J. Vac. Sci. Tech. A 7(6), 3325.
18. Smentkowski, V.S. and Yates, J.T., Jr. (1990) Surf. Sci. 232, 102.
19. Smentkowski, V.S., Ellison, M.D. and Yates, J.T., Jr. (1990) Surf. Sci. 235, 116.
20. Shaw, M.C. (1958/59) Wear 2, 217.
21. Smentkowski, V.S., Cheng, C.C. and Yates, J.T., Jr. (1989) Surf. Sci. 220, 307.
22. Linsebigler, A.L., Smentkowski, V.S., Ellison, M.D. and Yates, J.T., Jr. (1991) accepted, J. Am. Chem. Soc.
23. Linsebigler, A.L., Smentkowski, V.S. and Yates, J.T., Jr. submitted, Surf. Sci.
24. Smentkowski, V.S., Cheng, C.C. and Yates, J.T., Jr. (1989) Surf. Sci. Lett. 215, L279.
25. The $:CH_2$ radical has been reported to desorb near 200 K from an Al surface containing I_2CH_2 adsorbate. See Domen, K. and Chuang, T.J. (1987) J. Am. Chem. Soc., 109, 5288.
26. Smentkowski, V.S., Cheng, C.C. and Yates, J.T., Jr. (1990) Langmuir 6, 147.
27. Kubaschewski, O., Evans, E.L. and Alcock, C.B. (1967) Metallurgical Thermochemistry, Pergamon Press, New York.
28. Kelley, K.K. (1935) Bull-U.S. Bur. Mines, No. 383.
29. Schafer, H. (1955) Z. Anorg. Allg. Chem. 178, 300.
30. Schoonmaker, R.C. and Porter, R.F. (1960) J. Chem. Phys. 64, 86.
31. Smentkowski, V.S. and Yates, J.T., Jr. (1990) Surf. Sci. 232, 113.

LUBRICATION BY
LIQUIDS AND
MOLECULARLY-THIN
LAYERS

.... La fabrication des lubrifiants rappelle étonnamment la cuisine française: variété presqu'infinie, plus un art qu'une science, pratiquée en général au fond d'une cuisine.

ANON., quoted by C TROYANOWSKY in "Microscopic aspects of adhesion and lubrication", J-M GEORGES (ed), Elsevier 1982.

The first recorded tribologist - pouring lubricant (water) in front of the sledge in the transport of the statue of Ti (circa 2400 B.C.). [From D. Dowson, History of Tribology (Longman, London, 1979) p. 37, with permission.]

FRICTION and TRACTION IN LUBRICATED CONTACTS

Duncan DOWSON
Department of Mechanical Engineering
The University of Leeds
Leeds
LS2 9JT
England

1. LUBRICATION REGIMES

1. Fluid Film Lubrication

The phenomenon of fluid film lubrication was recognized in the 1880's through the pioneering experimental work of Petrov (1883) and Tower (1883) and the classical development of the hydrodynamic theory of lubrication by Osborne Reynolds (1886).   Petrov carried out an extensive study of friction in railway waggon axle boxes in the workshops of the St. Petersburg-Varsovie Railway and concluded that mediate friction was dependent upon the viscosity and not the density of the lubricant. He measured the friction torque acting on such bearings and found that it was determined by the shearing of a thin film of lubricant contained between a concentric journal and bearing. He also concluded that there was no slip between the lubricant and the bounding solids and this enabled him to write the following expression for the coefficient of friction in a full journal bearing.

$$\mu = \frac{4\pi^2 R^2 \ell}{c} \left(\frac{\eta \mathcal{N}}{P} \right) \tag{1}$$

- where; μ = coefficient of friction
              R = bearing radius
              ℓ = bearing length
              c = radial clearance
              η = viscosity of lubricant
              N = rotational speed of journal (revolutions/sec.)
              P = load

I. L. Singer and H. M. Pollock (eds.), Fundamentals of Friction: Macroscopic and Microscopic Processes, 325–349.
© 1992 Kluwer Academic Publishers. Printed in the Netherlands.

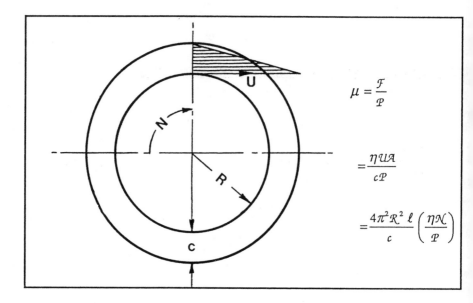

$$\mu = \frac{\mathcal{F}}{\mathcal{P}}$$

$$= \frac{\eta \mathcal{U} \mathcal{A}}{c \mathcal{P}}$$

$$= \frac{4\pi^2 \mathcal{R}^2 \ell}{c} \left(\frac{\eta \mathcal{N}}{\mathcal{P}} \right)$$

Figure 1   Petrov's Law

Equation (1) is one form of the equation frequently referred to as a representation of Petrov's Law. It can be seen that for a given bearing;

$$\mu = f \left(\frac{\eta \mathcal{N}}{\mathcal{P}} \right) \tag{2}$$

Tower (1883) also embarked upon the measurement of railway axle box friction, in this case in the Edgeware Road Station, Chapel Street Works of the Metropolitan Railway in London. He found that

> ....   *'all the common methods of lubrication are so irregular in their action that the friction of a bearing often varies considerably'*....

and proceeded to run his specially constructed testing mchine with the jounal immersed in oil. This immediately yielded steady and low coefficients of friction, often of order of magnitude 10^{-3}, but it also revealed, by chance, very high pressures in the running bearing. Tower concluded that;

> ....   *'the brass was actually floating on a film of oil, subject to a pressure due to the load'*

The concept of a rotating shaft separated from the bearing by a film of lubricant was thus established early in the 1880's by independent experimental investigations in Russia and the United Kingdom. Lord Rayleigh referred to Mr Tower's nd the ....*'coefficient as low as 1/1000'*... in his Presidential Address to the British Association or the Advancement of Science in Montreal, Canada in 1884. He also indicated that he xpected .... *'from Professor Stokes a further elucidation of the processes involved'*. Osborne Reynolds read two papers at the Montreal Meeting, one entitled *'On the Action of Lubricants'* nd the other *'On the Friction of Journals'*. There appear to be no records of these papers, but ome sixteen months after the Montreal meeting Reynolds (1886) submitted his classical paper o the Royal Society.

Reynolds, who had recently drawn attention to the distinction between laminar and turbulent low in pipes, concluded that the flow of lubricant in bearings was determined by viscous flow. He discussed the development of pressure in lubricating films through combined Poiseuille (pressure gradient) and Couette (surface motion) flows and introduced the important concept of he physical wedge.

The Reynolds equation of fluid-film lubrication, which determines the hydrodynamic pressure distribution in bearings, represents a combination of the equations of motion and continuity. It can be written as;

$$\frac{\partial}{\partial x}\left[\frac{\rho h^3}{12\eta}\left(\frac{\partial p}{\partial x}\right)\right] + \frac{\partial}{\partial y}\left[\frac{\rho h^3}{12\eta}\left(\frac{\partial p}{\partial y}\right)\right] = \frac{\partial}{\partial x}\left[\frac{\rho(u_1+u_2)h}{2}\right] + \frac{\partial}{\partial y}\left[\frac{\rho(v_1+v_2)h}{2}\right] + \frac{\partial}{\partial t}(\rho h)$$

(3)

- where;

h	= film thickness	
p	= pressure	
u_1, u_2	= surface velocities (x-direction)	
v_1, v_2	= surface velocities (y-direction)	
t	= time	
x, y	= coordinates	
ρ	= lubricant density	
η	= lubricant viscosity	

The terms on the right hand side of equation (3) indicate the physical actions responsible for the generation of pressure in a bearing. The first two represent *'entraining'* action, while the last one reflects *'squeeze'* film effects.

Analytical solutions are available for a number of simplified bearing configurations, but it is generally necessary to solve the Reynolds equation by numerical methods to provide reliable data for design purposes.

The zoologist William Bate Hardy drew attention to a totally different kind of lubrication,

> ....' *in which the solid faces are near enough together to influence directly the physical properties of the lubricant. This is the condition found with 'dry' or 'greasy' surfaces'...*

Hardy used the term *'Boundary Lubrication'* to describe the regime in which very thin films, of molecular proportions, are effective in reducing friction between opposing solids in sliding motion. He attributed this action to the role of absorbed molecular layers on the opposing solid surfaces and a series of papers by Hardy and Doubleday (1922(a),(b), 1923) are now recognized as classical contributions to the subject.

Friction experiments were carried out on glass, steel and bismuth in the presence of paraffin lubricants and their related acids and alcohols. It was found that as the concentration of ethyl alcohol vapour upon glass or steel increased, the coefficient of friction decreased from the dry contact level (0.74 - 0.93) to about (0.45 - 0.65). Furthermore, with saturated vapours, the coefficient of friction decreased linearly as the molecular weight of the lubricant increased, such that it could be expressed in the form;

$$\mu = b - a M \qquad (4)$$

- where;

μ	= coefficient of friction
M	= molecular weight of the lubricant
a	= a constant dependent upon the chemical composition of the lubricant
b	= a constant related to the nature of both the lubricant and the solid

Hardy offered a physical model to explain the role of molecular weight, or chain length, in which the molecules were orientated at right angles to the surface by the fields of attraction of the solids, as shown in Figure 2.

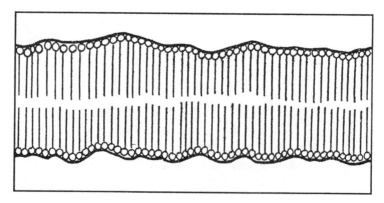

Figure 2   Sir William Bate Hardy's Concept of Boundary Lubrication

This model is sometimes likened to the pile of a carpet, with shear taking place between the ~ee ends of the long chain molecules. The effective thicknesses of boundary lubricating films ~e thus of molecular proportions (10^{-9} - 10^{-8} m).

In a *'Review of Existing Knowledge on Lubrication'* commissioned by the Department of cientific and Industrial Research (D.S.I.R. (1920)) in the United Kingdom, attention had ~ready been drawn to distinct modes or stages of lubrication, as shown in Table 1.

Stages of Lubrication	Laws	Coefficient of Friction
1. Unlubricated surfaces	Dry friction	0.10 - 0.40
2. Partially lubricated surfaces	Greasy friction	0.01 - 0.10
3. Completely lubricated surfaces	Viscous friction	0.001 - 0.01

Table 1.   D.S.I.R. Review of Existing Knowledge of Lubrication

The classification is neat and very little different from current understanding. By the mid-1920's the distinct modes of lubrication known as *'Fluid Film'* and *'Boundary'* were clearly ~dentified and although it is the former that is the subject of the present lecture, it is necessary to ~ppreciate the character of the latter. There is also an intermediate mode of lubrication between ~hese two regimes which is important in many lubricated tribological components.

1.3 Mixed Lubrication

The Couette expression for journal bearing friction (1) shows that the coefficient of friction in a fluid film bearing increases as either the viscosity or the speed increases and decreases as the applied load increases. In real bearings the shaft is located in an eccentric position, and the coefficient of friction is additionally a function of eccentricity ratio. Nevertheless, for a bearing of given geometry (radius, length and radial clearance), the coefficient of friction is a function of the lubricant viscosity, the rotational speed and the load (or mean pressure). The parameter ($\eta N/P$) shown in equation (1) is not dimensionless, but it can be written so by converting the load (P) to the mean pressure on the bearing (p). Hence, for a given bearing;

$$\mu = \phi \left(\frac{\eta \omega}{p} \right) \tag{5}$$

The quantity ($\eta\omega/P$) is sometimes known as the Gumbel number, and when multiplied b $(R/c)^2$, R being the bearing radius and (c) the radial clearance, as the Sommerfeld number.

Robert Henry Thurston, who became the first President of the American Society of Mechanica Engineers in 1880, drew attention to the practical limitation on the relationship (5) when h carried out a wide range of tests on the friction of lubricants in pendulum machines employin test bearings as fulcra, in which the load on the bearing was varied. He wrote (1885);

> .... *with increasing pressures, the limit of bearing power is attained or approached, and the friction must exhibit a change of law, the coefficient increasing, beyond that limit, as the intensity of pressure is augmented".*

Richard Stribeck (1902) carried out a similar range of experiments, in which both the load an the speed were varied. He confirmed the minimum point in the friction trace identified b Thurston.

The increase in the coefficient of friction as the load increased or the speed decreased furthe was associated with increasing asperity contacts. The rapid rise in coefficient of friction wit changes in load or speed bridged the relatively low minimum coefficient of friction associate with fluid-film lubrication (~ 0.001) and the higher, but steady, coefficient of boundar lubrication (~ 0.1). This region became known as the *Mixed Lubrication* regime since frictio was associated with contributions from both fluid-film and boundary actions.

1.4 The Stribeck Curve

The three main regimes of lubrication can be conveniently represented on a curve relatin coefficient of friction to the parameter ($\eta\omega/p$), as shown in Figure 3.

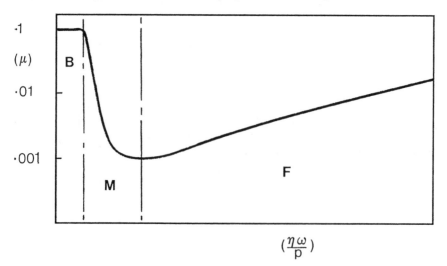

Figure 3 Stribeck Curve

It was Ludwig Gumbel who showed that Stribeck's separate friction traces could be condensed onto a single curve by the use of $(\eta\omega/p)$, and hence this dimensionless grouping is associated with his name.

In this lecture we are concerned primarily with the nature of friction in the fluid-film lubrication regime, to the right of the minimum in the Stribeck curve, but some reference will also be made to the mixed lubrication regime.

2. VISCOUS FLOW

2.1 Constitutive Equation. For a Newtonian (linear viscous) fluid, the shear stress (τ) is directly proportional to the shear rate of strain (γ).

Thus,

$$\tau_{xy} = \eta \left[\frac{\partial v}{\partial x} + \frac{\partial u}{\partial y} \right] etc. \tag{6}$$

- where; η is the **coefficient of viscosity.**

2.2 Continuity Equation.

The principle of conservation of mass can be written as;

$$\frac{\partial \rho}{\partial t} + \frac{\partial}{\partial x}(\rho u) + \frac{\partial}{\partial y}(\rho v) + \frac{\partial}{\partial z}(\rho w) = 0 \tag{7}$$

$$or, \frac{\mathcal{D}\rho}{\mathcal{D}t} + \rho\, div\ \mathbf{u} = 0 \tag{8}$$

For an incompressible fluid, this becomes,

$$\frac{\partial u}{\partial x} + \frac{\partial v}{\partial y} + \frac{\partial w}{\partial z} = 0 \tag{9}$$

2.3 Equations of Motion

If the principle of conservation of momentum is applied to an iso-viscous, incompressible, Newtonian fluid;

$$\rho \frac{Du}{Dt} = \rho X - \frac{dp}{dx} + \eta \nabla^2 u \; etc.$$ (10)

$$- where; \quad \nabla^2 = \frac{\partial^2 u}{\partial x^2} + \frac{\partial^2 u}{\partial y^2} + \frac{\partial^2 u}{\partial z^2}$$

If the Reynolds Number is small compared to unity, such that inertia forces are negligible compared to the viscous forces acting on the fluid, the equations of motion reduce to;

$$\left(\frac{\partial p}{\partial x} - \rho X \right) = \eta \nabla^2 u \; etc.$$ (11)

This equation represents a simple balance between the pressure and body forces acting on a fluid and the viscous force.

2.4 Slow Viscous Flow

Only three actions can cause a viscous fluid to flow;

- Motion of the Bounding Solid Surfaces - Couette flow
- Pressure differences in the fluid - Poiseuille flow
- Body forces - e.g. gravity.

Consider laminar flow between parallel flat plates separated by a distance (h) (Figure 4).

2.5 Couette Flow

If the two plates have velocities (U_1) and (U_2) in the x-direction and there are no pressure gradients, or body forces, the velocity distribution is linear as shown in Figure 4(a).

The viscous shear stress on any section is given by ($\eta \, \partial u / \partial y$) and hence the <u>viscous stresses acting on the lower and upper plates respectively</u> are;

$$\left(\tau_{\mp \frac{h}{2}} \right) = \pm \eta \left[\frac{U_2 - U_1}{h} \right]$$ (12)

2.6 Poiseuille and Body Force Flow

If the two plates are stationary, but flow is generated by either pressure gradient or body force actions, a parabolic velocity distribution arises as shown in Figure 3(b). In this case the viscous stresses acting on the lower and upper plates respectively are;

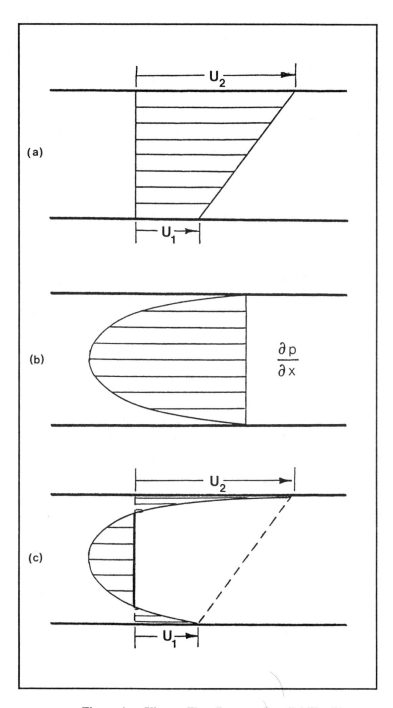

Figure 4.   Viscous Flow Between Parallel Flat Plates

$$\left(\tau_{\mp\frac{h}{2}}\right) = -\frac{h}{2}\left[\frac{\partial p}{\partial x} - \rho X\right] \tag{13}$$

2.7 Combined Flow

If both Couette and Poiseuille (or body force) flows take place simultaneously, the resultant velocity distribution can be found by superimposing the appropriate linear and parabolic velocity distributions as shown in Figure 3(c). It should be noted that reverse flow may take place in the film of fluid; a situation frequently encountered in bearings.

Likewise, the viscous stresses acting upon the solids can be found by combining (12) and (13) such that;

$$\left(\tau_{\mp\frac{h}{2}}\right) = \pm\,\eta\left[\frac{U_2 - U_1}{h}\right] - \frac{h}{2}\left[\frac{\partial p}{\partial x} - \rho X\right]$$

3. VISCOUS FRICTION and POWER LOSS in BEARINGS

The viscous force acting on the bearing surfaces can be obtained for any bearing configuration, by integrating the viscous stress ($\eta\,\partial u / \partial y$) over the full bearing area (A). Likewise, the power loss can be derived by multiplying the viscous force on a bearing surface by its velocity (U).

3.1 Parallel Surface Bearing.

Consider a parallel surface bearing of the form depicted in Figure 4(a). The viscous stresses on the lower and upper plates are given by equation (12) and the corresponding viscous forces are found by multiplying these stresses by the plane (x,z) area of the plates (A).

The power losses associated with each plate are;

$$E_1 = -\eta\left[\frac{U_2 - U_1}{h}\right]U_1\,A \qquad (a)$$

$$E_2 = +\eta\left[\frac{U_2 - U_1}{h}\right]U_2\,A \qquad (b)$$

$$\tag{15}$$

Note that if the lower surface moves with velocity (U) and the upper surface is stationary,

$$\mathcal{E} = \eta \frac{u^2 \mathcal{A}}{h} \qquad (16)$$

3.2 Plane Inclined Slider (or Tilting Pad) Bearing.

Consider flow in the (x,y) plane only in a bearing of the form shown in Figure 5. This implies no side-leakage in the z-direction.

The Reynolds equation (3) can be solved analytically for this configuration to reveal the pressure distribution. There will thus be combined flow (Couette and Poiseuille) at any section in the bearing, with a variation of viscous stresses along the bounding solid surfaces from inlet to outlet. The integration of these stresses allows the friction forces and the normal force components to be ascertained. It is found that (F_1) and (F_2) are different, but if the coefficient

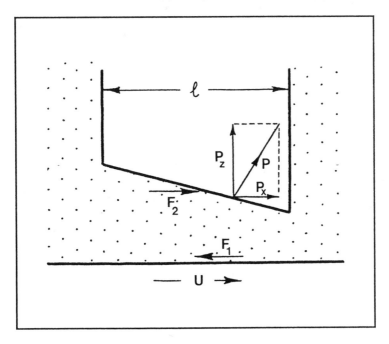

Figure 5.   Plane Inclined Surface Bearing

of friction (μ) is defined as (F_1/P_z), μ (ℓ/h_o) = ϕ (h_i/h_o), as shown in Figure 6.

Since the upper surface of the bearing is inclined to the (x,z) plane, the normal stresses generate components in both the (x) and the (y) directions. The latter provides the hydrodynamic lift to balance the applied load, whereas the former represents the balance between the two viscous friction forces acting on the runner and the bearing pad. It should be noted that, since inertia effects upon the fluid have been assumed to be negligible,

336

$$F_1 + F_2 + P_x = 0 \qquad\qquad (17)$$

3.3 Journal Bearings.

If the lubricant does not rupture due to gas or vapour release (cavitation), the pressure distribution is skew symmetric about the line of centres (the Sommerfeld (1904) solution). This ensures that the application of any steady load to the bearing will cause the journal centre to move within the bearing clearance at right angles to the load vector. If the eccentricity of the load (P) is (e), the moment (Pe) will represent the difference in the journal and bearing friction torques. Once again the friction stresses, forces and torques on the bearing and the shaft can readily be obtained from analytical solutions to the Reynolds equation. In general the shaft torque exceeds the bearing torque, typically by a factor of 6 at an eccentricity ratio of about 0.8 according to the Sommerfeld analysis.

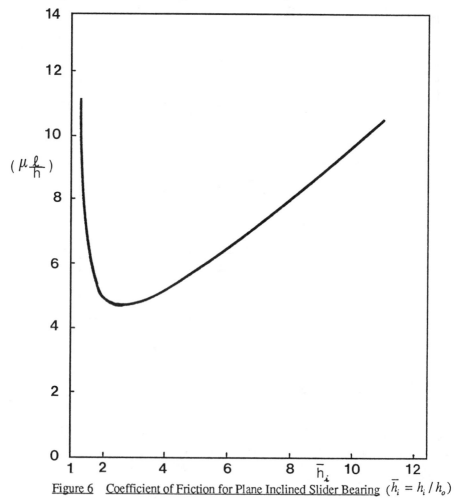

Figure 6   Coefficient of Friction for Plane Inclined Slider Bearing ($\overline{h}_i = h_i / h_o$)

In most convergent-divergent shaped bearings the lubricant film ruptures and in journal bearings this causes the shaft to move both along and perpendicular to the load vector. The attitude angle (Ψ) is defined as the angle through which the load vector has to be rotated in the direction of shaft rotation to become aligned with the line of centres. If the oil film in a journal bearing cavitates, the difference between shaft and bearing moments (see Figure 7) is thus (Pesin Ψ).

It is generally much easier to measure stationary bearing than rotating journal friction, but the former should not be used in power loss calculations.

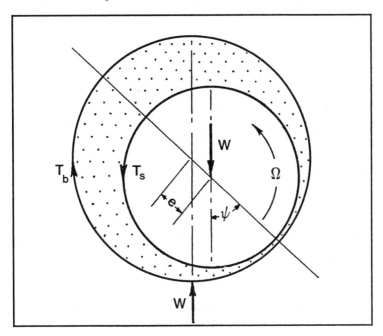

Figure 7.   Bearing and Journal Torques

3.4 Viscous Friction in Cavitation Zones

Dissolved gas is readily released from liquids whenever the pressure falls to the saturation pressure (generally atmospheric). This occurs in journal bearings shortly after the lubricant enters the divergent clearance space. In due course the full film reforms again. Within the cavitated zone the pressure is essentially constant at a value close to, but generally slightly less than, ambient. Visual observations suggest that the lubricant travels in thin streamers, extending from the journal to the bearing, under Couette action.

If the film ruptures along a line where the film thickness is (h*), continuity requires that the width of the bearing filled with lubricant under shear at any location where the film thickness is (h) is simply (h*/h) times the full width of the bearing.

4. ELASTOHYDRODYNAMIC CONJUNCTIONS

A major development in the understanding of fluid film lubrication which has taken place during the second half of the 20th century, has been the recognition of the beneficial influence of both elastic deformation of the solids and the increase in viscosity of the lubricant with pressure in highly stressed lubricated machine elements. It all started with an attempt to explain the indirect operating evidence of fluid film lubrication in gears, where conventional lubrication theory was failing to predict film thicknesses which were large compared to the known surface roughness of the gears. In due course it became clear that elasto-hydrodynamic action yielded separations of the solids which were orders of magnitude greater than those predicted by conventional rigid solid, iso-viscous lubricant hydrodynamic analysis.

The essence of elasto-hydrodynamic action is that the geometry of the lubricated components readily generates very high hydrodynamic pressures in local regions close to the point, or line, of closest approach.

These exceptionally high pressures cause local deformations, in regions of magnitude comparable to the dimensions of the Hertzian zone of dry contact, which in turn create a more conforming, effective fluid-film bearing. This leads to much greater film thicknesses than would otherwise be achieved between the original, rigid solids. If, in addition the lubricant viscosity increases with pressure, a doubly beneficial action takes place. A common representation of viscosity-pressure effects adopted in early elasto-hydrodynamic action is the Barus equation;

$$\eta = \eta_o \, e^{\,\alpha p} \tag{18}$$

- where the viscosity-pressure coefficient is of order 10^{-8} m^2/N at room temperature (typically 20-40 GPa^{-1}).

4.1 Elastohydrodynamic Film Thickness

The integration of elasticity theory and hydrodynamic analysis has revealed much about the very effective mode of fluid-film lubrication known as 'elastohydrodynamic'. Most of the studies have been concerned with film thickness predictions, since the essential requirement for the successful operation of highly stressed machine components is that the load transmission should be effected through lubricating films which are at least as thick as the composite roughness of the bounding solids. Representative equations for the minimum film thickness in line (Dowson (1968)) and elliptical (Chittenden et al (1985)) contact are;

$$\mathcal{H}_{min} = (h_{min} / \mathcal{R}) = 2.65 \; \mathcal{U}^{0.7} \, \mathcal{G}^{0.54} \; \mathcal{W}^{-0.13} \tag{19}$$

$$H_{min}^{\bullet} = 3.68\left[1-\exp\left[-0.67\left[\left\{(R_y/R_x)\cos^2\theta+\sin^2\theta\right\}/\left\{\cos^2\theta+(R_y/R_x)\sin^2\theta\right\}\right]^{2/3}\right]\right]$$

(20)

– where $H_{min}^{\bullet} = (h_{min}/R_e)/(U_e^{0.68}\ G^{0.49}\ W_e^{-0.073})$

with subscript (e) representing the direction of lubricant entrainment relative to the minor axis of the Hertzian contact ellipse.

The predictions of elasto-hydrodynamic lubrication (E.H.L.) film thickness and shape have been broadly confirmed by splendid experimental studies based upon capacitance or interferometry methods in both line and point contact configurations. Such confirmation engendered confidence in the theory and E.H.L. film thickness calculations are now well established features of the design of highly stressed machine elements such as **gears, ball** and **roller bearings** and **cams and followers.**

4.2 The Lambda (Λ) Ratio.

It has been found that the ratio of the calculated elasto-hydrodynamic film thickness to some composite measure of surface roughness has a profound effect upon the life of lubricated, highly stressed machine elements. The relationship is shown in Figure 8, which provides the basis for the design guidance that (Λ) ratios of 2 or 3 should be established for maximum life of the components.

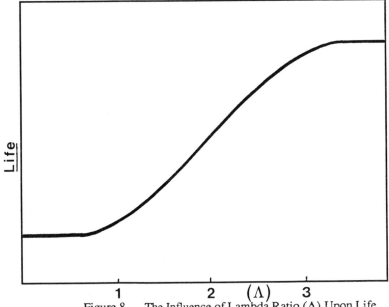

Figure 8.   The Influence of Lambda Ratio (Λ) Upon Life

340

4.3 Friction in Elasto-hydrodynamic conjunctions.

The rheological model of a Newtonian fluid, which proved to be so successful in predicting film thickness and the life of components operating in the elasto-hydrodynamic regime, proved to be totally inadequate in relation to the prediction of friction,traction and power loss.

The conditions experienced by the lubricant as it transverses elasto-hydrodynamic conjunctions are quite severe. The Hertzian contact zones have widths of order 10^{-3}m, maximum Hertzian pressures of order 3-4 GPa (up to 600,000 psi) , film thicknesses of about 10^{-7} - 10^{-6} m and entraining velocities of about 10 m/s. The transit times are thus about 10^{-4}; and the shear rates 10^7 to 10^8 s^{-1}.

Initially, a Newtonian model was assumed in the calculation of friction and traction, but this invariably and substantially overestimated friction. One reason for this was undoubtedly the influence of temperature upon viscosity and the simple Barus equation (18) had to be modified to include a viscosity-temperature coefficient. However, thermal effects alone proved to be incapable of accounting for the observed friction ~ shear rate characteristics in EHL conjunctions (e.g. in disc machines), as noted by Evans and Johnson (1986). Crook (1961) presented an impressive analysis of thermal effects in EHL traction.

The severity of the conditions in elasto-hydrodynamic conjunctions also promoted the idea that the elasticity of the fluid might be an important feature of its rheological response. Maxwell models for lubricants, with linear viscous and elastic elements in series, were proposed by Tanner (1960), Chow and Saibel (1971) and Dyson (1965). This yields an analysis restricted to relatively low shear stresses and strains. The shear strain rate expression is;

$$\dot{\gamma} = \frac{\dot{\tau}}{G} + \frac{\tau}{\eta}$$ (21)

- with the relaxation time being defined as (η/G).

A non-linear Maxwell model has been proposed by Hirst and Moore (1975) and by Johnson and Tevaarwerk (1977). In this representation the non-linear viscous term can be based upon the Rhee-Eyring fluid model and the shear rate can be written as;

$$\dot{\gamma} = \frac{\dot{\tau}}{G} + \frac{\tau_o}{\eta} \sinh\left(\frac{\tau}{\tau_o}\right)$$ (22)

- where (τ_0) is a reference (Eyring) shear stress.

It should be noted that for small values of (τ/τ_o), sinh (τ/τ_o) → (τ/τ_o) and (22) reduces to the linear small strain form (21). If $\tau \gg \tau_0$, sinh (τ/τ_0) ≈ (1/2) exp (τ/τ_0), and if the viscous term dominates;

$$\frac{\tau}{\tau_o} = \log_e \left(\frac{2\eta\dot{\gamma}}{\tau_o} \right) \tag{23}$$

- showing that the viscous shear stress (τ) is proportional to the logarithm of the rate of shear

Equation (22) can be written in dimensionless form as;

$$\frac{\eta\dot{\gamma}}{\tau_o} = \mathcal{D} \frac{\partial \bar{\tau}}{\partial x} + \sinh \frac{\tau}{\tau_o} \tag{24}$$

- where $\bar{\tau} = (\tau / \tau_o)$ and,

    $D = (\eta U/G\ell)$ the Deborrah number, representing the ratio of relaxation to transit times.

It is clear from (24) that the significance of visco-elastic behaviour in the lubricant is determined by the magnitude of the Deborrah number.

The three dominant lubricant properties responsible for the rheological behaviour of lubricants in elasto-hydrodynamic conjunctions are thus;

    G   elastic shear modulus
    η viscosity
    τ_0 reference (Eyring) stress

Typical values of (G) and (τ_0) for liquid lubricants are about 1 GPa, and 1 MPa respectively, while (η) is strongly dependent upon pressure and temperature. There is, however, another most influential lubricant characteristic to consider in elastohydrodynamic lubrication.

4.3.1. Solidification. The severity of the operating conditions in elasto-hydrodynamic conjunctions has already been noted. If Newtonian behaviour persisted the viscosity and the shear stress would be enormous. Indeed, it can readily be shown that the shear stresses would exceed the limits for the bounding metals, and it is therefore not surprising that alternative shear characteristics are imposed upon the lubricant.

One of the most significant findings in relation to lubricant behaviour in recent years has been that of a transition from a fluid to an amorphous solid under realistic elasto-hydrodynamic conditions. Smith (1959, 1960) first suggested that lubricants might experience a limiting shear stress, itself a function of pressure and temperature, and then deform as a plastic solid. The glass transition temperature at which solidification takes place is typically in the range 50°C-100°C at a pressure of 1 GPa. The simple linear-viscous (Newtonian) ~ plastic model of lubricant behaviour is shown in Figure 9.

 _ The limiting shear stress (τ_L) is found to be linearly related to pressure (Bair and Winer (1979); Höglund and Jacobson (1986) according to the relationship.

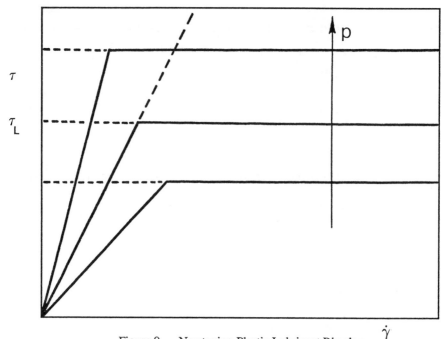

<u>Figure 9.</u>   <u>Newtonian-Plastic Lubricant Rheology</u>

$$\tau_L = \tau_o + \delta p \qquad (25)$$

$$or \qquad \overline{\tau} = \overline{\tau}_o + \delta P \qquad (26)$$

$$where, \quad \overline{\tau}_L = (\tau \, / E')$$
$$\overline{\tau}_o = (\tau_o / E')$$
$$P = (p / E')$$

Bair and Winer (1979) found (δ) to range from 0.05 to 0.1 while Jacobson (1985) suggested a wider range of 0.025 to 0.15. Jacobson and Jacobson and Hogland (1886) also found (τ_o) to be 1-5 MPa. Typical solidification pressures are of order 1 GPa.

At high pressures (GPa), $\tau_L \approx \delta p$ and (δ) thus represents a limiting value of the friction or traction in elasto-hydrodynamic conjunctions. The value of (δ) is related to the chemical tructure of the base oil (Högland 1989), and it appears to be practically independent of any dditives. It was reported that Napthemic base oils exhibited higher values of (δ) than either sters or paraffinic base oils and that poly-olefines produced the lowest values. The value of (δ) decreased as the viscosity increased.

. SUMMARY

The general concepts of friction in fluid-film bearings are well established and successful design procedures exist for power loss predictions in journal and thrust bearings. A Newtonian fluid model will normally suffice, if due account is taken of the influence of temperature, and perhaps pressure, upon lubricant viscosity. The possibility of film rupture (cavitation) and reformation in convergent-divergent bearings must also be considered.

A major extension of fluid-film analysis occurred in the second half of the 20th century with the recognition of elasto-hydrodynamic action. It is now recognized that many highly stressed, lubricated, machine elements such as gears, ball and roller bearings, certain traction drives, cams and followers and piston rings rely upon elasto-hydrodynamic action for their efficiency and durability.

The analysis of film thickness in elasto-hydrodynamic conjunctions based upon a simple Newtonian fluid model has been remarkably successful and confirmed by numerous experiments. However, the Newtonian model is quite inadequate for the prediction of friction or traction. Non -linear viscous (Eyring) effects have been introduced into visco-elastic (Maxwell) models of lubricant rheology, but the most significant feature of lubricant behaviour in many elasto-hydrodynamic conjunctions is solidification. The fluid solidifies to an amorphous solid at a glass transition temperature and then deforms like a plastic solid subjected to a limiting shear stress. The value of this limiting shear stress is linearly related to pressure by a limiting shear stress proportionality constant (δ), which ranges from about 0.02 to 0.15, but is typically 0.05 to 0.1.

The similarity between this friction or traction coefficient and the familiar coefficients of friction in boundary lubrication is consistent with a smooth transition from fluid-film to mixed or boundary lubrication.

The general form of a traction ~ slip velocity trace for an elasto-hydrodynamic conjunction is shown in Figure 10.

344

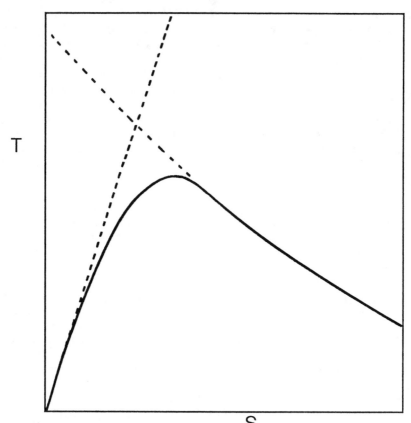

T

S

Figure 10.   Typical Traction ~ Slip Velocity Trace for Elasto-hydrodynamic Conjunctions.

References

Bair, S. and Winer, W.O. (1979), 'A Rheological Model for Elastohydrodynamic Contacts Based On Primary Laboratory Data', Trans. A.S.M.E., Vol. 101, pp 258-265.

Chittenden, R.J., Dowson, D., Dunn, J.F. and Taylor, C.M. (1985), 'A Theoretical Analysis of the Isothermal Elastohydrodynamic Lubrication of Concentrated Contacts. II. General Case, with Lubricant Entrainment Along Either Principal Axis of the Hertzian Contact Ellipse or at some Intermediate Angle', Proc. Roy. Soc. Lond., A397, 271-294.

Chou, T.S. and Saibel, E. (1971), 'The EHD Problem with a Viscoelastic Fluid', Trans A.S.M.E., Journal of Lubrication Technology, Vol. 93F, p. 26.

Crook, A.W. (1961), 'The Lubrication of Rollers, Part III. A Theoretical Discussion of Friction and the Temperatures in the Oil Film', Phil. Trans. R. Soc. Lond., A254, pp 237-258.

Dowson, D. (1968), 'Elastohydrodynamics', Proc. I.Mech..E., Vol. 182, Pt. 3A, pp 151-167.

D.S.I.R.   (1920), 'Report of the Lubricants and Lubrication Inquiry Committee', Advisory Council of the Department of Scientific and Industrial Research (HMSO, London).

Dyson, A. (1965), 'Flow Properties of Mineral Oils', in Elastohydrodynamic Lubrication Phil, Trans. Roy. Soc. Lond. A258 (No. 1093) p. 529.

Evans, C.R. and Johnson, K.L. (1986), 'Regimes of Traction in Elastohydrodynamic Lubrication', Proc. Inst. Mech. Engrs., Vol. 200, No. C5, pp 313-324.

Hardy, W.B. and Doubleday, I. (1922(a)), 'Boundary Lubrication - The Temperature Coefficient', Proc. Roy. Soc., A101, pp 487-492.

Hardy, W.B. and Doubleday, I. (1922(b)), 'Boundary Lubrication - The Paraffin Series', Proc. Roy. Soc., A 100, pp 550-574.

Hardy, W.B. and Doubleday, I., (1923), 'Boundary Lubrication - The Latents Period and Mixtures of Two Lubricants', Proc. Roy. Soc. A104, pp 25-39.

Hirst, W. and Moore, A.J. (1975), 'The Elastohydrodynamic Behaviour of Polyphenyl ether', Proc. Roy. Soc. Lond., A 344, pp 403-426.

Högland, E. (1989), 'The Relationship Between Lubricant Shear Strength and Chemical Composition of the Base Oil', WEAR, Vol . 130, pp 213-224.

Jacobson, B. (1985), 'A High Pressure-Short Time Shear Strength Analyser for Lubricants', Tribol. 107, pp 221-223.

Jacobson, B. and Högland, E. (1986), 'Experimental Investigation of the Shear Strength of Lubricant Subjected to High Pressure', Trans. A.S.M.E., Jnl. Lub. Technol. 108, No. 4, p.571.

Johnson, K.L. and Tevaarwerk, J.L. (1977), 'Shear Behaviour of Elastohydrodynamic Oil Films', Proc. Roy. Soc. Lond., A 356, pp 215-236.

Petrov, N.P. (1883), 'Friction in Machines and the Effect of the Lubricant', Inzh. Zh., St. Petersburg, 1, pp 71-140; 2, pp 227-279; 3, pp 377-436; 4, pp 535-564.

Reynolds, O. (1886), 'On the Theory of Lubrication and its Application to Mr Beauchamp Tower's Experiments, Including an Experimental Determination of the Viscosity of Olive Oil', Phil. Trans. Roy. Soc., Vol. 177, pp 157-234.

Smith, F.W. (1959), 'Lubricant Behaviour in Concentrated Contact Systems - The Castor Oil/Steel System', WEAR, Vol. 2, No. 4, pp 260-263.

Smith, F.W. (1960), 'Lubricant Behaviour in Concentrated Contact - Some Rheological Problems', Trans. A.S.L.E., Vol. 3, p. 18.

Sommerfeld, A. (1904), 'Zur hydrodynamischen Theorie der Schmiermittehreibung', Z. Math. Phys., 50, pp 97-155.

Stribeck, R. (1902), 'Die Wesentlichen Eigenschaften der Gleit und Rollenlayer', Z. Ver. dt. Ing., 46, No. 38, pp 1341-1348; 1432-1438; No. 39, 1463-1470.

Tanner, R.I. (1960), 'Full Film Lubrication Theory for a Maxwell Liquid', Int. Jnl. Mech. Sci., Pergamon Press, Vol. 1, pp 206-215.

Thurston, R.H. (1885), 'A Treatise on Friction and Lost Work in Machinery and Millwork', Wiley, New York, (7th Edn. 1903).

Tower, B. (1883), 'First Report on Friction Experiments (Friction of Lubricated Bearings)', Proc. Inst. Mech. Engrs., Nov. 1883, pp 632-659, Also, Jan. 1884, pp 29-35.

Discussion *following the lecture by D Dowson on "Friction and traction in lubricated contacts"*

A R SAVKOOR. Given that the coefficient of friction increases as the film gets thinner and thinner, what in practice will be its upper limit under these conditions?

D DOWSON. In general we are trying to operate machinery nearer to the friction minimum on the curve, and so certainly the coefficient of friction has come down with time, whereas when we were operating well up in the hydrodynamic region, most of the time unnecessary large film thicknesses were being generated. In manufacturing systems there is much interest in being able to operate with thinner films, and so the coefficient of friction has come down and down. If you look at the length of a bearing in a machinery design book of 50 years ago, you will find typically a square bearing, in other words a bearing whose length is as long as the diameter of the shaft. The length of the bearing was in fact introduced in order to make sure that you did have a great enough length of lubricant film. In recent years the width of the bearing has been reduced more and more, bringing us nearer to the bottom point of the curve and you will see now bearings which have a width which is perhaps a quarter or a fifth of the diameter, the so-called short bearings. So the power required to drive the machinery has been reduced enormously.

ANON. The rise in the coefficient of friction has been calculated purely on the basis of the elasto-hydrodynamic effect. Can you say something about the consequences for wear?

D DOWSON. There are some interesting implications when you are in the rising part of the curve in the mixed lubrication regime, without evidence of any wear. This mainly arises for two reasons. One if from micro-hydrodynamic effects: if you can have a film thickness which is nominally equally to, or a little bit less than, the undeformed value, then certainly with soft materials, you have effectively no wear at all. On the other hand, at the other end of the range you have in effect the imprint of a hard material on top of a soft material, as we heard last week. For example, if you rub a hard material against a soft material, the profile of the hard material imprints itself upon the soft material, and after some initial wear you do in fact have two very well conforming surfaces which subsequently operate in the fluid film region. There are quite a few instances like this.

J A GREENWOOD. Professor Dowson has discussed the importance of the lambda ratio in determining the state of friction, whether it is with fluid film lubrication or with asperity contacts. I think it is important to point out with regard to the lambda ratio, that is, the ratio of the film thickness to the surface roughness, that the work on it rather loses its impact when you remember that the people concerned have measured neither the film thickness nor the surface roughness. Apart from Winer or Thomas, there are really rather few people who have made any serious attempts to measure either of the two quantities.

D DOWSON. Dr Greenwood is quite right. On the slide that I presented, the ratio was written as *calculated* film thickness divided by composite surface roughness. You don't have to measure these quantities, because the relationship has already been demonstrated on full scale equipment. You calculate the properties, take a simple talysurf trace of the surface, and that is all you need. That is why in practice there are very few experimental determinations. Nevertheless attempts are being made now to model the lubrication of rough surfaces, first measuring the profiles in two dimensions and then using them in an EHL-type calculation.

U LANDMAN. This is a very practical question. What tricks can people use to control the temperature of the lubricant in experiments involving bearings? Is the temperature controlled through details of design, choice of materials, environment etc? What has been done with regard to the cooling of bearings?

D DOWSON. Nearly always it is the last of the things you have mentioned. There is far more lubricant there than you need; its job is to provide cooling. Very often it is said that we could make a lot of progress if we isolated the cooling fluid from the fluid that is really needed for lubricating the surface. But coming back to your question, it is mainly done by external cooling, air flow over the system, water jackets on engines, and a large amount of lubricant.

ANON. On your graph of the dimensions of the lubricating film we don't get to 10^{-9} m until about the present time perhaps. I wonder how you view the work of Zisman in the 1950's with Langmuir-Blodgett films on glass slides. Does that work fit onto your graph, or is your graph primarily for describing practical applications?

D DOWSON. Yes, I based it on the range of film thicknesses that are being talked about in terms of fluid-film-lubricated bearings, not on the properties of boundary lubricants that have been identified since the 1920's. What I want to emphasize is that people are measuring, in real systems, thicknesses several orders of magnitude smaller than was commonplace at the beginning of the century. It does seem to me that with the current implementation of records from full-scale laboratory experiments we are getting well down the curve, at any rate to 10-nanometre dimensions. However as an engineer I would say we are still talking about situations where the notional separation of the surfaces in our basic designs is still orders of magnitude greater than molecular dimensions.

B J BRISCOE. I want to make a point about the communality of response between nominally solid boundary lubricants and these fluids which exist under very high pressure. Professor Dowson has nicely illustrated this point today, showing that shear stress is a function of strain rate. After an initial elastic or viscous response, we move towards some form of plastic behaviour, and there will be a shear strength which will be a function of pressure. That is the point at which the model fluid experiments can be carried out: as a function of computed contact pressure, one gets for the oil a coefficient α of the order of 0.15 or even more than that. If one takes a solid lubricant like polyethylene, or indeed a stearate, you get pressure-dependences comparable to those obtained for mobile liquids. Thus, probably there is really no distinction between a material which is solid just because of absorption and confinement, and a liquid which is transformed by hydrostatic stress: the distinction between the liquid and the solid is now rather blurred. Professor Tabor did address this point about five years ago: if hydrocarbons are squashed together, in the majority of cases friction would just involve the rubbing of methyl groups, for example, over one another, and hence we would expect quite similar responses if we compare the behaviour of highly complex liquids with that of solids.

D TABOR. Professor Dowson's very fascinating historical contribution to the subject has received no comment so far. I would like to make just one light-hearted contribution on a point which has not yet been recorded in the annals of the history of lubrication. When the railway system in Europe began to spread about 150 years ago, the Russian railway authorities used as lubricant in their axle bearings mainly lard and fatty oils. The result was that whenever railway trucks stopped at village sidings, the peasants would come along and steal the lard and grease out of the axle boxes. This led to some fattening of the peasants but also to a very high rate of axle failure. In response, the railway authorities hit on a very neat solution, which was to mix soot with the lard that they put in their axle boxes. This had two beneficial effects. Firstly the peasants didn't steal it anymore, and secondly as it turned out, it provided better lubrication. I hope that you will be able to validate this next time you write a detailed account of lubrication practice in the railway system.

ADHESION, FRICTION AND LUBRICATION OF MOLECULARLY SMOOTH SURFACES

Jacob N. Israelachvili

Department of Chemical Engineering, and Materials Department,
University of California, Santa Barbara, California 93106, USA

ABSTRACT

The past few years have witnessed tremendous advances in experimental and theoretical techniques for probing both the static and dynamic the properties of surfaces at the ångstrom level. Here we review how these advances have significantly furthered our fundamental understanding of the following:

- ADHESION, particularly the processes that contribute to energy dissipation during adhesion and separation (loading-unloading cycles) in the absence of surface damage,

- INTERFACIAL FRICTION, particularly the relation between contact mechanics, adhesion and friction of unlubricated contacts, and

- LUBRICATION, particularly how the static and dynamic (e.g., viscous and rheological) properties of liquids in ultra thin films differ from the bulk liquid properties, and how this affects the shear behaviour and friction of lubricated contacts.

The emphasis will be on 'ideal' surfaces and interfaces – those that are molecularly smooth, films that are only a few molecular layers thick, and where no permanent damage or wear occurs during sliding.

PART I

REVERSIBLE AND IRREVERSIBLE ADHESION

Under ideal conditions the adhesion energy is considered to be a well-defined thermodynamic quantity. It is normally denoted by γ or W, and gives the work done on bringing two surfaces together or the work needed to separate two surfaces from contact. Under ideal, equilibrium conditions these two quantities are the same, but under most realistic conditions they are not: the work needed to separate two surfaces is always greater than that originally gained on bringing them together. An understanding of the molecular mechanisms underlying this phenomenon is essential for understanding many adhesion hysteresis effects, energy dissipation during loading-unloading cycles, contact angle hysteresis, and – ultimately – the molecular mechanisms associated with many frictional processes. We start by describing both the theoretical and experimental basis of this phenomenon, and how it arises even between perfectly smooth and chemically homogeneous surfaces.

1.1 Adhesion and wetting (contact angle) hysteresis

Most real processes involving adhesion and wetting are hysteretic or energy-dissipating even though they are usually described in terms of (ideally) reversible thermodynamic functions such adhesion free energy, reversible work of adhesion, surface tension, interfacial tension, etc. For example, the energy change, or work done, on

I. L. Singer and H. M. Pollock (eds.), Fundamentals of Friction: Macroscopic and Microscopic Processes, 351–385.
© 1992 *Kluwer Academic Publishers. Printed in the Netherlands.*

separating two surfaces from adhesive contact is generally not fully recoverable by bringing the two surfaces back into contact again. This may be referred to as *adhesion hysteresis*, and expressed as

$$W_R \quad > \quad W_A$$
$$\text{receding} \quad \text{advancing}$$
$$\text{(separating)} \quad \text{(approaching)}$$

or
$$\Delta W = (W_R - W_A) > 0 \tag{1}$$

where W_R and W_A are the adhesion energies for receding (separating) and advancing (approaching) two solid surfaces, respectively. Adhesion hysteresis is responsible for such phenomena as 'rolling' friction [1] and 'elastoplastic' adhesive contacts [2] during loading-unloading and adhesion-decohesion cycles.

Hysteresis effects are also commonly observed in wetting/dewetting phenomena [3]. For example, when a liquid spreads and then retracts from a surface the advancing contact angle θ_A is generally larger than the receding angle θ_R (cf. Fig. 1A). Since the contact angle, θ, is related to the liquid-vapor surface tension, γ, and the solid-liquid adhesion energy, W, by the Dupré equation (Fig. 1B):

$$(1 + \cos\theta)\gamma_L = W \tag{2}$$

we may conclude that *wetting hysteresis* or *contact angle hysteresis* ($\theta_A > \theta_R$) implies that either $\gamma_A > \gamma_R$ or that $W_R > W_A$. Since there is no reason nor any previous evidence for a liquid surface having different surface tensions for increasing and decreasing areas (so long as the liquid is pure), we may conclude that wetting or contact angle hysteresis actually implies adhesion hysteresis, viz. Eq. (1).

In all the above cases at least one of the surfaces is always a solid. In the case of solid-solid contacts, the hysteresis has generally been attributed to viscoelastic bulk deformations of the contacting materials or to plastic deformations of locally contacting asperities [1, 2]. In the case of solid-liquid contacts, hysteresis has usually been attributed to surface roughness or to chemical heterogeneity [3] as illustrated in Fig. 1C and 1D, though there have been reports of significant hysteresis on molecularly smooth and chemically homogeneous surfaces [4].

In this Part we shall focus on two other, less studied but more fundamental, mechanisms that can give rise to hysteresis. These may be conveniently referred to as (i) *mechanical hysteresis,* arising from intrinsic mechanical irreversibility of many adhesion/decohesion processes (see Fig. 1B *inset*, and Fig. 2), and (ii) *chemical hysteresis,* arising from the intrinsic chemical irreversibility at the surfaces associated with the necessarily finite time it takes to go through any adhesion/decohesion or wetting/dewetting process (Fig. 1E). Henceforth we shall use the term *approach/separation* to refer quite generally to any cyclic process, such as adhesion/decohesion, loading/unloading, advancing/receding and wetting/dewetting cycles.

As will be argued below, because of natural constraints of *finite time* and the *finite elasticity* of materials most approach/separation cycles are thermodynamically irreversible, and therefore energy dissipating. By thermodynamic irreversibly we simply mean that one cannot go through the approach/separation cycle via a continuous series of equilibrium states because some of these are connected via spontaneous—and therefore thermodynamically irreversible—instabilities or transitions. During such transitions there is an absence of *mechanical* and/or *chemical* equilibrium. In many cases the two are intimately

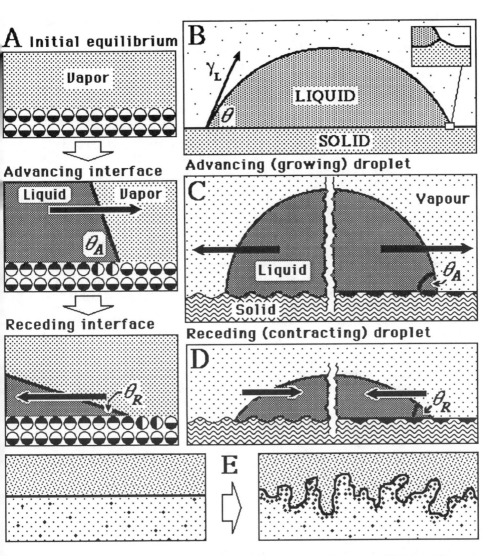

Fig.1. *Examples of wetting and contact-angle hysteresis. **A.** Solid surface in equilibrium with vapour. On wetting the surface, the advancing contact angle θ_A is observed; on dewetting it decreases to the receding angle, θ_R. This is an example of adhesion hysteresis during wetting/dewetting that is analogous to that during the loading/unloading of two solid surfaces. **B.** Liquid droplet resting on a flat solid surface. This is not a true equilibrium situation: at the three-phase contact boundary the normal liquid stress, $\gamma_L\sin\theta$, is balanced by high local stresses on the solid which induce elastic or plastic deformations (inset) and/or chemical rearrangements to relax these stresses. **C and D.** Contact angle hysteresis is usually explained by the inherent roughness (left side) or chemical heterogeneity (right side) of surfaces. **E.** Interdiffusion molecular reorientations and exchange processes at an interface may induce molecular or microscopic-scale roughness and chemical heterogeneity even though initially both surfaces are (initially) perfectly smooth and homogeneous.*

related or occur at the same time (and there may also be an absence of *thermal* equilibrium) but the above distinction is nevertheless a useful one since there appears to be two fairly distinct molecular processes that give rise to them. These will now be considered in turn.

1.2 Mechanical hysteresis

Consider two solid surfaces a distance D apart (Fig. 2) interacting with each other via an attractive potential and a hard-wall repulsion at some cut-off separation, D_o. Let the materials of the surfaces have a bulk elastic modulus K, so that depending on the system geometry the surfaces may be considered to be supported by a simple spring of effective 'spring constant' K_S [5]. When the surfaces are brought towards each other a mechanical instability occurs at some finite separation, D_A, from which the two surfaces jump spontaneously into contact (cf. lower part of Fig. 2). This instability occurs when the gradient of the attractive force, dF/dD, exceeds K_S [5]. Likewise, on *separation* from adhesive contact, there will be a spontaneous jump apart from D_o to D_R. Separation jumps are generally greater than approaching jumps.

Such spontaneous jumps occur at both the macroscopic and atomic levels. For example, they occur when two macroscopic (R≈1cm) surfaces are brought together in surface forces experiments [6]; they occur when STM or AFM tips approach a flat surface [5, 7], and they occur when individual bonds are broken (or 'popped') during fracture and crack propagation in solids [8]. But such mechanical instabilities will not occur if the attractive forces are weak or if the backing material supporting the surfaces is very rigid (high K or K_S). However, in many practical cases these conditions are not met and the adhesion/decohesion cycle is inherently hysteretic regardless of how smooth the surfaces, of how perfectly elastic the materials, and of how slowly one surface is made to approach the other (via the supporting material).

Thus, the adhesion energy on separation from contact will generally be greater than that on approach, and the process is unavoidably energy dissipative. It is important to note that this irreversibility does not mean that the surfaces must become damaged or even changed in any way, or that the molecular configuration is different at the end from what it was at the beginning of the cycle. Energy is always dissipated in the form of heat whenever two surfaces or molecules impact each other.

Before continuing with other types of hysteresis, it is worth mentioning what the true equilibrium situation is in the presence of mechanical instabilities. Referring to the lower part of Fig. 2, if the two surfaces are brought to some arbitrary separation, D', between D_A and D_R, and left to equilibrate at a finite temperature for a truly *infinite* length of time, the equilibrium separation will actually be a bimodal Boltzmann *distribution* of distances, peaking at D≈D' (the separated state) and D≈D_o (the contact state). Every now and then a spontaneous thermally induced fluctuation will occur, taking the surfaces from the separated to the contacting state, or vice versa. Over a long enough period of time the surfaces will have moved back and forth, sampling the whole of 'ergodic' space. The resulting *distribution* is what defines the equilibrium thermodynamic 'state' of the system.

It is also worth noting that (i) there is no one unique thermodynamically equilibrium separation, and (ii) if D' lies between D_A and D_R, the time-averaged distribution will center around two discrete separations D' and D_o which is analogous to a two-phase (e.g., solid-gas) system even though only two surfaces, or particles, are involved.

1.3 Chemical hysteresis

When two surfaces come into contact the molecules at the interfaces relax and/or rearrange to a new equilibrium configuration that is different from that when the surfaces were isolated (Fig. 1E). These rearrangements may involve simple positional and orientati-

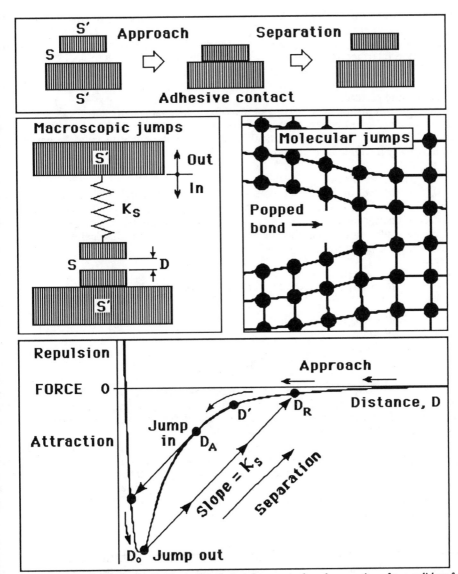

Fig.2. *Origin of mechanical adhesion hysteresis during the approach and separation of two solid surfaces.*
Top: The force between two real surfaces is never measured at the surfaces themselves, S, but at some other point, say S', to which the force is elastically transmitted via the material supporting the surfaces. Center (left): 'Magnet' analogy of two approaching surfaces, where the lower is fixed and where the other is supported at the end of a spring of stiffness K_S. Bottom: Force-distance curve for two surfaces interacting via an attractive van der Waals-type force-law, showing the path taken by the upper surface on approach and separation. On approach, an instability occurs at $D=D_A$, where the surfaces spontaneously jump into 'contact' at $D \approx D_O$. On separation, another instability occurs where the surfaces jump apart from $\sim D_O$ to D_R. Center (right): On the atomic level, the separation of two surfaces is accompanied by the spontaneous popping of bonds, which is analogous to the jump apart of two macroscopic surfaces.

onal changes of the surface molecules, as occurs when the molecules of two homopolymer surfaces slowly intermix by diffusion [9] or reptation [10]. In more complex situations new molecular groups that were previously buried below the surfaces may appear and intermix at the interface. This commonly occurs with surfaces whose molecules have both polar and nonpolar groups, for example, copolymer surfaces [11], surfactant surface [12,13] and protein surfaces [14]. All these effects act to enhance the adhesion or cohesion of the contacting surfaces.

What distinguishes chemical hysteresis from mechanical hysteresis is that during chemical hysteresis the chemical groups at the surfaces are different on separation from on approach. However, as with mechanical hysteresis, if the cycle were to be carried out infinitely slowly it *should* be reversible.

1.4 The JKR theory

Modern theories of the adhesion mechanics of two contacting solid surfaces are based on the JKR theory [16, 18]. In the JKR theory two spheres of radii R_1 and R_2, bulk elastic moduli K, and surface energy γ per unit area, will flatten when in contact. The contact area will increase under an external load or force, F, such that at mechanical equilibrium the contact radius a is given by (cf. Fig. 3)

$$a^3 = \frac{R}{K}\left[F + 6\pi RW + \sqrt{12\pi R\gamma F + (6\pi R\gamma)^2} \right]$$
(3)

where $R=R_1R_2/(R_1+R_2)$. Another important result of the JKR theory gives the adhesion force or 'pull off' force:

$$F_S = -3\pi R\gamma_S .$$
(4)

where, by definition, γ_S is related to the work of adhesion W, by $W=\gamma_S$. Note that according to the JKR theory a finite elastic modulus, K, while having an effect on load-area curve, has no effect on the adhesion force, F_S – an interesting and unexpected result that has nevertheless been verified experimentally [16, 17, 19-20].

Equation (3) is the basic equation of the JKR theory and provides a suitable framework for measuring the adhesion energies of contacting solids and for studying the effects of surface conditions and time on adhesion energy hysteresis. This can be done in two ways: first, by measuring how a varies with load (cf. Fig. 3) and comparing this with Eq. (3), and second, by measuring the 'pull off' force and comparing this with Eq. (4). We proceed to describe recent results of such experiments.[†]

[†] Of course, the JKR theory has been tested before, both for adhering and nonadhering surfaces [18, 24, 25], but space does not allow all these experiments to be reviewed here. For nonadhering surfaces, $\gamma=0$, and Eq. (3) reduces to the 'Hertzian' limit: $a^3=RF/K$. The inverse cubic dependence of a on F was verified by Horn et al. [25] for two microscopic, molecularly smooth curved surfaces of mica immersed in aqueous salt solution, where $\gamma=0$. Moreover, the measured a-F curves were found to be reversible (nonhysteretic) for increasing and decreasing loads. In contrast, for adhering mica surfaces in air ($\gamma>0$) there was a significant hysteresis in the a-F curves. This was attributed to viscoelastic effects in the glue supporting the mica sheets, though no systematic study was carried out to further investigate this phenomenon. We return to reconsider this matter in the light of the new results presented here.

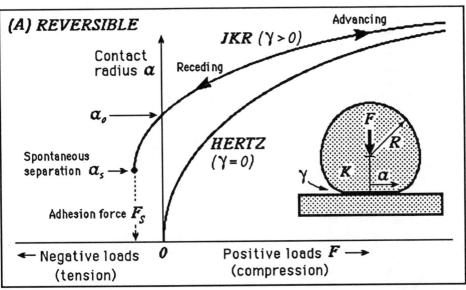

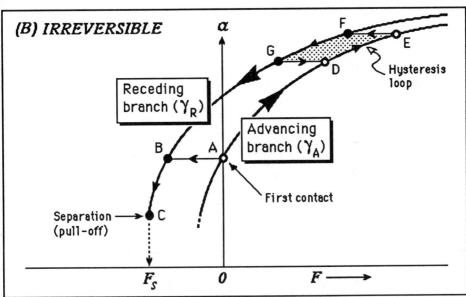

Fig.3 (A) *Reversible contact-radius versus load of nonadhesive Hertzian contact and adhesive JKR contact under ideal conditions. No hysteresis.* (B) *Irreversible a-F curves and the hysteresis loops they give rise to during an advancing-receding cycle (also known as compression-decompression, loading-unloading, and bonding-debonding cycles).*

1.5 Measurements of adhesion hysteresis between molecularly smooth surfaces

In our studies we have chosen to study these effects using molecularly smooth mica surfaces onto which well-characterized surfactant monolayers were adsorbed, either by adsorption from solution (sometimes referred to as 'self-assembly') or by the Langmuir-Blodgett deposition technique. Different types of surfactants and deposition techniques were used to provide surface-adsorbed monolayers with a wide variety of different properties such as surface coverage and phase state (solid, liquid or amorphous). Some of the surface used are shown schematically in Fig. 4.

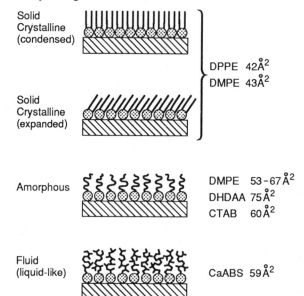

Fig. 4. *Likely chain configurations for monolayers in the crystalline, amorphous and fluid states (schematic). The first two phases shown are solid, the third is glassy or gel-like, and the last is liquid-like. The full names of the surfactants are: DPPE (di-palmitoyl-phosphatidyl-ethanolamine), DMPE (di-myristoyl-phosphatidyl-ethanolamine), DHDAA (di-hexadecyl-dimethyl-ammonium-acetate), CTAB (cetyl-dimethyl-ammonium-bromide), CaABS (calcium-alkyl-benzene-sulphonate). The values next to each surfactant gives its molecular area in the monolayer.*

The Surface Forces Apparatus technique [21-23] was used for measuring the adhesion or 'pull off' forces F_S, as well as the loading-unloading a-F curves for various surface combinations (cf. Fig. 5) under different experimental conditions. The pull-off method allows us to measure only γ_R, while the a-F curves give us both γ_A and γ_R. We may note that if all these processes were occurring at thermodynamic equilibrium, then γ_A and γ_R should be the same and equal to the well-known literature values of hydrocarbon surfaces, viz. $\gamma = 23$–31 mJ/m^2 [15]. Note, too, that the phase state of hydrocarbon chains has little effect on γ, as can be ascertained from the similar values for liquid hexadecane (27 mN m^{-1}) and solid paraffin wax (25-30 mN m^{-1}) at the same temperature of 25°C [15]. Thus, we would expect the *equilibrium* values of γ to fall within the range: 23-31 mJ/m^2.

Unless otherwise stated, all measurements were carried out after the monolayers had been allowed to equilibrate with an atmosphere of pure, dry nitrogen gas.

1.6 Effects of molecular relaxations at surfaces on adhesion hysteresis

Figure 6 shows the measured a-F curves obtained for a variety of surface-surface combinations. Both the advancing (open circles) and receding (black circles) points were fitted to Eq. (3). These fits are shown by the continuous solid lines in Fig. 6, and the corresponding fitted values of γ_A and γ_R are also shown.

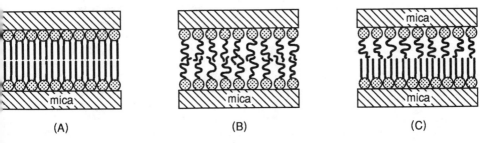

Fig. 5. *Schematics of likely chain interdigitations after two surfaces have been brought into contact.*
(A) Both surfaces in the solid crystalline state – no interdigitation.
(B) Both surfaces amorphous or fluid – interdigitation (entanglements) and disentangle-ments occur slowly for two amorphous surfaces and rapidly for two fluid surfaces. If the surfaces are separated sufficiently quickly, the effective molecular areas being separated from each other will be greater than the 'apparent' area, and the receding adhesion will be greater than the advancing adhesion.
(C) One surface solid crystalline, the other amorphous or fluid – no interdigitation.

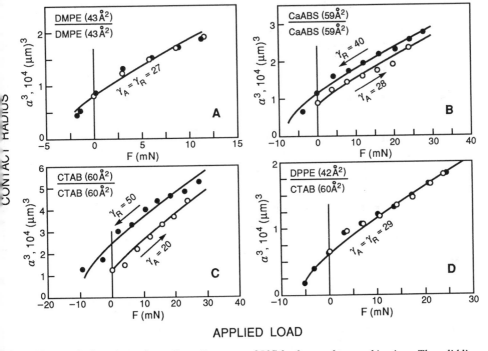

Fig. 6. *Measured advancing and receding a-F curves at 25 °C for four surface combinations. The solid lines are based on fitting the advancing and receding branches to the JKR theory, Eq. (3), from which the indicated values of γ_A and γ_R were determined. At the end of each cycle the pull-off force was measured. For the four cases shown here the following adhesion energy values were obtained based on Eq. (4): (A) crystalline on crystalline: $\gamma_R=28$ mJ/m^2, (B) fluid on fluid: $\gamma_R=36$ mJ/m^2, (C) amorphous on amorphous: $\gamma_R=44$-76 mJ/m^2, (D) amorphous on crystalline: $\gamma_R=32$ mJ/m^2. The equilibrium (literature) values for γ are in the range 23–31 mJ/m^2.*

As already mentioned, under ideal (thermodynamically reversible) conditions should be the same regardless of whether one is going up or down the JKR curve, as was shown in Fig. 3A. This was found to be the case for two solid crystalline monolayers (Fig. 6A) and almost so for the two fluid monolayers (Fig. 6B). The greatest hysteresis was found for two amorphous monolayers (Fig. 6C). However, no hysteresis was measured when an amorphous or a fluid monolayer was brought together with a solid crystalline monolayer (Fig. 6D). In all cases, the γ_R value determined independently from the measured pull-off force (see legend to Fig. 6) was found to be the same (to within $\pm 10\%$) with the value determined from the receding branch of the a-F curve.

Our results therefore provide a convincing test of the validity and inherent consistency of the two basic JKR equations, Eqs (3) and (4).§ They also show that experimental pull-off forces should, in general, be higher than given by the JKR theory, Eq. (3), unless the system is truly close to equilibrium conditions.

The data presented so far strongly suggests that chain interdiffusion, interdigitation or some other molecular-scale rearrangement occurs after two amorphous or fluid surfaces are brought into contact which enhances their adhesion during separation. The observation that two solid-crystalline or a crystalline and an amorphous surface do not exhibit hysteresis is consistent with this scenario, since only one surface needs to be frozen to prevent interdigitation from occurring with the other. All this is illustrated schematically in Fig. 5. The much reduced hysteresis between two fluid-like monolayers probably arises from the rapidity with which the molecules at these surfaces can disentangle (equilibrate) even as the two surfaces are being separated (peeled apart).

It appears, therefore, that the ability of molecules or molecular groups to interdiffuse, interdigitate and/or reorient at surfaces, and especially the relaxation times of these processes, determine the extend of adhesion hysteresis (chemical hysteresis). Little or no hysteresis arises between rigid, frozen surfaces since no rearrangements occur during the time course of typical loading-unloading rates. Liquid-like surfaces are likewise not hysteretic, but now because the molecular rearrangements can occur faster than the loading-unloading rates. Amorphous surfaces, being somewhat in between these two extremes are particularly prone to being hysteretic because their molecular relaxation times can be comparable to loading-unloading times (presumably the time for the bifurcation front to traverse some molecular scale length).

If this interpretation is correct it shows that very significant hysteresis effects can arise purely from surface hysteresis effects, which would be in addition to any contribution from bulk viscoelastic effects. The former involves molecular interdigitations that need not go much deeper than a few ångstroms from an interface. Some further experiments support the above picture.

1.7 Effects of time and temperature on adhesion hysteresis
The adhesion energy as determined from the pull-off force generally increased with the contact time for all the monolayers studied. This is shown in Fig. 7 for two amorphous monolayers of CTAB for which the effect was most pronounced. Notice how the

---

§ Earlier reports [19, 20] that the pull-off force is higher than $3\pi R\gamma$ and closer to $4\pi R\gamma$ must now be considered to be due to these surface energies being hysteretic due to the adsorption of a thin contaminant layer of organic material on the mica surfaces [23]. Hysteresis was indeed measured in these experiments but attributed to the glue supporting the surfaces. The present results show that the glue is not responsible for the hysteresis, which is due to surface effects, and that the unexpectedly higher values measured for F_s are indeed to be expected, but they do not correspond to any equilibrium values.

ysteresis increases as the more the monolayer goes into the amorphous, glassy state
15°C) and disappears once it is heated to above its chain-melting temperature (35°C).

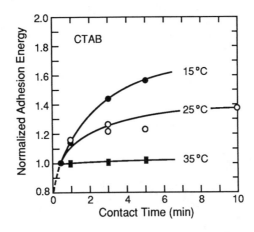

Fig.7. *Effect of contact time on the normalized adhesion energy of two CTAB (amorphous) monolayers at different temperatures (at 35°C CTAB is in the liquid state).*

Similarly, the rate at which two surfaces were loaded or unloaded also affected their adhesion energy. Again, for two amorphous CTAB surfaces, Fig. 8 shows that on slowing down the loading/unloading rate, the hysteresis loop becomes smaller and the advancing and receding energies, γ_A and γ_R, approach the equilibrium value.

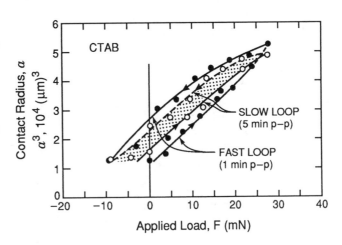

Fig.8. *Effect of advancing/receding rates on a-F curves for two CTAB monolayers at 25°C. By fitting the data points to Eq. (3) the following values were obtained. For the fast loop (1 minute between data points): $\gamma_A=20$ mJ/m², $\gamma_R=50$ mJ/m². For the slow loop (5 minutes between points): $\gamma_A=24$ mJ/m², $\gamma_R=44$ mJ/m².*

One should note that decreasing the unloading or peeling rate may sometimes act to *increase* the adhesion, since by decreasing the peeling rate one also allows the surfaces to remain longer in contact. In Fig. 8, the surfaces were first allowed to remain in contact for longer than was needed for the interdigitation processes to be complete (as ascertained from the contact time measurements of Fig. 7).

362

1.8 Effects of capillary condensation on adhesion hysteresis

Figure 9 shows that when liquid hydrocarbon vapour is introduced into the chamber and allowed to capillary condense around the contact zone, all hysteresis effects disappear. This again shows that by fluidizing the monolayers they can now equilibrate sufficiently fast to be considered always at equilibrium (like a true liquid). Such effects may be expected to occur with other surfaces as well, so long as the vapour condenses as a *liquid that wets* the surfaces.

Stick-slip adhesion, both on loading but particularly on unloading, was often found to occur with most of the monolayers studied, but was always eliminated when organic vapours were introduced.

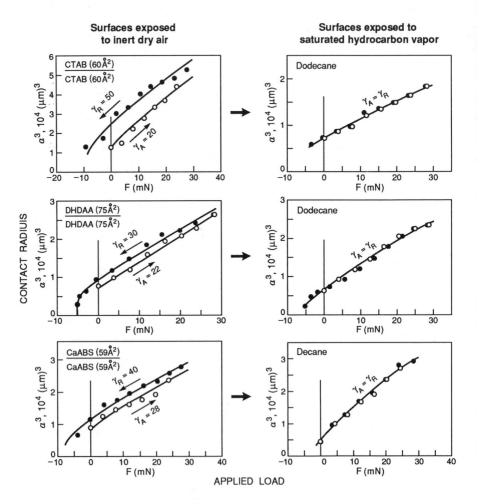

Fig.9. *Disappearance of adhesion hysteresis on exposure of monolayers to various organic vapors. The adhesion energies as measured by the pull-off forces were between 18 and 21 mJ/m² for the three systems shown.*

1.9 Summary of molecular origin of adhesion hysteresis

The above results show that the adhesion of two molecularly smooth and chemically homogeneous surfaces can be hysteretic due to structural and chemical changes occurring at the molecular (or even ångstrom) level. More specifically, adhesion hysteresis can arise from the finite elasticity of solid surfaces, the finite times of molecular rearrangements at interfaces, and the inevitably finite times of real loading/unloading (and wetting/dewetting) processes. Adhesion hysteresis increases with

(i) the ability of the molecular groups at the surfaces to reorient and interdiffuse across the contact interface which is often determined by the phase state of the surface molecules,

(ii) the time two surfaces remain in contact and the externally applied load during this time, and

(iii) the rate of approach and separation (or peeling) of the surfaces.

One consequence of adhesion hysteresis is that larger adhesion 'pull off' forces would generally be measured than expected from equilibrium surface energy values. The amorphous monolayers we studied provided the highest hysteresis observed, with cases where γ_R was more than twice the value of γ_A. One may interpret this as indicating that at the point of separation, the real surface area (at the molecular level) was more than twice the apparent surface area (also equal to the initial surface area). This is in contrast to the more common case, encountered in adhesion and friction phenomena, where the real area of contact is generally much less than the apparent area, due to asperity contacts.

For the molecularly smooth surfaces studied here, it appears that chemical hysteresis is far greater than mechanical hysteresis – the latter being capable of giving rise to energy dissipation effects equivalent to changing the surface energy by up to 50 mJ/m^2 even for 'low energy' organic surfaces.

Our results also question traditional explanations of hysteresis based purely on the *static* surface roughness and chemical heterogeneity of surfaces (cf. Fig. 1), and focuses more on the *dynamics* of these effects.

PART II

BOUNDARY AND INTERFACIAL FRICTION

Situations where atomically smooth, undamaged surfaces slide past each other may be referred to as 'interfacial' sliding. The surfaces may be separated by one or two semi-ordered layers of liquid (lubricated sliding). Such cases always involve adhesion, and even at zero applied load there is a finite area of contact and a finite friction. In the case of unlubricated sliding of 'dry' surfaces, the frictional force at low loads is described by the equation originally proposed by Bowden and Tabor: F=SA, where S is a critical shear stress and A is the molecular area of contact. Experiments show that the variation of A with applied load L is well described by the JKR theory, even during sliding and even at negative values of L. At higher loads there is an additional contribution to F that is proportional L, given by F=CL. This contribution is analogous to Amontons' law, but C has a different origin and exists even in the absence of adhesion. When damage occurs, Amontons' law F=μL, is now applicable where μ is the normal coefficient of friction. We discuss the factors that determine the magnitudes of S, C and μ for molecularly smooth crystalline surfaces like mica. Lubricated sliding is considered in PART III.

2.1 Normal friction, boundary friction and interfacial friction

We are still a long way from understanding what happens at the molecular level during the sliding of two surfaces past each other. Most frictional processes occur with the surfaces 'damaged' in one form or another [1]. This we shall refer to as 'normal' friction. In some cases, the surfaces will slide past each other while separated by large, almost macroscopically-sized particles of wear debris. In other cases, usually under high loads but also depending on the smoothness and hardness of the surfaces, the damage (or wear) may be localized within a much narrower interfacial region of plastically deformed nanometer-sized asperities.

There are also situations where sliding can occur between two perfectly smooth undamaged surfaces. Experimentally, it is usually difficult to unambiguously establish which type of sliding is occurring. The term 'boundary' friction elicits the idea that most of what is happening is restricted to a thin boundary region, which, as in a grain boundary, would typically not extend more than a few nanometers on either side of it. However, the term 'boundary lubrication' is more commonly used to denote the sliding of two surfaces separated by thin monomolecular layers of some suitable 'boundary' lubricant (e.g. a surfactant monolayer), though here too it is presumed that plastic deformations and damage of isolated asperity contacts are also occurring during sliding [1].

In contrast, there is currently no commonly accepted term to describe the sliding of two perfect, molecularly smooth, undamaged surfaces, whether in molecular contact or separated by molecularly thin films of liquid or lubricant fluids. Homola *et al.* [28] used the terms 'interfacial friction' or 'interfacial sliding' for such phenomena, which suggest that the frictional mechanism is now restricted to a molecularly thin interfacial region having a uniform gap thickness and a well-defined contact area. Clearly, the interaction forces associated with interfacial friction would be much more localized than in the case of boundary friction or normal friction.

Moreover, during interfacial sliding, the friction depends critically on the intermolecular forces between the surfaces, their area of contact, the precise distance between the two surfaces (at the ångstrom level), but it depends only weakly on the applied load and on the bulk viscosity of the lubricating liquid. In contrast, quite different parameters are important when sliding occurs between damaged surfaces separated by thicker layers of debris or bulk liquids.

These and other aspects of the differences between interfacial sliding and sliding in the presence of damage (wear) is the primary focus of PART II, as well as the elucidation of the nature of the transition between these two modes of sliding. These questions need to be answered if we are to gain a fundamental understanding of the relationships between adhesion, friction, lubrication and wear and the role of different intervening liquids (or lubricants) in these processes.

2.2 The Surface Forces Apparatus for studies of friction

Tabor & Winterton [24] and later Israelachvili & Tabor [25] developed the 'Surface Forces Apparatus' (SFA) for measuring the van der Waals forces between two mica surfaces as a function of their separation in air or vacuum. More sophisticated apparatuses were later developed for measuring the forces between surfaces in liquids [23]. In this type of apparatus the distance between the two surfaces can be controlled to better than 1Å, while the surface separation is measured to similar accuracy by an optical technique using multiple beam interference 'FECO' fringes. Additionally, the shapes of the FECO fringes give the surface profiles [22, 24, 26], so that the area of contact between two surfaces can be precisely measured, and any changes may be readily observed under both static and dynamic conditions (in real time) by monitoring the changing shapes of these fringes. Forces are measured from the deflection of a 'force-measuring spring'. Though the olecularly smooth surface of mica is the primary surface used in these measurements, it is ossible to deposit or coat each surface with surfactant layers, polymer and metal films, tc., prior to an experiment.

A new friction attachment was recently developed suitable for using with the SFA. This attachment (shown schematically in Fig. 10) allows for the two surfaces to be sheared ast each other at sliding speeds which can be varied continuously from 0.1-20 μm/sec while simultaneously measuring both the transverse (frictional) force and the normal load etween them. The externally applied load, L, can be varied continuously, and both ositive and negative loads can be applied. Finally, the distance between the surfaces D, their true molecular contact area A, their elastic (or viscoelastic or elastohydrodynamic) eformation, and their lateral motion can all be monitored simultaneously by recording the moving FECO fringe pattern using a video camera-recorder system [27, 28]

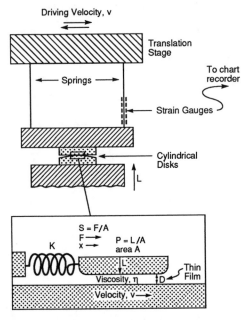

Fig.10.
Schematic drawing of the sliding attachment for use with the Surface Forces Apparatus.

2.3 Sliding of two mica surfaces in dry air

When two mica surfaces are brought into contact in air, a thin (approximately 5Å thick) interfacial layer of physisorbed mainly carbonaceous material separates the surfaces. The surfaces adhere to each other so that even under zero external load there is a finite area of contact. When subjected to an increasing external load, the area of contact increases. If the load is decreased to negative values, the contact area falls but remains finite until the pull-off force, F_s, is reached, at which point the surfaces spontaneously separate. This type of behavior is predicted by the JKR theory. During this process of compression and decompression the physisorbed layer between the surfaces is not squeezed out even under pressures up to 150 atm.

When sliding is initiated, the surfaces remain 'pinned' to each other until som
critical shear force is reached. At this point, the surfaces begin to slide past each other at
steady velocity equal to the driving velocity. The frictional force during steady sliding wi
be referred to as the kinetic friction.

It is found that during sliding the area-load relationship is still given by the JK
theory, Eq. (3). Additionally, the frictional force is not proportional to the load but
proportional to the area of contact, as predicted by Bowden & Tabors' model [1],

$$F = S_cA. \tag{5}$$

Since the area of contact and the applied load were found to follow the JKR theory, it
instructive to plot both the contact area and frictional force as a function of load on the sam
graph. This is done in Fig. 11. Note that there is a frictional force even at negative load:
where the surfaces are still sliding in adhesive contact. Hence, no constant 'coefficient c
friction' can be ascribed to this type of interfacial sliding. Instead, the 'critical shear stress
S_c, is found to be a constant for a particular system. In a number of different experiment
with two mica surfaces sliding in dry atmospheres, S_c was measured to be 2.5×10^7 N/m
and to be independent of the sliding velocity.

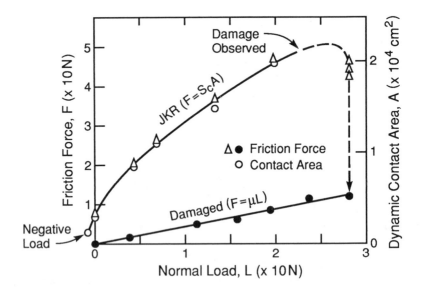

Fig.11. *Frictional force F and contact area A vs load L for two mica surfaces sliding i*
adhesive contact in dry air. The contact area is well described by the JKR theory, Eq. (3)
even during sliding, and the frictional force is found to be directly proportional this area
The vertical dashed line and arrow show the transition from 'interfacial' to 'normal' frictio
with the onset of wear (lower curve).

It is apparent that, especially at high loads, the frictional force is roughly roportional to the load, with the extrapolated line intercepting the load axis at some egative load. Such a model was proposed by Derjaguin [29] to describe interfacial friction the presence of adhesion. However, on closer scrutiny, it is clear that the data points do ot fall on a straight line. The friction is indeed proportional to the contact area which, specially at small and negative loads, shows no proportionality to the load.

.4 Sliding of two mica surfaces in humid air

In the presence of water vapour, the value of S_c falls by a factor of about 30 with icreasing relative humidity, even though the apparent contact area does not change and the eparation between the surfaces remains at 2.5-5Å. This effect is attributed to the repulsive ydration forces between the surfaces that arise when water condenses around the hydrated otassium ions on the mica surfaces [30].

Likewise, when interfacial sliding occurred with the surfaces totally immersed in a 0^{-2} M KCl salt solution, the friction was once again very low (Fig. 12). This is also ttributed to the existence of a purely repulsive short-range hydration force between the urfaces [30]. In particular, F was now found to be proportional to the load L, with a oefficient of friction of about 0.03 [28]. It appears that this type of friction follows Amontons' law but, as discussed below, it probably has a different origin from that of iormal' friction.

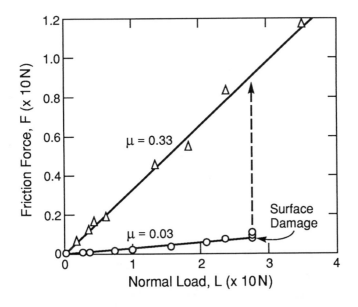

Fig. 12. *Mica surfaces sliding in 0.01 M KCl electrolyte where there is a strongly epulsive short-range 'hydration' force. During sliding the surfaces were separated by a 2.5Å water film. After damage occurred the surfaces were separated by a ~500Å–thick forest of mica flakes. Notice that in both cases sliding takes place in the absence of adhesion with the frictional force described by Amontons' law.*

It is interesting that a 2.5Å thick water film between two mica surfaces is sufficien to bring the coefficient of friction down to 0.02–0.03, a value the corresponds to th unusually low friction of ice. Clearly, a single monolayer of water can be a very goo lubricant (much better than any other liquid monolayer film between two mica surface discussed in PART III).

2.5 Sliding of surfaces in true molecular contact

Interfacial friction of two solid surfaces that are in true molecular contact durin sliding may called 'unlubricated interfacial sliding'. This is to distinguish this type c sliding from the case where the surfaces are separated by a one or two molecular layers c liquid ('lubricated interfacial sliding') or by some adsorbed, usually surfactant, monolaye ('boundary friction').

Two hydrocarbon surfaces in molecular contact will slide past each other withou any damage occurring if the sliding velocity and/or load are not too high. Such surfaces ca be produced by depositing various surfactant monolayers on mica surfaces. With calciun stearate monolayers in a dry environment, it was found that the friction can be dominate by stick-slip motion. Sliding can be smooth or intermittent, i.e. stick-slip, depending on th type of surfactant and the conditions of its deposition and the conformation of th molecules on the surface during sliding [32,33,34]. It is noteworthy that for sliding c undamaged calcium stearate monolayers the values of S_c are about $3-4 \times 10^6$ N/m². This i about an order of magnitude less than that between untreated mica surfaces in dr atmosphere (Fig. 13), but not as low as for mica surfaces in humid atmospheres or i water.

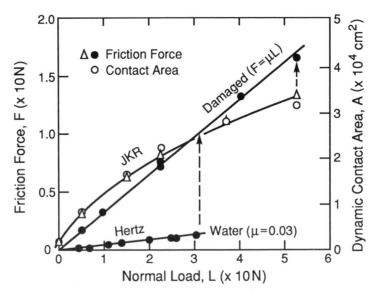

Fig.13. *Sliding of mica surfaces each coated with a monolayer of calcium stearate in th absence of damage (JKR 'boundary friction' curve) and in the presence of damag (Amontons' type 'damaged' curve). The lower (Hertz) line is for interfacial sliding with monolayer of water between the surfaces, as in Fig. 12, shown for comparison. In bot cases, after damage occurs, $\mu \approx 0.3$.*

Experiments currently in progress in this laboratory show that the friction and stick-slip of monolayers adsorbed on mica is very sensitive to the relative humidity, to temperature, and to the same factors that affect the adhesion hysteresis of surfactant-coated surfaces described in PART I.

2.6 Transition from interfacial friction to normal friction with wear

When damage occurs during sliding it is first seen as a small, highly localized, discontinuity (or spike) in the FECO fringes. This indicates that one or both of the surfaces have become torn so that a small mica flake now protrudes from one of the surfaces. This has been confirmed by electron micrographs of damaged regions taken at the end of an experiment. Damage usually occurs somewhere within the contact zone, though it can also occur at the periphery. Even though a flake trapped between the two surfaces may be very small (e.g. a few hundred ångstroms), it affects a much larger area by causing the surfaces to separate over a region of radius much greater than the size of the flake or asperity. Thus even one submicroscopic flake can cause a significant reduction in the area of contact and so in the friction.

Once damage occurs, it propagates rapidly throughout the contact zone. Secondary flakes form from the first and the real contact area rapidly falls. Within seconds, and well before the sliding surfaces have traversed one full contact diameter, there is an abrupt transition to 'normal' friction when the surfaces suddenly jump apart and sliding now proceeds smoothly with the surfaces separated by a 100-1,000Å gap of wear debris (mica flakes). At the transition, the friction changes abruptly from obeying $F = S_c A$ to obeying Amontons' law: $F = \mu L$, as shown in Figs. 11–13.

One remarkable feature of the results is that while the strength of interfacial friction, as reflected in the values of S_c, is very dependent on the type of surface, on the relative humidity, and on the nature of the liquid between the surfaces, this was not the case once the transition to 'normal' friction occurred, at which point all the friction coefficients were the same, i.e., insensitive to the type of coating on the mica surfaces or even to whether or not the surfaces were immersed in a liquid. This is illustrated in Figs. 11–13. Clearly, the mechanism and factors that determine the normal friction, at least of mica surfaces, must be quite different from those that govern interfacial friction. But one should point out that this effect probably only applies to certain types of brittle or layered materials. Preliminary experiments of the friction between silica surfaces, which are much more ductile, indicate a totally different mode of sliding, with no large flakes separating the two surfaces during sliding. In the case of silica, water and other surface-active components do have an effect on the friction coefficient and stick-slip behaviour.

2.7 Models of interfacial friction

A friction model was proposed by Tabor [31] and developed further by Sutcliffe *et al.* [43], McClelland [8], and Homola *et al.* [28] to explain the 'boundary' friction of two solid hydrocarbon surfaces sliding past each other in the absence of wear. In this model, the value of the critical shear stress, S_c, is calculated in terms of the energy needed to overcome the intermolecular forces between the surfaces as one layer is first slightly raised then slid across the other. The model applies equally well to the case of two sliding surfaces separated by a layer or more of liquid molecules.

This model is akin to pushing a cart over a road of cobblestones where the cartwheels represent the liquid molecules, and where the cobblestones represent the atoms of the surfaces over which the wheels have to roll before the cart can move. In this model the downward force of gravity replaces the attractive intermolecular forces between the two surfaces. When at rest the cartwheels find grooves between the cobblestones where they sit in potential energy minima and so the cart is at a stable equilibrium. A certain lateral force (the 'push') is required to raise the cartwheels against the force of gravity in order to initiate motion. Motion will continue as long as the cart is pushed, and rapidly stops once it is no longer pushed. Energy is dissipated by the liberation of heat (phonons, acoustic waves etc.) every time a wheel hits the next cobblestone. The Cobblestone model is not unlike the old Coulomb model of friction except that it is being applied at the molecular level and where the external load is replaced by attractive surface forces. Note that the model does not assume that all interfacial bonds are broken simultaneously over the whole contact area – just as for the cart moving on the cobblestone where the four wheels roll over individual stones more or less independently of each other – even though the overall motion is correlated on a larger scale.

2.8   Calculation of critical shear stress, S_c

We first consider the case of two surfaces sliding past each other whilst separated by a thin liquid film. When the two surfaces are brought into contact under some positive load the liquid molecules between them order themselves to fit snugly within the spaces between the atoms of the two surfaces, in an analogous manner to the self-positioning of the cartwheels on the cobblestone road. Hence, the free energy of the system is minimized.

A tangential force applied to one surface will not immediately result in the sliding of that surface relative to the other. The attractive van der Waals forces between the surfaces must first be overcome by the surfaces separating by a small amount. To initiate motion, let the normal distance between the two surfaces increase by a small amount, ΔD, while the lateral distance moved is Δd. If the force of adhesion between the surfaces is F_{ad}, then the interfacial energy change associated with the initiation of sliding is $\Delta D \times F_{ad}$. In a first approximation, this is expected to be some small fraction ε of the total adhesion energy $2\gamma A$, of the two surfaces, where A is the area of contact. In this simple model it is assumed that (i) the contribution of an external normal force to the work required to separate the surfaces is negligible in comparison to the internal van der Waals forces of the system, and (ii) the process is not reversible, with a fraction ε of the energy being dissipated as heat. If the friction force, i.e. the force needed to initiate sliding, is F_{fr}, then by equating the two energies, one obtains the condition necessary for sliding, viz.

$$\Delta d \times F_{fr} = \Delta D \times F_{ad} = (2\gamma A)\varepsilon. \tag{6}$$

Now, typically $\gamma \approx 25 \times 10^{-3}$ J/m^2 for a hydrocarbon or a van der Waals surface. Assuming that the surfaces move normally and laterally by about 1Å (i.e., $\Delta D \approx \Delta d \approx 10^{-10}$ m), and that 10% of the surface energy is lost every time the surfaces move across the characteristic length Δd [cf. McClelland, ref. 8], so that $\varepsilon \approx 0.1$, we obtain

$$S_c = F_{fr}/A \approx \frac{2\gamma\varepsilon}{\Delta d} \tag{7}$$

$$\approx 5 \times 10^7 \text{ N/m}^2 \quad \text{using the above parameters.}$$

This compares well with the experimental values of 2×10^7 N/m^2 for surfaces sliding of mica in air or separated by one molecular layer of cyclohexane. It is intuitively clear that the larger the separation, or the greater the number of liquid layers between the surfaces, the less the energy required to initiate sliding and, hence, the lower the friction. This is precisely what is observed.

2.9 Effect of external load on interfacial friction

When there is no interfacial adhesion S_c is zero. Thus, in the absence of any adhesive forces between the surfaces themselves, the only 'attractive' force that needs to be overcome for sliding to occur is the externally applied load or pressure (which has so far been ignored).

For a preliminary qualitative discussion of this question the magnitudes of the *externally* applied pressure to the *internal* van der Waals pressure between the two surfaces are compared. Assume a typical external pressure of $P_{ext}=100$ atm. The internal van der Waals pressure is given by $P_{int}=F_{ad}/A=A/6\pi D^3 \approx 10^4$ atm (using a typical Hamaker constant of $A=10^{-12}$ erg, and assuming $D\approx2$Å for the equilibrium interatomic spacing [16]). This implies that we do not expect the externally applied load to affect our measured values of S_c as long as these are due to attractive van der Waals forces, viz. as long as S_c is of the order of 10^7 N/m^2 or greater.

For a more general semi-quantitative analysis, again consider the Cobblestone model as used to derive Eq. (7) and now simply include an additional contribution to F_{ad}/A due to the externally applied pressure P_{ext} (equivalent to the action of gravity in the case of the cart being pushed over the cobblestones).

Thus:
$$\Delta d \times F_{fr}/A = \Delta D \times F_{ad}/A + \Delta D \times F_{ext}/A = \Delta D(P_{int}+P_{ext}) \qquad (8)$$

which gives the more general relation

$$S_c = F_{fr}/A = C_1 + C_2 P_{ext} \qquad (9)$$

where $P_{ext} = F_{ext}/A$ or L/A, and where C_1 and C_2 are constants.

The constant C_1 depends on the mutual adhesion of the two surfaces, while C_2 depends on the topography or atomic bumpiness of the surface groups – the smoother the surface groups the smaller the ratio $\Delta D/\Delta d$ and hence the lower the value of C_2. Equation (9) was previously derived by Briscoe & Evans [34] where the constant C_2 was interpreted in terms of two parameters Ω/ϕ whose physical significance are different from the interpretation proposed here.

In the absence of any attractive interfacial force, we have $C_1\approx0$, and the second term in Eq. (9) should dominate. Such situations typically arise when surfaces interact across a repulsive but molecularly thin liquid film, for example, when mica slides across a monolayer or two of water (Fig. 12). In such cases the total frictional force should be low and it should now increase *linearly* with the external load according to

$$F = C_2 P_{ext} A = C_2 L \qquad (10)$$

Note that this has the same form as Amontons' Law where C_2 is now analogous to a coefficient of friction, μ, which for sliding in 10^{-2} M KCl (cf. Fig. 12) was determined to be $C_2=0.025$.

It is important to note that the origin of this type of friction where Amontons' Law is obeyed is, nevertheless, quite different from the conventional explanation of 'normal' friction [ref. 1, Ch. III and VI] which is due to the shearing of *adhesive* junctions. Here, we do not have any adhesion between the surfaces.

The relative contributions of the two terms in Eq. (9) will depend very much on the nature and type of the system being studied. In many of the mica experiments described here, the external pressure P_{ext} did not exceed 100 atm (10 MPa) which is small relative to P_{int}. Briscoe and Evans [34] carried out experiments on surfactant-coated mica sheets (similar to those shown in Fig. 13), but they used surfaces of much smaller radii (R=0.05cm) and so managed to obtain pressures as high as 500 MPa. Interestingly, their results showed that for pressures above about 10-20 MPa, the shear stress does indeed increase linearly with P_{ext} as predicted by Eq. (10). From the slopes of their lines values for $\Delta D/\Delta d$ in the range $0.1 - 0.03$ may be deduced.

2.10 Summary of interfacial sliding

To summarize, it appears that during 'interfacial sliding' there are two major contributions to the total critical shear stress, which may be expressed as

$$S_{c\,(tot)} = S_{c\,(int)} + S_{c\,(ext)} = C_1 + C_2 P_{ext} \tag{11}$$

where the two terms refer to the internal (or interfacial force) contribution, the external (applied pressure) contribution. However, neither C_1 nor C_2 can be considered as truly constant since both can depend on P_{ext}. In particular, C_1 depends critically on the surface forces and so could vary, for example, with the number of liquid layers between two surfaces during sliding.

Equation (11) can be rewritten so as to give the frictional force, F_{fr}, in terms of the load, L. For high loads where $A \propto L^{2/3}$, we obtain

$$F_{fr} = AS_c = C_1 L^{2/3} + C_2 L \tag{12}$$

where C_1 and C_2 are constants. Note that the second term is proportional to L as in Amontons' Law for normal friction, but that it has a different origin since it does not require any interfacial adhesion nor the shearing of adhesive junctions.

When damage occurs, there is a rapid transition to 'normal sliding' in the presence of wear debris. Our observations [28] indicate that the mechanisms of interfacial friction and normal friction are vastly different on the submicroscopic and molecular levels. However, under certain circumstances both may appear to follow a similar equation (e.g. Amontons' law) even though the friction coefficients will be determined by quite different material properties in each case. Finally, if one considers the origin of Eqs (11) and (12) for interfacial friction, it can be seen that the second term may well apply to normal friction as well. If so, μ becomes identical to the C_2 of Eq. (12). This, in turn, is given by the simple ratio of two distance parameters $\Delta D/\Delta d$ representing the ratio of the vertical to horizontal displacements of the surfaces during sliding (as in Coulomb friction). It is possible that the invariant value for μ of about 0.3 obtained here for mica sliding on wear tracks under very different conditions is simply a manifestation of the mean geometry adopted by the torn up mica flakes. This mode of sliding (Coulombic friction) may very well occur in other brittle or crystalline systems.

PART III
PROPERTIES OF LIQUIDS IN MOLECULARLY THIN FILMS

The static and dynamic properties of ordinary liquids, when confined within a very thin film between two surfaces, can be very different from those of the bulk liquids. Films thinner than 10 molecular diameters can become much more viscous, and those below 4 molecular diameters can undergo a phase-transition into a liquid-crystalline or solid-like phase. Such 'liquid' films can now support a finite load and shear stress. These modified properties depend intimately on the geometry of the liquid molecules, and how these molecules can fit in between the two (atomically) corrugated surfaces. The tribological properties of such molecularly thin 'liquid' films are described in the light of recent experimental and theoretical (computer simulation) studies.

3.1 Introduction

When a liquid is confined between two surfaces or within any narrow space whose dimensions are less than 5 to 10 molecular diameters, both the static (equilibrium) and dynamic properties of the liquid, such as its compressibility and viscosity, can no longer be described even qualitatively in terms of the bulk properties. The molecules confined within such molecularly thin films become ordered into layers ('out-of-plane' ordering), and within each layer they can also have lateral order ('in-plane' ordering). Such films may be thought of as being more solidlike than liquidlike.

The measured forces between two solid surfaces across molecularly thin films exhibit exponentially decaying oscillations, varying between attraction and repulsion with a periodicity equal to some molecular dimension of the solvent molecules. Such films can therefore sustain a finite normal stress, and the adhesion between two surfaces across a film is 'quantized', depending on the number of layers between the surfaces [35] and of magnitude that depends also on the 'twist' angle between the two surface lattices [36]. The structuring of molecules in thin films and the oscillatory forces it gives rise to are now reasonably well understood, both experimentally and theoretically, at least for simple liquids [35-38].

Work has also recently been done on the dynamic, e.g., viscous or shear, forces associated with molecularly thin films. Both experiments [27,39,40] and theory [41,42] indicate that even when two surfaces are in steady state sliding they still prefer to remain in one of their stable potential energy minima, i.e., a sheared film of liquid can retain its basic layered structure, though its lateral in-plane ordering may be modified or lost. Thus, even during motion the film does not become totally liquidlike. Indeed, such films exhibit yield-points before they begin to flow. They can therefore sustain a finite shear stress, in addition to a finite normal stress. The value of the yield stress depends on the number of layers comprising the film and represents another 'quantized' property of molecularly thin films [27, 39-42].

The dynamic properties of a liquid film undergoing shear are very complex. Depending on whether the film is more liquidlike or solidlike, the motion will be smooth or of the 'stick-slip' type – the latter exhibiting yield points and/or periodic 'serrations' characteristic of the stress-strain behavior of ductile solids. During sliding, transitions can occur between n solidlike layers and (n-1) layers or (n+1) layers, and the details of the motion depend critically on the externally applied load, the temperature, the sliding velocity, the twist angle and the sliding direction relative to the surface lattices.

3.2   Molecular events during shear: stick-slip

Here we briefly review recent results on the shear properties of simple molecules in thin films and how these are related to changes in their molecular configurations. These have been studied using the Surface Forces Apparatus technique [27, 39,40,43] and computer simulations [41,42]. Figure 14 shows typical results for the friction measured as a function of time (after commencement of sliding) between two mica surfaces separated by n=3 molecular layers of the liquid OMCTS, and how the stick-slip friction increases to higher values in a quantized way when the number of layers falls to n=2 and then to n=1.

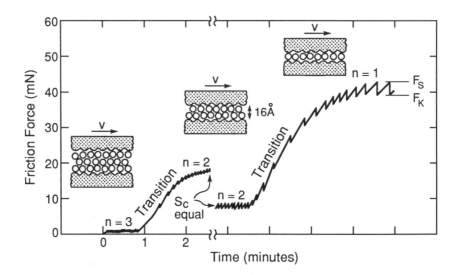

Fig.14. *Measured change in friction during interlayer transitions of the silicone octamethylcyclotetrasiloxane (OMCTS, an inert liquid whose quasi spherical molecules have a diameter of 8Å) [39]. In this system, the shear stress S=F_{fr}/A, was found to be constant so long as the number of layers n remained constant. Qualitatively similar results have been obtained with cyclohexane [27]. The shear stresses are only weakly dependent on the sliding velocity v. However, for sliding velocities above some critical value v_c, the stick-slip disappears and the sliding is now smooth or 'steady' at the kinetic value.*

With the much added insights provided by recent computer simulations of such systems [41,42,44] a number of distinct regimes can be identified during the stick-slip sliding that is characteristic of such films, shown in Fig. 15 (a) to (d).

Surfaces at rest — Fig. 15(a): With no externally applied shear force, solvent-surface interactions induce the liquid molecules in the film to adopt solidlike ordering [42]. Thus at rest the surfaces are stuck to each other through the film.

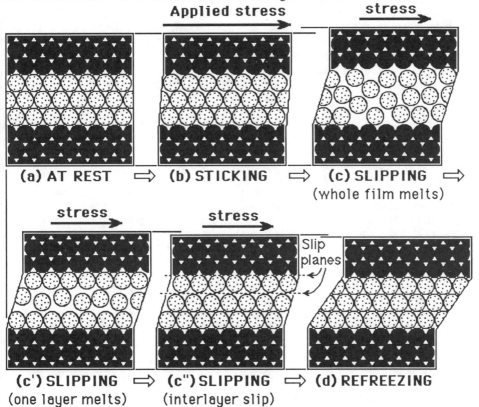

Fig.15. *Schematic illustration of molecular rearrangements occurring in a molecularly thin film of spherical or simple chain molecules between two solid surfaces during shear. Note that, depending on the system, a number of different molecular configurations within the film are possible during slipping and sliding, shown here as stages (c) – total disorder, (c') – partial disorder, and (c') – order persists even during sliding with slip occurring at a single slip-plane either within the film or at the film-solid interface. The configurations of branched-chained molecules is much less ordered (more entangled) and remains amorphous during sliding, leading to smoother sliding with reduced friction and little or no stick-slip.*

Sticking regime (frozen, solidlike film) — Fig. 15(b): A progressively increasing lateral shear stress, τ, is applied. This causes a small increase in the lateral displacement, x, and film thickness, D, but only by a small fraction of the lattice spacing or molecular dimension, σ. In this regime the film retains its solidlike 'frozen' state — all the strains are elastic and reversible, and the surfaces remain effectively stuck to each other.

Slipping and sliding regimes (melted, liquidlike film) — Fig. 15 (c) , (c'), (c'): When the applied shear stress has reached a certain critical value, the film suddenly melts (known as 'shear melting') and the two surfaces begin to slip rapidly past each other as the 'static shear stress' (τ_s) or 'static friction force' (F_s) has been reached (in the language of materials science the 'upper yield point' has been reached).

If the applied stress, τ, is kept at a constant value the upper surface will continue to slide indefinitely once it has settled down to some constant velocity. Even if the shear stress is reduced below τ_s, steady-state sliding will continue so long as it remains above some critical value, variously referred to as the 'dynamic shear stress' (τ_d), the 'dynamic friction force' (F_d), or the 'lower yield point'. The experimental observation that the static and dynamic stresses are different suggests that during steady-state sliding the configuration of the molecules within the film is almost certainly different from that during the slip.

Experiments with linear chain (alkane) molecules show that the film thickness remains quantized during sliding, so that the structure of such films is probably more like that of a nematic liquid crystal where the liquid molecules have become shear aligned in some direction enabling shear motion to occur while retaining some order within the film.

Computer simulations for simple spherical molecules [42] indicate that during the slip, the film thickness, D, is roughly 15% higher (i.e., the film density falls), and the order parameter drops from 0.85 to about 0.25. Both of these are consistent with a disorganized liquidlike state for the whole film during the slip, as illustrated schematically in Fig. 15(c). At this stage, we can only speculate on other possible configurations of molecules in the sliding regime. This probably depends on the shapes of the molecules (e.g., whether spherical or linear or branched), on the atomic structure of the surfaces, on the sliding velocity, etc. Figure 15 (c), (c') and (c') show three possible sliding modes wherein the molecules within the shearing film either totally melt, or retain their layered structure and where movement occurs either within one or two layers that have remained molten while the others are frozen, or where slip occurs between two or more totally frozen layers. Other sliding modes, for example, involving the movement of dislocations or disclinations are also possible, and it is unlikely that one single mechanism applies in all cases. Clearly, it would be very interesting to establish whether shearing films should be considered more like quasi-ordered liquids or like quasi-disordered solids.

Freezing transitions — Fig. 15(d): The slipping or sliding regime ends once the applied shear stress falls below, τ_d, when the film freezes and the surfaces become stuck once again. The freezing of a whole film can occur very rapidly. Depending on the system, freezing can occur after prolonged sliding or immediately after the slip. The latter case is particularly common whenever the stress is applied not directly at the two surfaces but transmitted through the material on either side of the surfaces. In such cases the stress on the surfaces relaxes elastically during the slip. If the slip is rapid enough and if the molecules in the film can freeze quickly, the slip will be immediately followed by a stick; and if the externally applied stress is maintained the system will go into a continuous 'stick-slip' cycle. On the other hand, if the slip mechanism is slow the surfaces will continue to slide smoothly and there will be no stick-slip. However, unless the liquid molecules are highly entangled or irregular in shape [39,40], there will always be a single stick-slip 'spike' on starting, as shown in Fig. 16.

3.3 Shearing experiments with non-spherical molecules

With more asymmetric molecules, such as branched isoparaffins and polymer melts, no regular spikes or stick-slip behaviour occurs at any speed since these molecules can never order themselves sufficiently to 'solidify'. Example of such liquids are perfluoropolyethers and polydimethylsiloxanes (PDMS).

A novel interpretation of the well known phenomenon of decreasing coefficient of friction with increasing sliding velocity has been proposed by Thompson and Robbins [42] based on their computer simulation which essentially reproduced the above scenario. This postulates that it is not the friction that changes with sliding speed v, but rather the time

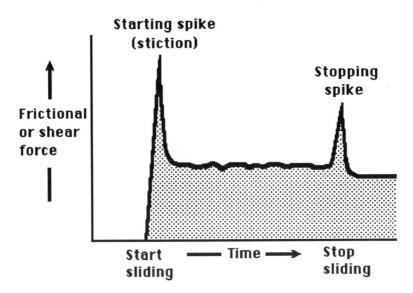

Fig.16. *Stiction is the high starting frictional force experienced by two moving surfaces which causes them to jerk forward rather than accelerate smoothly from rest. It is a major cause of surface damage and erosion. The figure shows a 'stiction spike' or 'starting spike' as well as a 'stopping spike'. The latter occurs when two sliding surfaces are brought to rest over a finite time during which the molecules in the film can freeze and stick before the surfaces stop moving.*

various parts of the system spend in the sticking and sliding modes. In other words, at any instant during sliding, the friction at any local region is always F_S or F_k, corresponding to the 'static' or 'kinetic' values. The measured frictional force, however, is the sum of all these discrete values averaged over the whole contact area. Since as v increases each local region spends more time in the sliding regime (F_k) and less in the sticking regime (F_S) the overall friction coefficient falls. Above a certain critical velocity v_c the stick-slip totally disappears and sliding proceeds at the kinetic value.

The above scenario is already quite complicated, and yet this is the situation for the simplest type of experimental system. The factors that appear to determine the critical velocity v_c depend on the type of liquid between the surfaces (as well as on the surface lattice structure). Small spherical molecules such as cyclohexane and OMCTS have been found to have very high v_c, which indicates that these molecules can rearrange relatively quickly in thin films. Chain molecules and especially branched chain molecules have been found to have much lower v_c, which is to be expected, and such liquids tend to slide smoothly rather than in a stick-slip fashion (see Table I). However, the values of v_c also depend on number of liquid layers comprising the film, the structure and relative orientation of the two surface lattices, the externally applied load, and of course on the stiffness of the spring (and in practice of the material of the surfaces).

Seven organic and polymeric liquids have been studied so far. These are listed in Table I, together with the type of sliding observed, the friction coefficient, and the bulk viscosity of the liquids (given for reference purposes).

Table I. Tribological characteristics of some liquids and polymer melts in molecularly thin films between two shearing mica surfaces.

Liquid (dry)	Short-range force	Type of sliding	Friction coefficient	Bulk viscosity (cP)
Spherical molecules				
cyclohexane (σ=5Å)	*Adhesive*	stick-slip	»1 (quantized)	0.6
OMCTS* (σ=9Å)	*Adhesive*	stick slip	»1 (quantized)	2.3
Chain molecules				
octane	*Adhesive*	stick slip	1.5	0.5
tetradecane	*Adhesive*	stick slip	1.0	2.3
octadecane (branched)	*Repulsive*	(stick slip)	0.35	5.5
PDMS* (M=3700, melt)	*Repulsive*	smooth	0.42	50
PBD* (M=3500, branched)	*Repulsive*	smooth	0.03	800
Water				
water (KCl solution)	*Repulsive*	smooth	0.01-0.03	1
hydrocarbon liquids (wet)	*Adhesive*†	smooth	0.03	~1

* PDMS: Polydimethylsiloxane, PBD: Polybutadiene, OMCTS: Octamethylcyclo-tetrasiloxane.

† The strong adhesion between two hydrophilic mica surfaces in wet hydrocarbon liquid is due to capillary forces, i.e., to the resolved Laplace pressure within the condensed water bridging the two surfaces. The direct force between the two surfaces across the liquid (water) is actually repulsive.

From the data of Table I (top part) it appears that there is a direct correlation between the shapes of molecules and their coefficient of friction. Small spherical or chain molecules have high friction with stick-slip because they can pack into ordered solidlike layers, whereas longer chained and branched molecules give low friction and smoother sliding. It is interesting to note that the friction coefficient generally decreases as the bulk viscosity of the liquids *increases*. This is because the factors that are conducive to low friction are generally conducive to high viscosity. Thus, molecules with side groups such as branched alkanes and polymer melts usually have higher bulk viscosities than their linear homologues for obvious reasons. However, in thin films the linear molecules have higher shear stresses. It is probably for this reason that branched liquid molecules are better lubricants—being more disordered in thin films because of this branching. In this respect it is important to note that if an 'effective' viscosity were to be calculated for the liquids of Table I, the values would be 10^6 to 100 times the bulk viscosities (10^6 for cyclohexane, 100 for PBD). This indicates that the bulk viscosity plays no direct role in determining the frictional forces in such ultrathin films. However, the bulk viscosity should give an indication of the *lowest* possible viscosity that might be attained in such films. Based on this hypothesis we may surmise that friction coefficients as low as 10^{-4}-10^{-3} might be attainable with the right system.

The only exception to the above correlations is water, which has been found to exhibit both low viscosity *and* low friction (cf. Fig. 12), yet water is essentially a small spherical molecule. In addition, the presence of water can drastically lower the friction and eliminate the stick-slip of hydrocarbon liquids when the sliding surfaces are hydrophilic. On the other hand, we have noted that with certain (hydrophobic) surfactant-coated monolayer surfaces and polymer melts the presence of water can act very differently, e.g. enhancing stick-slip. However, the results with monolayers and other surfaces are too few and too preliminary to allow us to draw any general conclusions at this stage.

3.4 Summary and conclusions

The static and dynamic properties of molecularly thin films between two solid surfaces cannot be described in terms of parameters or mechanisms applicable to the bulk liquids or solids. Such ultrathin films (typically less than 10-40Å thick) can support both normal and shear stresses, their effective viscosities can be 10^6 times the bulk values, molecular relaxations may take 10^{10} times longer, and their melting points are almost totally unrelated to the bulk values. Our results show that during normal sliding the films may undergo many freezing-melting cycles per second (solid-liquid transitions) — a mechanism that manifests itself macroscopically as stick-slip.

Most interestingly, experiments on a variety of simple liquids, linear chained hydrocarbons and branched chained polymer melts show that there is a direct correlation between the shapes of molecules and their effectiveness as lubricants (at least at low shear rates). Small spherical or simple chain molecules exhibit high friction with stick-slip because they can pack into well-ordered solid-like layers. In contrast, irregularly shaped molecules such as longer chained hydrocarbons or branched polymer liquids remain in an entangled, disordered, fluid-like state even in very thin films and these give low friction and smoother sliding. It is probably for this reason that irregularly shaped branched chain molecules are usually better lubricants. These findings also lead to the seemingly paradoxical conclusion that molecular entanglements which lead to a *high* viscosity in bulk liquids are conducive to *low* friction when the liquids are confined within a thin film. An understanding of these complex phenomena at the molecular level may help in the processing of better lubricants, more durable surfaces, and better materials.

ACKNOWLEDGEMENTS

I thank the ONR for supporting the research associated with the experimental techniques under grant number N00014-89-J-1101, and the DOE for supporting the research associated with the organic films under grant number DE-FG03-87ER45331.

REFERENCES

(1) Bowden, F. P.; Tabor, D. *Friction and Lubrication*, Methuen: 1967.
(2) Greenwood, J. A.; Johnson, K. L. *Phil. Mag. A* **1981**, *43*, 697. Michel, F. and Shanahan, M. E. R. *C. R. Acad. Sci. Paris* **1990**, *310 II*, 17. Maugis, D. *J. Materials Sci.* **1985**, *20*, 3041.
(3) Miller, C. A.; Neogi, P. *Interfacial Phenomena*, Marcel Decker: Basel, New York, 1985.

380

(4)   Schwartz, A. M. *J. Colloid Interface Sci.* **1980**, *75*, 404.

(5)   Pethica, J. B.; Sutton, A. B. *J. Vac. Sci. Technol. A* **1988**, *6*, 2490. Landman, U., Luedtke, W. D., Burnham, N. A., and Colton, R. J. *Science* **1990**, *248*, 454.

(6)   Horn, R.; Israelachvili, J. N. *Chem. Phys. Lett.*, **1980**, *71*, 192.

(7)   Weisenhorn, A. L.; Hansma, P. K.; Albrecht, T. R.; Quate, C. F. *Appl. Phys. Lett.* **1989**, *54*, 26. Hansma, P. K.; Elings, V. B.; Marti, O.; Bracker, C. E.; *Science,* **1988**, *243*, 1586.

(8)   Lawn, B. R; Wilshaw, T. R. *Fracture of Brittle Solids*; Cambridge Univ. Press: London, 1975. Sahimi, M.; Goddard, J.G. *Phys. Rev. B* **1986**, *33*, 7848. McClelland, G.M. in *Adhesion and Friction* (M. Grunze & H.J. Kreuzer, eds) Springer Series in Surface Science, Vol. 17, 1989, pp. 1-16.

(9)   Ellul, M. D.; Gent, A. N., *J. Polymer Science: Polymer Physics*, **1984**, *22*, 1953; **1985**, *23*, 1823. Shanahan, M. E. R.; Schreck, P.; Schultz, J. *C. R. Acad. Sci. Paris*, **1988**, *306 II*, 1325. Okawa, A., *B.Sc. Thesis*, Dept. Materials Science, University of Utah, June, 1983. Holly, F. J., Refojo, M. J. *J. Biomed. Materials Res.* **1975**, *9*, 315.

(10) Klein, J. *J. Chem. Soc. Faraday Trans. I*, **1983**, *79*, 99; *Makromol. Chem. Macromol. Symp.* **1986**, *1*, 125.

(11) Yasuda, H.; Sharma, A. K.; Yasuda, T. *J. Polym. Sci., Polym. Physics*, **1981**, *19*, 1285.

(12) Langmuir, I. *Science* **1938**, *87*, 493. Wasserman, S. R.; Tao, Y.-T.; Whitesides, G. M. *Langmuir*, **1985**, *5*, 1074.

(13) Chen, Y. L. E; Gee, M. L.; Helm, C. A., Israelachvili, J. N., McGuiggan, P. M. *J. Phys. Chem.* **1989**, *93*, 7057.

(14) Andrade, J. D.; Smith, L. M.; Gregonis, D. E. In *Surface and Interfacial Aspects of Biomedical Polymers*; Andrade, J. D., Ed.; Plenum: New York & London, 1985; Vol 1, p. 249.

(15) Zisman, W. A., *Ind. Eng. Chem.* **1963**, *55(10)*, 19. Fowkes, F.M. *Ind. Eng. Chem.* **1964**, *56*, 40. Zisman, W. A.; Fox, H. W., *J. Colloid Science* **1952**, *7*, 428.

(16) Israelachvili, J. N. *Intermolecular and Surface Forces*, Academic Press: New York, 1985. (2nd edition: 1991.)

(17) Johnson, K. L.; Kendall, K.; Roberts, A. D. *Proc. R. Soc. London A*, **1971**, *324*, 301.

(18) Pollock, H. M.; Barquins, M.; Maugis, D. *Appl. Phys. Lett.* **1978**, *33*, 798. Barquins, M.; Maugis, D. *J. Méc. Théor. Appl.*, **1982**, *1*, 331.

(19) Christenson, H. K.; Claesson, P. *J. Colloid Interface Sci.* **1990**, *139*, 589; Moy, E.; Neumann, A.W. *J. Colloid Interface Sci.* **1990**, *139*, 591

(20) Israelachvili, J. N.; Perez, E.; Tandon, R.K. *J. Colloid Interface Sci.* **1980**, *78*, 260.

(21) Chen, Y.L., Helm, C.A., Israelachvili, J.N., *J. Phys. Chem.* (in press)

(22) Horn, R. G.; Israelachvili, J. N.; Pribac, F. *J. Colloid Interface Sci.* **1987**, *115*, 480.

(23) Israelachvili, J. N.; Adams, G. E. *J. Chem. Soc. Faraday Trans. I* **1978**, *74*, 975; Israelachvili, J. N.; McGuiggan, P. M. *J. Mater. Res.* **1990**, *5*, 2223.

(24) D. Tabor and R.H.S. Winterton, *Proc. Roy. Soc. London,* **1969**, *A312*, 435.

(25) J.N. Israelachvili and D. Tabor, *Proc. Roy. Soc. London* **1972**, *A331*, 19.

(26) J.N. Israelachvili, *J. Colloid Interface Sci,* **1973**, *44*, 259;

(27) J.N. Israelachvili, P.M. McGuiggan and A.M. Homola, *Science,* **1988**, *240*, 189.

(28) A.M. Homola, J.N. Israelachvili, M.L. Gee and P.M. McGuiggan, *Trans. ASME, J. Tribology,* **1989,** *111,* 675. A.M. Homola, J.N. Israelachvili, P.M. McGuiggan, and M. L. Gee, *Wear* **1990,** *136,* 65.

(29) B.V. Derjaguin, *Wear,* **1988,** *128,* 19.

(30) J.N. Israelachvili, *Chemica Scripta* **1985,** *25,* 7.

(31) D. Tabor, in *Microscopic Aspects of Adhesion and Lubrication,* Paris, Societe de Chimie Physique, 1982.

(32) J.N. Israelachvili and D. Tabor, *Wear,* **1973,** *24,* 386.

(33) B.J. Briscoe, D.C. Evans and D. Tabor, *J. Colloid Interface Sci.,* **1977,** *61,* 9.

(34) B.J. Briscoe and D.C. Evans, *Proc. Roy. Soc. London,* **1982,** *A380,* 389.

(35) J. Israelachvili and P. McGuiggan, *Science* **1988,** *241,* 795.

(36) P. McGuiggan and J. Israelachvili, *J. Mater. Res.* **1990,** *5,* 2232.

(37) R. G. Horn, *J. Amer. Ceram. Soc.* **1990,** *73,* 1117.

(38) D. Henderson and M. Lozada-Cassou, *J. Coll. Int. Sci.* **1986,** *114,* 180.

(39) M. Gee, P. McGuiggan, J. Israelachvili and A. Homola, *J. Chem. Phys.* **1990,** *93,* 1895.

(40) A.M. Homola, H. V. Nguyen and G. Hadziioannou, *J. Chem. Phys.* (in press).

(41) M. Schoen, C. Rhykerd, D. Diestler and J. Cushman, *Science* **1989,** *245,* 1223.

(42) P. Thompson and M. Robbins, *Science* **1990,** *250,* 792.

(43) M.J. Sutcliffe, S.R. Taylor, and A. Cameron,*Wear* **1978,** *51,* 181.

(44) U. Landman, W.D. Luedtke, N.A. Burnham and R.J. Colton, *Science* **1990,** *248,* 454.

Discussion *following the lecture by J N Israelachvili on "Adhesion, friction and lubrication of molecularly smooth surfaces"*

K L JOHNSON. In one of your first slides you suggested that the popping of bonds is an irreversible processes, and yet your own results on the adhesion of two solid surfaces indicate complete reversibility on "coming on" and "coming off". Is there a discrepancy there?

J N ISRAELACHVILI. Initially, I was presenting an overview of the various energy dissipating processes that can occur during adhesion and debonding. These could be purely mechanical (as in the popping of bonds), or chemical, or viscoplastic, or some other mechanism. In our experiments the mechanical and viscoplastic contributions were negligible, and almost all of the hysteresis came from chemical effects (molecular rearrangements at surfaces). However, I do believe that bond popping is inherently energy dissipative and therefore irreversible in a thermodynamic sense. This type of contribution is perhaps more important in other situations, for example, when loading-unloading (advancing-receding) occurs via stick-slip motion.

M O ROBBINS. But wasn't one of your points, that you need a certain high strength of bond before it popped, liberating a lot of energy, rather than opening up smoothly? In other words, with the weak van der Waals forces operating between your weak hydrocarbon surfaces you should not really expect to have any abrupt popping.

J N ISRAELACHVILI. I agree.

U LANDMAN. Concerning the friction you measured with only a monolayer of water between the surfaces, you found a very low friction coefficient of 0.03 which is similar to that for sliding on ice. But you also said that in this case the friction-load and area-load curves did not obey the JKR theory. Does friction on ice obey the JKR theory?

J N ISRAELACHVILI. I am not aware of any tests of the JKR theory in relation to sliding on ice. In our experiments, we would actually not expect the data to obey the JKR theory because there was no adhesion (or only very weak adhesion) between the two mica surfaces across the water layer. We therefore expect to have the friction simply proportional to load (cf Equations 9 and 10) as was indeed found experimentally (cf Figure 12).

K L JOHNSON. Were the water films thick?

J N ISRAELACHVILI. No, one or two layers. But water behaves quite differently from other liquids - the surfaces have little or no adhesion, lower friction, and no stick-slip.

U LANDMAN. Why is there no adhesion in water?

J N ISRAELACHVILI. If the surfaces are rendered hydrophilic, for example by ion exchange as was the case here, there is a short-range oscillatory force superimposed on a monotonic "hydration" force. However, the adhesive minima of the oscillations were all above the zero-force line [cf Israelachvili and McGuiggan, Science (1988) 241, 795] and so there was no adhesion. With other, simpler liquids, the minima occur at negative force values (real adhesion) and the friction is therefore higher, has a F x Area component, and exhibits stick-slip.

K L JOHNSON. As a well-known colleague used to say, "water is the most common material with the most uncommon properties".

B R LAWN. Concerning mechanical and chemical hysteresis, Roger Horn and Doug Smith at NIST have recently studied the adhesion between dissimilar materials. In the case of mica and glass, charge transfer occurs on contact, and the adhesion increases dramatically to well above the values for mica-mica or glass-glass. In addition, the hysteresis is gigantic. Would you call that mechanical or chemical hysteresis?

J N ISRAELACHVILI. I would study this phenomenon more and do lots of experiments before I call it anything. [Added afterwards: this seems to be a case of chemical hysteresis, since ion or charge transfer is essentially the same as chemical transfer or molecular rearrangements.]

H M POLLOCK. In the cobblestone model, the four cartwheels will not lift off the ground at the same time. Likewise, the atoms at a shearing junction will not all move at the same time. Consequently, will your fudge factor, ϵ, be lower than it otherwise would be?

J N ISRAELACHVILI. I agree that in real situations, different regions within the shearing zone will not move (slide, slip or stick) at the same time. However, there must be some correlation between these motions via the backing material, but this is obviously a complex issue. Can you ask me simpler questions.

R S POLIZZOTTI. Do you see any evidence for vertical displacements at constant load (during sliding) that would show that the films are going through phase transformations?

J N ISRAELACHVILI. We have looked for this but have not seen it. If it's there, it's just beyond our measuring resolutions of about 1Å. Mark Robbins and Peter Thompson have concluded from their computer simulations that the thickness of a liquid film should be greater during motion than when at rest, i.e. that the film density falls during motion or slip, but I don't remember by how much.

M O ROBBINS. We saw an effect of about 10% in the two-layer film, which would correspond to less than 1Å - something that might be difficult to detect in these experiments.

J N ISRAELACHVILI. Someone in Professor Davis' group in Minnesota has recently developed a technique for analyzing fringe positions so that surface separations can be measured to about $\pm\ 0.1$ Å. We are hoping to use this method to look at film thickness changes during sliding experiments.

ANON. Have you analyzed the stick-slip spectrum you measure, such as the frequency response in relation to the stiffness of your springs?

J N ISRAELACHVILI. Not yet, though it should be very easy to do. Actually, only at high sliding velocities, just before the stick-slip disappears at $v = v_c$, does the stick-slip become erratic - with both the frequency and amplitude changing as v approaches v_c. I expect that this would be the most interesting regime to study.

I L SINGER. Did the hysteresis you measured go up at higher loads?

J N ISRAELACHVILI. Yes.

I L SINGER. Could you induce hysteresis with the solid-solid surfaces at higher loads?

J N ISRAELACHVILI. No, but maybe we didn't go high enough. Eventually even the solid surface groups break through in some way and become interdigitated; this would probably happen abruptly at some high load and would be seen as a yield or break point in the distance/force measurements. [Added later: Interdigitation could also occur after a sufficiently long time in contact.] With the amorphous surfaces, the hysteresis increased both with the load and contact time.

M O ROBBINS. Was there a load you had to exceed before you saw hysteresis?

J N ISRAELACHVILI. No.

M O ROBBINS. You always saw it?

J N ISRAELACHVILI. Yes. [Added later: Recall that according to the JKR theory, at the centre of the contact region the two surfaces experience a compressive pressure or load, even though the externally applied load is zero. We have tentatively concluded that most of the hysteresis comes from this central region.]

I L SINGER. Have you studied the adhesion or friction of non-organic solid films such as evaporated metal films?

J N ISRAELACHVILI. We are presently looking at sputtered alumina and silica films, as well as other sputtered or evaporated metal or metal-oxide films. The initial results indicate large quantitative and qualitative differences between the adhesion and friction of these surfaces, both relative to each other and relative to mica.

ANON. During sliding, do you ever squeeze out the last layer of liquid molecules between the surfaces?

J N ISRAELACHVILI. Yes, and when that happens the two surfaces come into strong adhesive contact and rapidly become damaged. The "interfacial" sliding then becomes replaced by "normal" sliding, with the two surfaces moving while separated by a forest of wear debris (torn up mica flakes). Such contacts and damage usually starts locally either at the middle or edge of the contact area where the two surfaces have "broken through" the last layer. Once damage is initiated it usually spreads rapidly throughout the contact zone. This whole damage initiation process is worth studying further.

E RABINOWICZ (written question): How many molecular layers must a liquid have to acquire the properties of a liquid? For example, if a water layer on a solid is 10 molecules thick, it presumably has surface tension. If a water layer is 0.1 molecules thick, it presumably has no surface tension. How about 0.3 molecules thick, or three? What determines the point at which liquid properties begin?

J N ISRAELACHVILI. In the case of a single surface exposed to water vapour, I would say 1-2 layers (even though the liquid properties may differ from those of the bulk). But for a liquid trapped between two solid surfaces the matter is more complex and ceases to be simply a question of "how many layers" or "how thick the film". It now depends on the details of exactly how two real (that is, structured) surfaces are positioned relative to each other and how the liquid molecules can fit between them. Thus, as the figure below shows, a film composed of simple spherical molecules that is normally in the liquid state in the bulk may be either in a solid-like or liquid-like state at the same time surface separation (B), or it may be solid-like at one separation but liquid *closer in* (A). Such effects appear to be important only at surface separations below about 5 molecular diameters. At larger separations, the film will behave as a liquid regardless of the details of how the two surfaces are positioned relative to each other.

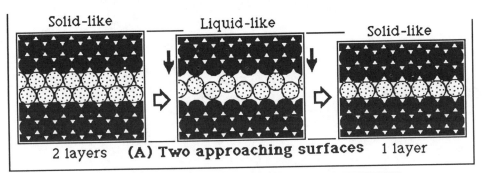

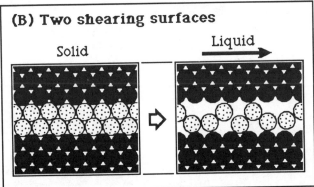

Illustrations of how liquid to solid-like transitions occur in molecularly thin films between two 'structured' solid surfaces.

MOLECULAR TRIBOLOGY OF FLUIDS

Steve Granick
University of Illinois at Urbana-Champaign
Department of Materials Science and Engineering
    and Materials Research Laboratory
105 South Goodwin, Urbana, IL   61801   U.S.A.

ABSTRACT. The rheology of fluids at interfaces differs considerably from that in the bulk. Differences depend in part on the nature of the surface (especially the strength and uniformity of adsorption) and in part on the chemical makeup of the fluids themselves. This review emphasizes the shear rheology of fluid films of thickness 15-100 Ångstroms. Differences from the bulk behavior are more prominent, the thinner the film. Ultimately a solid-like response to shear is observed. These findings show that one needs to develop a new intuition about the flow of even the simplest liquids, when one dimension of the flow is comparable to the size of the molecule. These new notions are not intuitive extrapolations of bulk properties.

Introduction

The science of tribology is poorly understood. This impedes both technological progress and fundamental understanding not only in tribology, but also in the many related areas that depend on successful solution of tribology problems. Without wishing to minimize the differences between these phenomena, it is certainly fair to say that they share many common questions. There are questions of intermolecular forces, of concentration and density profiles, of topography, chemical structure, and reactivity of surface species, of mass transport in the interface, and of evolution of all of these properties with time in contact.

The molecular tribology approach -- the study of tribology at atomic and molecular scales -- constitutes a new frontier of tribology research. In a major recent surge of activity, new experimental methods have been developed to measure dynamic interfacial forces in shear. Building in part on earlier friction studies which had been somewhat neglected [1-5], striking new findings have been obtained [6-45]. The new methods include the surface forces apparatus for measuring static interfacial forces [6], new molecular tribometers for measuring friction in shear [7-24], atomic force microscopy [25-28], use of UHV tribometers [29,30] and the quartz-crystal microbalance [31-33]. Theoretical calculations and molecular dynamics simulations are also emerging for friction in dry [34-39] and lubricated [40-45] sliding.

Development of methods for studying tribology has naturally accompanied developments in other areas of surface measurement. These may be classed roughly as microscopic and macroscopic. Among the microscopic methods of analysis are those based on surface spectroscopies, among them electron probe microanalysis, electron

I. L. Singer and H. M. Pollock (eds.), Fundamentals of Friction: Macroscopic and Microscopic Processes, 387–401.
© 1992 Kluwer Academic Publishers. Printed in the Netherlands.

spectroscopy for chemical analysis (ESCA), low energy electron diffraction (LEED), vibrational spectroscopies, and many others. Model microasperities have also been developed and studied, notably with atomic force microscopy (AFM). Much progress has been made with these modern methods. However, their application to the study of lubricants under externally driven flow is in its infancy.

In surveying the present situation, one is struck that one of the major difficulties in explaining tribology is the paucity of direct, experimental information concerning dynamic events on a microscopic scale. With macroscopic approaches one can measure, for example, the friction between surfaces (and other global properties) while the surfaces are actually in contact. Such macroscopic measurements are often difficult to interpret from a molecular point of view. Most of the questions posed in the preceding section are molecular in nature. With microscopic methods, one can examine surfaces either under unrealistically low rates of externally-driven flow, or before and after sliding has occurred. As such, these approaches yield incomplete information concerning the many interfacial situations where the presence of a lubricant, under extreme flow rates, is an essential part of the physical situation.

A further major problem is to characterize the true contact area and thickness of fluid films during flow. They can be gauged by resistance, capacitance, and other measurements, but of course many distributions of surface roughness are compatible with a given measurement. Furthermore, to the degree that a surface is rough, the surface separation and the thickness of the intervening film must be described by a distribution rather than any single number.

Consider first the canonical experiment depicted in Figure 1. Imagine that one takes a droplet of liquid, puts it between a ball and a table -- and lets the ball fall. Of course the liquid squirts out, initially rapidly, then slower and slower as the liquid thickness becomes less than the radius of the ball. This problem was solved over 100 years ago in a classic analysis of Reynolds [49]. Experiment shows that the film eventually stabilizes at a <u>finite</u> thickness of a few molecular diameters [46]. The liquid film supports the weight of the ball! Of course the heavier the ball, the less the ultimate thickness of the liquid film, but this is a general result: an extraordinarily large pressure is needed to squeeze out the final few layers of liquid between two solid surfaces. This experiment shows that when the thickness of a liquid film becomes comparable to molecular dimensions, classical intuition based on continuum properties no longer applies.

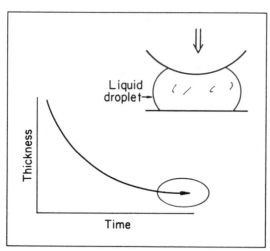

Figure 1. Hypothetical experiment showing that a liquid can support a normal load. A liquid droplet is placed between a ball and a flat surface. The graph, in which liquid thickness is plotted schematically against time after the ball has begun to fall, shows that the film thickness remains finite (a few molecular dimensions) even at equilibrium.

Atomically Smooth Surfaces -- A Model Approach

Striking new findings regarding the dynamics of fluid in intimate contact with solid surfaces have been obtained in recent years.

With rough surfaces, a fundamental difficulty is to separate the well-known origins of energy dissipation: (i) elastic and plastic deformation of the underlying substrate, (ii) wear, i.e. degradation of the surfaces and substrates in the course of sliding, and (iii) surface chemical and mechanical properties of the fluid itself. However, it is possible to separate these effects if atomically smooth surfaces are employed. This gives the situation shown schematically in Figure 2.

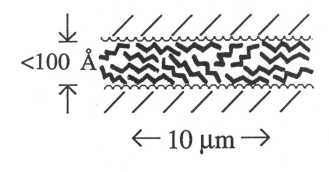

Figure 2. Schematic representation of the organization of chain molecules in a narrow gap. Mechanical deformations of the substrate are minimized, allowing study of the complex interplay of load, shear force, and molecular forces which occurs at a lubricated interface.

This idealized system constitutes a single asperity for the purposes of understanding the constraints on the flow of the fluid. As such, it provides a productive approach with which to address an array of open questions on a fundamental level.

Neglecting surface effects for the moment, recall that liquids can be distinguished from solids as sustaining no deformation at equilibrium. In responding to force, a liquid eases deformation by flowing at some rate. Shear deformation, in which one surface slides tangent to another, is especially simple to interpret. At sufficiently low shear rates (shear rate = velocity/film thickness), flow obeys Newton's law,

$$\sigma = \eta \, d\gamma/dt \qquad (1)$$

where σ is the shear stress (stress=force/area), $d\gamma/dt$ is the shear rate, $\dot{\gamma}$, and η is the viscosity, which has dimensions of mass length^{-1} time^{-1}. The viscosity of a liquid may be thought of qualitatively as its resistance to flow.

These notions become more subtle when the liquid film is so thin as to be anisotropic in the directions normal to and parallel to the surfaces. An example may be seen in Fig. 1; after the ball has fallen, the ultrathin liquid supports an equilibrium normal force. It acts as a solid in this direction.

What of sliding of one surface tangent to another? When dealing with an ultrathin film, analysis shows that if local viscosity coefficients could be measured, they would vary with distance across a thin liquid film just as the local density does [47]. A laboratory shear experiment averages over the width of the film. Nonetheless it is meaningful to define an effective viscosity based on Eq. 1. But when a liquid film is sufficiently thin, its response to a tangential force is that of a solid. Then one measures rigidity or yield stress.

The molecular rheometer used in these experiments has been described previously [7,9]. Although based on the use of atomically smooth mica substrates as are several other tribometers [1,3-5,8], its capabilities are quite different in that it permits travel over distances <u>small</u> compared to the area of contact. This, as will be seen below, has important consequences for investigating time-dependent reorganization of lubricant molecules.

The principle of the measurement was to confine ultrathin liquid films in a homebuilt surface forces apparatus, surrounded by a macroscopic droplet of the same liquid, between circular parallel plates of single crystals (usually of muscovite mica or thin films of other materials coated onto muscovite mica). At molecular thicknesses a liquid film supports a state of normal stress; the film thickness adjusts to externally applied normal pressure. The thickness at a given net normal pressure was measured to ±0.1 nm by optical interferometry between the back sides of the mica sheets. With this approach, my laboratory has studied viscous dissipation and elasticity of confined liquids using periodic sinusoidal oscillations over a range of amplitudes and frequencies [7,9-17]. Other workers have used a different apparatus to study the friction encountered during sliding at a constant speed [8,21-24]. Early measurements involved dry sliding [1-5]. The technical difficulties of these methods have been discussed elsewhere [48].

In this rapidly developing area, the experimental picture that has emerged to date is summarized in Figure 3. Energy dissipation of a fluid film is sketched schematically against thickness of the lubricant film. When the film is sufficiently thick (micrometers or more), it obeys well-known continuum relations. At smaller film thickness (on the order of nanometers) the effective viscosity rises, and the characteristic relaxation times are prolonged. At still smaller film thickness (on the order of Ångstroms to nanometers), the films become solid-like in the sense that sliding does not occur unless a certain shear stress (or "yield stress") is attained.

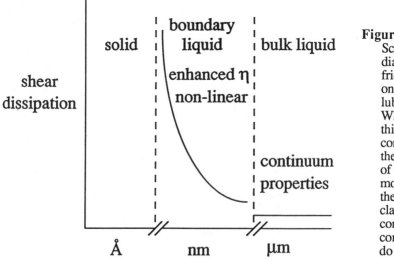

Figure 3. Schematic diagram of how friction depends on thickness of a lubricant film. When the thickness is comparable to the dimensions of the lubricant molecules themselves, classical continuum considerations do not apply.

Effective Viscosity of Simple Paraffin Liquids

Now we become quantitative. Figure 4 illustrates how the shear viscosity changes with film thickness for an experiment conducted with dodecane, $CH_3(CH_2)_{10}CH_3$, at 28°C. The limiting effective viscosity at low shear rate, $\eta°_{eff}$, is plotted against film thickness. At thickness 40 Å, $\eta°_{eff}$ was already 100 poise, significantly large than the 0.01 poise of the bulk liquid. Drainage to lesser thickness resulted in increase of the effective viscosity. At the thickness 26±1 angstroms, the effective viscosity seemed to diverge. To observe differences from the bulk behavior is qualitatively reasonable since the dodecane molecule is confined to a space comparable to its molecular dimensions; but the magnitude of the differences is spectacular.

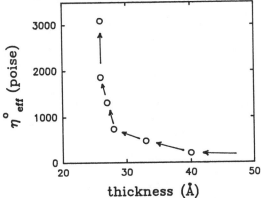

Figure 4. Effective limiting viscosity at low shear rate, plotted against liquid thickness, for dodecane confined between opposing mica plates. Temperature, 28°C. Arrows indicate direction of increasing compression. From ref. [11].

The simple flow properties of a beaker of dodecane turn complex under confinement. In Figure 5, one sees that whereas the effective viscosity was constant at low rates of deformation, as the rate of deformation increased, it decayed as a seeming power law in the effective strain rate (strain rate=velocity/film thickness). This nonlinear response set in at a deformation rate many orders of magnitude slower than that at which nonlinear effects would be observed in the bulk. Measurements of film thickness failed to detect any changes with shear (<0.1 nm); this does not rule out the possibility that the film thickness adjusted to the shear, possibly in such a way as to lower the mean density, but it puts the upper limit of 4% on possible shear-induced changes in the mean liquid density. Other control experiments showed reversibility when the velocity was raised and lowered.

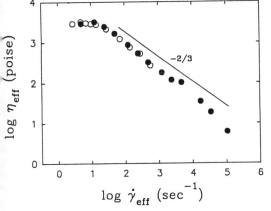

Figure 5. Log-log graph of effective viscosity as a function of effective strain rate, for dodecane confined between mica plates. Film thickness, 27 Å. Net normal pressure, 0.12 MPa. Temperature, 28°C. From ref. [11].

This is a central theme: that rheological relaxation of a trapped fluid is so much slower than in the bulk liquid. This same pattern of behavior under extreme confinement has been confirmed with a variety of other liquids. It may be related to the behavior expected of two-dimensional liquids [11], though explanations of these striking observations are still conjectural. These findings show that one needs to develop a new intuition about the flow of even the simplest liquids, when one dimension of the flow is comparable to the size of the molecule. These new notions are not intuitive extrapolations of bulk properties. Spectacular sensitivity to the speed of sliding is an essential feature.

Switch from Liquid to Solid Response

A solid-like response to shear stress further thickens the plot of this complex story. The rapid switching one could observe between liquid-like and solid-like response is illustrated in Figure 6 for an experiment with a siloxane liquid, octamethylcyclotetrasiloxane (OMCTS). The actual amplitude of oscillation during an experiment at 0.2 Hz is plotted against the elapsed time in data taken from an oscilloscope trace. The film thickness was two layers (1.8 nm) and the normal pressure was 3.6 MPa. A compliant response, on the left side of Figure 6, became increasingly ragged as the net normal pressure was slowly raised, then switched abruptly to a response characteristic only of the experimental apparatus (the mica and its underlying glue). After this point, a critical shear stress was required to initiate sliding. In Figure 6, one notes that the transition occurred in the space of less than 5 seconds. For longer-chain siloxane lubricants, however, the transition occurs over a period of more than 5 hrs [13]. The molecular dependence of the speed of the transition is striking. It may provide a means for the lubrication engineer to control over the emergence of this behavior.

Origin of static friction

When it is sufficiently thin, a confined liquid solidifies in the sense that it will not allow shear until a critical shear stress (yield stress) is exceeded. This occurs especially for films less than 4 molecular dimensions thick (i.e. less than 2 molecular dimensions per solid boundary) [8]. It is tempting to attribute the phenomenon simply to strong adsorption. However, the effect is also seen at larger thickness, up to 6 molecular dimensions [11,13]. Such yield stress behavior is familiar but not understood. It is the familiar fact of life, static friction.

What underlying organization of the liquid is responsible for observing a yield stress? Computer simulations show that under some circumstances (spherical liquid particles), the liquid particles form an epitaxial crystal on an underlying crystalline lattice [40-44]. The phenomenon appears, however, to be more general than this. In the laboratory, the phenomenon has also been observed using amorphous rather than crystalline confining solids [10] and a liquid that is believed to solidify to a glass rather than to a crystal [12,13]. Loss of fluidity may reflect vitrification imposed by the liquid's confinement.

The yield stress depends strongly on history [12,13]. The film thickness at solidification depends on the rate at which the liquid drop is thinned [13]. In addition, the yield stress grows over remarkably long times -- minutes to hours, depending on the liquid [13]. Figure 6 illustrates this strengthening in an experiment with a polymer glass-forming liquid (identified in the Figure legend). The yield stress on first measurement was approximately 1 MPa, but this value nearly tripled over a 4 hour interval. The sluggish ncrease in the yield stress implies that whatever the structure may be of the solidified state, apparently it is full of defects.

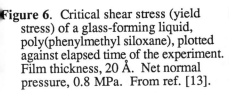

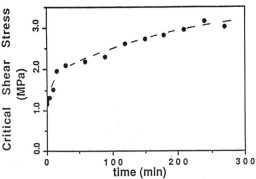

Figure 6. Critical shear stress (yield stress) of a glass-forming liquid, poly(phenylmethyl siloxane), plotted against elapsed time of the experiment. Film thickness, 20 Å. Net normal pressure, 0.8 MPa. From ref. [13].

This has implications for understanding static friction. Empirically, it has long been known that static friction usually increases with time after solids are left in contact. The traditional explanation is that this results entirely from the flattening of asperities as they conform, and resulting increases in the true area of contact between the opposing solids [49]. These experiments, in which the true area of contact could be measured directly, suggest that this is not the entire story. It seems that the buildup of static friction can also be an intrinsic phenomenon; in other words, that slow rearrangements of liquid molecules trapped between two solids can also result in static friction. A boundary has a profound influence on the structure and dynamics of fluids, actually leading to loss of fluidity when the fluid is sufficiently thin.

Generality of this Behavior

The conclusions, that effective viscosity is enhanced in an ultrathin lubricant film and that this is accompanied by an enormous prolongation of relaxation times, are also obtained if the surface lacks the long-range periodicity of a crystal and the adsorption to this surface is weaker.

Recently methods were developed to blanket mica with a securely attached, self-assembled organic monolayer [50,51]. The surface composition, composed of chemically reacted octadecylsilane (OTE) chains, is an array of closely-packed methyl groups. The methyl groups are amorphous over distances greater than 100 angstroms in films of this type [52]; liquids could not crystallize epitaxially onto these monolayers. In addition, the surface energy is only 22 mJ-m^{-2} [51], considerably less than that of freshly cleaved mica (200 to 400 mJ m^{-2}); liquids are expected to adsorb more weakly than onto mica. Thus, substantially weaker coupling to the surface is expected than in the case of mica. For dodecane confined between such layers, fundamentally the same effective shear viscosity was obtained as with mica [10].

Wear

On occasion, the solid surfaces perceptibly undergo wear in the course of these experiments. This event is easily detected; wear particles of mica are perceived as a jagged contour on the formerly-flat interference fringes used to view the zone of sliding contact. Wear is rare under the conditions of these experiments. It is more apt to occur under conditions of strong adhesive contact (critical shear stress response) than under conditions of viscous dissipation.

Conclusions and Engineering Significance

In summary, these experiments show the limits of linear and nonlinear viscous shear response of confined lubricating fluids. It emerges that a boundary can profoundly slow down the dynamics of the liquid state and enhance the effective viscosity. Interpretations are this stage are still conjectural. Among the possibilities to entertain are the importance of two-dimensional effects [53], of logjammed or vitrified states [12], and a breakdown of the normal hydrodynamic assumption of zero slip of fluid at a wall [54].

One tends to take friction, wear, and tear for granted. While tribology design and tribology-based applications are rooted in our economic life, too often the technologies and formulations are empirically-derived.

On the scientific side, appreciation is growing that scientific understanding is possible of these systems that are so complex and so far from equilibrium. Surfaces in sliding contact are more amenable to rational inquiry than might be feared. Tribology is becoming recognized as an area with many opportunities to do exciting and useful surface science.

The engineering significance is that the flow of these -- the simplest fluid lubricants -- under extreme confinement could not be understood simply by intuitive extrapolation of bulk properties. Features of specific engineering significance are:

(1) Solid lubrication need not necessarily reflect solid-solid contact. It takes an extraordinarily large pressure to squeeze the very last layers of liquid out from between two solid surfaces.

(2) Non-Newtonian microviscosity. Lubricants that would be Newtonian in the bulk are highly non-Newtonian in an ultrathin gap.

(3) Enhanced elasticity of longer-chain fluids. The elasticity of ultrathin films of lubricants, discussed elsewhere [55], may be related to stiction phenomena that are well-known to engineers.

(4) Time dependence of the static friction. Usually one considers that when the critical shear stress increases with elapsed time, the reason is the growth of the area of contact. However, slow molecular rearrangements of the fluid in the boundary layer can also contribute significantly.

(5) Solidification of ultra-thin fluid films can be suppressed by continuous sliding. The engineering implication is to better understand the conditions under which continuous operation of a piece of equipment may be desirable in order to impede stiction.

(6) <u>The pressure dependence of both the critical shear stress and the effective viscosity is enhanced over that for the bulk fluid</u>. This observation, discussed elsewhere [11,12,15], provides a means to control the friction during sliding.

(7) <u>Prospects for design of lubricant formulations</u>. The patterns described in this article appear to be general, but their magnitudes depend on particular properties of the lubricant (the molecular size, shape, polarity, and chemical composition), as well as on the strength of surface-fluid attraction.

Acknowledgments.

The author is indebted to his graduate students and postdoctoral coworkers, John Van Alsten, George Carson, Hsuan-Wei Hu, James Peachey, and John Peanasky, for collaborations and discussions regarding the work cited here. This work was supported by grants through the Tribology Program of the National Science Foundation and the Materials Research Laboratory at the University of Illinois, Grant NSF-DMR-89-20538.

References

1. Bailey, A.I. and Courtney-Pratt (1955), J.S., Proc. Roy. Soc. A<u>227</u>, 500.
2. Tabor, D. and Winterton, R.H.S. (1969), Proc. Roy. Soc. A<u>312</u>, 435.
3. Israelachvili, J.N. and Tabor, D. (1973), Wear <u>24</u>, 386.
4. Briscoe, B.J., Evans, D.C.B., and Tabor (1977), D., J. Coll. and Interface Sci. <u>61</u>, 9.
5. Briscoe, B.J. and Evans, D.C.B., Proc. Roy. Soc. (1982), A<u>380</u>, 389.
6. Israelachvili, J.N. and Adams, G.E., J. Chem. Soc. Faraday Trans. (1978), II<u>74</u>, 975.
7. Van Alsten, J. and Granick, S., Phys. Rev. Lett. (1988), <u>61</u>, 2570.
8. Israelachvili, J.N., McGuiggan, P.M. and Homola, A.M., Science (1988), <u>240</u>, 189.
9. Peachey, J., Van Alsten, J., and Granick, S., Rev. Sci. Instrum. (1991), <u>62</u>, 463.
10. Granick, S., Science (1991), <u>253</u>, 1374.
11. Hu, H.-W., Carson, G.A., and Granick, S. (1991), Phys. Rev. Lett. <u>66</u>, 2758.
12. Van Alsten, J. and Granick, S. (1990), Langmuir, <u>6</u>, 214.
13. Van Alsten, J. and Granick, S., Macromolecules (1990), <u>23</u>, 4856.
14. Van Alsten, J. and Granick, S. (1989), Trib. Trans. <u>32</u>, 246.
15. Van Alsten, J. and Granick, S. (1990), Trib. Trans. <u>33</u>, 436.
16. Carson, G.A., Hu, H.-W., and Granick, S. (1992), Tribology Trans., in press.
17. Granick, S., Materials Bulletin (1991), <u>16</u>(10), 33.
18. Georges, J.-M., Loubet, J.-L., and Tonck, A. (1988), C. R. Acad. Sci. (Paris) <u>306</u>, II871.
19. Tonck, A., Georges, J.-M., and Loubet, J.-L. (1988), J. Coll. Interface Sci. <u>126</u>, 150.
20. Georges, J.-M., Mazuyer, D., Tonck, A., and Loubet, J.-L. (1990), J. Phys.: Conden. Matter <u>2</u>, SA399.
21. Gee, M.L., McGuiggan, P.M., Israelachvili, J.N. and Homola, A. (1990), J. Chem. Phys. <u>93</u>, 1895.
22. Homola, A.M., Israelachvili, J.N., Gee, M.L. and McGuiggan, P.M. (1989), J. Tribology <u>111</u>, 675.

23. McGuiggan, P.M., Israelachvili, J.N., Gee, M.L., and Homola, A.M. (1989), Mater Res. Soc. Symp. Proc. 140, 79.
24. Homola, A.M., Israelachvili, J.N., McGuiggan, P.M. and Gee, M.L. (1990), Wea 136, 65.
25. Mate, C.M., McClelland, G.M., Erlandsson, R., and Chiang, S. (1987), Phys. Rev Lett 59, 1942.
26. Erlandsson, R., Hadziioannou, G., Mate, C.M., McClelland, G.M., and Chiang, S (1988), J. Chem. Phys. 89, 5190.
27. Mate, C.M., Lorenz, M.R., and Novotny, V.J. (1989), J. Chem. Phys. 90, 7550.
28. Blackman, G.S., Mate, C.M., and Philpott, C.M., Vacuum, in press.
29. Buckley, D.H. (1981), Surface Effects in Adhesion, Friction, Lubrication, and Wear Elsevier, Amsterdam.
30. Gellman, A.J., J. Vac. Sci. Tech., in press.
31. Watts, E.T., Krim, J., and Widom, A. (1990), Phys. Rev. B 41, 3466.
32. Krim, J., Watts, E.T., and Digel, J. (1990), J. Vac. Sci. Technol. A8, 3417.
33. Krim, J., Solina, D.H., and Chiarello (1991), Phys. Rev. Lett. 66, 181.
34. Landman, U., Luedtke, W.D., and Ribarsky, M.W. (1990), J. Vac. Sci. Technol A7, 2829.
35. Landman, U., Luedtke, W.D., Burnham, N.A., and Colton, R.J. (1990), Science 248, 454.
36. Landman, U., private communication.
37. Sokoloff, J.B. (1990), Phys. Rev. B 42, 760.
38. Zhong, W. and Tomanek. D. (1990), Phys. Rev. Lett. 64, 3054.
39. Sokoloff, J.B. (1991), Phys. Rev. Lett. 66, 965.
40. Schoen, M., Diestler, D.J., and Cushman, J.H. (1987), J. Chem. Phys. 87, 5464.
41. Schoen, M., Cushman, J.H., Diestler, D.J., and Rhykerd, C.L., Jr. (1988), J Chem. Phys. 88, 1394.
42. Schoen, M., Rhykerd, C.L., Jr., Diestler, D.J., and Cushman, J.H. (1989), Science 245, 1223.
43. Koplich, J., Banavar, J., and Willemsen, J. (1988), Phys. Rev. Lett. 60, 1282.
44. Thompson, P.A. and Robbins, M.O. (1990), Phys. Rev. A 41, 6830.
45. Thompson, P.A. and Robbins, M.O. (1990), Science 250, 792.
46. for a review, see J. N. Israelachvili, Intermolecular and Surfaces Forces (1985), Academic Press, New York.
47. I Bitsanis, J. J. Magda, M. Tirrell, H. T.Davis (1987), J. Chem. Phys. 87, 1733 (1987).
48. see appendix of H.-W. Hu and S. Granick, Macromolecules (1990), 23, 613; also see appendix of reference 9.
49. for a review, see J. Halling, Ed. (1975), Principles of Tribology, MacMillian Press, London.
50. G. Carson and S. Granick (1990), J. Mater. Sci. 5, 1745.
51. C. Kessel and S. Granick (1991), Langmuir 7, 532.
52. C.E.D. Chidsey, G.-Y. Liu, P. Rowntree, G. Scoles (1989), J. Chem. Phys. 91, 4421.
53. for a review, see J.-P. Hansen and I.R. McDonald, Theory of Simple Liquids, 2nd Ed. (1986), Academic Press, New York.
54. P. Debye and R.L. Cleland (1959), J. Appl. Phys. 30, 843.
55. H.-W. Hu, G. Carson, and S. Granick, submitted.

Discussion *following the lecture by S Granick on "Molecular tribology of fluids":*

U LANDMAN. The question of the relation between structure, amorphization, fluidity, and the like, is very interesting. Much is known in a related field, liquid phase epitaxy of a silicon. If you melt the top of a solid by irradiating it with photons, it recrystallizes explosively. Layering in front of the solid front is the first incipient act of crystallization. You can inhibit recrystallization (and form an amorphous film) by taking heat out faster than the solid front can move. Thus there is a competition between an external effect and crystallization. It could be that there are ways to measure these effects in your system, too.

S GRANICK. What I learn mostly from Uzi's interesting question is that there are parallels to other fields of endeavour and that we may learn more than just about tribology from this work. Not only that, but there are ways to design around these problems. That is, having recognised that lubrication has some essential physics in it, we can now worry about the mechanisms, so that the design engineer may profit from what may on first blush be seen as a purely academic exercise.

J N ISRAELACHVILI. Could you please clarify what were the ranges and amplitudes of the frequencies? These were all sinusoidal, I believe.

S GRANICK. The frequency varied from 0.026 Hz to 52 Hz. The amplitude varied from a minimum of one-half of an Angstrom to around 10 μm.

K L JOHNSON. You have mentioned what the repercussions were of these figures on the strain rates, but not on the strains. Have you an answer to that question off the top of your head? What would be range of strains be?

S GRANICK. The strains varied from a minimum of about 10%, where we know that we have a linear response, to a maximum of perhaps 2000%, where we don't.

K L JOHNSON. That's interesting; it means you're in the large strain range. You are not reproducing the sort of things that were done in the bulk by John Lamb and his collaborators in Glasgow, who were in the microstrain range. You really are in the nonlinear region.

S GRANICK. I'm sorry, Ken, that's not quite correct. We go into the high strain region but also into the low strain region. We can cover between both extremes smoothly, and see the variations.

K L JOHNSON. I beg your pardon. You're not restricted to molecular strains, where it was hard to imagine that molecular segments were moving really beyond their own equilibrium positions.

R POLIZZOTTI. I have several questions. We know from surface science that the lateral motion of molecules is much easier, in terms of the activation energy, than the desorption processes. It struck me in looking at your data that it might be possible to interpret the sum of the data with regard to localization shear at the interface between the mica and the liquid, corresponding to localized shear between alkyl chain terminated surfaces and the liquid. Is there anything in your data that would suggest that is not a problem? Are we really measuring the bulk properties of the 20 or 30Å film of liquid; or are we measuring properties of the shearing interface between the liquid and the solid surfaces?

U LANDMAN. It may not have a bulk.

S GRANICK. To avoid speculation, I think I have told you everything I know.

M O ROBBINS. I just want to say that in some simulations of polymer films at very high shear rates, you do get separation between the solid and the polymer. The polymers have gotten so entangled that the film can't respond internally. However, I wouldn't want to conclude that there is never shear in the film at the much lower experimental shear rates.[1]

S GRANICK. Let me give a less facetious but less well-founded response. Clearly, the film is inhomogeneous. Of that we are certain. We seek to characterize it by a single number, as an effective viscosity under certain conditions of strain rate. It's clearly an average. So you're really asking, where is that average taken. There is no contradiction with what you call surface science studies, by which I think you mean diffusion of small molecules on solids with ultrahigh vacuum on the other side, for these two reasons. First, the conditions of confinement are physically different. We have two solid surfaces squeezing down upon this film of macroscopic area - not single molecules diffusing about at a very forgiving vacuum interface.

The second reason we have a physically different situation is that these molecule are bigger, so that they adsorb more strongly. In the case of the chains - and do-decane qualifies as a chain in this respect - the overall sticking energy is large and you cannot imagine it moving about as a unit, as CO_2 might. You start scratching your head and imagining creeping and crawling caterpillar-like motions; something else, something really qualitatively different. I see no contradiction at all.

Whether the skip is at the walls, or in the center, or whether it is some mixed response, I hope that the simulations will tell us. They certainly have provided very provocative results already. Our findings are starting to coincide in a way I never expected a year or two ago. So we starting to have a confluence of results.

---

[1]Indeed, our work has progressed a great deal since last summer. We now know that the slip can occur entirely at the walls or throughout the film, depending on the wall-fluid coupling, chain length, and film thickness.

R S POLIZZOTTI. Your comment begs my second question. If you look at the lateral size of your contacts, they are much larger than those of the asperity contacts that Professor Greenwood demonstrated last week. I was wondering what is known about the effects of this.

S GRANICK. It's an excellent question. The activation volume we calculate is an uncertain quantity, but if we accept the argument, it amounts to a lateral extent of maybe 20 molecules. So that might suggest that an asperity only has to be larger than about 20 molecules to show these kinds of effects. It's only a suggestion but it's everything we know. In a sense, it's premature even to speculate.

J FERRANTE. Do you think that the diamond anvil cell would give one a way to know the structure of these films?

S GRANICK. The problem in making diffraction measurements on films this thin is to get enough diffraction signal. If you could find a way to increase the sensitivity you might do it in a diamond anvil cell, but the real problem is to get the sensitivity.

U LANDMAN. A cautionary word is in order here. Can you actually determine the effect of the glue, in back of your mica surfaces?

S GRANICK. In order to make these measurements cleanly, we require parallel plate geometry. To achieve this we need to use a material in back of the mica that is weaker than the interfacial liquid. We quantify the effect of the glue by measuring the system in dry contact and comparing with the response in the presence of liquid. The complex impedance of the glue is significantly larger than that of the liquid films I have been describing to you. We have looked at this question carefully.

J-M GEORGES. If I understand the design correctly, you have two springs in series (corresponding to effects of the glue and of the liquid). This seems difficult to interpret.

S GRANICK. This is not our picture. We need a dashpot (that is, a compliance), in addition to a spring, to account for the effect of the glue. A detailed description is found in Rev. Sci. Inst. 62, 463 (1991).

J N ISRAELACHVILI. I have a question about time effects. You were mentioning the long time it takes the films to equilibrate. Could it be that because you're going backwards and forwards over the same region, that's different from what we do, which is to have continued sliding in one direction?

S GRANICK. What are you suggesting is a strong possibility. It is true that it's difficult to find a numerical correspondence between the measurements you've made and ours.

ANON. Have you looked at water?

S GRANICK. Water is the obvious next system to look at. So far we've tried to simplify the situation as much as possible, and that has meant just working with monopolar liquids.

ANON. Are there any directional effects as you go along different directions on the mica?

S GRANICK. When the films are very thin the relative orientation of the mica seems to make a difference. Jacob Israelachvili has investigated this carefully. For the films I was emphasizing, where a liquid-like response is observed, we have seen no evidence of it. It's a lucky thing. Just imagine if the response depended on relative orientation; to find a generic response might be a hopeless task. One of the strongest pieces of evidence that this is not an essential part of the physics is that even with the amorphous surfaces I described, we see qualitatively the same effects.

D P LETA. You sparked my curiosity when you mentioned additives. Have experiments been done using large polymers at 0.1% concentration? These have a tremendous effect on lock-up.

S GRANICK. We have not yet done the experiment, but hope to do it.

J KRIM. When you tell us about the viscosity that has gone up by orders of magnitude, you obtain that number, I believe, by assuming no slip at the wall and uniform density. If I think about it, this gives a lower bound on the actual viscosity of what's in between. Do you agree?

S GRANICK. Yes. The effective viscosity is certainly not a macroscopic viscosity. The medium is not homogeneous. Transport coefficients certainly vary across the thickness of the film and maybe laterally across the area as well. But we can't measure any of that. We measure a force and a phase of that force. A convenient way to represent the data is in terms of a complex viscosity. It is a number for which we have some intuition from the bulk. It has practical significance as well because it gives tribologists a physical sense of how much mechanical energy needs to be pumped into a system to keep an excitation going.

J KRIM. Do you know that the liquid is deforming, rather than slipping at the wall?

S GRANICK. Not unequivocally. The former seems most likely for the thinnest films. The latter seems most likely for the thickest films.

P S THOMAS. I am concerned about the large pressure coefficient of critical shear stress that you measure. If you take the Eyring model, it falls out that the pressure coefficient is the ratio of the pressure activation volume to the stress activation volume. Your value of 20 seems unreasonable.

S GRANICK. How much physical meaning to assign to the Eyring model is a question already raised by Professor Tabor a week ago. He didn't know the answer and neither do I. The limitations of that model are well known. Among other things, the model rests on assuming an equilibrium between reactant and product. Let us clearly separate the experimental data, of which we are confident, from its interpretation, which involves a model with attendant assumptions and approximations.

I L SINGER. The critical shear stress that you measure - is it static friction?

S GRANICK. It is the force it takes to get motion started. Certainly when dealing with a glass in the bulk, it is true that forces depend on the rate of measurement. The same is probably true here.

I L SINGER. What is the temperature rise as you shear?

S GRANICK. A few tenths of a degree at most. Heat is efficiently dissipated because the surface-to-volume ratio is so large.

D TABOR. I have a very simple question for the simulators. Whenever I hear people here talking about their work, in the back of my head is the question - what is the mechanism by which energy is dissipated in the viscous process? In this case, can we consider a liquid as clusters of molecules, such that molecules are distorted by shear, rather in the way that I described for solids? I'd like to know if there is any evidence for this at all.

NEW APPROACHES AT THE NANO- AND ATOMIC SCALE

...... Despite the pleasing basic simplicity and consistency of the adhesion theory of friction, it has to be stretched until it is no longer simple, in order to take into account the many different sliding conditions which are possible. Alternatively, frictional energy input may first be stored as elastic strain energy and then converted into heat by repeated jumping of unstable interfacial atoms .....

J SKINNER, "An atomic model of friction", Central Electricity Generating Board report RD/B/N 3137 (1974).

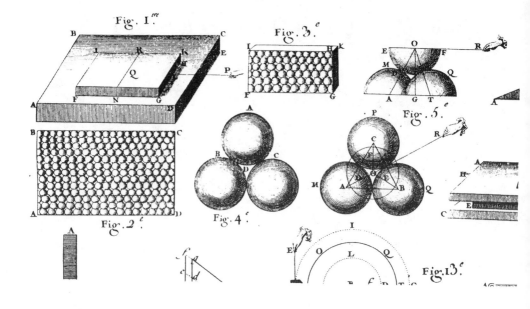

Belidor's representation of rough surfaces with ideal spherical asperities (Belidor, 1737). [From D. Dowson, <u>History of Tribology</u> (Longman, London, 1979) p. 165, with permission.]

FRICTION AT THE ATOMIC SCALE

GARY M. McCLELLAND
JAMES N. GLOSLI
IBM Research Division
Almaden Research Center
San Jose, CA 95120-6099 USA

ABSTRACT. This paper reviews simple theoretical models and recent experimental advances concerned with frictional processes at the atomic scale. Two simple models for wearless interfacial friction are discussed, the independent oscillator (IO) model, and the Frenkel-Kontorova model. The IO model is compared with molecular dynamics calculations between films of close-packed alkane molecules. Atomic force microscope studies of friction are discussed, and the role of other tip-based proximal probe techniques in understanding tribological processes is reviewed.

1. Introduction

Recent rapid progress in experimental, theoretical, and computational methods are providing new insight into atomic-level processes occurring at sliding interfaces. Experimental techniques are being developed which probe increasingly smaller and better defined interfaces. Simultaneously, theoretical approaches to atomic interactions and dynamics are being refined, and the computing power is becoming available to model ever larger systems for longer time periods. This article discusses three topics concerning an atomic-level approach to friction. To elucidate the mechanism of energy dissipation at sliding interfaces, two simple models of wearless friction are discussed, an independent oscillator model, and the Frenkel-Kontorova model. The concepts developed are illustrated by molecular dynamics calculations of interfacial sliding of alkane films. Atomic force microscope measurements of atomic-level friction are then described. Finally, results and prospects for related proximal probe techniques are presented.

2. Simple Models of Wearless Friction

At the present level of sophistication of computational hardware and experimental techniques, realistic atomic level simulations of sliding interfaces are now quite possible. Because such simulations can be very complex, it is still useful to understand simple models which can more clearly reveal basic physical principles. In this spirit, we review two simple models of wearless friction at a sliding interface, closely following an earlier discussion [1]. In most systems friction is accompanied by wear, and in adhesive friction and ploughing friction, energy is in fact dissipated not at the interface but rather in the bulk of the solids by movement of dislocations. However recent force microscope and surface force apparatus studies demonstrate convincingly that friction can occur without wear.

Figure 1 displays a model system containing a lower solid A and an upper solid B, which may have different lattice constants. For each solid, the layer of atoms furthest from the interface is at-

I. L. Singer and H. M. Pollock (eds.), Fundamentals of Friction: Macroscopic and Microscopic Processes, 405–425.

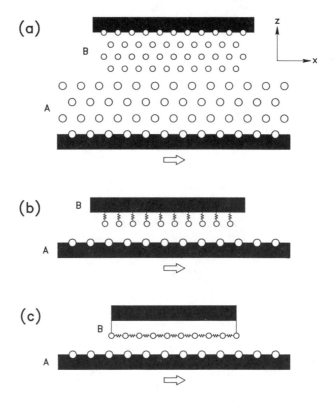

Figure 1. (a) Solid A sliding against solid B in two dimensions. (b) Independent oscillator model. (c) Frenkel-Kontorova model. From reference [1], used by permission.

tached to a rigid (black) support, and sliding is performed by moving support A toward the right while keeping the distance between the supports constant. Experimentally, it is the load rather than the distance which is normally specified, but the principal effect of load is to vary the contact area in the presence of surface roughness, an irrelevant effect for flat surfaces.

If the intrasolid bonds holding each solid together are covalent, ionic, or metallic, and the solids have no dangling bonds to bond across the interface, the interfacial interactions will be sufficiently weak that the solids are not disrupted during sliding. The static friction at the A-B interface is the force needed to translate solid A across B at an infinitesimally small velocity. The boundaries of the solids are attached to heat sinks at zero temperature, so that any energy released into vibration is conducted away.

In analyzing the occurrence of friction, we aim to discover by what mechanism infinitesimally slow motion of the A support is converted to rapid atomic vibration (heat). Since the sign of the frictional force depends on the direction of motion, and the frictional force itself is simply a function of the atomic positions, the change in the direction of the frictional force must arise from a difference in the atomic positions.

2.1 THE INDEPENDENT OSCILLATOR MODEL

Fundamental mechanisms of energy dissipation can be addressed through the independent oscillator (IO) model diagrammed in Fig. 1b, in which solid A has been simplified to a single row of rigidly

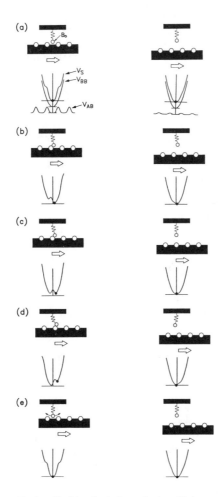

Figure 2 Motion of an atom B_0 described by the independent oscillator model. Left and right columns represent strong and weak interfacial interactions, respectively. The top panel represents the relevant potentials. Subsequent panels show the response of B_0, represented by a black dot on V_S, to gradual sliding of Λ, determined by the potential sum V_S plotted below each diagram. From reference [1], used by permission.

connected atoms. The row of interfacial B atoms do not interact with each other, but they feel an interfacial potential from the Λ atoms and are connected by a single flexible bond to a rigid support representing the remainder of the B solid. To account for dissipation of any excitation of the interfacial B atoms, these bonds are allowed to lose energy into the B support. The IO model is obviously a great simplification, but it contains all the major elements of the wearless frictional problem: movable surface atoms, a periodic corrugated interfacial structure, and a mechanism to dissipate vibrational energy created at the surface. The analysis presented here is similar to that of Tomlinson [2], who recognized long ago the central role of mechanical adiabaticity in understanding friction. In an unpublished manuscript Skinner has also lucidly discussed many of these issues [3].

The dynamics of the IO model are now analyzed in terms of the potential energy curves V_{AB} and V_{BB} which govern the lateral motion of a particular B atom B_0 as A slides by it (Fig. 2). V_{BB} is a simple parabolic potential well with curvature k_{BB}, representing the strong bond connecting B_0 to solid B. We initially assume V_{AB} has periodic sharp repulsive peaks for each A atom. As shown on the left side of Fig. 2, the net potential affecting B_0 is the sum $V_S = V_{AB} + V_{BB}$; the changing shape of this sum as A is advanced along x governs the motion of B_0. These potentials are presented as effective potential functions of a single coordinate which lies mostly in the x direction along the path of B_0. The sum V_S is presented on the left side of Fig. 2 for sequential positions of A as it is moved right; at each A position, B_0 rests at a local minimum of the potential. If A is slid infinitesimally slowly so that V_S changes infinitesimally slowly, the principal of adiabatic invariance requires that the action of classical mechanics for B_0 motion is conserved [4]. (Slow sliding in this case simply means slow compared to the rate at which strain equilibrates in the solid, which is the speed of sound.) Thus B_0 remains in the changing minimum of V_S without becoming vibrationally excited. The critical exceptions occurs at a point between Fig. 2 d and e, where a local minimum disappears. Here B_0 must fall abruptly to the potential bottom, becoming vibrationally excited. This vibrational energy is then dissipated irreversibly in the solids. The potential V_{AB} has "plucked" the harmonic V_{BB} bond. It is just at this plucking where the strain energy put into the B bond by translation is converted into vibrational motion (heat).

The directionality of the frictional force arises from the double minimum in V_S induced by V_{AB}. For example, for the A position diagrammed in Fig 2c, if A has been moving to the right as just described, atom B_0 is pushed to the right, exerting a force on solid B to the right. On the other hand, if A has been moving to the left, B_0 finds itself in the left minimum, reversing the resulting forces.

The mechanism we have just outlined for energy dissipation cannot occur for sufficiently weak interfacial forces, a case diagrammed on the right side of Fig 2. The interfacial potential used on the right side is identical to that on the left side, but reduced by a factor of 5. In the schematic of the interface, this is suggested by drawing the interfacial atoms farther apart. As a consequence of this interaction reduction, the shape of the total effective potential is qualitatively changed. For any position of the A solid diagrammed in Fig. 2a-e, the total potential has only a single minimum. There are no metastable local minima which can disappear with sliding and lead to plucking as there are for the stronger interacting case. Thus as solid A is slid along infinitesimally slowly, atom B_0. remains in the single minimum of the total potential, the position of which varies smoothly as A slides. In this case there is no mechanism for energy dissipation, and (at infinitesimally low velocity) *the friction vanishes.*

Of course there is a conservative lateral force exerted on B through B_0, but unlike a frictional force, this force is independent of the direction of motion, because the B_0 atom occupies the same position in V_S (the unique minimum) for both sliding directions.

The requirement just deduced that to dissipate energy more than one local minimum must arise in V_S is equivalent to requiring that V_S have a local maximum. This can only occur if the downward curvature at the maximum of V_{AB} exceeds the curvature of the bond V_{BB}:

$$k_{BB} < -\frac{\partial^2 V_{AB}}{\partial^2 x_{B_0}}. \tag{1}$$

For the IO model this is the simple criterion for whether or not frictional dissipation can occur.

From another point of view, it is clear that while sliding two solids across each other, breaking interfacial bonds (even van der Waals bonds) and distorting intrasolid bonds at the interface requires energy. For the weak interaction adiabatic case, potential energy is returned through the relaxation of the distortion and the reversible formation of new bonds. For the strong interaction "plucking" case satisfying Eq. 1, the potential energy is not returned; the new bonds are formed and the re-

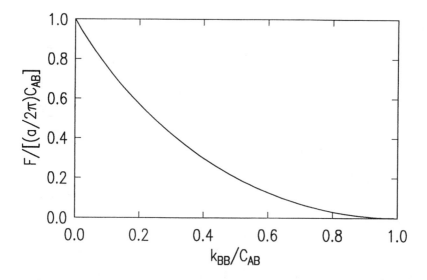

Figure 3 Normalized frictional force in the independent oscillator model *vs.* the ratio k_{BB}/C_{AB} of the intrasolid to interfacial interaction strengths.

laxation occurs instantaneously, so that the energy is not regained as mechanical energy but dissipated to the rest of the solid as phonons.

It is important to recognize that the bending of the B_0 atom can represent a variety of different types of degrees of freedom in the solid. The key feature of this model is the metastability caused by a strong interaction interacting with a weak spring. Hirano and Shinjo [5] have explored in considerable detail potential features leading to plucking in more realistic models of sliding interfaces. We have also recently discussed the possible role of trapped interfacial atoms or molecules at the sliding interface by a scheme very similar to that of Fig. 2 [1]. The molecule is assumed to move in the slowly varying potential formed by combining the potentials of the surfaces. Even for weakly bound molecules, friction can arise from sudden dissipative motions of the interfacial molecules when the maxima in the molecule-surface potential is more sharply curved than the minima.

To further analyze the effect of changing potential shapes on the frictional force, suppose that the interfacial potential is

$$V_{AB} = C_{AB}\left(\frac{a_A}{2\pi}\right)^2 \cos\left[\frac{2\pi}{a_A}(x_{B_0} - \delta_A)\right], \tag{2}$$

which has a period a_A and a maximum downward curvature of magnitude C_{AB}. Whenever C_{AB} is less than k_{BB}, there is no local maximum in V_S and no dissipation of frictional energy. As C_{AB} increases from zero, the friction remains zero until it reaches k_{BB}, at which point the friction starts to increase with C_{AB}.

Figure 3 presents as a function of k_{BB}/C_{AB} the frictional force (for a single B_0 atom) of the IO model. This friction is simply the energy dissipated per pluck divided by the lattice constant, or equivalently the lateral force averaged over the A motion. The friction reaches its maximum for very weak levers $k_{BB}/C_{AB} \to 0$, where its value is $(a_A/2\pi)C_{AB}$, which is simply the maximum force at the most sloped part of the V_{AB} potential. In this limit of small k_{BB}, the B_0 bond is stretched many lattice constants a_A before it plucks, and the lateral force is not reduced significantly during the pluck. One might naively expect that maximum frictional force would correspond to releasing all the potential energy difference from the maximum of V_{AB} to its minimum, which would give an average friction of $(a_A/2\pi^2)C_{AB}$. The actual value is a factor of π larger than this, because when B_0 plucks, it is accelerated by its bond force as well as the interfacial force.

The preceding discussion has assumed that the sliding solids are at zero temperature, so that the interfacial atoms always seek their lowest potential energy. At higher temperatures, the friction will often be lower, because thermal activation will allow the interfacial atoms to pass over potential maxima at lower strain. Indeed, if the interface is always at thermal equilibrium, there can be no friction, because at thermal equilibrium the position of the interfacial atoms and the shear force will be independent of the direction of sliding.

Although frictional energy is only dissipated for small k_{BB}, each B atom is always subject to an interfacial force. The total force exerted across the interface is simply the sum of these forces for all B_0. Suppose that the two solids are incommensurate so that the lattice constants a_A and a_B do not form the ratio of two simple rational numbers. Then the average over all these atoms will simply be the average interfacial interaction shown in Fig. 3, and will be constant with respect to sliding. For the zero-friction adiabatic case, more precisely, the lateral force will be conservative, on the order of that for a single B_0.

Consider now the case when the interfacial atoms of A and B have identical spacings. Then the interfacial force of all the atoms acts in phase so that even in the adiabatic case the lateral force to slide one solid over the other can be very large, alternating in sign. Actually in this case the flexibility of the underlying solid, which we have ignored until now, can have a drastic effect. To analyze this effect, consider two interacting three-dimensional solids, with a circular contact area of radius R containing N interfacial A and B atoms, all interacting in phase. In the model we have explored so far, in which all B_0 bonds would bend in parallel, we could just as well model a single atom with force constant Nk_{BB} and interfacial interaction NV_{AB}. However in series with the individual B_0 bonds is another spring formed by the elasticity of the solid B, which from continuum contact mechanics is known to vary proportionally to the contact radius, thus being proportional to $\sqrt{N}$. Since the interfacial interaction varies as N, as N becomes large the interfacial force will always become larger than the internal force constant arising from the continuum contact elasticity. Due to this effect, a extended commensurate contact will always exhibit plucking behavior, dissipating energy, regardless of the relative values of the interfacial and bond forces. We have described this effect in a very simplified way, assuming that the B atoms move together with a uniform distortion of the solid. In practice the deformation of the B solid leading to dissipation will be much more complex. The point is that motions involving cooperative movement of the B atoms must be taken into account, especially for commensurate solids.

2.2 THE FRENKEL-KONTOROVA MODEL

As the simple argument about elastic continuum deformations demonstrates, it is important to consider relative motion of the atoms at the interface. The interfacial B atoms interact not only with the atoms of the A solid but also with other atoms of B. In particular, the force to move N neighboring B atoms together is much less than N times the force required to move a single B atom. Interactions between B atoms were first included in a model described by Frenkel and Kontorova [6] and soon developed by Frank and van der Mewe [7] and later by many others [8]. Recently Sokoloff has further discussed the application of this model to mechanical friction [9].

The Frenkel-Kontorova (FK) model, diagrammed in Fig. 1c, concerns the mechanics of a row of B atoms constrained to a line, with each atom interacting with its neighbors through a harmonic potential. The total B potential is thus

$$V_{BB} = \frac{k_{BB}}{4} \sum_n (x_{Bn} - x_{Bn-1} - a_B)^2. \tag{3}$$

The B atoms are required to have an average spacing a_B, which might be ensured by anchoring the ends of the B chain to the rigid support (Fig. 1c). To facilitate comparison to the IO model discussed above, the normalization of V_{BB} has been chosen so that if all atoms but one are held fixed at their equilibrium position, the force constant for moving the atom is $\partial^2 V_{BB}/\partial^2 x_{Bn} = k_{BB}$. Each B atom interacts with the rigid B solid with the same sinusoidal potential Eq. 2 used for the IO model. Note that while the IO model does not promote the coherent motion of nearby B atoms, in the FK model, coherent motion occurs more easily than it would in a real solid, where the interfacial atoms are bound to atoms further into the solid as well as to other interfacial atoms.

In the FK model, a competition occurs between the B - B interactions, which tend to keep the B atom spaced by a_B, and the A - B interactions, which tend to space the B atoms at the minima of the A - B potential, which are spaced by a_A. Consider how the spacings of the B atoms depend on the relative values of k_B and C_{AB}. When C_{AB} is zero, the B atoms space themselves evenly by a_B. For most cases in which $0 < C_{AB} << k_{BB}$, the B atoms are slightly displaced from equal spacings of a_B, with the displacement depending on the local phase of the A and B positions. For $C_{AB} >> k_{BB}$, however, a distinctly different behavior is found. In this case, the B atoms are forced to be locally in phase with the A lattice, lying in the minima of the potential Eq. 2. Between such regions are misfit dislocations with no atoms or more than one atom in a minima, allowing the average spacing of B to be a_B. In this case the lattices are said to be "pinned" together.

Our interest is in the force per B atom required to move the B lattice with respect to the A lattice. In the case of strong interactions ($C_{AB} > k_{BB}$), this force is substantial, but the force per atom may be considerably less than the force required to move a single B atom in the A potential. This phenomenon, which is familiar from the mechanics of dislocation motion in a perfect solid, occurs because each atom does not pluck simultaneously, and the force to move an interfacial atom over a potential barrier is partially supplied by its neighbors. For smaller values of C_{AB}, the force required per atom to move the B layer is zero for an infinite system [8]. In other words, there is no friction (at zero velocity) between A and B. For a finite system, the force needed to translate all of A is conservative (independent of the direction of motion), oscillating about 0 as A is translated, with a magnitude about that of the force required to translate a single atom of A along B. This force is very similar to that which would be found if the B atoms were held rigidly to a spacing a_B as B was moved.

Summarizing, as in the IO model, as the ratio of the interaction to the spring potentials is increased, a transition from free sliding to frictional dissipation occurs at some critical value $(C_{AB}/k_{BB})_c$. This critical value depends strongly on a_B/a_A. If a_B/a_A is near a simple rational number, V_{AB} acts in phase over many atoms of the chain, so that $(C_{AB}/k_{BB})_c$ is quite small. The highest values of $(C_{AB}/k_{BB})_c$ range up to $0.2\pi^2/4 \simeq 0.5$ [10], and occur for a_B/a_A values far from a simple rational number. This case corresponds to the least amount of correlated motion among the B atoms, and it is expected to agree most closely with the corresponding critical value from the IO model, which is unity. The fact that a smaller $(C_{AB}/k_{BB})_c$ is obtained for the FK than the IO model accords with the fact that, if, as in the FK model, B atoms neighboring a particular B atom relax in response to its motion, the effective force constant for the motion of that B atom will be reduced.

Considering the significant differences between the two models, the agreement within a factor of two of the parameter values for the onset of friction is satisfying. The FK model produces much richer phenomena, because it allows V_{AB} to locally distort the B solid into registry with it. But as models for the interface between two sliding solids, the IO and FK models seem equally inadequate, the IO model ignoring the interaction of interfacial atoms with their neighbors along the interface, the FK model ignoring interactions with atoms one layer away from the interface.

2.3 MOLECULAR DYNAMICS SIMULATIONS

To demonstrate some of the simple concepts just discussed, we present molecular dynamics calculations of the friction between close-packed alkane films, which will be described in more detail

412

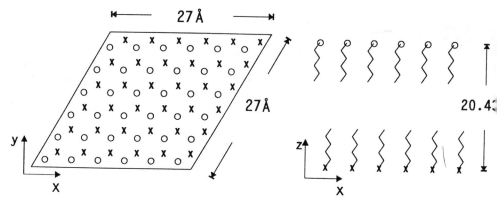

Figure 4. Schematic of a molecular dynamics simulation of the friction between two close-packed alkane monolayers.

elsewhere [11]. In these calculations, alkane chains of 6 carbons are bound to two substrates of fixed separation (Fig. 4). Each carbon with its attached hydrogen atoms is represented as a single atom, which interacts with atoms of other chains by Lennard-Jones interactions. The computational cell contains 72 chains, and periodic boundary conditions in the x and y directions are used. Internally, the chain bonds can bend and twist in a realistic potential, but stretching is not allowed. The simulations are performed near zero net load, i.e. where attractive and repulsive interfacial forces balance.

Fig. 5a displays the shear stress as one solid is slid past the other in the x direction at 20 K and the low velocity of 2.3 m/s. The shear stress exhibits a distinctive sawtooth pattern with a period of the lattice constant .45 nm. Averaging the shear stress, we find the frictional shear strength to be 50 ± 3 MPa. In the sawtooth, the force changes gradually from the sign aiding the sliding motion to that opposing it, then suddenly reverses direction. An examination of the chain configurations shows that this sudden change is associated with a sudden plucking motion of the chains. For close-packed chains at these low temperatures, this motion is particularly simple, as the chain behaves nearly as a rigid body. Two plucks are observed per lattice period, because the interfacial potential passes through two minima for each displacement by a lattice constant, due to the triangular structure of the interface.

The fact that the plucking motion is responsible for energy release in this model is supported by the plot of the cumulative thermal energy flow out of the system into the thermostat during sliding (Fig. 5b). Almost all of the energy flow occurs at the plucks, in agreement with the simple IO model described above. At 300 K, the friction is reduced to about half its 20 K value due to thermal activation.

The dependence of the friction on the interfacial interaction (analogous to C_{AB} in the above discussion) has been determined by scaling the interfacial Lennard-Jones well depth. As for the IO and FK models, we find a threshold below which the plucking motion disappears when the LJ interaction has been reduced to 40 % below the nominal bulk value used for the calculations of Fig. 5. Below this interaction strength, there is a small amount of dissipation by a mechanism akin to viscous dissipation in simple fluids. For this low interaction case, around 100 K, where the orientational order of the chains melts out, the friction is enhanced by an interaction with rotational motion of the chains about their axes.

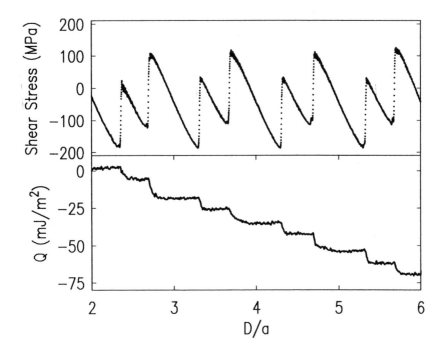

Figure 5. Interfacial shear stress (a) and cumulative heat flow (b) as a function of sliding distance D/a, normalized by the lattice constant $a = 0.45$ nm.

3. Atomic Force Microscope Measurements of Atomic-Scale Friction

Because frictional force depends critically on contact area, surface roughness is an extremely important parameter in determining frictional forces. The idea of simplifying interfacial contact by replacing one surface by a single tip has a long history [12], and some work has used quite well characterized surfaces [13].

The potential of the single asperity approach was considerably enhanced in 1986 by the invention by Binnig, Quate, and Gerber of the atomic force microscope (AFM) [14], a device in which a sample is positioned by piezoelectric translators with respect to a sharp tip mounted on a flexible cantilever. When the sample is moved with respect to the tip, the cantilever deflection records the changing force between the tip and surface. If the sample position is scanned perpendicular to the tip axis and the sample position along the axis is adjusted to keep the normal force between the tip and sample constant, a constant-force topographic profile is obtained which may show atomic resolution [15].

In addition to its usefulness as a profilometer, the AFM can also probe the mechanics of the interaction between two solids at an unprecedentedly fine level. Piezoelectrics can easily be actuated to a precision of a small fraction of an Ångstrom, and the AFM can be made insensitive to external vibrations by miniaturization and spring suspension. The fundamental limit to the force detection sensitivity is set by thermal noise, which for a typical lever is less than 10^{-11} N when averaged over a 1 ms time constant. This can be compared with the strength of a single chemical bond, which

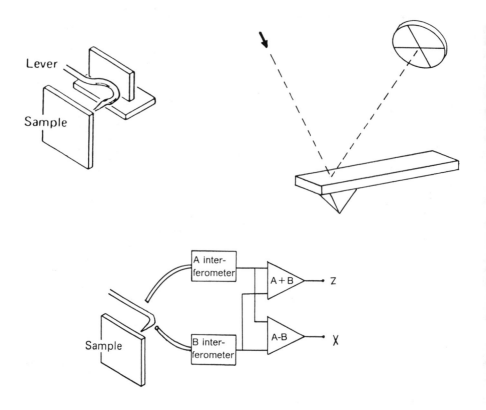

Figure 6. Three designs for simultaneously detecting the motion of an AFM tip perpendicular and parallel to a sample. (a) capacitance [16], (b) laser beam deflection [17,18], (c) dual fiber interferometry

is typically 10^{-8} N (10^{-6} gram force). In the first AFM, tunneling from the lever to a second tip was used to sense the lever deflection, but this method has now been largely supplanted by more reliable optical interferometric and deflection techniques.

For AFM studies of friction, it is important to measure both the force parallel (friction) and the force normal (load) to the surface. For this purpose, several methods, diagrammed in Fig. 6, have been devised. The first used two capacitance sensing plates located near the lever [16]. Marti *et al.* [17] and Amer *et al.* [18] have devised an optical lever deflection technique in which a bending and a torsional mode of a microfabricated lever are sensed simultaneously by a quadrant photodiode detector. A final bidirectional detector method, which we have recently used in our lab, uses two optical fiber interferometers of the type developed by Rugar *et al.* [20] to measure the lever deflection along two orthogonal directions angled 45° with respect to the surface normal. This geometry is adopted to avoid the close tolerance which would be required if one fiber were brought in parallel to the surface. The frictional and normal forces are measured from the difference and sum of the two interferometer signals. Of these three methods, the laser deflection method seems the most generally useful, because it requires only a single laser beam, which may be focussed by a lens remote from the lever, and because the bidirectional capability may be added to a standard monodirectional force sensor by simply replacing a two-element photodiode by a four-element photodiode.

In 1987 the first atomic-scale AFM friction experiments were reported by Mate *et al.* [21] for a 300 nm radius etched tungsten tip sliding on graphite in air. In this experiment, the motion of

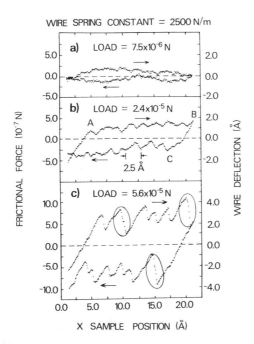

WIRE SPRING CONSTANT = 2500 N/m

Figure 7. Frictional force in air between a tungsten tip and a graphite surface as a function of sample position for three different loads. From reference [21], used by permission.

the lever parallel to the surface was monitored by a simple (non-fiber) interferometer. The load was not measured directly, but rather deduced by assuming that deformation of the graphite sample was negligible, so that after the sample had made contact with the lever and advanced by a distance z, the loading force was $k_\ell z$, where k_ℓ is the lever force constant.

Figure 7 presents for three different loads the variation of tip-surface lateral force as the surface is scanned 20 Å back and forth across the interface at a frequency of 10 Hz. At the lowest load, the hysteresis associated with the frictional force is barely visible above the instrumental noise. At the intermediate load distinct continuous features are observed, while at the largest load, a sawtooth pattern is clearly visible. By rastering the scan in both y and x directions, it was determined that the periodicity of the friction is that of the graphite lattice, a periodicity which is not so apparent in single directional scans along a direction other than a lattice vector. The sawtooth pattern of Fig. 7 is clearly reminiscent of the stick-slip friction commonly observed in macroscopic friction, in which the tip alternately sticks to the surface (the regions of unity slope) and then slides abruptly (the vertical regions). Classic stick-slip behavior occurs when kinetic friction is less than static friction, but this is not the case for the AFM experiments, where the frictional force is observed not to depend on velocity. Remember that the average tip velocity in AFM experiments is remarkably low, $\simeq 10^{-7}$ m/s, so little velocity effect is expected.

The stick-slip behavior observed here and in other AFM experiments arises from the interaction of the compliant lever with the tip-surface interaction. To analyze this behavior, recognize that for a tip of effective mass m_t the tip acceleration a_t is governed by the force F_t which is the sum of the force exerted by the lever and the sample:

$$m a_t = F_t = -k_\ell x_t + F_{ts}(x_t - x_s). \tag{4}$$

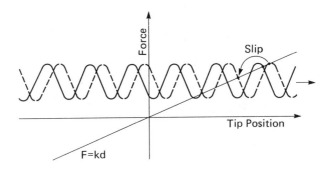

Figure 8. Phenomenological model of tip motion determined by the force exerted by the lever (straight line) and sample (curved lines) as a function of tip position. From reference [21], used by permission.

Here $F_{ts}(x_t - x_s)$ is the force exerted by the sample on the lever for tip and sample positions x_t and x_s, respectively. Since the sample is moved by the piezoelectric drivers slowly compared to the lever resonant frequency, the lever is usually at mechanical equilibrium $a_t = 0$, so that

$$k_\ell x_t = F_{ts}(x_t - x_s). \tag{5}$$

The result of this equation can be seen in the construction of Fig. 8, where the left hand side of Eq. 5 is represented by a straight line and the right hand side by a periodic function. The force on the tip is zero wherever these two lines cross, and this point represents a stable equilibrium whenever $\partial F_t/\partial x_t < 0$, which means that, going from left to right, the surface curve intersects the lever line from above. In the figure, this type of intersection is seen where the lever line intersects the solid curve. Moving the sample to more positive x is equivalent to moving F_{ts} to the right. The tip simply follows the changing intersection point. For some shapes F_{ts}, moving the sample rightward will result in the F_{ts} becoming tangent to $k_\ell x_t$, (see the dashed curve in Fig. 8) and then losing the intersection point all together. At this point the resultant force accelerates the tip until it comes to equilibrium at the next intersection, a process marked "slip" in Fig. 8.

This model predicts that if the tip-surface force is small enough so that $\partial F_{ts}/\partial x_t$ is always less than k_ℓ, the tangency and disappearance of the intersection point and the resultant slip will not occur. This is in agreement with the data displayed in Fig. 7, in which the lateral force shows no slips at lower loads.

There is a direct correspondence between the model just presented for stick-slip motion of the AFM lever and the independent oscillator model for friction discussed above. The interfacial B_0 atom of the IO model corresponds to the tip, the V_{BB} potential to the lever potential, and the interfacial interaction V_{AB} to the *phenomenological* tip-surface interaction F_{ts}. For describing the AFM mechanics, the tip-surface interaction need not be conservative (i.e. its average is non-zero in Fig. 8); our description of the slips requires only that it be well-defined for a particular direction of surface motion.

A surprising feature of the friction observations on graphite is that atomic scale structure is observed, even at loads above 10^{-5} N, where a continuum model of the contact of the tip with the compliant graphite surface predicts a contact 100 nm across. A possible explanation of this phenomenon is a suggestion by Pethica to explain STM imaging mechanisms [19], that a graphite flake may have broken off the surface to adhere to the tip. If this tip maintained its orientation with respect to the surface, the periodicity of the surface could be observed even for large contacts.

Another possibility is that the tip and surface make contact at only a few nm-scale asperities on a rough tip, so that the surface corrugation is not entirely averaged out.

The height of the friction loops displayed in Fig. 7 is equal to twice the frictional force. By recording this height as a function of load, it was found that the frictional force is rather closely proportional to load, with a friction coefficient of 0.01. Since the proportionality of frictional force with load in macroscopic systems undergoing elastic contact is thought to result from the statistics of multiasperity contact [22], single asperity contact in an AFM need not follow this proportionality. In fact Hertzian elastic contact of a single spherical tip and a surface gives a contact area which increases by only the 2/3 power of the load. The proportionality we observe may result from multiasperity contact on a tip with only nm-scale roughness, but it is also may result from the variation with load of the interfacial shear strength.

Erlandsson et al. reported a similar study of friction of a tungsten tip on a mica surface [23]. As they recorded friction loops while approaching the sample to the surface, a sudden increase of the frictional force was observed, which they attributed to a "snapping-in" of the tip by the attractive tip-surface force, perhaps mediated by water condensation at the interface from the ambient air. While subsequently withdrawing the sample, the friction was found to persist to negative applied loads, undoubtedly because of the loading effect of the attractive force. As for graphite, the friction exhibited the periodicity of the mica surface, and was proportional to the load, but in this instance the coefficient of friction was 0.1.

Marti et al. [17] used a bidirectional AFM to probe the friction between a Si_3N_4 cantilever and a mica surface at much lower loads, 2×10^{-8} N, than used in previous AFM friction studies. While scanning, they observed a lateral force variation of 10^{-9} N with the lattice periodicity, and were able to correlate this variation with topographic features. Normal and lateral force images of a wear indentation made by the AFM show greater contrast in the frictional than in the normal force; steps are observed clearly in the frictional force. This observation shows the utility of lateral forces for surface imaging, independent of their value for understanding friction.

Meyer and Amer have used a bidirectional laser deflection AFM to probe the friction of Si and Si_3N_4 tips on clean NaCl surfaces in UHV [18]. At a load of 10^{-8} N, the lateral force was about 1×10^{-9} N during long 700 nm scans across the surface. When the tip went up over a single step edge at the surface, the frictional force increased by about 4×10^{-10} N over a region of about 20 nm, a distance presumably reflecting the tip shape. No corresponding feature is observed in the frictional force as the tip moves down a step. The authors speculate that the height of the feature observed in the frictional force is related to the energy required to raise the tip over the step against the load.

In our lab we have recently studied the frictional force of hydrogen-terminated diamond (111) and (100) surfaces against diamond tips formed by chemical vapor deposition at loads of 10^{-8} - 10^{-7} N [24]. Periodic frictional features with the diamond lattice spacing are observed, and the frictional force increases very slowly at loads $> 10^{-8}$ N. We find that in many cases the non-conservative hysteresis component in the lateral force is smaller than the conservative reversible component.

As emphasized in Section 2 above, under conditions where the surface atoms are bound strongly to the solids which support them and the interfacial interactions are weak, the non-conservative friction component of the lateral force should disappear, leaving only a reversible conservative component. To the authors knowledge, this phenomenon has yet not been observed by AFM.

The only AFM work concerning metal friction is an ultrahigh vacuum study by Cohen et al. [25] of the friction of iridium tips on gold using a bidirectional capacitance-sensing AFM. By approaching the sample to the surface while oscillating the sample back and forth 10 nm parallel to the sample surface, friction loops were generated while the loading behavior of the tip on the sample was recorded. In these experiments, no slips were recorded in the lever motion, but the resolution of the AFM was limited. Fig. 9 displays the normal deflection of the lever and the friction (deduced from the height of the friction loops) as the sample is first pushed toward and then withdrawn from

418

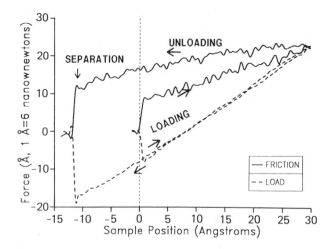

Figure 9. Friction and loading forces between an iridium tip and a gold (111) surface as the sample is pushed into the tip while simultaneously oscillating in the sample plane. Positive sample position is moving toward the tip. From reference [25], used by permission.

the tip. Although the observed friction does not depend linearly on load, the effective friction co-efficient obtained from the slope of the plot of friction with respect to load ranges from 0.4 to 1.0. These plots display a clear hysteresis, with the friction being much larger during unloading than during loading, suggesting that the tip penetrates plastically into the surface so that the effective contact area is greater during unloading than during loading. However the normal force ("load") curve itself invalidates this interpretation, because the curve is nearly elastic, showing a hysteresis no greater than 0.1 nm. Any rearrangement of the interface is thus on the atomic level and does not involve tip penetration.

As exemplified by the work of Kaneko *et al*, AFM friction experiments can give insight into the practical problems of tribology of magnetic disk surfaces [26].

4.0 Other Applications of Proximal Probes to Nanotribology

Since atomic resolution was first achieved by an STM 8 years ago [27], the development and use of "proximal probe" techniques has been rapid. These instruments employ the positioning and feedback methods of the STM, but sense a quantity other than tunneling current, such as force, thermal conductivity, ionic conductivity, electrical potential, light transmission, or capacitance. A useful compilation of literature on this subject can be found in proceedings of the annual STM meetings [28].

The use of the AFM to study lateral forces associated with friction have been described above. However the AFM has been used principally as a high resolution profilometer; indeed it works by the same principle as the commercial profilometers long used for tribology studies. For studies of rough surfaces with steep slopes it is important to use a tip which is not only atomically sharp but also has sharp sides; otherwise, at sharp deep features on the surface, it will be the tip, not the surface, which is profiled. It is now possible with a single AFM to characterize a sample 10 μm across at atomic resolution. This 10 μm scan range overlaps the resolution of optical interferometers. Together, these two instruments can characterize the roughness of a sample from the millimeter to the Ångstrom scale using forces too small to perturb even a single chemical bond. As emphasized in the paper by Greenwood in this volume, surface topography plays an essential role in the contact mechanics of friction, and complete understanding of this contact requires knowledge over a complete range of distance scales. Meyer *et al.* [29] have used force microscopy to characterize the topographic properties of carbon coatings on disk surfaces for magnetic recording, in an effort to correlate this behavior with the frictional behavior. Blackman and Bhushan have combined AFM and optical profilometry of recording heads and disks to understand contact mechanics in disk drives [30].

The fact that while measuring topography, an AFM tip can be scanned repeatedly across the same surface region while the atomic structure remains unchanged is very significant, as it indicates that friction can occur without wear at the atomic level. The fact that images which apparently resolve atoms can be recorded at loads of 10^{-7} N, which is more than enough to break a chemical bond, has led to speculation that the tip sometimes breaks off a flake from the surface, which in turn slides over the surface during imaging [19]. While imaging in air, it is also possible that the tip is mostly supported and lubricated by an absorbed contamination layer, leaving the tip atom to extend through the layer to do the imaging [31].

In purely topographic AFM imaging recording forces normal to the surface, anomalous frictional effects are often observed [32,29,33]. A featureless region appearing immediately after the scanning surface changes direction is attributed to the tip sticking to the surface. Sometimes "grooving" is observed, in which two scans along x with slightly different y values are drastically different. This may occur when the tip has been trapped in one row of an atomic lattice, and then suddenly slips to the next row. Such effects occur by a mechanism analogous to the stick-slip effects observed in the friction studies described above. It is likely that lateral forces affect AFM topographs more often than is realized. Routinely recording images during both forward and reverse scans might help resolve this issue, but only for non-reversible (non-conservative) lateral forces. Because AFMs based on laser deflection can be modified very easily to measure both force components, many groups will soon record both normal and lateral forces while imaging. In fact a commercial AFM is now available with simultaneous normal and lateral force-sensing capability. As Meyer *et al.* has emphasized in this volume, frictional forces may in some cases be a more sensitive structural probe than normal forces.

In a natural extension of classic scanning electron microscope studies of wear, force microscopes have been used to probe wear mechanisms by imaging surfaces after sliding. The AFM tip can be used both as the "pin" in a miniature "pin on flat" experiment and as the imaging tool. Blackman *et al.* [34] have used non-destructive attractive tip-surface forces to image Langmuir-Blodgett (LB) films of cadmium arachidate on which a 1000 Å radius tungsten tip has been slid at 10^{-7} N. They found a monolayer to be undamaged by sliding, although it was swept clean of "molecular debris." However when a thicker film of five layers was used, the first four layers were easily worn away, leaving the last layer tenaciously bound to the surface. Albrecht [32] has described similar studies on self-assembled films, which show that the wear can be eliminated by chain polymerization (cross-linking) at the base. These results correlate well with more macroscopic studies, which suggest that one layer provides all the lubricating action [35]. During AFM imaging the layered dry lubricant material $NbSe_2$, Lieber and Kim have observed the growth of wear "pits" only one layer deep and several nanometers in diameter over periods of minutes [36].

Despite the recent excitement over tunneling and force microscopes, field ion microscopy (FIM), the first technique to image surfaces with atomic resolution [37] should not be forgotten. In this instrument, a large positive voltage is applied to a sharp 100 nm radius conductive tip. The resultant strong field at the surface atoms ionizes the surrounding imaging gas, and the positive ions are drawn to a nearby fluorescent screen, giving a direct projection image of the tip [38]. A unique feature of the FIM is that it provides a method for preparing a perfect tip in a vacuum by field evaporation, heating, thermal evaporation, ion bombardment and reactive etching. By sequential field evaporation of atomic layers, the tip structure can be profiled with atomic resolution, and even the chemical identity of individual atoms can be determined. Well before the advent of the STM and AFM, several groups used the FIM tip as a model system for single asperity surface contact [38,39,40]. In a study of contact of tungsten and platinum tips touched to platinum surfaces at loads of $\simeq 10^{-6}$ N, Walko found that tip damage extends to a depth of 0.5-1.5 nm [40]. By determining as a function of load the area over which atomic-level damage occurs, it was determined that the contact followed Hertz's equations for elastic contact. With the recently developed STM and AFM technology, the opportunities for applying the FIM to study nm-scale frictional events are evident. In such studies, an AFM tip would be imaged by FIM, slid across a surface while monitoring the force, and then imaged again to detect any changes. The combination of FIM with scanning probe techniques was first advocated by Fink, who demonstrated that monatomic and triatomic tips are easily fabricated [41].

The very careful surface force apparatus experiments reviewed elsewhere in this volume have provided a comprehensive picture of the mechanical properties of ultra-thin liquid films. Scanning probe techniques can complement this picture by providing structural information. For example, Mate and his coworkers have used an AFM to profile the distribution of perfluoropolyether polymer liquid films as thin as 2 nm on silicon and gold surfaces [42]. As the AFM tip approaches the surface, contact with the liquid film is signalled by the sudden onset of an attractive meniscus force, while the later onset of repulsive force indicates contact with the underlying solid surface. By recording the height difference between these two measurements, the film thickness can be profiled, and a transition between wetting for thin films and dewetting for thick films is observed. During pulloff of the tip from the surface, the distance from the solid surface at which the liquid film breaks away from the surface determines the disjoining pressure of the film, which is a measure of the film's tendency to thin or thicken [43]. This point can occur tens of nm from the surface, but if the lubricant is bonded to the surface by heating to prevent flow along the surface, the attractive force extends only several nm above the surface [44]. This group has also studied the penetration of the tip through one or more LB layers, a process which may occur gradually or suddenly [44]. A wide variety of adhesive tip-surface interactions are observable for different materials [45].

Only a decade ago, it was unimaginable that lubricant layers between solid surfaces could be imaged at atomic resolution, but now STM imaging of organic layers is routine; the STM can simply penetrate through a liquid to the other surface. A beautiful study of the dynamics of ordered monolayers of didodecylbenzene adsorbed from a solution on graphite shows that the motion of grain boundaries on a 100 ms time scale is associated with diffusing free volume [46]. By fabricating a tip which just penetrates through an ultra-smooth solid surface, it seems quite possible to image during sliding lubricant layers of the type studied in the surface force apparatus.

Because force microscopes are capable of measuring charges as small as an elementary charge [47], they are an ideal tools for probing tribocharging, the process by which charge is transferred between two surfaces during the repeated asperity contact caused by sliding. This process is understood to involve equilibration of the Fermi levels during metal contact, but its mechanism during insulator contact is not understood. By AFM imaging, Terris et al. find that when a 0.1 μm radius Ni tip is touched to a polymethyl methacrylate surface, a charged region 10 μm in diameter is formed containing both positive and negative charges, which are stable on the surface for days before decaying away [48]. Further work found that the charge transferred at a single spot is independent of the number of contacts and depends on the bias voltage of the tip [49].

5. Discussion

The potential of tip-based mechanics experiments is most dramatically illustrated by the recent work of Eigler and Schweizer [50], who used a cryogenic STM to precisely and reproducibly position Xe atoms on particular lattice sites of a single-crystal nickel surface. By any standard, this experiment is a milestone of modern physical science. In a sense it is a wear experiment, because material (individual Xe atoms) were removed from particular areas on the surface. Although forces were not measured in this particular experiment, such measurements are quite possible.

What is surely most lacking in AFM measurements to date is characterization of the tip. While for topographic imaging it is sufficient to somehow simply generate a sharp tip, in order to quantitatively compare measured forces with theory, an experiment must be well defined. Ironically, more traditional experiments with larger tips and loads suffer much less from this problem, because the contact mechanics may not depend critically on atomic-scale surface roughness and chemical integrity. From this standpoint, surface force apparatus experiments, which use extended atomically flat mica surfaces, have a substantial advantage over the present state of force microscopy. As discussed in section 4, complete characterization of AFM tips by field ion microscopy would be a very valuable procedure.

ACKNOWLEDGEMENT. We have enjoyed collaborating with S. R. Cohen, S. Chiang, R. Erlandsson, G. J. Germann, C. M. Mate, and G. Neubauer. We thank Farid Abraham for interesting us in alkane films, and for much help and advice in setting up the molecular dynamics calculations. This work was partially supported by the Office of Naval Research contract N00014-88-C-0419 and the Air Force Office on Scientific Research contract F49620-89-C-0068.

REFERENCES

1. G. M. McClelland, in *Adhesion and Friction*, M. Grunze and H. J. Kreuzer, eds. Springer Series in Surface Science **17** (Springer Verlag, Berlin, 1990), p. 1.

2. G. A. Tomlinson, *Phil. Mag. Series 7* **7**, 905 (1929).

3. J. Skinner, "An Atomic Model of Friction," unpublished manuscript, Central Electricity Generating Board Report number RD/B/N3137, Berkeley Nuclear Laboratories (1974).

4. see, for example H. Goldstein, *Classical Mechanics*, 2nd ed. (Addison-Wesley, Reading, 1980).

5. M. Hirano and K. Shinjo, *Phys. Rev. B* **41**, 11837 (1990).

6. Y. I. Frenkel and T. Kontorova, *Zh. Eksp. Teor. Fiz.* **8**, 1340 (1938).

7. F. C. Frank and J. H.van der Mewe, *Proc. R. Soc.* **198**, 205, 216 (1949).

8. for a review, see P. Bak, *Rep. Prog. Phys.* **45**, 587 (1982).

9. J. B. Sokoloff, *Surf. Sci.* **144**, 267 (1984); Phys. Rev. Lett. **66**, 965 (1991); *Phys. Rev. B* **42**, 760 (1990);

10. S. Aubry, *Ferroelectrics* **24**, 53 (1980); M. Peyrard and S. Aubry: *J. Phys.* **C16**, 1593 (1983).

11. J. N. Glosli and G. M. McClelland, to be published.

12. D. Maugis, G. Desalos-Andarelli, A. Heurtel, and R. Courtel, *ASLE Trans.* **21**,1 (1976); R. Feder and P. Chaudhari, *Wear* **19**, 109 (1972); G. Andarelli, D. Maugis and R. Courtel, *Wear* **23**, 21 (1973); N. Gane and J. Skinner *Wear* **25**,381 (1973); G. M. Pharr and W. C. Oliver *J. Mater. Res.* **4**, 94 (1989); N. Gane and F.P. Bowden, *Jour. Appl. Phys.* **39**, 1432 (1968); N. Gane, *Proc. Roy. Soc. Lond. A* **317**, 367 (1970); N. Gane and J.M. Cox, *Phil. Mag.* **22**, 881 (1970);

13. Q. Guo, J.D.J. Ross, and H.M. Pollock, *Mat. Res. Soc. Symp. Proc.* **140**, 51 (1989); M.D. Pashley, J.B. Pethica, and D. Tabor, *Wear* **100**,7 (1984); D. Maugis and H.M. Pollock, *Acta Metallurgica* **32**, 1323 (1984).

14. G. Binnig, C. F. Quate, and Ch. Gerber, *Phys. Rev. Lett.* **56**, 930 (1986).

15. for a review see D. Rugar and P. Hansma, *Phys. Today* **43**, 23 (1990).

16. G. Neubauer, S.R. Cohen, G.M. McClelland, D.E. Horn, and C.M. Mate, and C. M. Mate, Rev. Sci. Instrum. **61** 2296 (1990).

422

17.   O. Marti, J. Colchero and J. Mlynek, *Nanotechnology* **1**, 141 (1990).

18.   G. Meyer and N. M. Amer, *Appl. Phys. Lett.* **57**, 2089 (1990).

19.   J.B. Pethica, *Phys. Rev. Lett.* **57**,3235 (1986).

20.   D. Rugar, H.J. Mamin, R. Erlandsson, J.E. Stern, and B.D. Terris, *Rev. Sci. Inst.* **59**, 2337 (1988); D. Rugar H. J. Mamin and P. Günther, *Appl. Phys. Lett.* **55**, 2588 (1989).

21.   C.M. Mate, G.M. McClelland, R. Erlandsson, and S. Chiang, *Phys. Rev. Lett.* **59**, 1942 (1987).

22.   J. A. Greenwood and J. H. Tripp, *J. Appl. Mech.* March 1979 p. 153; J. A. Greenwood, *J. Lubrication Tech.* Jan. 1967, p. 81; J. A. Greenwood and J. B. P. Williamson, *Proc. Roy. Soc. Lond.* **295**, 300 (1966).

23.   R. Erlandsson, G. Hadziioannou, C.M. Mate, G.M. McClelland, and S. Chiang, *J. Chem. Phys.* **89**, 5190 (1988).

24.   G. J. Germann, G. M. McClelland, G. Neubauer, S. R. Cohen, H. Seki, and Y. Mitsuda, to be published.

25.   S.R. Cohen, G. Neubauer, and G.M. McClelland, *J. Vac. Sci. Technol. A* **8**, 3449 (1990).

26.   R. Kaneko, K. Nonaka, and K. Yasuda, *J. Vac. Sci. Technol. A* **6** 291 (1988); R. Kaneko, *J. Microscopy* **152**, 363 (1988).

27.   G. Binnig, H. Rohrer, Ch. Gerber, and E. Weibel, *Phys. Rev. Lett.* **50**, 120 (1983).

28.   *J. Microscopy* **152** (1989); *J. Vac. Sci. Tech. A* **8**, no. 1 (1990); *J. Vac. Sci Tech. B* **9**, No. 2, Part 2 (1991).

29.   E. Meyer, H. Heinzelmann, P. Grütter, Th. Jung, H.-R. Hidber, H. Rudin, and H.-J. Güntherodt, *Thin Solid Films*, **181**, 527 (1989).

30.   B. Bhushan, and G. S. Blackman, *Trans. ASME J. Tribology*, to be published.

31.   H.J. Mamin, E. Ganz, D.W. Abraham, R.E. Thomson, and J. Clarke, *Phys. Rev. B* **34**, 9015 (1986).

32.   T.R. Albrecht, Ph.D. Dissertation, Stanford University, (1989).

33.   E. Meyer, H. Heinzelmann, D. Brodbeck, G. Overney, R. Overney, L. Howald, H. Hug, T. Jung, H. R. Hidber, and H.-J. Güntherodt, *J. Vac. Sci Technol. B* **9**, 1329 (1991).

34.   G. S. Blackman, C. M. Mate, and M. R. Philpott, *Vacuum* **41**, 1283 (1990).

35.   V. Novotny, J. D. Swalen, and J. P. Rabe, *Langmuir* **5**, 485 (1989); V. DePalma and N. Tillman, *Langmuir* **5**, 868 (1989); O. Levine and W.A. Zisman, *J. Phys. Chem.* **61**,1068 (1957).

36.   C. M. Lieber and Y. Kim, to be published.

37.   E.W. Müller, *Z. Physik* **131**, 136 (1951).

38.   E.W. Müller and T.T. Tsong, *Field Ion Microscopy* (Elsevier, New York, 1969).

39.   D. H. Buckley, *Surface Effects in Adhesion, Friction, Wear, and Lubrication,* (Elsevier, Amsterdam, 1981).

40.   R. J. Walko, *Surface Sci.* **70**, 302 (1978).

41.   H.-W. Fink, *IBM J. Res. Develop.* **30**, 460 (1986); *Physica Scripta* **38**, 260 (1988).

42.   C. M. Mate, M. R. Lorenz, and V. J. Novotny, *J. Chem. Phys.* **90**, 7550 (1989).

43.   C. M. Mate and V. J. Novotny, *J. Chem. Phys.* **94**, 8420 (1991).

44.   G. S. Blackman, C. M. Mate, and M. R. Philpott, *Phys. Rev. Lett.* **65**, 2270 (1990).

45.   N. A. Burnham, D. D. Dominguez, R. L. Mowery, and R. J. Colton, *Phys. Rev. Lett.* **64**, 1931 (1990).

46.   J. P. Rabe and S. Bucholz, *Phys. Rev. Lett.* **66**, 2096 (1991).

47.   C. Schönenberger and S. F. Alvarado, *Phys. Rev. Lett.* **65**, 3162 (1990).

48.   B. D. Terris, J .E. Stern, D. Rugar, and H. J. Mamin, *Phys. Rev. Lett.* **63**, 2669 (1989);

49.   F. Saurenbach and B. D. Terris, *Trans. IEEE-IAS*, Jan/Feb. 1992, to be published.

50.   D. M. Eigler and E. K. Schweizer, *Nature* **344**, 524 (1990).

Discussion *following the lecture by G M McClelland on "Friction at the atomic scale"*

M O ROBBINS. Models of dissipation in other systems including charge-density-wave conduction, contact-line motion and flux flow may teach us something about friction. For example, the Tomlinson model arises in a variety of contexts including Fisher's mean field theory of charge-density wave conduction and my work with Joanny on contact-line motion. One finds a number of results which you didn't mention which may be relevant. The first is that displacement at constant force is never adiabatic - it always produces "pops". (This is not surprising if one thinks of constant force as the weak spring limit.) The second is that the variation of force with velocity at low velocities is different for constant force, weak springs and strong springs. In fact one finds different power laws relating the change in force to velocity. It would be interesting if these conclusions could be tested in your experiments. Any failure would help to devise improved models.

G M McCLELLAND. These are all interesting issues. The typical velocity used in the AFM experiments is about 10^{-6} cm/s. For most systems, it may not be possible to move fast enough to observe a velocity dependence. [For further comments, see page.]

J B SOKOLOFF. (Physics Department, Northeastern University, Boston, MA 02115, USA): written contribution entitled

Theory of atomic level sliding friction

In reference [1], one of the simplest models for friction between perfect crystalline surfaces was studied analytically. The model consists of a harmonic solid, one of whose sides interacts with a sliding periodic potential; the side opposite this side is held stationary. It is illustrated in figure 1. The equations of motion for $u_{\vec{R}}$, the displacement of the atom at point $\vec{R}$ along the direction of sliding in this model, are then

$$m\ddot{u}_{\vec{R}} = -\alpha(4u_{\vec{R}} - \sum_{\vec{a}_1} u_{\vec{R}+\vec{a}_1}) - \beta(2u_{\vec{R}} - \sum_{\vec{a}_2} u_{\vec{R}+\vec{a}_2}) - m\gamma\dot{u}_{\vec{R}}$$
$$-\delta_{Z,Nc}\lambda_0 sin[(2\pi/a)(x_{\vec{R}}^0 + vt + u_{\vec{R}})] \tag{1}$$

where β and $\vec{a_2}$ are the force constant and a nearest neighbour distance along the c-axis and α and $\vec{a_1}$ are the force constant and nearest neighbour distance perpendicular to the c-axis, $\vec{R} = (X, Y, Z)$ (where $X = n_1 a$, $Y = n_2 a$, and $Z = n_3 c$ where c and a are the lattice constants along and perpendicular to the c-axis respectively and n_1, n_2 and n_3 are integers), Nc (where N is an integer) is the Z value for the plane of atoms in contact with the sinusoidal potential (i.e. at the slip plane), $\lambda_0 \sin[(2\pi/a)(x^0_{\vec{R}} + vt + u_{\vec{R}})]$ is the periodic potential, $x^0_{\vec{R}}$ is the initial position of the atom at point $\vec{R}$, and v is the sliding velocity. The term $-m\gamma \dot{u}_{\vec{R}}$ represents the damping of the vibrational modes. Following reference [1], this equation may be formally solved for $u_{\vec{R}}$ in terms of the sinusoidal potential using the Green's function for equation (1) and the average force of friction F_{av} determined by setting the rate of work done on the crystal by the sinusoidal potential equal to $F_{av} v$. F_{av} was determined at high speeds using lowest order perturbation theory, which means that we neglect the term proportional to $u_{\vec{R}}$ in the argument of the potential, in which case the model reduces to a collection of harmonically driven harmonic oscillators, which can be solved analytically[1]. The speed required for this approximation to be valid can be estimated by the method itself by calculating the root mean square value of $u_{\vec{R}}$ (i.e. the time average value of $u^2_{\vec{R}}$) and determining whether or not it is much smaller than a lattice spacing. Second, equation (1) will be integrated numerically.

When equation (1) was solved numerically with $\gamma = 0$, the force that must be applied to the sinusoidal potential to slide it at uniform speed v was found to oscillate between positive and negative values comparable in magnitude to λ_0, averaging to a value of F_{av} much smaller than λ_0 [2], because a harmonic oscillator in a periodic driving force will generally not absorb from the driving force. In order for there to be friction, γ must have a value comparable to the phonon mode spacing. When such a value of γ is used in the simulations, we obtain excellent agreement with the perturbation theory result for F_{av} for values of v for which the root mean square atomic displacement is small compared to a lattice constant.

Perturbation theory calculations done in reference [1] for a sinusoidal potential incommensurate with the crystal gives a value which is a factor of 10^{-14} smaller than the value of σ_{fric} found in the last section for the commensurate case for $v = 0.03v_p$ if we take $\gamma = 0.1vQ$ and Q, the difference between the periods of the crystal and the potential, of the order of $0.1(2\pi/a)$.

As we have seen in the last two sections, without defects, the force of friction between two ideal crystalline surfaces will be either much too large or too small. The value of σ_{fric} calculated for the case of a potential commensurate with the crystal containing a concentration of one dislocation line for every 10,000 atomic distances along the direction of sliding at the interface (needed to make the Peierls stress for this case comparable to experimentally observed values) is of the order of $10^7 \, N/m^2$ at $v = 1 \, cm/s$, which is comparable to the value observed for MoS_2 [2].

References

1. J B Sokoloff, Phys. Rev. B42 (1990) 760.

2. I L Singer, R N Bolster, J Wegand, S Fayeulle and B C Stupp, Appl. Phys. Lett. 57 (1990) 995.

Figure

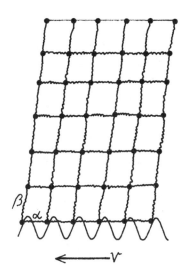

1. The model for sliding friction which is studied in this contribution is illustrated in this figure. Black dots are atoms, and the curly lines that connect them indicate that they interact harmonically. The vertical and horizontal force constants β and α are labelled. The atoms in the top layer are connected by solid lines to indicate that they are held fixed in place. The sinusoidal potential moves to the left with speed v, as indicated.

FRICTION FORCE MICROSCOPY OF LANGMUIR-BLODGETT FILMS

E. Meyer, R. Overney, L. Howald, D. Brodbeck,
R. Lüthi and H.-J. Güntherodt
Institut für Physik, Universität Basel,
Klingelbergstrasse 82,
4056 Basel, Switzerland

ABSTRACT. In friction force microscopy both lateral and normal deflections of a cantilever-type spring are simultaneously monitored. Measurements on microfabricated structures and on mica demonstrate the performance of the instrument on the micron scale and on the atomic scale. Langmuir-Blodgett films are well-known for their exceptional tribological properties. Multilayers of Cd-arachidate are investigated. Lubrication is observed on the microscopic scale.

1. Introduction

The inventions of the scanning tunneling microscope (STM) [1] and the atomic force microscope (AFM) [2, 3] have stimulated the development of a variety of scanning probe microscopes. In all these microscopes a fine probing tip is scanned in x- and y-directions along the sample surface. Scales from a few hundred microns down to the atomic scale can be covered by the use of piezo-electric transducers. During the scan process the specific interactions (tunneling current in STM, force between probing tip and sample in AFM) are used as input signals for the feedback loop which controls the z-motion of the probing tip or sample. The signals can be digitized as a function of the lateral position and images are displayed by a computer.

The friction force microscope (FFM) [4] is one of the latest developments of scanning probe microscopes. Here, both the lateral and the normal deflections of a cantilever-type spring (called "lever") are simultaneously measured. Similar to the AFM the normal deflection of the lever is used as input parameter for the feedback-loop. The technique is rather versatile and offers many different operation modes. Instead of measuring in the equiforce mode (force is kept constant by the feedback loop) the sample can be scanned at constant height and variable deflection images are created. This mode is often used for atomic scale images. Furthermore "spectroscopic" modes, where force-distance curves [9] are performed at different places, might give additional information.

In this contribution we present the realization of a FFM which is based on laser beam deflection [5, 6], where the lateral force is measured by the torsion of the lever. Other designs based upon optical interferometry [4] and capacitance methods [7] are

I. L. Singer and H. M. Pollock (eds.), Fundamentals of Friction: Macroscopic and Microscopic Processes, 427–436.
© 1992 *Kluwer Academic Publishers. Printed in the Netherlands.*

shortly discussed.

First results on mica and on magneto optical discs (MOD) demonstrate the performance of the instrument on the atomic scale as well as at the micron scale. A more systematic study on Langmuir-Blodgett (LB) films of Cd-arachidate shows the potential of this technique. Comparison between the arrangement of the molecules on 2- and 4-layer films allow some conclusions about the inter- and intralayer interactions. By increasing parameters such as the load and the scan speed the onset of plastic deformation is observed. Comparison between the friction image and the topography reveals that the uncovered parts of the substrate give rise to larger frictional forces than the covered ones. This is a demonstration of lubrication on the microscopic scale. In the friction image small features are found, which are not visible in the topography. These features are interpreted as chemical inhomogeneities in the film which are not observable by topographical measurements.

2. Experimental

Up to now two different designs of FFM have been introduced: The design used in our laboratory is based upon laser beam deflection [5, 6]. A laser beam is reflected off the rear side of a microfabricated cantilever. The deflection of the reflected beam is then measured by a quadrant detector. The normal deflection of the lever causes a difference between the upper and the lower segments of the photodiode whereas the torsion of the lever causes a difference between the left and the right segments.

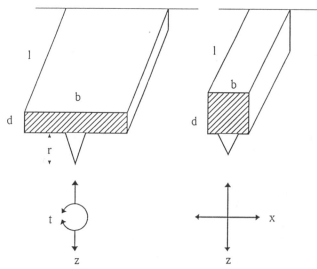

Fig. 1. Perspective view of two different levers, representing two different designs of FFM. On the left: Simultaneous measuring of torsion and z–bending by beam deflection. On the right: x– and z–deflection are measured simultaneously but separately by two sensors.

The spring constant of a rectangular beam for the z-bending is given by

$$c_z = \frac{Ebd^3}{4l^3} \tag{1}$$

and the spring constant for the torsion is approximated by

$$c_t = \frac{Gbd^3}{3lr^2} \tag{2}$$

where E is the Young's modulus, $G = \frac{E}{2(1+\sigma)}$ the shear modulus and σ the poisson ratio. In order to be sensitive to torsion the thickness d can be reduced and the tip length r increased (cf. Fig. 1). An alternative design needs two separated deflection sensors for the x- and z-direction. The additional spring constant for the x-deflection is given by

$$c_x = \frac{Eb^3d}{4l^3} \tag{3}$$

As deflection sensors optical interferometry [4] and capacitance methods [7] have been implemented. Another solution is to use two tunneling microscopes for each direction. For these designs cantilevers with quadratic or spherical cross section are best suited.

Spring constants from commercially available microfabricated cantilevers are given in table 1. The Si_3N_4 levers have a typical width-thickness ratio of 10 to 30 which results in 100 to 1000 times stiffer spring constants in the lateral direction compared to the normal direction. Therefore these levers are well suited for torsion. An angle resolution of 10^{-7}rad is achieved with our set-up and lateral forces on the order of 10^{-10}N can be measured in the lateral direction. A critical aspect of the FFM technique is the coupling between the lateral and normal forces. Careful alignment of the detectors and the lever minimizes this cross-talk.

Table 1. Spring constants of typical rectangular Si_3N_4-levers. Due to the high stiffness in the lateral direction, the levers are suited for torsion.

Lever $(l \times b \times d)\ \mu m$	Si_3N_4 N/m		
	c_z	c_x	c_t
$100 \times 10 \times 0.6$	0.21	58	114
$100 \times 20 \times 0.6$	0.41	462	228
$200 \times 20 \times 0.6$	0.05	58	114

3. Results and Discussion

On mica atomic scale protrusions are found with a spacing of 5.2Å which corresponds well to the spacing between the six-fold rings of the SiO_4-tetrahedra. The variation of the frictional forces are found to be about 10^{-9}N. As seen in Fig. 2 the contrast in the friction image is even better than in the topography. These results confirm previous FFM-measurements [6, 8]. It is also in agreement with previous AFM measurements, where it has been found that atomic scale features on layered materials such as graphite or mica are often dominated by frictional forces [9, 10]. It has been concluded that the contact is not formed by a single atom. Multiple tip imaging occurs amplifying the variations of the friction whereas the corrugation of normal forces remains approximately constant.

Fig. 2. FFM of mica. On the left side the image of the normal deflection (topography) and on the right side the image of the lateral deflection (friction) are displayed. The spacing between the protrusions is 5.2Å. The variations of the frictional force are about 10^{-9}N. The load is 10^{-8}N.

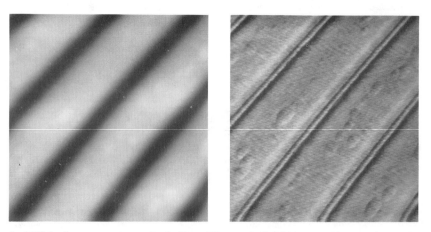

Fig. 3. FFM of a magneto optical disc. The period of the tracks is 1.6μm. A strong resemblance to the derivative of the topography is found in the friction image.

In order to test the microscope on a micron scale, measurements have been performed on magneto optical discs (MOD). The spacing between the tracks is 1.6µm. The structure is well resolved on both the friction and the topography image (cf. Fig. 3). In the friction image small hills are easily observed which can be hardly seen in the topography. The lateral component of the force can be used to increase the contrast of topographical information. The lateral forces are found to be large in areas where the component of the local gradient along the scan direction is large. Therefore the friction image represents a derivative of the topography. On the MOD no features could be found which are not related to the topography.

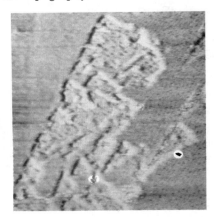

Fig. 4. FFM of a 2-layer Cd-arachidate film $(6.0 \times 6.0 \mu m^2)$. The steps are monomolecular (27Å) and are preferentially oriented in the $\langle 1\bar{2}1 \rangle$ directions of the Si(111) wafer. At the step edges the strongest variations of the lateral force occur.

In the second part results on Cd-arachidate films prepared by the LB-technique are presented. Ultrathin soap films are known to be good boundary lubricants. One monolayer can reduce the friction on metals by a factor of 10 and the wear by a factor of 10000 or more [11]. Previous AFM measurements have revealed the molecular structure of LB-films [12, 13]. Here, a more systematic study of Cd-arachidate is presented [14]. Comparisons between the topography of 2-layer and 4-layer films show characteristic differences.

On the 2-layer films we mainly observe monolayer steps (27Å) whereas on the 4-layer films predominantly bilayer steps (54Å) are observed. We conclude that the interaction between the hydrophobic tails is smaller than the interaction between the hydrophilic heads. Furthermore the interaction between the substrate and the first layer seems to be even stronger, stabilizing the monolayer. This conclusion is also supported by our intentional scratching on the surface.

The lateral arrangement of 4-layer films shows no preferential orientation. The steps are curved, which indicates that a liquid-like condensation of particles occurs during the formation of the films. As seen in Fig. 4 the steps on the 2-layer films are found to be oriented in the $\langle 1\bar{2}1 \rangle$ directions of the Si(111) wafer demonstrating again the strong interaction between the substrate and the film. Similar arrangements have been observed by LEEM-images [15]. There it has been found that the 7×7 reconstruction preferentially starts to grow at the upper side of steps. The steps themselves are oriented in the $\langle 1\bar{2}1 \rangle$ directions. A similar mechanism seems to be effective in the case of LB-films. On the unreconstructed Si(111)1×1 surface the molecules align along the steps forming the observed boundaries.

Fig. 5. AFM of a 4-layer Cd-arachidate film ($2.5 \times 2.5 \mu m^2$). A rectangular hole has been created by increasing the load on the order of 10^{-7}N and the scan speed to 5μm/sec. Damage was found to start at the pores. In the vicinity of the hole the density of pores is decreased.

The films were found to be stable with forces of 10^{-8} to 10^{-7}N. Assuming an approximate contact area of 10^{-16}m^2 (as estimated from the resolution at step edges and pores) a contact pressure of 100 to 1000 MPa is calculated. Surface force apparatus measurements [16] have shown that LB-films can withstand pressures of 100 to 400 MPa [11, 17, 18]. By simultaneously increasing the force and the scan speed (0.5μm/sec to 5μm/sec) the films can be damaged. It is found that the damage often starts at imperfections like pores or steps and then extends over the whole scan area. Afterwards the rectangular holes, which simply reflect the scan area during plastic deformation, could be imaged with smaller force and appropriate scan speed (cf. Fig. 5). In the surroundings of the holes the density of pores is found to be decreased. This indicates that pores play an important role in the response of the films to mechanical stress. During the application of pressure the pores fill up. If several holes are created in close vicinity, coalescence of the holes is observed on a time scale of hours.

Detailed comparison between the topography and friction images reveal certain interesting aspects. In Fig. 6a an area is shown where several bilayer steps are seen. The uncovered substrate is on the lowest level. Such places, where the bare substrate appears are rarely observed on the films. Figure 6a is therefore a rather atypical image for the films. However, we learn from this example that the friction on the substrate is increased compared to the friction on the covered parts by an order of magnitude. The difference between the higher levels (2-,4-,6-layer) is found to be less pronounced. These measurements are the first observation of lubrication on the microscopic scale which show clearly the contrast between friction with and without a boundary lubricant. In Fig. 6b features are visible (dark stripes running from the upper right to the lower left) which are not observable in the topography. We interpret these features as inhomogeneities of the films (small inclusions of water or organic contaminants).

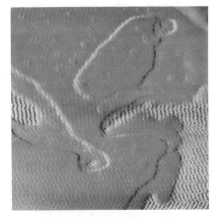

Fig. 6. FFM images of a 4-layer Cd-arachidate film ($3.0 \times 3.0 \mu m^2$).
(a) Several bilayer steps can be observed in the topography. In the friction image the difference between higher levels (2-layer, 4-layer and 6-layer) are small. However, the difference between the higher levels and the lowest level (substrate) is significant. On the higher, covered levels the variations of the frictional forces are about 10^{-9}N whereas on the substrate the variations are increased to about 10^{-8}N.

(b) In the friction image features are observable (dark stripes running from upper right to the lower left) which are not seen in the topography. The features are interpreted as chemical inhomogeneities.

Therefore FFM might even have the potential to reveal chemical inhomogeneities in ultrathin films. Figure 7 shows an atomic scale image of the 4-layer films. The separation between the protrusion is ≈ 5Å, which is in agreement with the intermolecular spacing. The molecular structure is also observable in the friction. However, the contrast of the friction is not as strong (compared to the topography) as in the case of mica which again might be related to lubrication of the LB-film.

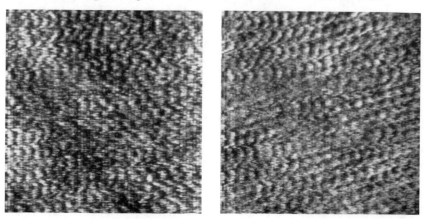

Fig. 7. FFM of a 4-layer Cd-arachidate film. The protrusions are 5Å apart which corresponds well to the intermolecular distances. The load is approximately $5 \cdot 10^{-9}$ N.

4. Summary

In summary, it has been shown that the FFM technique is a new generation of the family of scanning probe microscopes which can provide useful information about tribological properties. The results on mica and on MOD have shown the good performance of the instrument. The results on the LB-films have demonstrated the versatility of the instrument. From topographical images and nanoindentations conclusions about the inter- and intralayer interactions could be drawn. The comparison between topography and friction has revealed lubrication on the microscopic scale. Furthermore features in the friction image are observed which are not visible in the topography. These features are interpreted as chemical inhomogeneities. We think that this new technique will give impacts on the field of tribology (e.g. boundary lubrication).

5. Acknowledgements

We wish to thank T. Wagner, H. Schier, S. Roth for many stimulating discussions and the preparation of samples. This work was supported by the Swiss National Science Foundation and the Kommission zur Förderung der wissenschaftlichen Forschung.

References

[1] Binnig, G., Rohrer, H., Gerber, C. and Weibel, E. (1982) 'Surface Studies by Scanning Tunneling Microscopy', *Phys. Rev. Lett.* **49**, 57-60.

[2] Binnig, G., Quate, C. F., and Gerber C. (1986) 'Atomic force microscopy', *Phys. Rev. Lett.* **56**, 930-933.

[3] For recent reviews see e.g.:
Rugar, D. and Hansma, P. (1990), 'Atomic force microscopy', *Physics Today* **43**, 23-30.
Meyer, E. and Frommer, J. (1991), 'Forcing surface issues', *Physics World* **4**, 46-49.

[4] Mate, C. M., McClelland, G. M., Erlandsson, R. and Chiang, S. (1987), 'Atomic-scale friction of a tungsten tip on a graphite surface', *Phys. Rev. Lett.* **59**, 1942-1945.

[5] Meyer, G. and Amer, N. M. (1990), 'Simultaneous measurement of lateral and normal forces with an optical-beam-deflection atomic force microscope', *Appl. Phys. Lett.* **57**, 2089-2091.

[6] Marti, O., Colchero, J. and Mlynek, J. (1990), 'Combined scanning force and friction microscopy of mica', *Nanotechnology* **1**, 141-144.

[7] Neubauer, G., Cohen, S. R., McClelland, G. M., Horne, D. and Mate, C. M. (1990), 'Force microscopy with a bidirectional capacitance sensor', *Rev. Sci. Instr.* **61**, 2296-2308.

[8] Erlandsson, R., Chiang, S., McClelland, G. M., Mate, C. M. and Hadziioannou, G. (1988), 'Atomic scale friction between the muscovite mica cleavage plane and a tungsten tip', *J. Chem. Phys.* **89**, 5190-5193.

[9] Meyer, E., Heinzelmann, H., Grütter, P., Jung, T., Weisskopf, T., Hidber, H.-R., Lapka, R., Rudin, H. and Güntherodt, H.-J. (1988), 'Comparative study of lithium fluoride and graphite by atomic force microscopy', *J. of Microscopy* **152**, 269-280.

[10] Meyer, E., Heinzelmann, H., Brodbeck, D., Overney, G., Overney, R., Howald, L., Hug, H., Jung, T., Hidber, H.-R. and Güntherodt, H.-J. (1991), 'Atomic resolution on the surface of LiF(100) by atomic force microscopy', *J. Vac. Sci. Techn. B* **9**, 1329-1332.

[11] Briscoe, B. J. and Evans, D. C. B. (1981), 'The shear properties of Langmuir-Blodgett layers' *Proc. Roy. Soc. A* **380**, 389-407.

[12] Meyer, E., Howald, L., Overney, R. M., Heinzelmann, H., Frommer, J., Güntherodt, H.-J., Wagner, T., Schier, H. and Roth, S. (1991), 'Molecular-resolution images of Langmuir-Blodgett films using atomic force microscopy', *Nature* **349**, 398-399.

436

[13] Weisenhorn, A. L., Drake, B., Prater, C. B., Gould, S. A. C., Hansma, P. K., Ohnesorge, F., Egger, M., Heyn, S.-P. and Gaub, H. E. (1990), 'Immobilized proteins in buffer imaged at molecular resolution by atomic force microscopy', *Biophys. J.* **58**, 1251-1258.

[14] For a description of the preparation see : Schreck, M., Schmeisser, D., Göpel, W., Schier, H., Habermeier, H. U., Roth, S. and Dulog, L. (1989), 'Interaction of metals with cadmium arachidate Langmuir-Blodgett films studied by x-ray photoelectron spectroscopy', *Thin Solid Films* **175**, 95-101.

[15] Telieps, W. and Bauer, E. (1985), 'The $(7\times7)\leftrightarrow(1\times1)$' phase transition on Si(111)', *Surface Science* **162**, 163-168.

[16] Israelachvili, J. N. (1985), *Intermolecular and Surface Forces*, Academic Press.

[17] Israelachvili, J.N. and Tabor, D. (1973), 'The shear properties of molecular films', *Wear* **24**, 386-390.

[18] Bailey, A. I. and Courtney-Pratt, J. S. (1954), 'The area of real contact and the shear strength of monomolecular layers of a boundary lubricant', *Proc. Roy. Soc. A* **227**, 500-515.

COMPUTATIONAL TECHNIQUES IN TRIBOLOGY AND MATERIAL SCIENCE AT THE ATOMIC LEVEL

J. FERRANTE
N.A.S.A. Lewis Res. Ctr.
MS 5-9
21000 Brookpark Rd.
Cleveland, OH 44135
U.S.A.

and

G. BOZZOLO
Analex Corp. and N.A.S.A. Lewis Res. Ctr.
MS 5-9
21000 Brookpark Rd.
Cleveland, OH 44135
U.S.A.

ABSTRACT. Computations in tribology and material science at the atomic level present considerable difficulties. In this paper computational techniques ranging from first-principles to semi-empirical and their limitations are discussed. Example calculations of metallic surface energies using semi-empirical techniques are presented. Finally, application of the methods to calculation of adhesion and friction are presented.

I. Introduction

Computational material science or the theory of materials has recently come of age. Calculation of properties of real materials at the atomic level such as grain boundary or dislocation energies or the dynamics thereof, which in the recent past have seemed intractable, now have some hope for realistic modeling. An even more startling assertion is that modeling of tribological phenomena is also now feasible. Tribology is a particularly difficult field for theoretical studies at the atomic level because of the different possible materials in contact under high loads often with high degrees of disorder, but as Landman's etal have shown (paper in this proceedings) a great deal of progress has been made in approaching problems of practical interest. This paper is concerned with developments which have enabled this progress. We concentrate on two methods for determining energetics, the embedded atom method (EAM) and equivalent crystal theory (ECT). We concentrate on EAM because it is the method of choice at present and has successfully been used to treat many problems concerning defect energetics. We include ECT, because it is a new method which has the capability of treating a wider class of materials with good quantitative agreement for defect energetics in situations where there

I. L. Singer and H. M. Pollock (eds.), Fundamentals of Friction: Macroscopic and Microscopic Processes, 437–462.
© 1992 *Kluwer Academic Publishers. Printed in the Netherlands.*

is a large deviation from equilibrium. An example of how one might calculate surface energy, adhesive energy and energy as a function of position for sliding one metal over another will be given with the intent of encouraging the unfamiliar reader to apply these relatively simple techniques to problems of interest.

II. Background
First, we address the issue of what has changed in atomistic modeling in order to enable examination of problems of interest in tribology. First principles (ab-initio) approaches have dealt with perfect single crystals in which lattice periodicity [1] reduces the calculation to a single unit cell. The results of these calculations were impressively successful, in that accurate agreement with experiment was obtained in predicting the band structure, transport and magnetic properties of solids. Properties of interest to the material scientist, however, are generally related to "defects" in atomic structure which involve a partial loss of periodicity and thus require calculations over a large number of atoms, for example, dislocations, grain boundaries and interfaces between different materials. The breakdown of periodicity and the involvement of many atoms clearly complicates such problems For example, we show a twist boundary for an fcc(100) interface (fig.1). One can see that the geometry is quite complex and that there are many non-equivalent atoms which must be considered in the calculation. Until recently, the lack of adequate affordable computing power combined with lack of sufficiently efficient methods for performing quantum mechanical (ab-initio) calculations made such complex problems difficult. For example, it only recently [2,3] became possible to predict which simple structure-fcc, bcc or hcp-for an elemental metal had the lowest energy.

The question arises, why aren't all calculations performed using first-principles methods? In order to answer this question we again refer to fig. 1. The rigid twist boundary defect shown is not the minimum energy configuration, consequently a complex search must be performed to find the minimum energy structure. Consequently, a first-principles calculation would have to be performed for each configuration in order to determine the optimum structure. Finally, if temperature and dynamic effects are to be included the situation becomes even more complex. The complexity and cpu time requirements lead one to seek alternate approaches to treat such problems. Having outlined the problem we now describe some approaches.

III. Approaches for Describing the Interactions
This section gives a general discussion of approaches for calculating defect energies. The order will be inverted in that it will start with the most general, proceed to the least general and build to better approximations from the least general. We address four approaches, first-principles calculations, pair-potentials, many-body potentials and semi-empirical methods. The purpose of these discussions will be to give a qualitative understanding of the methods and to provide a

Fig. 1. A 4 °rigid twist angle for
an fcc(100) surface.

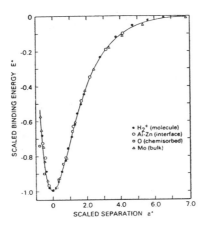

Fig. 2. Scaled binding energy as a
function of separation for the
four systems shown.

starting point for a literature search should a more comprehensive understanding be desired.

A. First-Principles Calculations

This is the most fundamental approach for calculating material properties. In principle, the only inputs are fundamental constants and the atomic number of the atoms of interest, and it is not materials limited. Recently, techniques have been developed to treat some of the problems discussed i.e., loss of periodicity, energy minimization, and dynamic effects. The loss of periodicity is handled by a trick called "super cells" [4]. A quasi-periodicity is produced in which a large cell which mimics the defect is repeated throughout the solid. The hope is that if the cell can be made sufficiently large, the energetics will converge to that of the real defect. Of course in highly disordered systems such as one might find in tribology this approach will not work as well. Carr and Parrinello [5] have developed a new technique which allows minimization of the total energy as a function of the electronic degrees of freedom and also allows the study of dynamic and temperature dependent effects. However, first-principles calculations remain cpu-time-intensive.

We now proceed to give a description of first-principles calculations. These normally involve solving the Kohn-Sham (KS) equations(the Schroedinger equation in a form appropriate to these problems) within the local density approximation (LDA) [6] with Poisson's equation. This is a set of equations using the mean-field approximation which is written as a set of one-electron equations where the electron is moving in the mean field of all of the other particles. The objective of the calculation is to obtain the self-consistent electron density distribution The total energy is expanded in terms of the electron density, since it can be shown that the total energy is a function of the electron density. Thus, once the electron density distribution is known, the total energy for a given configuration of atoms can be determined.

Although the first-principles methods are based on the firmest theoretical foundations, there are still many approximations used to get a solution, such as the mean-field approximations and the LDA. Only ab-initio methods can be used to calculate the electronic structure of the solids and are material independent of the methods discussed.

B. Pair-Potentials

In this section we will start with a more general discussion following Carlsson and Ashcroft [7], thus it will lead naturally to the following sections. What is the functional form of the cohesive energy $E_{coh}(R_1, \ldots, R_N)$ in a system consisting of N atoms with nuclei at sites $\{R_i\}$? A simple guess for expressing the cohesive energy would be a cluster expansion in pair and higher-order interactions

$$E_{coh} = E_1 + E_2 + E_3 + \ldots$$

$$= E_1 + (1/2) \sum_{i,j}' V_2(\mathbf{R}_i,\mathbf{R}_j) + (1/6) \sum_{i,j,k}' V_3(\mathbf{R}_i,\mathbf{R}_j,\mathbf{R}_k) + \ldots \quad (1)$$

where the prime indicates $i \neq j$ and E_1 represents external forces. The pair-potentials refers to keeping only the V_2 term in the expansion which dominates when the higher order terms are small in magnitude compared to the pair term. In practice, however, the application of pair-potentials hasn't followed from such an approach, but simply assuming some form for V_2 and then empirically fitting to some physical parameters. For example, suppose one could assume a Morse potential, in order to describe the interactions,

$$V_2 = D \left(\exp(-2\alpha(r_{i,j}-r_0))-2 \exp(-\alpha(r_{i,j}-r_0))\right) \quad (2)$$

the unknown parameters in eq. 2 could be selected to give the correct cohesive energy, bulk modulus and equilibrium lattice constant, etc., depending on the number of fitting parameters in the expressions. Once these parameters and the defect geometry are established, the energy is calculated by performing a sum over all pairs of atoms. For some solids such as rare gas solids this approach might be a sufficient. In metals, where volume dependent terms are important [8,9], it is not. There are other deficiencies. For example, it requires that the elastic constants satisfy $C_{12} = C_{44}$ [1] which seldom occurs in real metals. In addition, they require that the vacancy formation energy equals the cohesive energy which is also not the case. The vacancy formation energy is typically of the order of one third of the cohesive energy [7,8]. The principal advantage in pair-potentials is simplicity, however they may be useful in predicting trends when stearic effects dominate.

C. Many-Body Potentials

Many-body potentials are constructed in order to correct for the outlined deficiencies in pair potentials and treat situations where directional bonding is important such as in covalent solids such as Si [10]. These potentials correspond to keeping higher order terms in equation 1 which include terms which depend on bond angle in the bulk solid. The unknown constants which arise are determined from experimental properties as in pair potentials. This approach adds a fair amount of complexity in computations as compared to pair potentials, however they are still substantially simpler than first-principles methods for complex structures.

D. Semi-Empirical Methods

This category of approaches tackles the many-body problem in a different fashion. The term semi-empirical refers to determining a functional form for the cohesive energy based on some physical model, but including some parameters determined by fitting to experimental.

Once these constants are determined, the form is used to calculate the energy, dynamic behavior or other properties of interest such as defect energies. This approach has proven to be extremely effective in accomplishing its goals. This section concentrates on two methods, the embedded atom method [11,12,13] (EAM) and equivalent crystal theory [14,15,16] (ECT). The next section concentrates on giving examples from EAM and ECT. ECT treats a wide class of materials with a high degree of quantitative agreement with experiment and first-principles calculations. EAM has been applied to a wide variety of problems with considerable success and is currently the method of choice for metals. We should mention there is a class of approaches based on tight-binding theory [1] such as the method of Finis and Sinclair [8,17], however these will not be discussed here.

a. Universality of Binding Energy Relations

We digress for the moment to discuss an issue which will be important in the discussions of EAM and ECT. Rose, Smith and Ferrante [18,19] (RSF) discovered that there were similarities between certain binding energy relations, i.e. binding energy versus distance. They found that the functional form of the scaled binding energy curves describing cohesion, adhesion, chemisorption and diatomic molecules for the case of no charge transfer was the same to a high degree of accuracy (fig.2). Stated more formally, we write the energy versus separation as

$$E(R) = \Delta E \, E^*(a^*) \tag{3}$$

where

$$a^* = (R-R_e)/l \tag{4}$$

ΔE is the binding energy, a^* is a scaled length, R is the distance between particles, R_e is the equilibrium distance and l is a scaling length. RSF found that a simple functional form was an accurate representation of the function $E^*(a^*)$ first proposed by Rydberg for diatomics

$$E^*(a^*) = - (1+a^*) \exp(-a^*) \tag{5}$$

and the length scaling was chosen as

$$l = (\Delta E/(d^2E/dR^2)_{R_e})^{-1/2} \tag{6}$$

This result was dubbed the "universal binding energy relation" (UBER). The application of this result to EAM and ECT will be discussed in the next sections.

b. EAM

The embedded atom method can be considered as a breakthrough in atomistic simulations of metals. It has been successfully applied to problems such as phonons, liquid metals, defects, fracture, surface structure, surface segregation etc [20]. It is based on the effective medium theory of Stott and Zaremba [21] or the quasi-atom approach of Norskov and Lang [22] in which the energy to embed an atom in jellium is calculated. The total energy of the system is written as

$$E_{tot} = \sum_i^N E_i \qquad (7)$$

where

$$E_i = F_i(\rho_{h,i}) + (1/2) \sum_{j \neq i}^N \Phi(R_{ij}) \qquad (8)$$

and $\rho_{h,j}$ is the total electron density at atomic site i due to all the other atoms in the system, F_i is the embedding energy for placing an atom at site i and $\Phi(R_{ij})$ is a short-range pair-interaction representing the core-core repulsion of atoms i and j separated by R_{ij}. N is the total number of atoms in the system. The host electron density at site i is approximated by the linear superposition of all of the other atomic densities. The pair repulsion is given by

$$\Phi(R_{ij}) = Z_i(R_{ij}) \, Z_j(R_{ij})/R_{ij} \qquad (9)$$

where Foiles, Baskes and Daw (FBD) have parameterized a simple function for the effective charge Z(R) as

$$Z(R) = Z_0(1+\beta R^\nu) \exp(-\alpha R) \qquad (10)$$

in order to guarantee the short range behaviour, Z_0 is the nuclear charge. The embedding energy $F(\rho)$ is determined by requiring that it agree with the cohesive energy as described by the UBER (eq. 5) for the case of an isotropic expansion or contraction giving

$$F(R) = E_{coh} (1+\nu x) \exp(-\nu x) - \phi \qquad (11)$$

where Φ is the second term in eq. 8, E_{coh} is the cohesive energy, $x=(r/r_0-1)$ and $\nu =3 (V_0 B_0/E_{coh})^{1/2}$ V_0, B_0 and r_0 are the equilibrium volume, bulk modulus and Wigner-Seitz radius, respectively. The unknown constants are again obtained by fitting to such properties as the vacancy formation energy, elastic constants, and diatomic molecule parameters. It is unnecessary to perform the fitting for many materials since they are provided by FDB in their papers. Another point to emphasize is that once the embedding energy is specified for a given metal the same function is then used for any defect. Alloys have been treated in EAM [23] by keeping the same embedding function for a given metal and constructing the density by the overlap of atomic densities for all species present. For example, in a Ni-Cu alloy, the geometry would be specified, then at an Ni site, the overlap contribution to the electronic charge density from both species at the

Ni site is calculated. Finally, that density is used in the Ni embedding energy function to determine the contribution for the Ni atom at that site. A geometric mean of the two pair repulsion terms is used for the different atom interaction, [24]

c. ECT

Equivalent Crystal Theory is a new technique developed by Smith, Banerjea, and Co-workers [25] which also uses the UBER in order to implement the method. The technique gives quantitatively accurate values for the surface energy and surface relaxation of metals and covalently bonded solids. The application of EAM to bcc metals [26] and to covalently bonded solids is limited [27]. ECT is based on the fact that the UBER gives the ground state energy distance relation for a solid.

Consider a single crystal of an elemental solid. Next introduce a defect or an array of lattice defects into the crystal. The total energy E(*defect*) of the crystal containing the defects is equal to energy of an *ideal* single crystal (the equivalent crystal), E(*crystal*) plus a perturbation series, where the perturbing potential is the difference between the array of ion core potentials of crystal containing defects and those of the single crystal:

$$E(defect) = E(crystal) + \text{Perturbation Series} \tag{12}$$

E(*crystal*) is given by eq. 5 where $a^* = (r_{WS}-r_{WSE})/l$, r_{WS} is the Wigner-Seitz radius of equilibrium value $[3/(4\pi r_{WSE}^3) \equiv$ bulk atom density], $l = [\Delta E/(12\pi B r_{WSE}]^{1/2}$ and B is the bulk modulus of the ground state single crystal.

The "equivalent crystal" is a conceptually ideal crystal of the material which is either expanded or contracted from the ground state crystal and thus its energy is given by the UBER at some different scaled distance, a_0^*. The procedure is to find the value of a_0^* for which the perturbation series = 0. If this can be accomplished then

$$E(defect) = \Delta E\ E^*(a_0^*) \tag{13}$$

which can be evaluated simply by using the UBER. In order to implement ECT, we must have a method to represent the perturbation series. The perturbation series has matrix elements which are integrals over products of the density and the difference in potential between the perturbed and unperturbed state. These integrals are approximated by

$$g \propto R^p \exp(-\alpha R) \tag{14}$$

for nearest neighbors and

$$g \propto R^p \exp(-\alpha R) \exp(-R/\lambda) \tag{15}$$

for next nearest neighbors and beyond, where p = 2n-2, n = principal quantum number and λ = electronic screening length. It is assumed that the form of the electron density is that of the highest partially occupied s-orbital. Screening is introduced for next nearest neighbors by addition of the term $\exp(-R/\lambda)$. The calculation proceeds by solving the perturbation equation on a site by site basis, so that the energy change in creating a defect is written as the sum over the changes at each non-equivalent site given by

$$\delta E = \sum_{1=1}^{s} [E_1(defect) - E_1(crystal)] \qquad (16)$$

where the E_1's are the energies associated with site 1 for the defect

crystal and the ground state crystal, respectively. In many defects, for example a surface, where a high degree of symmetry is maintained, the value of s in eq. 16 can be quite small.

We now present an older version of ECT which is all that is needed to calculate surface energy for an example calculation which follows. A more general method used in determining surface relaxation will be outlined after this discussion. We now introduce the actual working equations for application of the perturbation equation [32] for $\delta E=0$ for a given site 1 to next nearest neighbors

$$b1 \; R_1^p(1) \; e^{-\alpha R_1(1)} + b_{,2} \; R_2^p \; e^{-\alpha R_2(1)} \; e^{-R_2(1)/\lambda}$$

$$= \sum_{\text{defect n}} R_i'^p(1) \; e^{-R_i(1)} + \sum_{\text{defect nn}} R_i'^p \; e^{-\alpha R_i'(1)} \; e^{-R_i'(1)/\lambda} \qquad (17)$$

where b1 and b2 are the number of nearest(next) neighbors respectively; and R_1 and R_2 are the nearest(next) neighbor distances, respectively.

We solve this equation for R_1, the equivalent crystal nearest neighbor

distance, which has a simple geometric relationship to the Wigner-Seitz radius and any of the neighbor distances, since we are dealing with a perfect crystal. To restate, an atom in a defect in the material experiences an environment as if it were in a perfect crystal, which is expanded or contracted from its equilibrium state to a new nearest neighbor distance given by R_1. The expression for the surface energy,

for which we will give an example later, is given by

$$\sigma = (\Delta E/A) \sum_{1=1}^{s} [F^*(a_{01}^*)] \qquad (18)$$

where

$$F^*(a^*) = 1 + E^*(a^*) = 1 - (1+a^*) \exp(-a^*) \qquad (19)$$

where A is the surface area, s = number of layers with energy from the bulk ($\cong 6$). ECT in this form has essentially only one fitting parameter, α, which is obtained from the vacancy formation energy. The other physical parameters such as the cohesive energy, bulk modulus and equilibrium lattice parameter appear explicitly in the model.

Since the purpose of this presentation is to give an overview of the methods, we will not go into details for the modification of ECT as presented so that a wider class of materials and relaxation of atomic positions can be treated. We will only indicate the general procedures leaving the details to reading ref.25. The defect formation energy ε_i is written in terms of four components in the spirit of the many-body expansion given in eq. 1 in which bond angle and bond compression terms are included in the expansion. In this formulation there are four fitting parameters compared to one, α, presented in the overview. These four parameters are obtained from the vacancy formation energy and the elastic constants. The other input parameters remain as before. The procedure is more complicated but not difficult, because each perturbation is treated as independent. There are now four equations similar to eq.17 but no more difficult to solve for the a 's. A comparison of various results EAM and ECT will be shown in the next section. Finally, alloys [28] are treated by an extension of the basic ECT ideas in which the α's reflect whether the neighbors are of type A or B.

IV. Examples of Surface Energy Calculations

We now give an example of a surface energy calculation for a rigid fcc (100) metal surface using EAM and ECT to demonstrate the relative simplicity of such calculations. Each atom in each plane is equivalent to all other atoms in the plane, therefore it is only necessary to evaluate the EAM or ECT equations for a single atom in each plane The only remaining question is how many planes are needed for convergence.

Consider EAM, the first step is to pick an atom in the plane, say atom A, and then evaluate the electron density at A from all of the other atoms in the crystal. The density is evaluated from analytic expressions [12]. Once the density is known, one examines the value of the embedding function, $F(\rho)$ eq.11, for the given metal at that density. Then one need only evaluate the pair term in eq. 8 using the parameters given in ref. 11 and then evaluate the contribution to the energy for this atom from eq. 7. The calculation is finished for the top plane. The same calculation is repeated for a single atom in the second plane. This procedure is continued in each plane until there is no change in the energy from the bulk value. In order to calculate the surface energy, the bulk energy for each atom is subtracted from the EAM energy then divided by the area of a primitive cell and then summed over each non-equivalent site ie.

$$\sigma = \sum_i^s (E_i^{plane} - E^{bulk})/A \tag{20}$$

We now outline the procedure for doing the same calculation with ECT. Again we start with the same picture, an atom in the surface plane. We will only concern ourselves with up to next nearest neighbors in this example. Referring to eq. 17, the perfect fcc equivalent crystal has 12 nearest-neighbors and 6 next nearest-neighbors, therefore b1 = 12 and b2 = 6. An atom in the free surface has lost four nearest-neighbors and one next nearest neighbor, thus we can rewrite eq. 17 for the surface atom as

$$12 \; R_1^p(1) \; e^{-\alpha R_1(1)} + 6 \; R_2^p(1) \; e^{-\alpha R_2(1)} \; e^{-R_2(1)/\lambda}$$

$$= 8 \ R_1^{'P}(1) \ e^{-\alpha R_1^{'}(1)} + 5 \ R_2^{'P}(1) \ e^{-\alpha R_2^{'}(1)} \ e^{-R_2^{'}(1)/\lambda} \tag{21}$$

An atom in the second plane keeps all of its nearest-neighbors and loses one next nearest-neighbor, therefore we have

$$12 \ R_1^P(2) \ e^{-\alpha R_1(2)} + 6 \ R_2^P(2) \ e^{-\alpha R_2(2)} \ e^{-R_2(2)/\lambda}$$
$$= 12 \ R_1^{'P}(2) \ e^{-\alpha R_1^{'}(2)} + 5 \ R_2^{'P}(2) \ e^{-\alpha R_2^{'}(2)} \ e^{-R_2^{'}(2)/\lambda} \tag{22}$$

where now the arguments of R refer to the equivalent atom in each plane. R_2 is simply a geometrical factor times R_1 and R' is known since we are specifying the geometry of the defect. The values of α and λ for a number of elements are given in the ECT papers. Thus, we are simply left with the problem of solving each transcendental equation for R_1 for each non-equivalent atom in the surface region.

Once this is done we simply substitute the values into eq. 18 and evaluate the energy.. It is conceivable that for high degrees of symmetry and truncating at next-nearest neighbors that the calculation can be done on a programmable hand calculator.

Let's consider how many terms must be kept for the surface energy calculation. Note that the cpu time consuming step in EAM is overlapping the densities, and in ECT it is solving the transcendental equation. Table I shows the contributions to the surface energy for EAM and ECT for a Ni(100) surface. In either case, three planes are sufficient to evaluate the energy. Table II shows the ECT values for an Al(210) surface in the case of the higher index planes, lower lying planes can have neighbors in the surface layer and thus more planes must be kept in the calculation.

In table III we show a comparison between ECT, EAM, First-principles and experiment for the surface energies. ECT gives quantitatively accurate agreement with first-principles calculations (typically < 10% disagreement) whereas EAM gives about 40% disagreement. In addition, ECT can be used for any crystal structure whereas EAM has been primarily applied to fcc metals although it has been applied to bcc and hcp structures in some cases. In table IV we show a comparison for surface relaxation. Again ECT reproduces the magnitudes and trends quite well. The more general form of ECT must be used for this calculation [25]. Predicting surface relaxations is a very stringent test, since the energy differences are very small. In table V we show the results of using the more general form of ECT applied to Si.

It is evident that EAM underestimates the surface energy. Our opinion is that this underestimation is caused by the fact that overlapping atomic electron densities does not include relaxation of the electron gas i.e. allowing it to redistribute in order to minimize the energy. At a free surface there are large deviations from equilibrium. Thus for a free surface, simple overlap of free atom

Table I Planar Contribution to the Surface Energy using EAM and ECT for Ni(100).

Plane	EAM	ECT
1	.290 eV/atom	.654 eV/atom
2	.0182	1.25×10^{-6}

Table II Planar Contribution to the Surface Energy using ECT for Al(210).

Plane	ECT
1	1.25 eV/atom
2	.404
3	.0551
4	1.25×10^{-6}

Table III Rigid and Relaxed Surface Energies in erg/cm^2 from ECT, First-principles and EAM (other) Calculations [33].

Element	Crystal face	ECT rigid	ECT relaxed	LDA rigid	Other
Cu	(111)	1830	1780	2100	1170[i]
	(100)	2380	2320	2300	1280[i]
	(110)	2270	2210		1400
Ag	(111)	1270	1230		620[i]
	(100)	1630	1600	1650'	705'
	(110)	1540	1510		770
Ni	(111)	2400	2320		1450
	(100)	3120	3040	3050	1580
	(110)	2980	2910		1730[i]
Al	(111)	920	860		
	(100)	1290	1220		
	(110)	1310	1230	1100'	
Fe	(110)	1820			
	(100)	3490		3100'	1693
W	(110)	3330			
	(100)	5880		5200	2926

Table IV Percentage Changes in Interlayer Spacing due to Relaxation, references for each value are given in ref. 33.

Element	$\Delta d_{n,n+1}$	ECT	EAM	Experiment	Technique
Cu(110)	Δd_{12}	−7.7%	−4.9%	(−8.5±0.6)% (−7.5±1.5)%	LEED Ion scattering
	Δd_{23}	+3.4%	+0.2%	(+2.3±0.8)% (+2.5±1.5)%	LEED Ion scattering
Cu(100)	Δd_{12}	−3.7%	−1.4%	(−2.1±1.7)% (−1.1±0.4)%	LEED LEED'
	Δd_{23}	+1.9%	−0.3%	(+0.45±1.7)% (+1.7±0.6)%	LEED' LEED'
Cu(111)	Δd_{12}	−3.1%	−1.40%	(−0.7±0.5)%	LEED'
	Δd_{23}	+1.9%	−0.05%		
Ag(110)	Δd_{12}	−6.0%	−5.7%	−5.7% (−7.8±2.5)%	LEED' Ion scattering
	Δd_{23}	+2.8%	+0.3%	+2.2% (+4.3±2.5)%	LEED Ion scattering'
Ag(100)	Δd_{12}	−3.0%	−1.90%		
	Δd_{23}	+1.7%	−0.05%		
Ag(111)	Δd_{12}	−2.5%	−1.30%		
	Δd_{23}	+1.6%	−0.04%		
Ni(110)	Δd_{12}	−7.6%	−4.87%	(−8.7±0.5)% (−9.0±1.0)%	LEED Ion scattering
	Δd_{23}	+3.4%	+0.57%	(+3.0±0.6)% (+3.5±1.5)%	LEED· Ion scattering
Ni(100)	Δd_{12}	−3.7%	−0.002%	(−3.2±0.5)%	Ion scattering'
	Δd_{23}	+2.0%	−0.001%		
Ni(111)	Δd_{12}	−3.1%	−0.05%	(−1.2±1.2)%	LEED·
	Δd_{23}	+1.9%	+0.00%		
Al(110)	Δd_{12}	−10.4%	−10.4%	(−8.6±0.8)% (−8.5±1.0)%	LEED LEED
	Δd_{23}	+4.7%	+3.1%	(+5.0±1.1)% (+5.5±1.1)%	LEED\ LEED'
Al(100)	Δd_{12}	−4.9%			
	Δd_{23}	+1.8%			
Al(111)	Δd_{12}	−3.9%		(+0.9±0.7)%	LEED
	Δd_{23}	+2.5%			

450

Table V Surface Energies in ergs/cm^2 for Si(100). The (2x1)
Values are for Symmetric Dimer Formation.

Surface	ECT	Ref. a	Ref. b	Ref. c	Ref.d
Ideal	2850	2740			2390
(1x1)	2820	2690			1970
(2x1)	1550		1590	1610	1910

a. M.T. Yin and M. Cohen, Phys. Rev. B **24**, 2303 (1981).
b. K.C. Pandey in **Proceedings of the seventh International Conference
on the Physics of Semiconductors**, eds. D.F. Chadi and W.A. Harrison
(Springer-Verlag, New York, 1985), p. 55.
c. N Roberts and R.J. Needs, J. Phys. Cond. Mat. **1**, 3139 (1989).
d. M.I. Baskes, J.S. Nelson, and A.F. Wright, Phys. Rev. B **40**, 6085
(1989).

Element	E_m	σ_m	σ_w
Ag	0.292	6.19	3.32
Ni	0.411	12.3	3.94
Cu	0.334	9.36	3.64
Al	0.225	4.34	—

Table VI Energy Barrier to Slip E_m (eV/atom), maximum slip stress σ_m
(10^{10} dynes/cm^2) and measured [a] maximum whisker strength σ_s (10^{10}
dynes/cm^2)

a. **Whiskers**, ed. J. Gordon Cook (Mills and Boon Limited, London, 1972)
pp. 46-55.

electron densities [9] may be a poor representation of the actual density distribution, whereas for an internal interface such as a grain boundary, overlap is a less severe approximation and the energies obtained with EAM may be quite good. ECT has, in some sense, electronic relaxation built into the procedure of finding the equivalent crystal and thus may avoid this difficulty. The effect of not relaxing the density is less severe than might be expected since the energy depends variationally on errors in the electron density. Thus, errors in the density only cause second order errors in the energy. In this section, it is hoped that we have communicated that the use of EAM and ECT for many problems of interest can be relatively simple and they have the promise of giving reasonable results.

V. **Friction and Other Stuff**

Although these proceedings involve friction, we are going to avoid the real issue of how do you model the loss processes and instead present results of another problem which is a bound on the friction process, i.e. sliding one single crystal surface over another.

We start with a general discussion of the adhesion by referring back to fig. 2. Here we show first a solid which we separate along some plane. If we calculate the total energy at each separation, we define the adhesive binding energy as

$$E_{AD}(a) = (E(a)-E(\infty))/(2 A) \tag{23}$$

where a is the separation between surfaces and A is the cross-sectional area. In fig. 3a we show what general shape one would expect the curves to have. If we take the derivative of this curve we expect the tensile stress to be the curve shown in fig. 3b. In fig. 5 we show the results of a first-principles, jellium calculation performed by Ferrante and Smith [29] solving the Kohn-Sham equations for this approximation. We can see that the curves for different metals in contact have large binding energies and have the same features as same metal contacts. In fig. 5 we show that all of these curves scale onto one binding energy relation, the UBER described earlier.

We now show adhesion calculations using ECT. A rigid adhesion calculation for ECT or EAM is essentially as simple as the surface energy calculation. The only difference is that the other half space is present and its position is changed. In fig. 6a we show [30,31] the results of an adhesion calculation for the (111) surface of a number of fcc metals and in 6 b,c we show the adhesive energy curves for the (110) surface of two bcc metals. The corresponding scaled binding energy curves are presented in figs. 7 a and b along with a fit to the Rydberg function. We can see that the curves scale. However, the fit to the Rydberg function is not as good, because of an ambiguity in the identification of nearest neighbors and next-nearest neighbors for the screening. Fig. 8, gives the force of separation for (110) iron and tungsten along corresponding to fig. 8b. ECT can be formulated completely in terms of force, therefore these calculations can be done directly. These calculations are relatively trivial compared to the

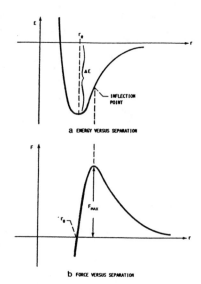

Fig. 3. Example of a binding energy
curve: (a) binding energy vs separation, (b) force vs separation

ADHESIVE BINDING ENERGIES

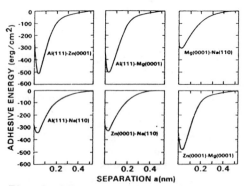

Fig. 4. Adhesive binding energy vs
the separation between the
surfaces indicated.

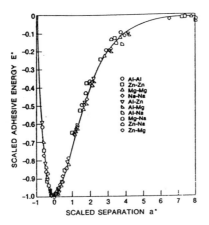

**Fig. 5. Scaled adhesive binding
energies vs scaled separation
for the interfaces indicated.**

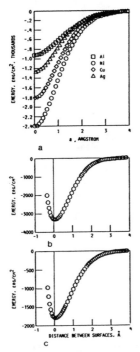

**Fig. 6. Rigid adhesive binding
energy vs separation using ECT:
(a) fcc(111), Al, Ni, and Ag,
(b) bcc(110), W, (c) bcc(110) Fe.**

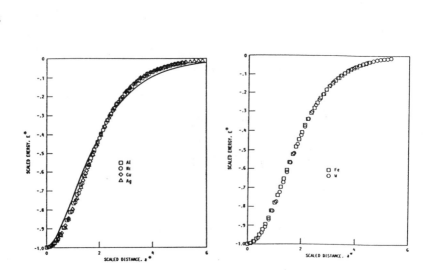

Fig. 7. Scaled adhesive energies
from fig. 8, (a) Al, Ni and Ag,
(b) W, (c) Fe.

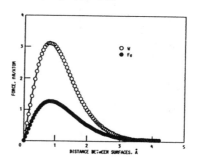

Fig. 8. Rigid adhesive binding force
vs separtion for W(110) and Fe(110).

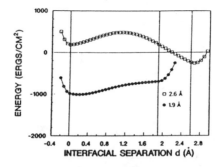

Fig. 9. Total energies as a function
symmetric movement of only the
surface layers for Ni(100) for
two rigid separations.

jellium or fully three dimensional first-principles results presented.

An interesting aspect of the adhesion calculation which was ignored in the previously relates to instabilities which can occur as surfaces approach one another called "avalanche". Our work on this was prompted by a paper by Pethica and Sutton [32]. We followed up on their result using ECT [33] applied to planar surfaces. The assumption that adhesion could take place with the half spaces essentially rigid with possibly some small relaxation is clearly not correct. We found, as did Pethica and Sutton, that the surfaces will snap together at a separation depending on the thickness of the metals in contact. In retrospect this result is not too surprising since the spring constant for the restoring force depends on the length of the spring. In fig. 9 we show the energy for moving one surface layer as a function of separation. We can see that at a separation of 2.6 Angstroms there is a barrier to moving one layer from each surface , since the restoring force is greater than the adhesive attraction. At 1.9 Angstroms this barrier is removed, since the situation is reversed. In fig. 10 we show the adhesive energy as a function of separation for allowing an increasing number of surface layers to relax. We can see that as the number of planes relaxed (length of the solid) is increased the "avalanche" occurs at larger separations. In fig. 11 we show how the separation at which avalanche occurs depends on the number of layers relaxed and the rigid separation. This example demonstrates that these calculational methods enable examination of phenomena which are difficult to observe experimentally.

We now address some ECT results that model friction more closely, that of the slip of one half-space over another [34]. We have used ECT to calculate the energy and force as a function of tangential position for slip on an fcc (100) surface for a number of metals. This calculation was performed under zero applied load conditions, that is, as a force tangential to the surface was applied the two half-spaces were allowed to move apart at the separation plane in the direction normal to the (100) surface. The numerical results were fitted to a Fourier series for the energy modulated by exponential term to represent the variation in force normal to the surfaces

$$\sigma(\mathbf{r}) = \sum_{i=0}^{2} \Delta_i \ (1+z_i+\beta_i z_i^3) \ e^{-z_i} \ H_i(x,y) \tag{24}$$

where

$$z_i = (z-z_0^{(i)})/l_i \tag{25}$$

$$H_0(x,y) = 1 - H_1(x,y) \tag{26}$$

$$H_1(x,y) = \cos(2\pi x/a) \ \cos(2\pi y/a) \tag{27}$$

$$H_2(x,y) = \cos(4\pi x/a) + \cos(4\pi y/a) - 2 \ H_1(x,y) \tag{28}$$

where a is the lattice parameter and Δ_i, β_i and l_i are fitting constants from the ECT calculations and are given in ref. 34. Eqs.

456

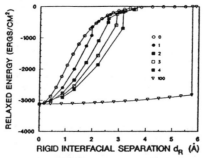

Fig. 10. Relaxed and rigid adhesive
binding energies for Ni(100)
crystal surfaces in registry
allowing different numbers of
surface layers to move on each

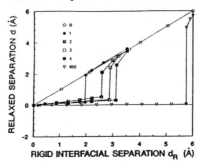

Fig. 11. Relaxed interfacial separation, d, as a function of rigid
(unrelaxed separation, d_R for different number of layers allowed to
relax.

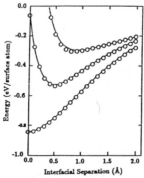

Fig. 12. Total cleavage energy vs
interfacial separation for Ag(100)
for different positions of the
surface

24-28 which can be thought of as an extension of the Frenkel model [1] is an excellent fit to the numerical results, with an average deviation of order 10^{-4} eV per surface atom for all metals treated. This is illustrated in Figs. 12 and 13 where the numerical results of equations 24-28 are plotted as solid curves. Fig. 12 shows ideal adhesion or cleavage as a function of z at fixed (x,y) for the Ag(100) interface. The bottom curve corresponds to (x,y) = (0,0), the middle to (a/4,0) and the top to (a/2,0).

Eqs. 24-28 and fig. 13 apply to arbitrary slip directions where the repeat distance can be many times the lattice constant a. Fig. 13 shows zero-load results for two illustrative slip directions, the lower curve shows results for slip along the x-axis and in the upper at an angle of arctan(3/4) relative to the x-axis. Fig. 13 also shows that the energy for all four metals scales onto one curve where the scaling is given by dividing the energy by the surface energy, Δ_1, and the

distance by the lattice parameter, a. Since the energy scales, there appears to be a universal form for ideal slip. This universality leads to useful rules of thumb for slip, analogous to the Griffith criterion for cracks. We find that the average energy barrier to slip is E_m = .36 Δ_1 and the average maximum slip stress is σ_m = 1.45Δ_1/a for

fcc(100).

Fig. 14 provides an overall picture of the stress profile. for an Ag interface sliding on the (100) surface at zero load. The size of the arrow head and the length of the vector are both proportional to the magnitude of the slip stress. In table VI we show the numerical results for the height of the energy barrier and the maximum shear stress in the direction of minimum to slip along with a comparison with whisker strengths. Although the theoretical results are larger, they are of the same order of magnitude as the experimental results. Since in the fcc system slip occurs on (111) planes, the results are in reasonable agreement. The ability to map the entire energy surface and then analyze the results for other relationships shows the power of the semi-empirical methods. What we have presented here is not friction, however, since there are no energy dissipation mechanisms in the calculation. In addition, at this stage of the calculation we have ignored relaxation of the atomic positions and any dynamic effects that will occur during sliding. However, the minimum stress to initiate sliding is a number of interest and the fact that scaling was found may mean that there are some underlying simplicities in the mechanisms. We should mention that Zhong and Tomanek [35] have performed a first-principles calculation similar to the above for sliding at a Pd-graphite interface.

One final topic we should address is ceramic and ionic compounds, since tribology has a wide range of materials involved in sliding contact and ceramic compounds are becoming increasingly important in newer high-temperature engines. At present pair-potentials [36,37] are the forms most often used to model interfaces and other defects. This approach suffers from the same difficulties previously described for

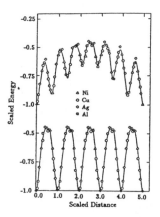

Fig. 13. Scaled total energy vs scaled
slip distance for two different slip
directions on (100) interfaces of
Ni, Ag, Cu, and Al.

Fig. 14. A map of slip stress for Ag
crystals sliding on (100) surfaces
for zero load.

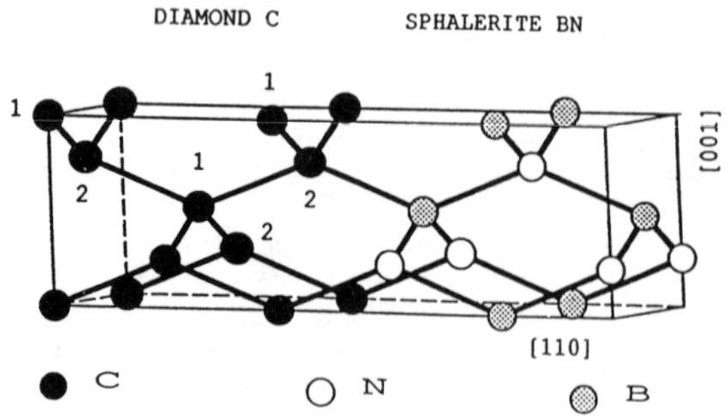

Fig. 15. Crystal structure of the $C_2(110)$
and BN(001) 1x1 superlattice [39].

pair-potentials with the added complexity of having the possibility of having net charges at interfaces. Development of new approaches to treat these problems is definitely a goal for the future.

First-principles methods are presently being applied to treat the problem of ceramic interfaces but with the limitation that super-cells are used. As example of such calculations we show a super-cell (fig. 16) structure used to calculate the interfacial energy for a diamond C/sphalerate Boron-Nitride interface [38] using a linear-muffin-tin-orbital-approach (LMTO). They obtain adhesive energies ranging from .89 to .95 eV/unit cell area depending on what structure is assumed for the interface. These calculations do not include any relaxation of the structure. These structural relaxations can be done if the changes are mainly local which is probably the case. The first-principles calculations give promise for providing information concerning interfacial energies, at least, for the materials combinations of interest in tribology.

At present there are no semi-empirical approaches for non-metals. Recently, we have proposed [39,40] an expression for the total energy for situations with charge transfer such as ionic solids with the hope that a method could be developed that would treat ceramic materials. The expression we propose is given by

$$E(R) = -C (1+a^*) e^{-a} - \delta Z^2 f(R)/R \qquad (29)$$

where the first term allows for the possibility of both covalent bonding and the repulsion and the second the coulombic attraction due to charge transfer, where C is the well depth of the Rydberg part a^* is now $(R-R_E)/l$ and R_E is the distance at which the minimum occurs for the Rydberg part, δZ is the charge transfer and $f(R)$ is a function that accounts for the crossing of the energy from two ions to two neutrals at some distance. This result can be contrasted with a Born-Mayer potential [1] which is a similar expression , but has a repulsive exponential core and thus does not allow for the possibility of covalent bonding. Fig. 16 shows a fit of this function to the potential energy curves from first-principles calculations for three ionic molecules, AlCl, AlF and LiF. In addition to the fits being quite good the values obtained for the charge transfer, the relative covalent contribution and the spectroscopic constants from the global fits are quite reasonable. The hope in pursuing these studies is that a semi-empirical method can be developed similar to EAM or ECT which treats non-metal defects and metal-non-metal interfaces.

VI. Concluding Remarks

In this presentation we have tried to give a description of the various techniques presently being used to approach problems in computational material science. We have also indicated that such calculational techniques are now sufficiently accurate, along with improvements in computer speed to enable serious consideration for attacking problems in tribology for certain classes of materials. At present it seems unlikely that the configurational energy minimization

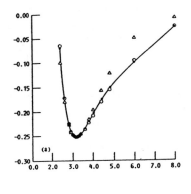

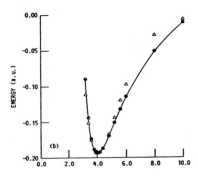

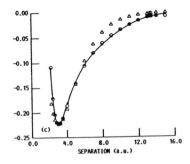

Fig. 16. a comparison between first-
principles diatomic molecule potential curves (0), the Rydberg
function, (Δ), and (●) eq. 29
(a) AlF, (b) Alcl, and (c) LiF.

and dynamics needs for modeling tribological problems will be amenable to first-principles approaches because of the complexity of the problem and the computer time needed to treat each configuration. A more likely scenario will be the development of semi-empirical approaches which accurately reproduce static defect energies as compared to first-principles calculations and thus can be applied to problems in tribology.

References

1. Ashcroft, N.W. and Mermin, N.D., **Solid State Physics,** pub. Holt, Rinehart and Wilson, New York, 1976.
2. Skriver, H.L., Phys. Rev. Lett. **49,** 1968 (1982).
3. Yin, M.T., and Cohen, M.L., Phys. Rev. B **26,** (1982).
4. Vicenzo, D.P., Alehand, D.L., Schlutter, M., and Wilkins, J.W., Phys. Rev. Lett. **56,** 1925 (1986); See also Payne, M.W., Bristowe, P.D., and Johannapoulis, J.D., Phys. Rev. Lett. **58,** 1348 (1987).
5. Carr, R., and Parrinello, M. Phys. Rev. Lett. **55,** 2471 (1985).
6. Kohn, W., and Sham, L.J., Phys. Rev. A **140,** 1133 (1985).
7. Carlsson, A.E. and Ashcroft, N.W., Phys. Rev. B **27,** 2101, (1983).
8. Finnis, M.W. and Sinclair, J., Phil. Mag. A **50,** 45 (1984).
9. Smith, J.R. and Ferrante J., Phys. Rev. B **31,** 5262 (1988).
10. Stillinger, F. H. and Weber, T.A., Phys Rev. B **31,** 5262 (1982).
11. Foiles, S.M., Baskes, M.I. and Daw, M.S., Phys. Rev. B **33,** 7983 (1986).
12. Foiles, S. M., Phys. Rev. B **32,** 3409 (1985).
13. Daw, M.S., Phys. Rev. B **39,** 7441 (1989).
14. Smith, J.R. and Banerjea, A., Phys. Rev. Lett. **59,** 2451, (1987).
15. Smith, J.R. and Banerjea, A., Phys. Rev. B **37,** 1041 (1988).
16. Smith, J.R., Perry, T.A. and Banerjea, A., in **Atomistic Simulations Beyond Pair-Potentials,** eds. V. Vitek and D.J. Srolovitz (Plenum Press, N.Y. 1989) p. 279.
17. Sutton, A.P., Finnis, M.W., Pettifor, D.G., Ohta, Y., J. Phys. C **21,** 35 (1988).
18. Rose, J.H., Smith, J.R. and Ferrante, J., Phys. Rev. B **28,** 1935 (1983).
19. Rose, J.H., Ferrante, J. and Smith, J.R., Phys. Rev. Lett. **47,** 675 (1981).
20. Foiles, S.M., Phys. Rev. B **32,** 7658 (1985).
21. Stott, M.J. and Zaremba, E., Phys. Rev. B **22,** 1564 (1980).
22. Norskov, J.R. and Lang, N., Phys. Rev. A **26,** 2857 (1982).
23. Foiles, S.M. and Daw, M.S., J. Mat. Res. **2,** 5 (1987).
24. Johnson, R.A., Phys. Rev. B **39,** 12554 (1989).
25. Smith, J.R., Perry, T.A., Banerjea, A., Ferrante, J. and Bozzolo, G., Phys Rev. B **44,** 6444 (1991).
26. Williams, F. and Massobrio, C., Phys. Rev. B **43,** 11653 (1991).
27. Baskes, M.I., Phys. Rev. Lett. **59,** 2666 (1987).
28. Bozzolo, G. Ferrante, J. and Smith, J.R., Rapid Comm. Phys. Rev. B

462

45, 493 (1992)

29. Ferrante J. and Smith, J.R., Phys. Rev. B **31**, 3427 (1985).

30. Banerjea, A., Ferrante, J. and Smith, J.R., J. Phys: Cond. Mat. **2,** 8841 (1990).

31. Banerjea, A., Ferrante, J. and Smith, J.R., **Fundamentals of Adhesion,** (ed. L.-H Lee, Plenum, New York, 1991) p. 325.

32. Pethica, J.B. and Sutton, A., J. Vac. Sci. Tech. **6**, 2494 (1988).

33. Smith, J.R., Bozzolo, G., Banerjea, A. and Ferrante, J., Phys. Rev. Lett. **63**, 305 (1989).

34. Bozzolo, G., Ferrante, J. and Smith, J.R., Scripta Met. et Mat. **25,** 1927 (1991).

35. Zhong, W. and Tomanek, D., Phys. Rev. Lett. **64**, 3054 (1990).

36. Catlow, C.R.A., Diller, K.M. and Nagett, M.J., J. Phys. C: Solid State **10**, 1395 (1977).

37. Duffy, D.M., J. Phys. C.: Solid State **19**, 4393 (1986).

38. Lambrecht, W.R.L. and Segall, B., Phys. Rev. B **40**, 7793 (1989).

39. Ferrante, J., Schlosser, H., and Smith, J.R., Phys. Rev. A **43**, 3487 (1991).

40. Smith, J.R., Schlosser, H., Leaf, W. Ferrante, J. and Rose, J.H., Phys. Ref. A **39**, 925 (1989).

MOLECULAR DYNAMICS SIMULATIONS OF ADHESIVE CONTACT FORMATION AND FRICTION

Uzi Landman, W. D. Luedtke and Eric M. Ringer
School of Physics
Georgia Institute of Technology
Atlanta, GA 30332 , USA.

ABSTRACT. Investigations using large-scale molecular dynamics simulations of atomistic mechanisms of adhesive contact formation, friction, and wear processes occurring as a consequence of interactions between material tips and substrate surfaces reveal the energetics and dynamics of jump-to-contact, elastic, plastic and yield processes, connective neck formation, wetting, reconstruction, atomic scale stick-slip, and materials transfer and wear phenomena. Results are presented for several tip and substrate materials, including metallic (nickel and gold), ionic (CaF_2), covalent (Si), and thin alkane (n-hexadecane) films adsorbed on a metal (gold) surface and interacting with a metal (nickel) tip.

I. INTRODUCTION

Understanding the atomistic mechanisms, energetics, and dynamics underlying the interactions and physical processes that occur when two materials are brought together (or separated) is fundamentally important to basic and applied problems such as adhesion [1-8], contact formation [3-18], surface deformations [7,8,17-25], materials elastic and plastic response characteristics [3,7,8,18-25], materials hardness [26-28], microindentation [6,11, 27-30], friction and wear [17,20,31,32], and fracture [33,34]. These considerations have motivated for over a century [1,3,18-21] extensive theoretical and experimental research endeavors of the above phenomena and their technological consequences. Most theoretical approaches to these problems, with a few exceptions [7,8,15-17], have been anchored in continuum elasticity and contact mechanics [18-26]. Similarly, until quite recently [35-38] experimental observations and measurements of surface forces and the consequent materials response to such interactions have been macroscopic in nature.

The study of frictional (or tribological) phenomena has a long and interesting history [39]. Leaping over centuries of empirical observations we start with the classical friction law presented by Amontons in 1699 and extended by Coulomb in 1781 (although it was actually known to da Vinci in the fifteenth century) which asserts that relative sliding of two bodies in contact will occur when the net tangential force reaches a

I. L. Singer and H. M. Pollock (eds.), Fundamentals of Friction: Macroscopic and Microscopic Processes, 463–510.

critical value proportional to the net force pressing the two bodies together. Furthermore, this proportionality factor, the friction coefficient, is independent of the apparent contact area, and depends only weakly on the surface roughness.

The first to introduce the notion of cohesive forces between material bodies in contact, and their contribution to the overall frictional resistance experienced by sliding bodies, was Desagulier's whose ideas on friction are contained in a book published in 1734, entitled "A Course of Experimental Philosophy". His observations which introduced adhesion as a factor in the friction, additional to the idea of interlocking asperities favoured in France, were conceived in the context of the role of surface finish where he writes "---the flat surfaces of metals or other Bodies may be so far polish'd as to increase Friction and this is a mechanical Paradox: but the reason will appear when we consider that the Attraction of Cohesion becomes sensible as we bring the Surfaces of Bodies nearer and nearer to Contact".

While it was recognized for many years that the Amontons-Coulomb laws of friction are applicable only to the description of friction between effectively rigid bodies and gross sliding of one body relative to another, the concepts of stress and the elastostatics came only later, in the writings of Cauchy, Navier and others, and the formulation by Hertz in 1881 of elastic contact mechanics [18]. Extensions of Hertz's theory to the contact of two elastic bodies including the influence of friction of the contact interface were made first by Cattaneo [40] and independently later by Mindlin [41].

Not attempting to give here a complete historical account, we note the growing realization since the beginning of the century of the role of adhesive interactions, plastic deformation and yield in determining the mechanical response and friction between bodies in contact. In particular the notion that the yield point of a ductile metal is governed by shear stress [22]; either the absolute maximum (Tresca criterion) or the octahedral shear stress (von Mises criterion). The relationship between the interfacial adhesive formation and shearing of intermetallic junctions and friction, was succinctly summarized by Tabor and Bowden as follows [42]: "Friction is the force required to shear intermetallic junctions plus the force required to plow the surface of the softer metal by the asperities on the harder surface."

The first successful theory of the contact mechanics of adhesive contact was formulated [23] by Johnson, Kendall and Roberts (JKR) who observed, during an investigation into the friction of automobile windscreen wipers, that in a situation of pressing together two bodies, when one or both surfaces is very compliant, the radius of the contact circle exceeded the value predicted by Hertz; moreover, when the load was removed a measurable contact area remained, and it was necessary to apply a tensile force to separate the surfaces. The JKR theory considers the adhesion between the two interfacing bodies simply as a change in surface energy only where they are in contact (i.e., infinitely short-range attractive forces). An alternative formulation [24] by Derjaguin, Muller and Toporov [DMT] on the other hand asserts that the attractive force between the solids must have a finite range, and in their original formulation they assume this interaction to act in a region just outside the contact zone. (In addition the DMT theory assumes that the deformed shape of the surfaces is Hertzian, i.e., unaffected by the surface forces). More complete formulations, which allow solid-solid interactions to be a prescribed function of the local separation between the surfaces have also been suggested [43,44]. Furthermore, Maugis and Pollock [12] have investigated, in the context of metal microcontacts, the development of plastic deformation and adherence under zero applied load by considering the influence of surface forces, and have derived conditions for ductile or brittle modes of separation after an elasto-plastic or full plastic contact.

The quest to understand and observe natural phenomena on refined microscopic scales has led to the development of conceptual and technological devices allowing the interrogation of materials with increasing resolution. On the experimental front, the importance of investigating single asperity contact in order to study the fundamental micromechanical response of solids has been long recognized. Such conditions are usually associated (i.e., assumed to be valid) for tip on flat configurations, with a tip radius of 1-2 μm or less [11,45-47]. This may very well be the case for clean metal contacts [48-50]. Indeed, evidence for continuous contact over an entire tip of several thousand Angstroms radius was given first by Pollock et al. [10].

The recent emergence and proliferation of surface force apparatus (SFA [36], of scanning tunneling microscopy (STM) [37], and of the related atomic force microscopy (AFM) [35] broaden our perspectives and abilities to probe the morphology, electronic structure, and nature of interatomic forces in materials, as well as enhance our ability to manipulate materials on the atomic scale [38].

On the theoretical front, recent advances in the formulation and evaluation of the energetics and interatomic interactions in materials [7,51], coupled with the development and implementation of computational methods and simulation techniques [7,8,52], open new avenues for investigations of the microscopic origins of complex materials phenomena. In particular, large-scale molecular dynamics computer simulations, which are in a sense computer experiments, where the evolution of a system of interacting particles is simulated with high spatial and temporal resolution by means of direct integration of the particles' equations of motion, have greatly enhanced our understanding of a broad range of materials phenomena.

Although our knowledge of interfacial processes occurring when two material bodies are brought together has significantly progressed since the original presentation by Heinrich Hertz before the Berlin Physical Society in January 1881 of his theory of the contact of elastic bodies [18], full microscopic understanding of these processes is still lacking. Moreover, it has been recognized that continuum mechanics is not fully applicable as the scale of the material bodies and the characteristic dimension of the contact between them are reduced [25,45]. Furthermore, it had been observed [20,28] that the mechanical properties of materials exhibit a strong dependence on the size of the sample (small specimens appear to be stronger than larger ones). Since the junctions between contacting solids can be small, their mechanical properties may be drastically different from those of the same materials in their bulk form. Consequently, the application of the newly developed theoretical and experimental techniques to these problems promises to provide significant insights concerning the microscopic mechanisms and the role of surface forces in the formation of microcontacts and to enhance our understanding of fundamental issues pertaining to interfacial adherence, microindentation, structural deformations, and the transition from elastic to elastoplastic or fully developed plastic response of materials. Additionally, studies such as those described in this paper allow critical assessment of the range of validity of continuum-based theories of these phenomena and could inspire improved analytical formulations. Finally, knowledge of the interactions and atomic-scale processes occurring between small tips and materials surfaces, and their consequences, is of crucial importance to optimize, control, interpret, and design experiments employing the novel tip-based microscopies [6,7,14,15,17,35-38,53-58].

In an attempt to address the above issues on an atomistic level, we have embarked on a series of investigations [7,8,17] of the energetics, mechanisms and consequences of interactions between material tips and substrate surfaces, using molecular dynamics simulations. Since material phenomena and processes are governed by the nature and magnitude of bonding and interatomic interactions, as well as by other materials characteristics (such as thermodynamic state, structure, and degree of compositional and structural perfection) a comprehensive study of any class of phenomena (interfacial processes in particular) requires systematic investigations for a range of material dependent parameters.

In this paper we review our studies of adhesive interactions and their tribological consequences in several materials systems. Following a brief discussion in Section II of the methodology of molecular dynamics simulations we present in Section III results of our investigations for several materials starting with metallic systems, including our recent results for a nickel tip interacting with an alkane film adsorbed on a gold surface, followed by studies of an ionic tip and surface system (CaF_2) and ending with our results for a covalently bonded system (Si).

II. METHODOLOGY

Prior to presentation of our results we provide in this section pertinent details of our simulation studies.

Molecular dynamics (MD) simulations consists of integration of the equations of motion of a system of particles interacting via prescribed interaction potentials [52]. The interatomic interactions that govern the energetics and dynamics of the system are characteristic to the system under investigation. Thus, for an ionic material (e.g., CaF_2) the energy may be described rather adequately as a sum of pair-wise interactions between the ions, with the potential $V_{\alpha\beta}(r)$ between ions of types α and β taken to be

$$V_{\alpha\beta}(r) = z_\alpha z_\beta e^2/r + A_{\alpha\beta}e^{-r/\rho_{\alpha\beta}} - C_{\alpha\beta}/r^6 , \qquad (1)$$

where, the first two terms correspond to the coulomb and overlap-repulsion contributions, and the last term represents the van der Waals dispersion interaction; the charge on ions of type α, in units of the electron charge e, is denoted by z_α. In our simulations of CaF_2 [59] we have employed a parameterization of the potential in Eq. (1) determined partly by fitting to selected experimental data for the low temperature bulk crystal (such as structure, lattice parameters, cohesive energy, elastic constants and defect formation energies) and partly by appeal to quantum-mechanical calculations [60]. To describe covalently bonded materials (e.g., silicon) requires, in addition to pair-interactions, three-body terms due to the directional nature of the bonds [61].

The nature of cohesion in metals, is rather different. Here, the dominant contribution to the total cohesive energy of the system (E_{coh}) is due to the electronic distribution interacting with the metal ions embedded in it. Based on the philosophy of density-functional theory [62], a description of metallic systems which is amenable to molecular dynamics simulations is provided by the embedded atom theory (EAM) [63,64]. The basic feature of this method is that the effect of the surroundings on each atom in the system can be described in terms of the average electron density which other atoms in the system provide around the atom in question. The electronic structure problem is then converted to that of embedding an atom in a homogeneous electron gas, which can be described in terms of a universal density dependent energy function. Thus the density dependent term gives rise to many-body interactions.

In the EAM the cohesive energy E_{coh} of the metal is written as

$$E_{coh} = \sum_i \left\{ F_i [\sum_{j \neq i} \rho_j^a(R_{ij}) + \frac{1}{2} \sum_{j \neq i} \phi_{ij}(R_{ij})] \right\} \qquad (2)$$

where ρ^a is the spherically averaged atomic electron density and R_{ij} is the distance between atoms i and j. In EAM the embedding function, F, and the pair-repulsion between the partially screened ions, ϕ, are determined by choosing functional forms which meet certain general requirements, and fitting parameters [65] in these functions to a number of bulk equilibrium properties of the solid, such as lattice constant, heat of sublimation, elastic constants, vacancy-formation energy, heat of solution (for alloys) etc. The EAM has been used with significant success in studies of metallic systems in various thermodynamic states and degrees of aggregation [63-66].

In our simulations of tip-substrate systems the surface part of the system is modeled by a slab containing n_d layers of dynamic atoms, with n atoms per layer, exposing an (hkℓ) surface plane, and interacting with n_s layers of the same material and crystallographical orientation. The surface atoms interact with a dynamic crystalline tip arranged initially in a pyramidal (tapered) geometry with the bottom layer (closest to the substrate surface) consisting of n_1 atom, the next layer consisting of n_1 > n_1 atoms and so on. In addition the tip interacts with a static

468

holder, made of the same material as the tip, consisting of n_h atoms located in $n_{h\ell}$ layers. This system is periodically replicated in the two directions parallel to the surface plane, and no boundary conditions are imposed in the direction normal to the surface.

The simulations were performed at 300K with temperature control imposed only on the deepest layer of the dynamic substrate, (i.e., the one closest to the static substrate). No significant variations in temperature were observed during the simulations. The equations of motion were integrated using a fifth-order predictor-corrector algorithm with a time step Δt ($\Delta t = 3 \times 10^{-15}$ fs for the metallic systems and 1×10^{-15} fs for the ionic ones).

Following equilibration of the system at 300K with the tip outside the range of interaction, the tip was lowered slowly toward the surface. Motion of the tip occurs by changing the position of the tip-holder assembly in increments of 0.25 Å over 500 Δt. After each increment the system is fully relaxed, that is, dynamically evolved, until no discernable variations in system properties are observed beyond natural fluctuations.

Analysis of the phase-space trajectories generated during the simulations allows determination of energetic, structural and dynamical properties. For example, the kinetic temperature (T) of a set of N particles is defined via the relation $(3N/2) k_B T = \sum_{i=1}^{N} m_i \vec{v}_i{}^2/2$, where $\vec{v}_i$ is the velocity vector of a particle with mass m_i and k_B is the Boltzmann constant. Another quantity, of particular interest in studies of the mechanical properties of materials, is the stress tensor and individual atomic contributions to it can be derived most generally from the Lagrangian of the system [67,68]. For the particular case of pair interactions the matrix of the stress tensor $\underset{\sim}{\sigma}$ is given in dyadic tensor notation by

$$\overleftrightarrow{\sigma} \equiv \sum_i \overleftrightarrow{\sigma}_i = \sum_i [m_i \vec{v}_i \vec{v}_i + \frac{1}{2} \sum_{j \neq i} \chi(r_{ij})(\vec{r}_i - \vec{r}_j)(\vec{r}_i - \vec{r}_j)]\Omega_i^{-1} , \qquad (3)$$

where Ω_i is the volume per particle i, $\vec{v}_i$ is the velocity vector of the particle, $\vec{r}_i$ is it's position vector and $\chi(r) = -r^{-1}\dfrac{dV(r)}{dr}$, where $V(r)$ is the pair-potential. The expression for potentials beyond pair-interactions are somewhat more complicated [61b]. From the atomic stresses, invariants of the stress tensor can be calculated, in particular the second invariant of the stress deviator, J_2, which is proportional to the stored strain energy and is related to the Von Mises shear strain-energy criterion for the onset of plastic yielding [19,22], is given by

$$J_2 = \frac{1}{2} \text{Tr}[\underset{\sim}{\Gamma} \bullet \underset{\sim}{\Gamma}^T] \, , \qquad\qquad (4a)$$

where Tr denotes the trace of the matrix product in square brackets, $\underset{\sim}{\Gamma}^T$ is the transpose of the matrix $\underset{\sim}{\Gamma}$ defined as

$$\underset{\sim}{\Gamma} = \underset{\sim}{\sigma} - p \underset{\sim}{1} \, , \qquad\qquad (4b)$$

where the hydrostatic pressure $p = \frac{1}{3} \text{Tr} \underset{\sim}{\sigma}$, and $\underset{\sim}{1}$ is the unit matrix.

III. CASE STUDIES

In this section results of several investigations are summarized. We start with simulations of the approach to adhesive contact followed by retraction (pull-off) of a clean Ni tip interacting with a clean gold substrate (Ni/Au). Simulations for the same system but when the tip is lowered beyond the point of contact formation are also described (nano indentation). Subsequently the materials in the tip and substrate are interchanged, i.e., a clean gold tip is lowered toward a clean nickel substrate (Au/Ni). We also give results for a modified nickel tip, (i.e., the one obtained at the end of the Ni/Au nano-indentation, where the bottom of the tip is coated by an epitaxial gold layer) interacting with an undamaged gold substrate. Following the discussion of inter-metallic interactions between a tip and a clean surface we present our recent results for a nickel tip interacting with an alkane (n-hexadecane) liquid film adsorbed on a gold surface. Next, consequences of the interactions between a CaF_2 surface and a tip made of the same material are described for both vertical and tangential displacements of the tip relative to the surface. Finally studies of an atomic-scale stick-slip phenomenon occuring upon sliding a silicon tip on a silicon surface are presented.

1. Clean Nickel Tip/Gold Surface [7a]

Simulated force versus distance curves for the system are shown in Fig. 1 as well as the calculated potential energy versus distance (Fig. 1c). Results for tip-to-sample approach followed by separation are shown, for adhesive contact (Fig. 1a) and indentation (Figs. 1b and 1c) studies. In these simulations the substrate consists of $n_s = 3$, $n_d = 8$, $n = 450$ atoms/layer exposing the (001) face. The tip consists of a bottom layer of 72 atoms exposing a (001) facet, the next layer consists of 128 atoms and the remaining six layers contain 200 dynamic atoms each. The static holder consists of 1176 atoms arranged in three (001) layers. This gives

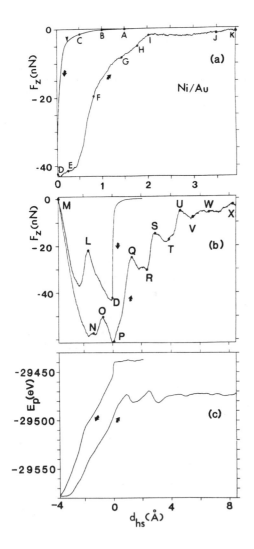

Fig. 1. Calculated force on the tip atoms, F_z, versus tip-to-sample distance d_{hs}, between a Ni tip and an Au sample for: (a) approach and jump-to-contact followed by separation; (b) approach, jump-to-contact, indentation, and subsequent separation; d_{hs} denotes the distance between the rigid tip-holder assembly and the static substrate of the Au surface (d_{hs} = 0 at the jump-to-contact point marked D). The capital letters on the curves denote the actual distances, d_{ts}, between the bottom part of the Ni tip and the Au surface; in (a): A = 5.7 Å, B = 5.2 Å, C = 4.7 Å, D = 3.8 Å, E = 4.4 Å, F = 4.85 Å, G = 5.5 Å, H = 5.9 Å, I = 6.2 Å, J = 7.5 Å, and K = 8.0 Å; in (b); D = 3.8 Å, L = 2.4 , M = 0.8 Å, N = 2.6 Å, O = 3.0 Å, P = 3.8 Å, Q = 5.4 Å, R = 6.4 Å, S = 7.0 Å, T = 7.7 Å, U = 9.1 Å, V = 9.6 Å, W = 10.5 Å, and X = 12.8 Å. (c)Potential energy of the system for a complete cycle of the tip approach, jump-to-contact, indentation, and subsequent separation. Forces in units of nanonewtons, energy in electron volts and distances in angstroms.

the tip an effective radius of curvature of ~ 30 Å.

The simulations correspond to a case of a rigid cantilever and therefore the recorded properties of the system as the tip-holder assembly approaches or retracts from the sample portray directly consequences of the interatomic interactions between the tip and the sample. The distance scale that we have chosen in presenting the calculated results is the separation (denoted as d_{hs}) between the rigid (static) holder of the tip and the static gold lattice underlying the dynamic substrate. The origin of the distance scale is chosen such that d_{hs} = 0 after jump-to-contact occurs ($d_{hs} \geq 0$ when the system is not advanced beyond the JC point and $d_{hs} < 0$ corresponds to indentation).

Since the dynamic Ni tip and Au substrate atoms displace in response to the interaction between them, the distance d_{hs} does not give directly the actual separation between regions in the dynamic tip and substrate material. The <u>actual</u> <u>relative</u> <u>distances</u>, d_{ts}, between the bottom part of the tip (averaged z-position of atoms in the <u>bottom-most</u> layer of the tip) and the surface (averaged z-position of the <u>topmost layer</u> of the Au surface, calculated for atoms in the first layer <u>away</u> <u>from</u> <u>the</u> <u>perturbed</u> <u>region</u> in the vicinity of the tip) are given by the letter symbols in Fig. 1(a,b) Note that the distance between the bottom of the tip and the gold atoms in the region <u>immediately</u> <u>underneath</u> it may differ from d_{ts}. Thus for example when d_{hs} = 0 (point D in Figs. 1a,b) the tip to <u>unperturbed</u> gold distance, d_{ts}, is 3.8 Å, while the average distance between the bottom layer of the tip and the adherent gold layer in <u>immediate</u> <u>contact</u> with it is 2.1 Å.

Tip-Substrate Approach

Following an initial slow variation of the force between the Au substrate and the Ni tip we observe in the simulations the onset of an instability, signified by a sharp increase in the attraction between the two (see Fig. 1a as well as Figs. 1b and 1c where the segments corresponding to lower- ing of the tip up to the point D describe the same stage as that shown in segment AD in Fig. 1a.) which is accompanied by a marked decrease in the potential energy of the system (see sudden drop of E_p in Fig. 1c as d_{hs} approaches zero from the right). We note the rather sudden onset of the instability which occurs only for separations d_{hs} smaller than 0.25 Å (marked by an arrow on the curve in Fig. 1a). Our simulations reveal that in response to the imbalance between the forces on atoms in each of the materials and those due to intermetallic interactions a jump-to- contact (JC) phenomenon occurs via a fast process where Au atoms in the region of the surface under the Ni tip displace by approximately 2 Å toward the tip in a short time span of ~ 1 ps. After the jump-to-contact occurs the distance between the bottom layer of the Ni tip and the layer of adherent Au atoms in the region immediately underneath it decreases to 2.1 Å from a value of 4.2 Å. In addition to the adhesive contact formation between the two surfaces an adhesion-induced partial wetting of the edges of the Ni tip by Au atoms is observed.

   The jump-to-contact phenomenon in metallic systems is driven by the marked tendency of the atoms at the interfacial regions of the tip and substrate materials to optimize their embedding energies (which are density dependent, deriving from the tails of the atomic electronic charge densities) while maintaining their individual material cohesive binding (in the Ni and Au) albeit strained due to the deformation caused by the atomic displacements during the JC process. In this context we note the difference between the surface energies of the two metals, with the one for Ni markedly larger than that of Au, and the differences in their mechanical properties, such as elastic moduli, yield, hardness, and strength parameters (for example, the elastic moduli are 21×10^{10} and 8.2×10^{10} N/m^2 for Ni and Au, respectively [69]).

Further insight into the JC process is provided by the local hydrostatic pressure in the materials (evaluated as the trace of the atomic stress tensors [67]) shown in Fig. 2a after contact formation (i.e., point D in Fig. 1a). The pressure contours reveal that atoms at the periphery of the contact zone (at $X = \pm 0.19$ and $Z = 0.27$) are under extreme tensile stress ($- 10^5$ atm $= - 10^{10}$ N/m^2 $= -10 GPa$). In fact we observe that the tip as well as an extended region of the substrate in the vicinity of the contact zone are under tension. Both the structural deformation profile of the system and the pressure distribution which we find in our atomistic MD simulations are similar, in general terms, to those described by certain modern contact mechanics theories [19-22] where the influence of adhesive interactions is included.

Tip-Substrate Separation After Contact
_____

Starting from contact the force versus distance (F_z vs. d_{hs}) curve exhibits a marked hysteresis seen both experimentally and theoretically (Fig. 1a) as the surfaces are separated [7a]. We remark that, in the simulation and the measurements [7a], separating the surfaces prior to contact results in no hysteresis. The hysteresis is a consequence of the adhesive bonding between the two materials and, as demonstrated by the simulation, separation is accompanied by inelastic processes in which the topmost layer of the Au sample adheres to the Ni tip. The mechanism of the process is demonstrated by the pressure contours during liftoff of the tip shown in Fig. 2b, recorded for the configuration marked G (d_{ts} = 5.5 Å in Fig. 1a). As seen the maximum tensile stress is located near the edges of the adhesive contact. We further observe that the diameter of the contact area decreases during lifting of the tip, resulting in the formation of a thin "adhesive neck" due to ductile extension, which stretches as the process continues, ultimately breaking at a distance d_{ts} of ~ 9-10 Å. The evolution of adhesion and tear mechanisms which we observe can be classified as mode-I fracture [33], reemphasizing the importance of forces operating across the crack in modeling crack propagation [33,34].

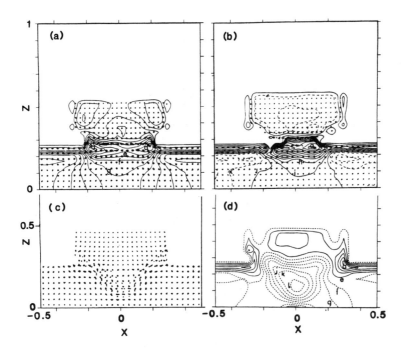

Fig. 2. Calculated pressure contours and atomic configurations viewed along the [010] direction, in slices through the system. The Ni tip occupies the topmost eight atomic layers. Short-time atomic trajectories appear as dots. Distance along the X and Z directions in units of X = 1 and Z = 1 corresponding to 61.2 A each. Solid contours correspond to tensile stress (i.e., negative pressure) and dotted ones to compressive stress. (a) after jump-to-contact (point D in Fig. 1a). The maximum magnitude of the tensile (i.e., negative pressure, 10GPa, is at the periphery of the contact, (X,Z) = (± 0.19,0.27). The contours are spaced with an increment, Δ, of 1 GPa. Thus the contours marked e, f and g correspond to -6, -5 and -4 GPa, respectively. (b) During separation following contact, (point G in Fig. 1a). The maximum tensile pressure (marked a), ~ -9GPa, is at the periphery of the contact at (X,Z) equal to (0.1,0.25) and (-0.04,0.25). Δ = 0.9 GPa. The marked contours h, i, j and k correspond to -2.5, -1.6, -0.66 and 0.27GPa, respectively. (c) Short-time particle trajectories at the final stage of relaxation of the system, corresponding to point M in Fig. 1b, (i.e., F = 0). Note slip along the [111] planes in the substrate. (d) Pressure contours corresponding to the final configuration shown in (c). Note the development of compressive pressure in the substrate which maximizes in the region of the contour marked ℓ (8.2GPa). The increment between contours Δ = 1.4GPa. The contours marked a and e correspond to -6.4GPa and -1.1GPa, respectively, and those marked f and g to 0.2 and 1.6GPa.

Indentation

We turn now to theoretical results recorded when the tip is allowed to advance past the jump-to-contact point, i.e., indentation (see Figs. 1(b,c), and Fig. 3). As evident from Fig. 1b, decreasing the separation between the tip and the substrate causes first a decrease in the magnitude of the force on the tip (i.e., less attraction, see segment DL) and an increase in the binding energy (i.e., larger magnitude of the potential energy, shown in Fig. 1c). However, upon reaching the point marked L in Fig. (1b) a sharp increase in the attraction occurs, followed by a monotonic decrease in the magnitude of the force till $F_z = 0$ (point M in

Fig. 1b) at $d_{ts} = 0.8$ Å. The variations of the force (in the segment DLM) are correlated with large deformations of the Au substrate. In particular, the nonmonotonic feature (near point L) results from tip-induced flow of gold atoms which relieve the increasing stress via wetting of the sides of the tip. Indeed the atomic configurations display a "piling-up" around the edges of the indenter due to atomic flow driven by the deformation of the Au substrate and the adhesive interactions between the Au and Ni atoms. Further indentation is accompanied by slip of Au layers (along (111) planes) and the generation of interstitial defects. In addition, the calculations predict that during the indentation process a small number of Ni atoms diffuse into the surrounding Au, occupying substitutional sites. Furthermore the calculated pressure contours at this stage of indentation, shown in Fig. 2d, demonstrate that the substrate surface zone in the vicinity of the edges of the tip is under tensile stress, while the deformed region under the tip is compressed with the maximum pressure (8.2GPa) occurring at about the fifth Au layer below the center of the Ni tip-indenter. The general characteristics of the pressure (and stress) distributions obtained in our indentation simulations correspond to those associated [12,19,23] with the onset and development of plastic deformation in the substrate.
   Experimentally, advancing the sample past the contact point is noted by the change in slope of the force as the increasing repulsive forces push the tip and cantilever back towards their rest position. We remark that the calculated pressures from the simulations compare favorably with the average contact pressure of ~ 3 GPa determined experimentally [7a] by dividing the measured attractive force by the estimated circular contact area of radius 20 nm.

Tip-Substrate Separation After Indentation

Reversal of the direction of the tip motion relative to the substrate from the point of zero force (point M in Fig. 1b) results in the force- and potential energy-versus distance curves shown in Figs. 1b and 1c. The force curve exhibits first a sharp monotonic increase in the magnitude of the attractive force (segment MN in Fig. 1b) with a corresponding increase in the potential energy (Fig. 1c). During this stage the response of the system is mostly elastic accompanied by the generation of a small number of vacancies and substitutional defects in the substrate. Past this stage the force and energy curves versus tip-to-sample separa-

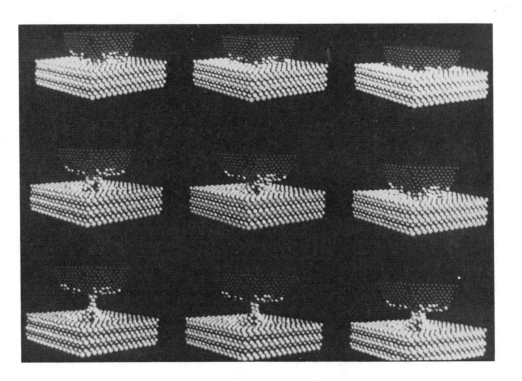

Fig. 3. Sequence of atomic configuration starting from a Ni tip
         indented in a Au (001) substrate (top right) and during the
         process of retraction of the tip (from right to left)
         accompanied by formation of a connective neck.

tion exhibit a nonmonotonic behavior which is associated mainly with the
process of elongation of the connective neck which forms between the
substrate and the retracting tip.

To illustrate the neck formation and elongation process we show in
Fig. 4 a sequence of atomic configurations corresponding to the maxima in
the force curve (Fig. 1b, points marked O, Q, S, U, W and X). As evident, upon increased separation between the tip-holder and the substrate
a connective neck forms consisting mainly of gold atoms (see atomic

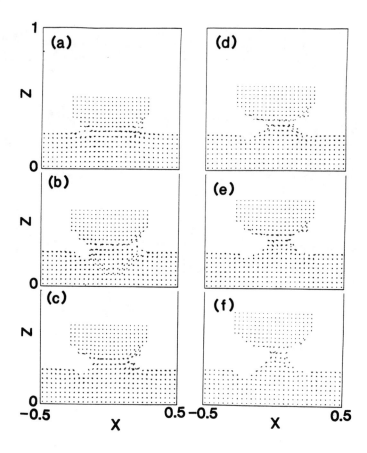

Fig. 4. Atomic configurations in slices through the system illustrating the formation of a connective neck between the Ni tip and the Au substrate during separation following indentation. The Ni tip occupies the topmost eight layers. The configurations (a-f) correspond to the stages marked O, Q, S, U, W, and X in Fig. 1b. Note the crystalline structure of the neck. Successive elongations of the neck, upon increased separation between the tip-holder assembly and the substrate, occur via structural transformation resulting in successive addition of layers in the neck accompanied by narrowing (i.e., reduction in cross-sectional area of the neck). Distance in units of X and Z, with X = 1 and Z = 1 corresponding to 61.2 Å.

configurations shown in Fig. 3). The mechanism of elongation of the neck involves atomic structural transformations whereby in each elongation stage atoms in adjacent layers in the neck disorder and then rearrange to form an added layer, i.e., a more extended neck of a smaller cross-sectional area. Throughout the process the neck maintains a layered

crystalline structure (see Figs. 4) except for the rather short structural transformation periods, corresponding to the sharp variations in
the force curve, (see segments PQ, RS, TU and VW in Fig. 1b) and the
associated features in the calculated potential energy shown in Fig. 1c
where the minima correspond to ordered layered structures after the
structural rearrangements. We note that beyond the initial formation
stage, the number of atoms in the connective neck region remains roughly
constant throughout the elongation process.

Further insight into the microscopic mechanism of elongation of the
connective neck can be gained via consideration of the variation of the
second invariant of the stress deviator, J_2, which is related to the von

Mises shear strain-energy criterion for the onset of plastic yielding
[19,22,26]. Returning to the force and potential energy curves shown in
Figs. 1b and 1c, we have observed that between each of the elongation
events (i.e., layer additions, points marked Q, S, U, W and X) the
initial response of the system to the strain induced by the increased
separation between the tip-holder and the substrate is mainly elastic
(segments OP, QR, ST, UV in Fig. 1b, and correspondingly the variations
in Fig. 1c), accompanied by a gradual increase of $\sqrt{J_2}$, and thus the

stored strain energy. The onsets of the stages of structural rearrangements are found to be correlated with a critical maximum value of $\sqrt{J_2}$ of

about 3 GPa (occuring for states at the end of the intervals marked OP,
QR, ST and UV in Fig. 1b) localized in the neck in the region of the
ensuing structural transformation. After each of the elongation events
the maximum value of $\sqrt{J_2}$ (for the states marked Q, S, U, W and X in Fig.

1b) drops to approximately 2 GPa.

In this context, it is interesting to remark that the value of the
normal component of the force per unit area in the narrowest region of
the neck remains roughly constant ($\sim$ 1 GPa) throughout the elongation
process, increasing by about 20% prior to each of the aforementioned
structural rearrangements. This value has been estimated both by using
the data given in Figs. 1b and the cross sectional areas from atomic
configuration plots (such as given in Fig. 4), and via a calculation of
the average axial component (zz element) of the atomic stress tensors
[51] in the narrow region of the neck. We note that the above observations constitute atomic-scale realizations of basic concepts which
underlie macroscopic theories of materials behavior under load [18-24].

A typical distribution of the stress, $\sqrt{J_2}$, prior to a structural

transformation is shown in Fig. 5 (shown for the state corresponding to
the point marked T in Fig. 1b). As seen, the maximum of $\sqrt{J_2}$ is localized

about a narrow region around the periphery in the strained neck.
Comparison between the atomic configuration at this stage (see Fig. 5, or
the very similar configuration shown in Fig. 4c) and the configuration
after the structural transformation has occurred (see Fig. 4d, corresponding to the point marked U in Fig. 1b) illustrates the elongation of
the neck by the addition of a layer and accompanying reduction in areal
cross section. We note that as the height of the connective neck increases the magnitude of the variations in the force and potential energy

478

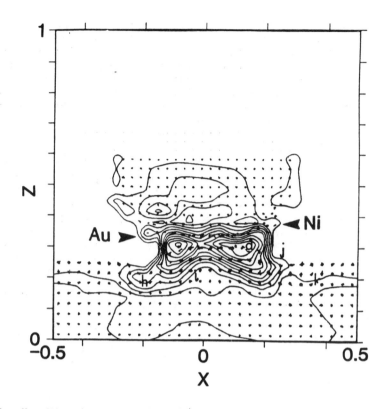

Fig. 5. Von Mises' shear stress ($\sqrt{J_2}$) corresponding to the configuration marked T in Fig. 1b (that is, just before the structural transformation resulting in the configuration (d) in Fig. 4). The proximal interfacial layers of Ni and Au are marked by arrows. The maximum contours (2.9 GPa, marked a) occur on the periphery of the neck (X, Z) = ($\pm$01, 0.3). The increment between contours is 0.2 GPa. The contours marked h, i, j, and k correspond to 1.1, 0.9, 0.7, and 0.5 GPa, respectively. Distance along X and Z in units of X = 1 and Z = 1 corresponding to 61.2 Å.

during the elongation stages diminishes. The behavior of the system past the state shown in Fig. 4f (corresponding to the point marked X in the force curve shown in Fig. 1b) is similar to that observed at the final stages of separation after jump-to-contact (Fig. 1b), characterized by strain induced disordering and thinning in a narrow region of the neck near the gold covered bottom of the tip and eventual fracture of the neck (occuring for a tip-to-substrate distance $d_{ts} \simeq 18$ Å), resulting in a Ni tip whose bottom is covered by an adherent Au layer.

The theoretically predicted increased hysteresis upon tip-substrate separation following indentation, relative to that found after contact (compare Figs. 1a and 1b), is also observed experimentally [7a]. In both theory and experiment the maximum attractive force after indentation is roughly 50% greater than when contact is first made. Note however that the nonmonotonic features found in the simulations (Fig. 1b) are not discernible in the experiment which is apparently not sufficiently sensitive to resolve such individual atomic-scale events when averaged over the entire contact area.

2. Gold Covered Ni Tip/Au Surface [8]

Having described in the previous section the processes occurring as a result of the interaction between a clean Ni tip with a gold surface, we turn next to a system where a Ni tip "wetted" by a gold monolayer is lowered and subsequently retracted from a clean initially undamaged Au(001) surface.

As aforementioned the retraction of the Ni tip from the gold surface after formation of an adhesive contact between the two, is accompanied by wetting of the tip by gold atoms. The gold coated Ni tip used in the present simulations was obtained following a slight indentation (see Section III.1).

The force vs. distance curve obtained for the system, along with the one corresponding to the clean Ni tip (see also Fig. 1a) shown in Fig. 6 and inspection of atomic configurations reveal that while the adhesive interaction is reduced for the coated tip, jump-to-contact instability, formation of an adhesive contact and hysteresis during subsequent retraction occur in both cases. However we should note that in the present case while a connective neck, made solely of substrate gold atoms, is formed during retraction of the tip, (of dimensions similar to that formed upon indentation, see Fig. 3) insignificant transfer of atoms from the surface to the tip occurs upon complete separation (i.e., while gold wets by adhering to a bare nickel tip, no wetting occurs for a gold covered tip).

3. Clean Gold Tip/Nickel Surface [7b]

In the studies discussed in Section III.1 the tip material (i.e.,Ni) was the harder one, characterized additionally by a larger surface energy. In this section studies where the materials composing the tip and surface are interchanged, i.e., a Au tip interacts with a Ni(001) surface. In these investigations the substrate consists again of n_s = 3 and n_d = 8 layers, exposing Ni(001) surface, but with a larger number of atoms per layer (n = 800) than in the previous studies to allow for spreading of the gold tip (see below). The gold tip, exposing a Au(001) facet, is prepared in a similar manner to that described in Section III.1.

Contact Formation and Separation

The simulated force on the tip and potential energy versus distance curve

480

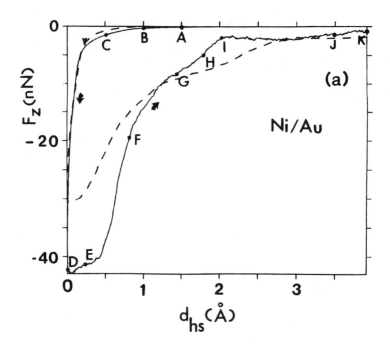

Fig. 6. Calculated force on the tip atoms, F_z, versus tip-to-sample distance, d_{hs}, for the case of a gold coated Ni tip approaching and then being retracted from a clean Au (001) surface (dashed line), as well as for the case of a clean Ni tip (see Fig. 1a).

are shown in Fig. 7, and atomic configurations of the system (i.e., short-time atomic trajectories, shown for a slice of 11 Å in width through the system) recorded at selected stages during the tip-to-substrate approach and subsequent separation are shown in Fig. 8.

Following an initial slow variation of the force between the Ni substrate and the Au tip we observe the onset of an instability, signified by a sharp increase in the attraction between the two (see segment AB in Fig. 7a) which is accompanied by a marked decrease in the potential energy of the system (see sudden drop of E_p in Fig. 7b as d_{hs} approaches zero). We note the rather sudden onset of the instability which occurs only for separations d_{hs} smaller than 0.25 Å (marked by an arrow on the curve in Fig. 7a and corresponding to d_{ts} = 4.2 Å). Our simulations reveal that in response to the imbalance between the forces on atoms in each of the materials and those due to intermetallic interactions a JC phenomenon occurs via a fast process where Au tip atoms

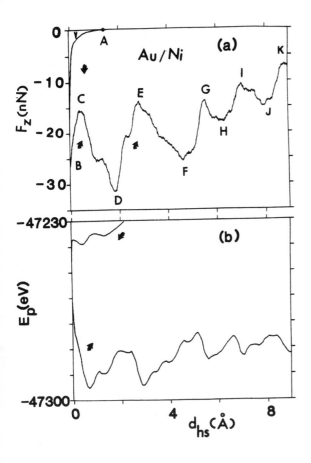

Fig. 7: Calculated force F_z (in a) and potential energy of the system E_p (in b), versus tip-to-sample distance, d_{hs}, between an Au tip and a Ni sample for approach and jump-to-contact followed by separation. d_{hs} denotes the distance between the rigid tip-holder assembly and the static substrate of the Ni surface (d_{hs} = 0 at the jump-to-contact point, marked B). The distance between the bottom layer of the tip and the top layer of the sample is 4.2 Å at the onset of the jump-to-contact i.e., last stable point upon tip lowering, (marked by an arrowhead in (a)). That distance is 2.1 Å after contact formation (point marked B). The points marked C, E, G, I and K correspond to ordered configurations of the tip each containing an additional layer. Force in units of nN, energy in eV and distance in Å.

displace by approximately 2 Å toward the surface in a short time span of ~ 1 ps (see Figs. 8a and 8b where the atomic configurations before and after the JC are depicted). After the JC occurs the distance between the bottom layer of the Au tip and the top layer of the Ni surface decreases to 2.1 Å from a value of 4.2 Å. The response in this system should be contrasted with that observed by us in simulations of a Ni tip approaching a Au surface, (see Section A.1) where the JC phenomenon involved a bulging of the Au surface underneath the tip.

Contours of the local pressure (evaluated as the trace of the atomic

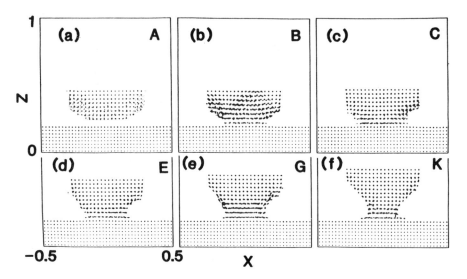

Fig. 8: Atomic configurations in a slice, in the XZ plane (i.e., containing the [100] and [001] directions), of width 11 Å through the system
(in the [010] direction) illustrating the atomic arrangements in the
system before and after jump-to-contact (in a and b, respectively) and
for the ordered configurations during the tip elongation processes which
occur upon retraction of the tip from the point of adhesive contact (in
c-f). The capital letters identify the corresponding points on the F_z
vs. distance curve given in Fig. 7a. Note the crystalline structure of
the neck. Successive elongations of the neck, upon increased separation
between the tip-holder assembly and the substrate, occur by way of a
structural transformation resulting in successive addition of layers in
the neck, accompanied by narrowing (that is, reduction in cross-sectional
area of the neck). Distance in units of X and Z, with X = 1 and Z = 1
corresponding to 70.4 Å.

stress tensors) and the square root of J_2 (second invariant of the stress
deviator), after JC had occurred, are shown in Fig. 9a and 9b, respectively. We observe that the pressure on the periphery of the tip is large
and negative (tensile), achieving values up to -1 x 10^5 atm = -10^{10} N/m =
10 GPa, while the middle core of the tip is under a small compressive
pressure. The $\sqrt{J_2}$ maximizes in the interfacial contact region achieving
a value of 2.0 x 10^4 atm = 2 GPa.
    As a consequence of the formation of the adhesive contact between
the two materials, the interfacial region of the gold tip exhibits large
structural rearrangements, both in the normal and lateral directions.

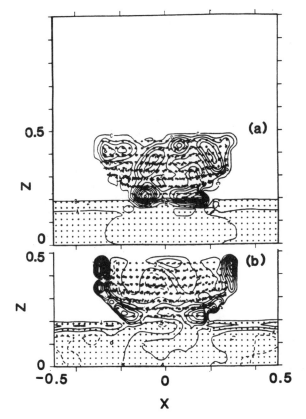

Fig. 9: Contours of the von Mises' shear stress ($\sqrt{J_2}$, in a) and pressure (in b) after the formation of adhesive contact between the Au tip and the Ni surface (corresponding to point B in Fig. 7a). Note that $\sqrt{J_2}$ maximizes at the periphery of the contact achieving a value of 2 GPa. The increment between the contours in a is 0.17 GPa. The hydrostatic pressure (in b) is tensile and large on the periphery of the tip and achieves a value of ~ -5 GPa at the peripheral region of the contact. The increment between contours in 1 GPa. Distance along the X and Z directions in units of X = 1 and Z = 1 corresponding to 70.4 $\mathring{A}$.

The atomic configuration in the three layers of the tip closest to the Ni surface are shown in Fig. 10. The structure of the proximal Au layer (large dots in Fig. 10a, superimposed on the atomic structure of the underlying Ni (001) surface) exhibits a marked tendency towards a (111) reconstruction (in this context we remark that only a small tendency towards a (111) structure was observed by us for a patch of a gold monolayer deposited on a Ni (001) surface.) Furthermore, the (111) reconstruction extends 3-4 layers from the interface into the Au tip. Accompanying the epitaxial surface structural rearrangement in the gold, which is partially driven by the lattice constant mismatch between gold and nickel, an increase in interlayer spacing occurs (d_{12} = 2.44 $\mathring{A}$, d_{23} = d_{34} = d_{45} = 2.53 $\mathring{A}$, d_{56} = d_{67} = 2.2 $\mathring{A}$ and d_{78} = 2.1 $\mathring{A}$, compared to the interlayer spacing between (001) layers in the bulk gold of 2.04 $\mathring{A}$ where d_{nn+1} is the spacing between layers n and n+1, and layer number 1 corresponds to the Au layer proximal to the Ni topmost surface layer). The (111) reconstruction and expanded interlayer spacings in the interfacial region of the Au tip persist throughout the separation process.

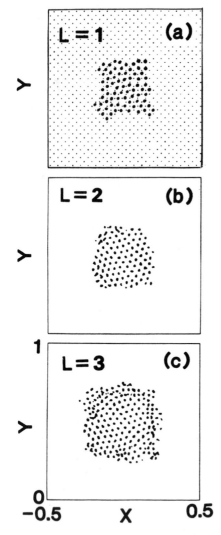

Fig. 10: Atomic configurations in layers of the Au tip closest to the Ni substrate after the formation of the adhesive contact, (point B in Fig. 7a). L = 1 corresponds to the proximal layer (large dots in a) of the Au tip in contact with the Ni (001) substrate (small dots in a). Note the (111) reconstruction of the gold layers. L = 2 and L = 3 are the two layer above the proximal layer. Distance along the X ([100]) and Y ([001]) directions in units of 70.4 Å.

Starting from contact the force versus distance (F_z versus d_{hs}) curve exhibits a marked hysteresis as the tip is retracted from the substrate (see Fig. 7a). We remark that separating the surfaces prior to jump-to-contact results in no hysteresis. The hysteresis is a consequence of the adhesive bonding between the two materials and, as demonstrated by the simulation, separation is accompanied by inelastic processes and the formation of an extended gold connective neck (see Fig. 8(c-f)).

The hysteresis in the force vs. distance curve exhibits marked variations (see Fig. 7a) which portray the atomistic processes of gold neck formation and elongation. In this context we remark that our

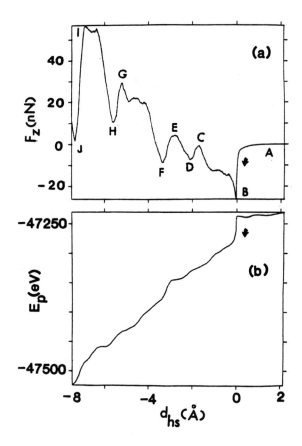

Fig. 11: Calculated force F_z (in a) and potential energy of the system E_p (in b), versus tip-to-sample distance, d_{hs}, for a Au tip lowered towards a Ni (001) surface. The point marked B (d_{hs} = 0) corresponds to the adhesive contact. d_{hs} < 0 corresponds to continued motion of the tip toward the surface past the contact point. The points marked E, G and I correspond to ordered configurations of the compressed tip, each containing one atomic layer less than the previous one (starting with 8 layer at point E). Force in units of nN, energy in eV and distance in Å.

earlier simulations (and accompanying AFM experiments) of a nickel tip interacting with a gold surface have also shown a marked hysteresis upon separation from adhesive contact. However in that case the variation of the force F_z vs. distance curve was monotonic, and the extension of the connective neck was rather limited. Variations similar to those shown in Fig. 7a were observed by us before (Section III.1) only upon tip-sample

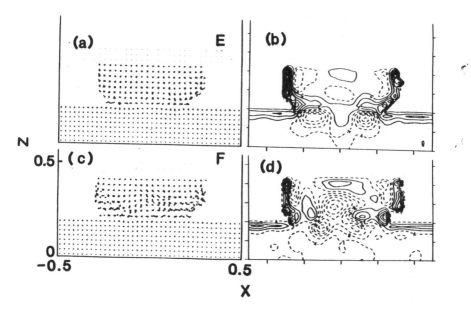

Fig. 12: Atomic configurations and pressure contours in an 11 Å slice through the system, corresponding to the points marked E and F in Fig. 11a. Note that in (a) the tip consists of 8 Au layers, while in (b), the peripheral region rearranges to form 7 layers, while the core region consists still of 8 layers, thus containing an interstitial defect layer. In (b), corresponding to point E, the compressive pressure (dashed contours) concentrates and maximizes just below the Ni surface, achieving a value of 6 GPa. The increment between contours is 1.2 GPa. Solid contours correspond to tensile pressure. The pressure contours during the intermediate state between ordered configurations (point F in Fig. 11(a)) exhibit concentration of compressive pressure in the core defect region (achieving a maximum value of 4.8 GPa inside the tip) and just below the surface. The increment between contours is 1 GPa.

separation following a slight indentation (see Fig. 1b) of the Au surface by the Ni tip.

The maxima in Fig. 7a (points marked C, E, G, I and K) are associated with ordered structures of the elongated connective neck, each corresponding to a neck consisting of one more layer than the previous one. Thus the number of Au layers at point C is 9 (one more than the original 8-layer tip), 10 layers at point E, etc. (see Fig. 8).

The mechanism of elongation of the neck is similar to that discussed before (see Section III.1) involving atomic structural transformations whereby in each elongation stage atoms in adjacent layers in the neck disorder and then rearrange to form an added layer, that is, a more extended neck of a smaller cross-sectional area. Throughout the process the neck maintains a layered crystalline structure except for the rather short structural transformation periods, corresponding to the sharp variations in the force curve, (see segments BC, DE, FG, HI and JK in

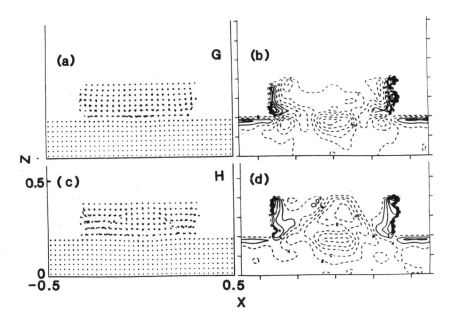

Fig. 13: Same as Fig. 12 for the configurations marked G and H in Fig. 11a. The ordered configuration marked G contains 7 layers in the Au tip and evolves from configuration F (see Fig. 12c). Note the increase in contact area. Configuration H contains 6 layers at the outside region of the tip and an interstitial defect layer at the tip core region. As in Fig. 12 the compressive pressure (dashed contours) maximizes just below the surface region for the ordered configuration (in b) achieving a value of 7.6 GPa). During the intermediate stage (in d) compressive pressure concentrates near the surface region (where it achieves a maximum value of 7.7 GPa) and in the core of the tip (where a maximum value of 6.2 GPa is achieved). The increment between contours is 1.5 GPa.

Fig. 7a and the associated features in the calculated potential energy shown in Fig. 7b where the minima correspond to ordered layered structures after the structural rearrangements.

As we mentioned before, associated with the formation of the adhesive neck and throughout the elongation stage the interfacial region of the neck is structurally reconstructed laterally (i.e., (111) reconstruction in layers) accompanied by expanding interlayer spacings compared to that in Au (001). The depth of the reconstructed region tends to be larger during the stretching stages, prior to the reordering which results in an additional layer. The reconstruction and increased interlayer spacings are related to the coupled effects of lattice constant mismatch between Au and Ni and the fact that the tensile stress in the normal direction acts on interfacial layers whose areas are smaller than those in the region of the tip close to the static holder, resulting in a larger strain in the former region.

Tip Compression

Having discussed the processes of contact formation and pulloff, we show
in Fig. 11 the force and potential energy curves obtained when the motion
of the Au tip towards the Ni substrate is continued past contact formation (point marked B, $d_{hs} = 0$, in Fig. 11. In this Figure $d_{hs} < 0$
corresponds to continued motion towards the surface). The force curve
exhibits nonmonotonic variations and reverses sign, signifying the onset
of a repulsive interaction.

We observe that a continued compression of the tip results in an increase in the interfacial contact area between the tip and the substrate,
and a "flattening" of the tip. To illustrate the process we show in
Figs. 12 and 13 atomic configurations and pressure contours of the system, corresponding to the points marked E, F, G and H on the force curve
(Fig. 11a). As can be seen the number of layers in the Au tip corresponding to points E and G (Figs. 12a and 13a, respectively) decreases
from 8 to 7 (contact areas $\sim$ 1100 $\overset{o}{A}^2$ and 1750 $\overset{o}{A}^2$, respectively). The
pressure contours corresponding to these configurations show tensile
(negative) pressure on the sides of the tip and a concentration of
compressive pressure in the interfacial substrate region.

The evolution of the structural transformation, induced by continued
compression, between the ordered structures, involves the generation of
interstitial-layer partial dislocations in the central region (core) of
the tip and subsequent transformation, as may be seen from Figs. 12c and
13c (the latter one rearranges to a 6-layer tip upon continued compression (point marked I in Fig. 11a, with a contact surface area of $\sim$ 2400
$\overset{o}{A}^2$). In the course of lowering of the tip (between ordered configurations) the outer regions of the tip rearrange first to reduce the number
of crystalline layers leaving an interstitial-layer-defect in the core of
the tip which is characterized as a high compressive pressure region (see
Figs. 12d and 13d). Continued lowering of the tip results in a "dissolution" (or annealing) of the interstitial-layer defect which is achieved
by fast correlated atomic motions along preferred (110) directions (see
Fig. 14, showing short-time trajectories in layers 1-4 of the tip,
recorded in the segment FG of Fig. 11a). Accompanying the annealing of
the defect the compressive pressure transfers from the tip to the
interfacial substrate region. The initial atomic rearrangement of the
peripheral region of the tip and the eventual expulsion of the remaining
interstitial layer atoms from the core, contribute to the increase of the
contact area between the tip and the substrate.

4. Nickel Tip/Hexadecane Film/Gold Surface [8]

Our discussion up to this point was confined to the interaction between
material tips and bare crystalline substrates. Motivated by the fundamental and practical importance of understanding the properties of adsorbed molecularly thin films and phenomena occuring when films are
confined between two solid surfaces, pertaining to diverse fields [70-73]
such as lubrication, prevention of degradation and wear, wetting, spreading and drainage, we have initiated most recently investigations of such

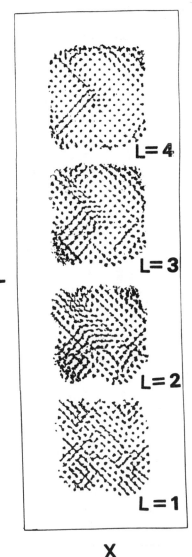

Fig. 14. Atomic trajectories in the interfacial layers of the compressed Au tip, recorded during part of the segment FG in Fig. 11(a), i.e., between configurations starting from one which contains a core interstitial defect layer and ending in an ordered configuration [compare Figs. 12(c) and 13(a)]. L = 1 corresponds to the proximal gold layer. The trajectories illustrate the atomic mechanism by which the core interstitial layer defect is expelled.

systems [74]. Among the issues which we attempted to address are the structure, dynamics, and response of confined complex films, their rheological properties, and modifications which they may cause to adhesive and tribological phenomena, such as inhibition of jump-to-contact instabilities and prevention of contact junction formation. Furthermore, these studies are of importance in light of recent AFM experiments on adsorbed polymeric films [75]. In the following we highlight certain of our results.

490

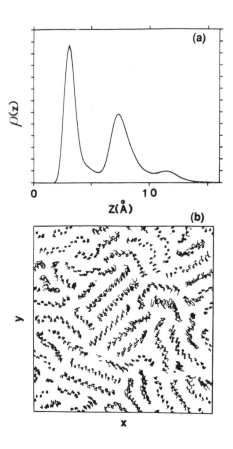

Fig. 15. Density profile (in a) and top view of the first molecular layer adsorbed on the surface (in b) for a thin film of n-hexadecane adsorbed on a Au (001) surface at 300K.

Distance in units of Å. The dimension along the x and y

axis in (b) is 61.2 Å. The origin of the z axis (normal to the surface) is at the average position of the centers of the gold atoms in the top-most surface layer.

The molecular film which we studied, n-hexadecane ($C_{16}H_{34}$) is model-ed by interaction potentials developed by Ryckart and Bellemans [76] and which have been employed before in investigations of the thermodynamic, structural and rheological properties of bulk liquid n-alkanes [77]. In this model the CH_2 and CH_3 groups are represented by pseudo-atoms of mass 2.41×10^{-23} grams, and the intermolecular bond lengths are fixed at 1.53 Å and the bond-angles at 109° 28'. A 6-12 Lennard-Jones (LJ) potential describes the intermolecular interaction between sites (pseudo-atoms) in different molecules, and the intramolecular interactions between sites more than three apart. The LJ potential well-depth parameter $\epsilon_2 = 6.2 \times 10^{-3}$ eV, and the distance parameter $\sigma_2 = 3.923$ Å. The range of the LJ interaction is cut-off at 9.8075 Å. An angle dependent dihedral poten-tial is used to model the effect of missing hydrogen atoms on the molecular conformation.

The substrate (Au(001)) and tip (Ni) which we use are described using the EAM potentials as in our aforementioned studies of Ni/Au (001) (see Section A.1). The interaction between the n-hexadecane molecules and the metallic tip and substrate is modeled using a LJ potential with $\epsilon_3 = 3\epsilon_2 = 18.6 \times 10^{-3}$ eV and $\sigma_3 = 3.0715$ Å. The cutoff distance of the molecule-surface interaction is 7.679 Å. We remark that the choice of ϵ_3 corresponds to an enhanced adsorption tendency of the alkane molecules onto the metals, which is a reasonable assumption, based on theoretical estimates obtained using the theory of dispersion interactions [70].

All details of the simulation pertaining to the metallic tip and substrate are as those given in Section III.1. The hexadecane film is composed of 73 alkane molecules (1168 pseudo-atoms) equilibrated initially on the Au (001) surface at a temperature of 300K. The constrained equations of motion for the molecules are solved using a recently proposed method [78], employing the Gear 5-th order predictor corrector algorithm.

The equilibrated adsorbed molecular film prior to interaction with the Ni tip is layered (see Fig. 15a) with the interfacial layer (the one closest to the Au (001) substrate, see Fig. 15b) exhibiting a high degree of orientational order. The molecules in this layer tend to be oriented parallel to the surface plane.

Lowering of the (001) faceted Ni tip to within the range of interaction causes first adherence of some of the alkane molecules to the tip resulting in partial "swelling" of the film [75,79] and a small attractive force on the tip. Continued approach of the tip toward the substrate causes "flattening" of the molecular film, accompanied by partial wetting of the sides of the tip, and reduced mobility of the molecules directly underneath it (see short time trajectories shown in Fig. 16a, corresponding to a distance $d_{ts} = 9.5$ Å between the bottom layer of the tip and the top-most layer of the Au (001) surface). The arrangement of molecules in the interfacial layer of the film is shown in Fig. 17a. At this stage the tip experiences a repulsive force $F_z = 2$ nN.

Continued lowering of the tip induces drainage of the second layer molecules from under the tip, increased wetting of the sides of the tip, and "pinning" of the hexadecane molecules under it. Side views for several tip-lowering stages are shown in Fig. 16(b-d), corresponding to $d_{ts} = 6.5$ Å (in b), 5.1 Å (in c), and 4.0 Å (in d and e). (The corresponding recorded forces on the tip, after relaxation, for these values of the tip-to-surface separations are 0 nN, 25nN, and -5nN, respectively.) Note that for $d_{ts} = 5.1$ Å and $d_{ts} = 4.0$ Å the region of the surface of the gold substrate directly under the tip is deformed and the above d_{ts} values represent averages over the whole surface area (in this context we mention that for $d_{ts} = 5.1$ Å the average pressure in the contact area between the tip and the sample is ~ 2 GPa). We also remark that we have

492

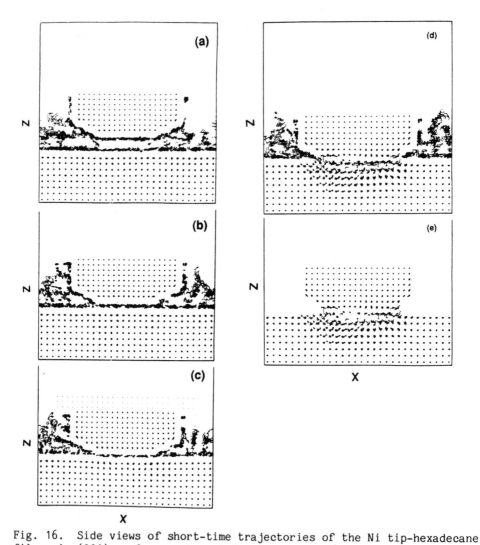

Fig. 16. Side views of short-time trajectories of the Ni tip-hexadecane film - Au (001) surface system at four stages of the tip lowering process: (a) d_{ts} = 9.5 Å, (b) d_{ts} = 6.5 Å, (c) d_{ts} = 5.1 Å, (d,e) d_{ts} = 4.0 Å, where in (d) both the metal atoms and alkane molecules are displayed, and in (e) only the metal atoms are shown. The values of d_{ts} are average distances between atoms in the bottom layer of the nickel tip and those in the topmost layer of the gold substrate.

observed that during the later stages of the tip-lowering process, drainage of entangled, or "stapled", molecules from under the tip is assisted by transient local inward deformations of the substrate which apparently

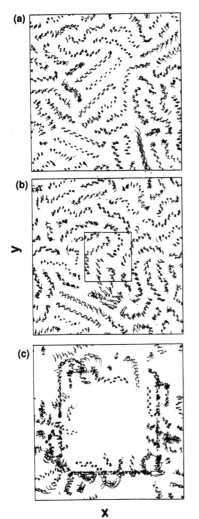

(a)

(b)

y

(c)

x

Fig. 17. (a) Arrangement of molecules in the first interfacial layer adsorbed on the Au (001) substrate for d_{ts} = 9.5 Å and in the first molecular layer (b) and the region above it (c), at d_{ts} = 6.5 Å. Note that molecules above the first layer drained from under the tip. The inner marked square in (b) denotes the projected area of the bottom layer of the tip.

lower the barriers for the relaxation of such unfavorable conformations of the confined alkane molecules. The arrangement of molecules in the first adsorbed alkane layer and in the region above it for d_{ts} = 6.5 Å is shown in Fig. 17(b,c). The molecules are oriented preferentially parallel to the surface, particularly in the region under the tip (exhibiting in addition a reduced mobility).

Comparison of the response of the system with that described in Section III.1 for the bare gold surface (see Fig. 1a) reveals that while in the latter case the force between the tip and the substrate is attractive throughout (and remains attractive even for a slight indentation of the surface, see Fig. 1b), the overall force on the tip in the presence

494

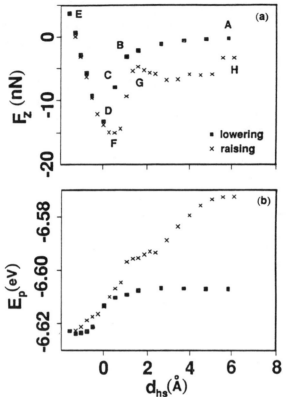

Fig. 18. Calculated force on the tip atoms, F_z, in (a), and potential energy per atom in the tip, E_p, versus tip-to-sample distance d_{hs}, for a CaF_2 tip approaching (filled squares) and subsequently retracting (crosses) from a CaF_2 (111) surface. The distance from the bottom layer of the tip to the top-most surface layer, d_{ts}, for the points marked by letters is: A (8.6 Å), B (3.8 Å), C (3.0 Å), D (2.3 Å), E (1.43 Å), F (2.54 Å), G (2.7 Å) and H (3.3 Å). Distance in A, energy in eV, and force in nN.

of the adsorbed alkane film is repulsive for relative tip-to-substrate distances for which it was attractive in the other case (except for the initial stages of the tip approach process). However we note for the smallest tip-to-substrate separation which we investigated here (average distance d_{ts} = 4.0 Å before relaxation) the onset of intermetallic contact formation, occuring by displacement of gold atoms towards the nickel tip accompanied by partial drainage of alkane molecules, resulting in a

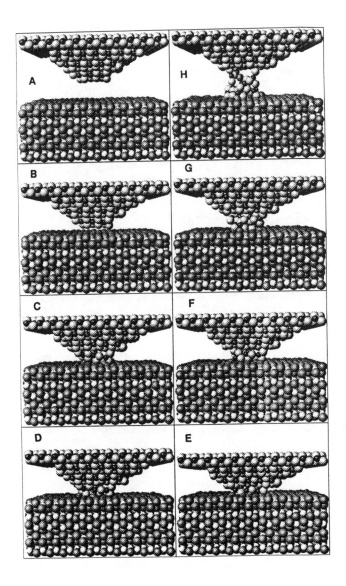

Fig. 19. Atomic configurations corresponding to the marked points in Fig. 18a. Small and large balls correspond to Ca^{+2} and F^- ions, respectively. The images were obtained for a cut in the middle of the system.

net attractive force on the tip of about -5nN, (the intermetallic contribution to this force is about -20nN and the alkane repulsive contribution
is about 15nN).

From these preliminary results we conclude that lowering of a faceted nickel tip towards a gold surface covered by a thin adsorbed n-hexadecane film results first in small attraction between the film and the tip
followed, upon further lowering of the tip, by ordering (layering) of the
molecular film. During continued approach of the tip toward the surface
the total interaction between the tip and the substrate (metal plus film)
is repulsive, and the process is accompanied by molecular drainage from
the region directly under the tip, wetting of the sides of the tip, and
ordering of the adsorbed molecular monolayer under the tip. Further
lowering of the tip is accompanied by inward deformation of the substrate
and eventual formation of intermetallic contact (occuring via displacement of surface gold atoms towards the tip) which is accompanied by
partial molecular drainage and results in a net attractive force on the
tip. The implications of these results to the analysis of AFM measurements of the thickness of adsorbed films [75], and the dependence of the
results on the extent of the film, and on the nature of the adsorbed
molecular film and its interaction with the substrate and tip, are
currently under investigation in our laboratory.

5. CaF_2 Tip/CaF_2 Surface [8]

In the previous sections results pertaining to intermetallic contacts
were discussed. Here we turn to results obtained in simulations of a
CaF_2 tip interacting with a CaF_2 (111) surface. As remarked in Section
II, the nature of bonding in ionic materials is different from that in
metallic systems, including long-range Coulombic interactions.

In these simulations the substrate is modeled by 3 static layers
interacting with 12 layers of dynamic atoms, with 242 Ca^{+2} cations in
each calcium layer and 242 F^- anions in each fluorine layer, exposing the
(111) surface of a CaF_2 crystal (the stacking sequence is ABAABA... where
A and B correspond to all F^- and all Ca^{+2} layers, respectively. The top
surface layer is an A layer). The CaF_2 tip is prepared as a (111)
faceted microcrystal containing nine (111) layers, with the bottom layer
containing 18 F^- anions, the one above it 18 Ca^{+2} cations followed by a
layer of 18 F^- anions. The next 3 layers contain 50 ions per layer, and
the 3 layers above it contain 98 ions in each layer. The static holder
of the tip is made of 3 CaF_2 (111) layers (242 ions total). The system
is periodically replicated in the two directions parallel to the surface
plane and no periodic boundary conditions are imposed in the normal, z,
direction. The long range Coulomb interactions are treated via the Ewald
summation method and temperature is controlled to 300 K via scaling of
the velocities of atoms in the three layers closest to the static
substrate. The integration time step $\Delta t = 1.0 \times 10^{-15}$ s, and motion of

the tip occurs in increments of 0.5 Å over a time span of 1 ps (10^{-12} s). As before after each increment in the position of the tip-holder assembly the system is allowed to dynamically relax.

Curves of the average force, F_z, on the tip atoms recorded for the fully relaxed configurations, versus distance d_{ns}, are shown in Fig. 18a along with the corresponding variations in the potential energy of the tip atoms (Fig. 18b). From Fig. 18a we observe following a gradual increase in the attraction upon approach of the tip to the surface, the onset of an instability marked by a sharp increase in attraction occuring when the bottom layer of the tip approaches a distance $d_{ts} \sim 3.75$ Å from the top layer of the surface. This stage is accompanied by an increase in the interlayer spacing in the tip material, i.e., tip elongation, and is reminiscent of the jump-to-contact phenomena which we discussed in the context of intermetallic contacts, although the elongation found in the present case (~ 0.35 Å) is much smaller than that obtained for the metallic systems.

Decreasing the distance between the tip-holder assembly and the substrate past the distance corresponding to maximum adhesive interaction (which occurs at $d_{ts} \sim 2.3$ Å) results in a decrease in the attractive interaction, which eventually turns slightly repulsive (positive value of F_z), accompanied by a slight compression of the tip material. Starting from that point ($d_{hs} = 26.5$ Å, $d_{ts} \sim 1.4$ Å) and reversing the direction of motion of the tip-holder assembly (i.e., detracting it from the surface) results in the force curve denoted by crosses in Figs. 5a and b.

As clearly observed from Fig. 18a, the force versus distance relationship upon tip-to-substrate approach and subsequent separation exhibits a pronounced hysteresis. The origin of this behavior, which is also reflected in the tip potential-energy versus distance curve shown in Fig. 15b, is a plastic deformation of the crystalline tip, leading to eventual fracture. At the end of the lifting processes part of the tip remains bonded to the substrate. Atomic configurations corresponding to those stages marked by letters on the force curve (Fig. 18a) are shown in Fig. 19.

Tip Sliding

Starting from the tip-substrate configuration under a slight attractive load (see point marked by an arrow in Fig. 18a ($h_{ts} = 1.7$ Å, $F_z = -3.0$ nN) lateral motion of the tip parallel to the surface plane is initiated by translating the tip-holder assembly in the $\langle \overline{1},1,0 \rangle$ direction in increments of 0.5 Å followed by a period of relaxation, while maintaining the vertical distance between the tip-holder and the substrate at a constant value. This then corresponds to a constant-height scan in the language of atomic-force microscopy. We have also performed constant-

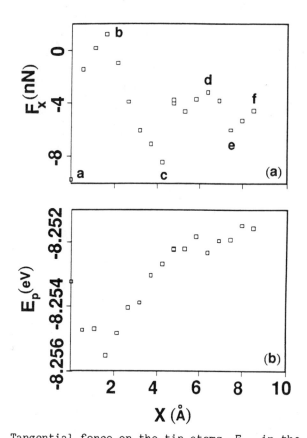

Fig. 20. Tangential force on the tip atoms, F_x, in the $\langle 1\overline{1}0 \rangle$ direction, and per-ion potential energy in the tip, E_p, versus distance (X, along

the $\langle 1\overline{1}0 \rangle$ direction), calculated for a CaF_2 tip translated at constant

height parallel to a CaF_2 (111) substrate surface in the $\langle 1\overline{1}0 \rangle$ direction. Note the oscillatory character of the force curve, portraying an atomic stick-slip process. Note also the increase in E_p with translated

distance. The marked points in (a) correspond to minima and maxima of

the F_x curve along the $\langle 1\overline{1}0 \rangle$ (X) direction. Distance in $\overset{\circ}{A}$, energy in eV and force in nN.

load simulations which will not be discussed here.

The recorded component of the force on the tip atoms in the direction of the lateral motion, as a function of the displacement of the tip-holder assembly, is shown in Fig. 20a and the corresponding potential

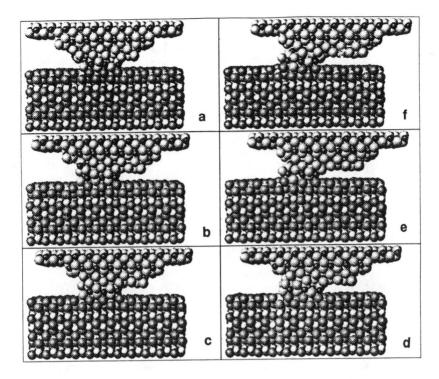

Fig. 21. Atomic configurations corresponding to the marked points in Fig. 20a. Note the interlayer slip occuring in the tip as the tip-holder assembly is translated from left to right. The bottom three layers of the tip adhere to the surface leading to an adhesive wear process of the sliding tip.

energy of the tip atoms is given in Fig. 20b. As seen, the force on the tip exhibits an oscillatory variation as a function of lateral displace-ment which is a characteristic of atomic-scale stick-slip behavior. In-spection of the atomic configurations along the trajectory of the system reveals that the lateral displacement results in shear-cleavage of the tip. The sequence of atomic configurations shown in Fig. 21 reveals that the bottom part of the tip remains bonded to the substrate, and sliding occurs between that portion of the tip material and the adjacent layers. This result indicates that under the conditions of the simulation (i.e., small load), atomic layers of the tip may be transferred to the substrate upon sliding resulting in tip wear. From the average value of the re-corded variation in the tangential force on the tip (see Fig. 20a), and the contact area we estimate that the critical yield stress associated with the initiation of slip in the system is ~ 9 GPa, in good corres-pondence with other simulations of shear deformations of perfect bulk crystalline CaF_2.

6. Si Tip/Si Surface [17a,c)]

In our earlier studies [17(a,c)] we have investigated the interaction
between silicon tips and silicon surfaces (i.e., a case of reactive tip-
substrate system). Our simulations, in both the constant-tip height and
constant-force scan modes, revealed that the local structure of the surface can be stressed and modified as a consequence of the tip-substrate
dynamical interaction, even at tip-substrate separations which correspond
to weak interaction. For large separations these perturbations anneal
upon advancement of the tip while permanent damage can occur for smaller
separations. For this system (employing the interatomic potentials
constructed by Stillinger and Weber [61a], which include two- and three-
body interactions reflecting the directional bonding character in covalent materials) we did not find long-range elastic deformations, which may
occur in other circumstances (such as a graphite surface, [56]) depending
upon the elastic properties of the material and the nature of interactions. Furthermore, we found [17(a,c)] that the characteristics of the
data depend upon the geometry of the scan, the degree of perfection of
the substrate, and the temperature.

In the following we focus on consequences of the interaction of a
large dynamic tip (consisting of 102 silicon atoms, arranged and equilibrated initially in four layers and exposing a 16 atom (111) facet)
scanning a substrate surface consisting of 6 (111) layers of dynamic Si
atoms (100 atoms per layer), at 300K.

Lateral scans in both a constant-height and constant-force modes
were performed. In constant-force scanning simulations, in addition to
the particle equations of motion, the center of mass of the tip-holder
assembly, Z, is required to obey $M\ddot{Z} = (\vec{F}(t) - \vec{F}_{ext}) \cdot \hat{Z} - \gamma\dot{Z}$ where $\vec{F}$ is the
total force exerted by the tip-atoms on the static holder at time t,
which corresponds to the force acting on the tip atoms due to their
interaction with the substrate, $\vec{F}_{ext}$ is the desired (prescribed) force
for a given scan, γ is a damping factor, and M is the mass of the holder.
In these simulations the system is brought to equilibrium for a prescribed value of $F_{z,ext}$, and the scan proceeds as described above while
the height of the tip-holder assembly adjusts dynamically according to
the above feedback mechanism.

In Figs 22(a) - 22(d) and Fig. 23 we show the results for a
constant-force scan, for $F_{z,ext} = -13\ \epsilon/\sigma$ (corresponding to -2.15×10^{-8}
N, i.e., negative load). Side views of the system trajectories at the
beginning and end stages of the scan are shown in Figs. 22(a) and 22(b),
and 22(c), respectively. As seen, the tip-substrate interactions induce
local modifications of the substrate and tip structure, which are
transient (compare the surface structure under the tip at the beginning
of the scan [Fig. 22(a)], exhibiting outward atomic displacements of the
tip-layer atoms, to that at the end of the scan [Fig. 22(c), where that
region relaxed to the unperturbed configuration]. The recorded force on
the tip holder along the scan direction (x) is shown in Fig. 22(d) and in
Fig. 23(a), exhibiting a periodic modulation, portraying the periodicity

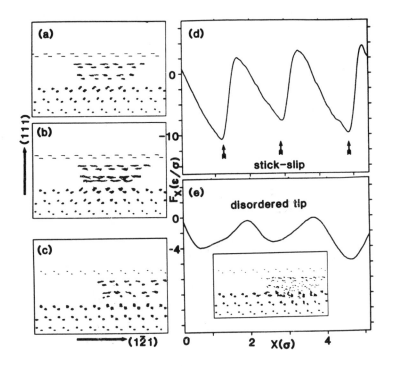

Fig. 22. (a)-(c) Particle trajectories in a constant-force simulation, $F_{z,ext} = -13.0$ (i.e., -2.15×10^{-8} N), viewed along the $(10\bar{1})$ direction just before (a) and after (b) a stick-slip event and towards the end of the scan (c), for a large, initially ordered, dynamic tip. (d) The recorded F_x, exhibiting stick-slip behavior. (e) The F_x force in a constant-force scan ($F_{z,ext} = 1.0$) employing a glassy static tip, exhibiting the periodicity of the substrate. Shown in the inset are the real-space trajectories towards the end of the scan, demonstrating the tip-induced substrate local modifications. Distance in units of $\sigma = 2.095$ Å, and force in units of $\epsilon/\sigma = 1.66 \times 10^{-9}$ N.

of the substrate. At the same time the normal force F_z fluctuates around the prescribed value [Fig. 23(c)] and no significant variations are observed in the force component normal to the scan direction [F_y in Fig. 23(b)].

Most significant is the stick-slip behavior signified by the asymmetry in F_x [observed also in the real-space atomic trajectories in Figs. 22(a) and 22(b)]. Here, the tip atoms closest to the substrate attempt to remain in a favorable bonding environment as the tip-holder

502

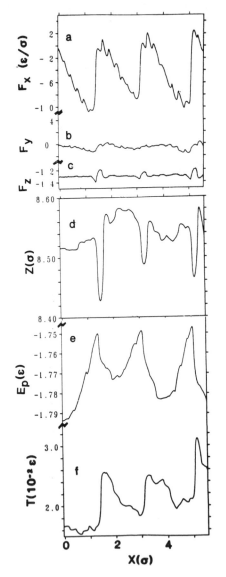

Fig. 23. Constant-force scan
simulation at $F_{z,ext}$ = - 13 (i.e.,
negative load of - 2.15 x 10^{-8} N)
employing a large (102 atoms)
faceted, ordered dynamical tip.
(a)-(c) Variations of the force F_x
(along the scan direction) and of
the force components normal to it,
F_y and F_z. Forces in the units of
1.66 x 10^{-9} N. Note that the
recorded force in the z direction
fluctuates around the prescribed
value; (d) center-of-mass height of
the tip-holder assembly; (e) potential energy E_p of the tip atoms; (f)
kinetic temperature of the tip
atoms. Note the discontinuous
variation and asymmetry in F_x,
signifying stick-slip behavior,
and the accompanying variations in
E_p and T. Distance in units of σ =
2.095 Å, energy in units of ϵ = 50
kcal/mole, and temperature in units
of 10^{-2} ϵ = 250K.

assembly proceeds to scan. When the forces on these atoms due to the
other tip atoms exceed the forces from the substrate, they move rapidly
by breaking their current bonds to the surface and forming new bonds in a
region translated by one unit cell along the scan direction. The detailed energetics of the atomic-scale stick-slip phenomenon can be elucidated
from the variations in the potential and kinetic energies of the tip
atoms along the scan, shown in Fig. 23(e) and 23(f), respectively. As
seen, during the stick stage, the potential energy of the strained bonds

between the tip and substrate atoms increases. The slip stage is signified by a discontinuity in the force along the scan direction [F_x, in
Fig. 23(a)], and by a sharp decrease in the potential energy [Fig.
23(e)], which is accompanied by a sudden increase in the kinetic temperature of the tip atoms [Fig. 23(f)] as a result of the disruption of
the bonds to the substrate and rapid motion of the tip atoms to the new
equilibrium positions. We note that the excess kinetic energy (local
heating) acquired by the tip during the rapid slip, dissipates effectively during the subsequent stick stage, via the tip to substrate
interaction [see the gradual decrease in T in Fig. 23(f), following the
sudden increase]. In this context we remark on the possibility that in
the presence of impurities (such as adsorbed molecules) the above
mentioned transient heating may induce (activate) interfacial chemical
reactions.

   We note that our constant-force simulation method corresponds to the
experiments in Ref. 32 in the limit of a stiff wire (lever) and thus the
stick-slip phenomena which we observed are a direct consequence of the
interplay between the surface forces between the tip and substrate atoms
and the interatomic interactions in the tip. The F_x force which we
record corresponds to the frictional force. From the extrema in F_x
[Figs. 22(d) or 23(a)] and the load ($F_{z,ext}$) used we obtain a coefficient
of friction $\mu = |F_x|/|F_{z,ext}| = 0.77$, in the range of typical values
obtained from tribological measurement in vacuum, although we should
caution against taking this comparison rigorously.

   Results for a constant-force scan at a positive load ($F_{z,ext} = 0.1$,
i.e., 1.66×10^{-10} N), employing the large faceted tip, are shown in Fig.
24. As seen from Fig. 24(d), the center-of-mass height of the tip and
holder assembly from the surface, Z, exhibits an almost monotonic
decrease during the scan, in order to keep the force on the tip atoms
around the prescribed value of 0.1 [see Fig. 24(e)]. At the same time
the potential energy of the tip increases. This curious behavior
corresponds to a "smearing" of the tip as revealed from the real-space
trajectories shown in Figs. 24(a) - 24(c). Comparison of the atomic
configurations at the beginning [Fig. 24(a)], during the scan [Fig. 24(b)
corresponding to X ~ 1σ, demonstrating a slip] and towards the end of the
scan [Fig. 24(c)] shows that as a result of the interaction between the
tip and substrate atoms, the bottom layer of the tip adheres to the
substrate and thus in order to maintain the same force on the tip holder
throughout the scan (as required in the constant-force scan mode) the tip
assembly must move closer to the substrate. These simulations demonstrate that in reactive tip-substrate systems, even under relatively
small loads, rather drastic structural modifications may occur, such as
"coating" of the substrate by the tip (or vice versa).

   The frictional force obtained in simulations employing a disordered
rigid 102-atom tip, prepared by quenching of a molten droplet, scanning
under a load $F_{z,ext} = 1.0$ are shown in Fig. 22(e). The significance of
this result lies in the periodic variation of the force, reflecting the

504

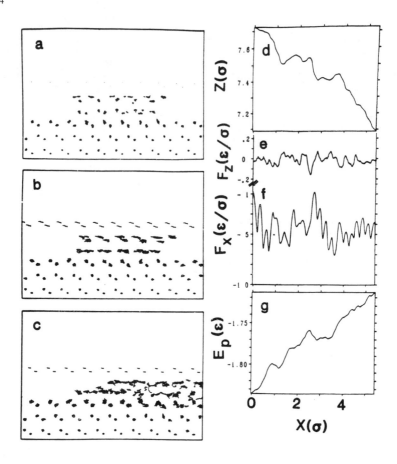

Fig. 24. Constant-force scan simulation at $F_{z,ext}$ = 0.1 (i.e., 1.66 x 10^{-10} N), employing a large (102 atoms), initially ordered dynamic tip. (a)-(c) Real-space particle trajectories at selected times during the scan, beginning (a), middle (b), and end (c), respectively. Note that the bottom layer of the tip adheres to the substrate surface (c); (d) Center-of-mass height of the tip holder assembly during the scan, as a function of scan distance, σ = 2.095 Å. Note the decrease in height associated with the adherence of the bottom tip atoms to the substrate [see (c)]. (e), (f) Normal force F_z and tangential force in the direction of the scan F_x during the scan. (g) Potential energy of the tip atoms E_p during the scan.

atomic structure of the substrate. This demonstrates that microscopic investigations of structural characteristics and tribological properties of crystalline substrates are not limited to ordered tips [32].

Acknowledgement

This work was supported by the U. S. Department of Energy, the Air Force Office of Scientific Research and the National Science Foundation. Simulations were performed on the CRAY Research, Inc. computers at the National Energy Research Supercomputer Center, Livermore, CA, through a grant from DOE, and at the Pittsburgh Supercomputer Center.

REFERENCES
1. D. Tabor, J. Coll. Interface Sci. 58, 2 (1977); M. D. Pashley and D. Tabor, Vacuum 31, 619 (1981).
2. N. Gane, P. F. Pfaelzer, and D. Tabor, Proc. R. Soc. London A 340, 395 (1974).
3. H. M. Pollock, Vacuum 31, 609 (1981).
4. D. Maugis, Le Vide 186, 1 (1977).
5. R. G. Horn, J. N. Israelachivili, and F. Pribac, J. Coll. Interface Sci. 115, 480 (1987) and references therein.
6. N. A. Burnham and R. J. Colton, J. Vac. Sci. Technol. A7, 2906 (1989).
7. (a) U. Landman, W. D. Luedtke, N. A. Burnham and R. J. Colton, Science 248, 454 (1990). (b). U. Landman and W. D. Luedtke, J. Vac. Sci. Technol. 9, 414 (1991).
8. U. Landman, W. D. Luedtke and E. M. Ringer, Wear (to be published).
9. D. Tabor and R. H. S. Winterton, Proc. R. Soc. London A 312, 435 (1969).
10. H. M. Pollock, P. Shufflebottom, and J. Skinner, J. Phys. D 10, 127 (1977); H. M. Pollock, ibid. 11, 39 (1978).
11. N. Gane and F. P. Bowden, J. Appl. Phys. 39, 1432 (1968).
12. D. Maugis and H. M. Pollock, Acta Metall. 32, 1323 (1984), and references therein.
13. U. Durig, J. K. Gimzewski, and D. W. Pohl, Phys. Rev. Lett. 57, 2403 (1986); U. Durig, O. Zuger, and D. W. Pohl, J. Microsc. 152, 259 (1988).
14. J. M. Gimzewski and R. Moller, Phys. Rev. B 36, 1284 (1987).
15. J. B. Pethica and A. P. Sutton, J. Vac. Sci. Technol. A6, 2494 (1988).
16. J. R. Smith, G. Bozzolo, A. Banerjea, and J. Ferrante, Phys. Rev. Lett. 63, 1269 (1989).
17. (a) U. Landman, W. D. Luedtke, and M. W. Ribarsky, J. Vac. Sci. Technol. A7, 2829 (1989); Mater. Res. Soc. Symp. Proc. 140, 101 (1989); see also (b) M. W. Ribarsky and U. Landman, Phys. Rev. B 38, 9522 (1988); (c) U. Landman, W. D. Luedtke and A. Nitzan, Surf. Sci. 210, L177 (1989).
18. H. Hertz, J. Reine Angew. Math. 92, 156 (1882); also in Miscellaneous Papers (Macmillan, London, 1896), p. 146; see review by K. L. Johnson, Proc. Instrm. Mech. Engrs. 196, 363 (1982).
19. G. Dieter, Mechanical Metallurgy (McGraw-Hill, New York, 1967).
20. E. Rabinowicz, Friction and Wear of Materials (Wiley, New York, 1965).
21. S. P. Timoshenko and J. N. Goodier, Theory of Elasticity (McGraw-Hill, New York, ed. 3, 1970).

506

22. K. L. Johnson, Contact Mechanics (Cambridge Univ. Press, Cambridge, 1985).
23. K. L. Johnson, K. Kendall, and A. D. Roberts, Proc. R. Soc. London A 324, 301 (1971).
24. B. V. Derjaguin, and V. M. Muller, Yu. P. Toporov, J. Coll. Interface Sci 53, 314 (1975); V. M. Muller, and B. V. Derjaguin, Yu. P. Toporov, Colloids Surfaces 7, 251 (1983).
25. P. A. Pashley, Colloids Surfaces 12, 69 (1984).
26. D. Tabor, The Hardness of Metals (Clarendon Press, Oxford, 1951).
27. J. B. Pettica, R. Hutchings, and W. C. Oliver, Philos. Mag. A48, 593 (1983).
28. N. Gane, Proc. R. Soc. London A 317, 367 (1970), and references therein.
29. P. J. Blau and B. R. Lawn, Eds., Microindentation Techniques in Materials Science and Engineering (American Society for Testing and Materials, Philadelphia, 1985).
30. M. F. Doerner and W. D. Nix, J. Mater. Res. 1, 601 (1988).
31. See articles in Mater. Res. Soc. Symp. Proc. 140, 101 (1989), edited by L. E. Pope, L. L. Fehrenbacher and W. O. Winer (Materials Research Society, Pittsburgh, 1989); F. P. Bowden and D. Tabor, Friction and Lubrication Solids (Clarendon Press, Oxford, 1950).
32. C. W. Mate, G. M. McClelland, R. Erlandsson, and S. Chiang, Phys. Rev. Lett. 59, 1942 (1987).
33. R. Thomson, Solid State Phys. 9, 1 (1986).
34. B. R. Lawn, Appl. Phys. Lett. 47, 809 (1985).
35. G. Binning, C. F. Quate, Ch. Gerber, Phys. Rev. Lett. 56, 930 (1986).
36. J. N. Isrealachvili, Acc. Chem. Res. 20, 415 (1987); Proc. Nat. Acad. Sci. U.S.A. 84, 4722 (1987); J. N. Israelachvili, P. M. McGuggan, and A. M. Homola, Science 240, 189 (1988).
37. G. Binning, H. Rohrer, Ch. Gerber, and E. Weibel, Phys. Rev. Lett. 50, 120 (1983).
38. See reviews by: P. K. Hansma and J. Tersoff, J. Appl. Phys. 61, R1 (1986); R. J. Colton and J. S. Murday, Naval Res. Rev. 40, 2 (1988); J. S. Murday and R. J. Colton, Mater. Sci. Eng. B, in press; J. S. Murday and R. J. Colton in Chemistry and Physics of Solid Surfaces, VIII, R. Vanselow and R. Howe, Eds., Springer Ser. Surf. Sci. (Springer, Berlin, 1990).
39. D. Dowson, History of Tribology (Longman, London, 1979).
40. C. Cattaneo, Rend. Accad. Naz. dei Lincei, Ser. 6, fol. 27 (1938); Part I, pp. 342-348, Part II, pp. 434-436; Part III, pp. 474-478.
41. R. D. Mindlin, J. Appl. Mech. 16, 259 (1949); see also J. L. Lubkin, in Handbook of Engineering Mechanics, ed. W. Flugge, (McGraw-Hill, NY, 1962).
42. F. P. Bowden and D. Tabor, Friction (Anchor Press/Doubleday, Garden City, NY, 1973), p. 62.
43. V. M. Muller, V. S. Yushchenko and B. V. Derjaguin, J. Coll. Interface Sci. 77, 91 (1980); ibid. 92, 92 (1983).
44. B. D. Hughes and L. R. White, Quat. J. Mech. Appl. Math. 32, 445 (1979); A. Burgess, B. D. Hughes and L. R. White.
45. M. D. Pashley, J. B. Pethica and D. Tabor, Wear 100, 7 (1984).
46. J. Skinner and N. Gane, J. Appl. Phys. D: Appl. Phys. 5, 2087 (1972).

47. D. Maugis, G. Desatos-Andarelli, A. Heurtel and R. Courtel, ASLE Trans. 21, 1 (1976).
48. J. B. Pethica and W. C. Oliver, Physica Scripta T19, 61 (1987).
49. J. B. Pethica, Phys. Rev. Lett. 57, 323 (1986).
50. Q. Guo, J. D. J. Ross and H. M. Pollock, Mater. Res. Soc. Proc. 140, 51 (1989).
51. See articles in Atomistic Simulations of Materials, Beyond Pair Potentials, V. Vitek and D. J. Srolovitz, Eds. (Plenum, New York, 1989); Many Body Interactions in Solids, edited by R. M. Nieminen, M. J. Puska, and M. J. Manninen (Plenum, New York, 1989).
52. See reviews by F. F. Abraham, Adv. Phys. 35, 1 (1986); J. Vac. Sci. Technol. B2, 534 (1984); U. Landman, in Computer Simulation Studies in Condensed Matter Physics: Recent Developments, D. P. Landau, K. K. Mon, and H. B. Schuttler, Eds. (Springer, Berlin, 1988), p. 108.
53. F. F. Abraham, I. P. Batra, and S. Ciraci, Phys. Rev. Lett. 60, 1314 (1988).
54. R. J. Colton et al., J. Vac. Sci. Technol. A6, 349 (1988).
55. D. Tomanek, C. Overney, H. Miyazaki, S. D. Mahanti, and H. J. Guntherodt, Phys. Rev. Lett. 63, 876 (1989).
56. J. M. Soler, A. M. Baro, N. Garcia, and H. Rohrer, ibid. 57, 444 (1986); see comment by J. B. Pethica, ibid., p. 3235.
57. N. A. Burnham, D. D. Dominguez, R. L. Mowery, and R. J. Colton, Phys. Rev. Lett. 64, 1931 (1990).
58. W. Zhong and D. Tomanek, Phys. Rev. Lett. 64, 3054 (1990).
59. E. Ringer and U. Landman (to be published).
60. C. R. A. Catlow, M. Dixon, and W. C. Mackrodt in Computer Simulations of Solids, Lecture Notes in Physics, Vol. 166 (Springer, Berlin, 1982), p. 130; see also M. Gillan in Ionic Solids at High Temperatures, edited by A. M. Stoneham, (World Scientific, Singapore, 1989), p. 57.
61. (a) F. H. Stillinger and T. A. Weber, Phys. Rev. B31, 5262 (1985); (b) U. Landman, W. D. Luedtke, M. W. Ribarsky, R. N. Barnett and C. L. Cleveland, Phys. Rev. B37, 4637 (1988).
62. P. Hohenberg and W. Kohn, Phys. Rev. B 136, 864 (1964).
63. See review by M. Baskes, M. Daw, B. Dodson, and S. Foiles, Mater. Res. Soc. Bull. 13, 28 (1988).
64. S. M. Foiles, M. I. Baskes, and M. S. Daw, Phys. Rev. B 33, 7983 (1986).
65. The parameterization used in our calculations is due to J. B. Adams, S. M. Foiles, and W. G. Wolfer, J. Mater. Res. Soc. 4, 102 (1989).
66. E. T. Chen, R. N. Barnett, and U. Landman, Phys. Rev. B 40, 924 (1989); ibid. 41, 439 (1990); C. L. Cleveland and U. Landman, J. Chem. Phys. 94, 7376 (1991); R. N. Barnett and U. Landman, Phys. Rev. B44, 3226 (1991); W. D. Luedtke and U. Landman, Phys. Rev. B (rapid communication), 44, 5970 (1991).
67. T. Egami and D. Srolovitz, J. Phys. 12, 2141 (1982).
68. M. Parrinello and A. Rahman, Phys. Rev. Lett. 45, 1196 (1980).
69. Mater. Eng. 90, C120 (1979).
70. (a) J. N. Israelachvili, Intermolecular and Surface Forces, (Academic Press, London, 1985); (b) R. G. Horn, J. Am. Ceram. Soc. 73, 1117

(1990); (c) R. J. Hunter, <u>Foundations of Colloid Science</u>, Vols. 1 and 2 (Oxford University Press, Oxford, 1987 and 1989); (d) <u>Thin Liquid Films</u>, edited by I. B. Ivanov (Marcel Dekker, New York, 1988).

71. D. Y. Chan and R. G. Horn, J. Chem. Phys. <u>83</u>, 5311 (1985).
72. J. N. Israelachvili, P. M. McGuiggan and A. M. Homola, Science <u>240</u>, 189 (1988); A. M. Homola, J. N. Israelachvili, P. M. McGuiggan and M. L. Gee, Wear <u>136</u>, 65 (1990).
73. J. Van Alsten and S. Granick, Phys. Rev. Lett. <u>61</u>, 2570 (1988); H.-W. Hu, G. A. Carson and S. Granick, Phys. Rev. Lett. <u>66</u>, 2758 (1991); S. Granick, Science <u>253</u>, 1374 (1991).
74. For simulations of the structural and dynamical properties of thin alkane films confined between two solid boundaries and the dynamics of film collapse upon application of load see M. W. Ribarsky, and U. Landman, J. Chem. Phys. (1991); for simulations of metal tips interacting with thin alkane films adsorbed on metal surfaces see W. D. Luedtke, U. Landman, M. W. Ribarsky, T. K. Xia and O. Yang, J. Chem. Phys. (1991); see also reference 8.
75. C. M. Mate, M. R. Lorenz, and V. J. Novotny, J. Chem. Phys. <u>90</u>, 7550 (1989); C. M. Mate and V. J. Novotny, J. Chem. Phys. <u>94</u>, 8420 (1991).
76. J. P. Ryckaert and A. Bellmans, Discuss. Faraday Soc. <u>66</u>, 96 (1978).
77. J. H. R. Clark and D. Brown, J. Chem. Phys. <u>86</u>, 1542 (1987); R. Edberg, G. P. Morriss, and D. J. Evans, J. Chem. Phys. <u>86</u>, 4555 (1987).
78. R. Edberg, D. J. Evans, and G. P. Morriss, J. Chem. Phys. <u>84</u>, 6933 (1986).
79. M. L. Forcada, M. M. Jakas and A. Gras-Marti, J. Chem. Phys. <u>95</u>, 706 (1991).

Discussion *following the lecture by U Landman on Molecular dynamics simulation of adhesion, contact formation and friction*

J N ISRAELACHVILI. Each step in your simulation run takes a few femtoseconds, so that the whole thing is over in a few nanoseconds?

M O ROBBINS. Yes, if there were something happening on a one microsecond time-scale, would you see it? The answer has to be no. What you would see is some relaxation over the calculated interval towards what one may only hope represents the final state. Slower processes would not show up, and cannot be excluded. We are not yet at the level where simulations can cover all the relevant time-scales.

U LANDMAN. The temperature is 300 K in all the simulations described. I let each step equilibrate, so don't associate any particular time with each step. We wanted a succession of quasi-equilibrated states, performed as slowly as possible, not continuously at any given speed. If there is something happening on the one microsecond time scale, then we would aim at a succession of very small changes, starting from non-interacting to fully-interacting systems.

As regards relaxation processes that cover more than several decades of vibrational frequency, I cannot speak about them.

J B PETHICA. If you look at actual stick-slip, one of the parameters that really matters is how far up the slope you go before you get the slip. I was wondering if this question of local temperature or whatever you call it is actually important, since one of the things that determine whether or not you go over that hill is how much activation energy is lying around. So, in other words, the fact that you have a temperature that is artificially a long way from the equilibrium may actually be very important in determining the value of the friction coefficient that you get locally.

U LANDMAN. The temperatures that I have shown here are for the whole tip. As I have commented already, the concept of temperature here is just an artificial representation; maybe I should have described it just in terms of energy.

D TABOR. I wanted to mention the beautiful experiment which Bowden proposed. He had a marvelous idea for determining the temperature of a crack as it travels through a brittle solid. He said if you get the crack going through calcium carbonate, the surface temperature of the crack could be determined from the amount of carbon dioxide which is liberated in the time that the crack propagates. He did that calculation: it came out as 2000° C, but the velocity only allows every surface atom to make one vibration, so it is not a real temperature. However this temperature really is playing a part in the mechanical decomposition of calcium carbonate, so my question is: even if what you quote is not the real temperature, maybe it does tell us something about the activity of the molecular groups, even if it is not equilibrium temperature as understood by thermodynamicists.

U LANDMAN. As regards energy that is available to do something chemically, the chemical bond doesn't carry a thermometer. It likes activation energy, but is not interested in whether or not it is equipartitioned. So I agree and applaud the idea that you have mentioned.

J FERRANTE. How far can you push the concepts of elasticity, and how strong will the binding be at the interface? Are you just looking for an overall picture of the effects of the softness of the material and the binding at the interface, or can you deal with particular materials?

U LANDMAN. Details of interfacial energies, hardness and so on, quantitatively speaking, are important. It is clear though that ionic bonding gives very different results to metallic bonding. Ionic bonding shows brittleness.

ANON. Are the elastic properties that you measure here, namely the response of the material to the applied load, consistent with the idea of literally compressing the hexadecane with the molecule itself?

U LANDMAN. The potential energy stored in the hinterland is only very mildly affected. It is the *inter* molecular component which is the major contribution to the stored energy.

J-M GEORGES. Does the squeezing of two or three layers depend upon the elasticity of the substrate?

U LANDMAN. In this simulation we assumed that the substrate has an infinitely high static hardness. In later simulations we used substrates with alkanes deposited onto them, and found some differences in the results.

ANON. A question regarding the orientation of the molecules at the surface: how sensitive was that to whether or not the surface was strongly absorbing?

U LANDMAN. It was more orientated parallel to the surface in the case of the strongly absorbing layer than in the other case. In most cases the majority of the configurations show end enhancement effects. The ends like to adsorb more than the middle does. This has been seen in true polymers by others. The amount of configuration entropy that you loose be detaching the sides only decides what happens to the ends of the molecules; it's smaller than the amount of entropy that you loose by rotating the middle.

ANON. As regards the work on hexadecane, you have invested a lot of money in these simulations. If you had done it on the cheap would it have made much difference?

U LANDMAN. There are people who have done work with cheaper models. I haven't tried those potentials. Simulations that allow one to use combined segments in polymers, or taking for example every four atoms as one unit, give results that are very different from what we get in some ways, but similar in other ways admittedly.

THE INDENTATION AND SCRAPING OF A METAL SURFACE: A MOLECULAR DYNAMICS STUDY[1]

J. BELAK and I.F. STOWERS
University of California
Lawrence Livermore National Laboratory
P.O.Box 808
Livermore, CA 94550
USA

ABSTRACT: The molecular dynamics computer simulation method is used to study the indentation and scraping of a clean metal surface by a hard diamond tool. Both two and three dimensional models are considered. The embedded atom method is used to represent a metallic work material. To make connection with macroscopic continuum models, we calculate the stress field from the atomistic computer experiment. The agreement is excellent during the initial elastic indentation. However, the onset of plastic deformation occurs at a much higher yield stress in the atomistic simulations. This enhanced hardness, with shallow indentations, corresponds to the theoretical yield stress required to create dislocations and leads to a dramatic increase in specific cutting energy for small depths of cut. Remarkably, the range of plastic deformation in our three dimensional simulations is only a few lattice spacings, unlike the several hundred lattice spacings dislocations readily propagate in our two dimensional simulations. This suggests that dislocations are not a very efficient mechanism for accommodating strain at this nanometer length scale in three dimensions.

1. Introduction

A fundamental understanding of tribological phenomena, such as friction and wear, is important in many engineering applications. For example, state-of-the-art single point diamond turning machines, such as the large optics diamond turning machine operated by the Precision Engineering Program at LLNL, routinely achieve depths of cut as small as a few nanometers[1]. At this length scale, the nature of elastic and plastic deformation is very sensitive to fundamental processes such as interatomic motion. Typical cutting speeds are about 1m/s, while speeds of several hundred meters per second are used for special applications. Thus, the entire chip formation process occurs on time scales much less than 1µs.

These observations suggest that a molecular dynamics (MD) model, in which explicit account is taken of interatomic motion, may yield new and useful insight into the cutting process. We have

---

1. Work performed under the auspices of the U.S. Department of Energy by the Lawrence Livermore National Laboratory under contract No. W-7405-ENG-48.

I. L. Singer and H. M. Pollock (eds.), Fundamentals of Friction: Macroscopic and Microscopic Processes, 511–520.

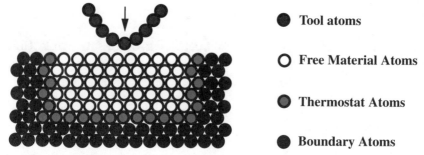

Figure 1. A cartoon illustrating the boundary conditions used in our molecular dynamics simulations of the indentation and cutting of a clean metal surface. The dark circles represent boundary atoms. These atoms are fixed during an indentation computer experiment and propagate to the right at a constant velocity during cutting. Immediately inside the boundary, we place a region of thermostat atoms (gray circles). The remaining atoms are free from further constraint. The computer experiment consists of moving the tool into the work material at a constant velocity until reaching the desired depth of cut (indentation).

developed a steady state MD model of an idealization of this process known as orthogonal cutting[2]. During orthogonal cutting, all of the work material motion is constrained to two spatial dimensions. There is no flow of material orthogonal to the cutting direction. This represents reality when the cutting direction is orthogonal to the rake face of the tool and the width of the material being cut is much greater than the depth of cut. Recently, we have extended this model to three spatial dimensions and have studied the indentation and scraping of a hard asperity across a clean metal surface. The purpose of the present paper is to demonstrate the connection between the microscopic atomistic modeling and macroscopic continuum mechanics modeling, illustrate an enhanced hardness effect that manifests itself on this nanometer length scale, and discuss some differences that occur between MD modeling in two and three spatial dimensions.

2. A Molecular Dynamics Model of Indentation and Scraping

Molecular dynamics modeling is very simple in principle. Given the positions of all atoms, calculate the force on each atom due to its neighbors and advance the positions according to the Newtonian equations of motion. Numerically, a finite difference integration scheme, such as predictor-corrector or leap-frog, is used. Further details may be found elsewhere in these proceedings[3]. In our simulations, we employ an embedded atom method[4] to express the forces between atoms in a metallic work material. The potential parameters due to Holian *et. al*[5] are used in our two dimensional simulations and the parameters for copper due to Oh and Johnson[6] are used in our three dimensional simulations. We approximate the interaction between the copper atoms in the work and the carbon atoms in the diamond tool using a Lennard-Jones potential model. The only adhesive interaction included in this model is the long-ranged van der Waals attraction.

Shown in Figure 1 is a cartoon illustrating the boundary conditions used in our molecular dynamics simulations. The MD simulation cell is a box surrounding these atoms. We define three geometrical regions within the work material. At the outmost, within two lattice spacings of the MD cell, we define a region of boundary atoms. These atoms are fixed during an indentation simulation and propagate to the right at a constant velocity during a cutting or scraping simulation. Dur-

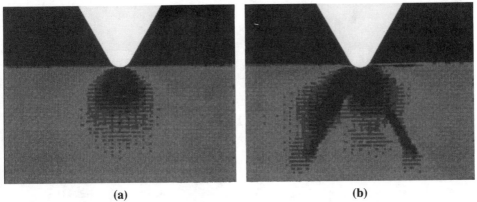

| (a) | (b) |

Figure 2. Two frames from our molecular dynamics simulation of surface indentation in two dimensions. The work material contains 43,440 atoms in 160 layers. The tool edge radius is about 5 nm and the indentation rate is about 1 m/s. The atoms are shaded by the local value of the deviatoric stress field. (a) An elastic indentation of three layers. (b) A plastic indentation of six layers.

ing cutting, atoms are continuously inserted from the left while atoms that leave the simulation cell are thrown away, producing a steady state flow. Next to this boundary region, we place a region of thermostat atoms. These atoms are maintained at room temperature. Their role is to draw away heat generated at the tool tip, effectively emulating a larger material. The remaining work atoms are free from further constraint. We model the diamond tool as infinitely hard. The simulated carbon atoms do not vibrate. The extension of this model to three dimensions is straightforward. We use the rigid boundary conditions in our 3D indentation simulations and periodic boundary conditions orthogonal to the cutting direction in our 3D cutting simulations. The three dimensional tool is generated by cleaving along three {111} planes in the cubic diamond lattice, producing a triangular (Berkovich-type) tip. The first five layers of tip-most carbon atoms are removed to make a blunted tip as shown in Figure 6.

3. Results

3.1 Two Dimensional Simulations

Shown in Figure 2 are two snapshots from our MD simulation of the indentation of a metallic 2D crystal. The length scale is comparable to recent experimental studies using thin films[7]. The indentation is performed at a slow constant velocity (~1 m/s). The atoms in the figure are shaded by the local expectation value of the deviatoric stress field[8]. In two dimensions, the deviatoric stress is one half the difference between the principle stresses and represents a shearing stress. To make the connection between the inherently macroscopic concept of stress and our microscopic MD model, we divide the simulation cell into many small sub-cells, each containing on average about 10 atoms. For a fixed indentation, we evaluate the ensemble average of the microscopic stress tensor for each of these sub-cells. Each atom in the sub-cell is assigned this value of stress. The resulting stress field for an elastic indentation of three layers is shown in Figure 2(a). The circular region of constant stress is the Hertzian stress field, well known in the study of contact mechanics[9]. To make a quantitative comparison with the Hertzian theory, we show in Figure

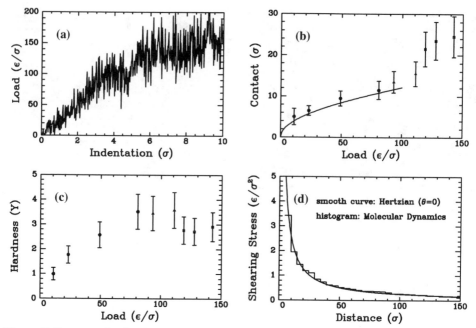

Figure 3. Some results for the indentation simulation described in Figure 2. (a) The load on the tool as a function of indentation. The units are the energy scale (ε) and the length scale (σ) of the 2D Holian potential model[5]. To make comparison with measurements on copper, we use $\varepsilon/k_B = 4560$ K and $\sigma = 2.3$ Å. (b) The 2D contact length between the tool tip and the metal surface as a function of load. The circles, triangles, and squares represent the elastic response, initial plastic response, and full plastic response, respectively. The smooth curve is the result from 2D Hertzian contact mechanics. (c) The calculated hardness (load/contact), in units of our calculated Theoretical yield strength, as a function of load. Symbols have the same meaning as in (b). (d) The deviatoric stress field along the axis of symmetry ($\theta = 0$) compared with the Hertzian result.

3(d) a plot of the deviatoric stress along the axis of symmetry. The histogram is the MD result and the smooth curve is the Hertzian result expected for the average load at this indentation (three layers). The agreement is excellent. Upon further indentation, the surface yields by creating dislocations along the easy slip planes at ±60° (see Figure 1). These 2D dislocations are single dislocation edges, unlike the dislocation loops that form in 3D. The active slip bands are clearly observable for the plastic indentation of six layers shown in Figure 2(b).

Shown in Figure 3(a) is the instantaneous normal force (load) on the tool. The short time scale fluctuations are due to the rapid motion of the surface atoms repeatedly colliding against the tool atoms. During the initial elastic indentation, the load rises linearly with increasing indentation. At an indentation of about 3.5 layers (1 layer ~ 1 σ), the surface yields through the creation of a single dislocation edge on one of the 60° slip planes. The load decreases slightly with further indentation as we work the active slip band. At an indentation of about 5 layers, the load begins to rise as we build up stress on the other side that has not yet yielded. After an indentation of about 6 layers, that side yields as well (see Figure 2(b)) and further indentation causes the load to rise only slightly.

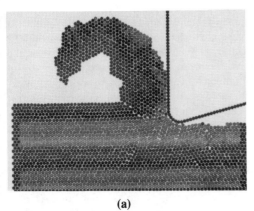
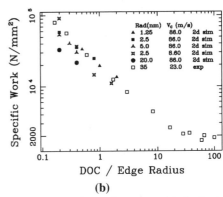

	Rad(nm)	V_c (m/s)	
▲	1.25	86.0	2d sim
■	2.5	86.0	2d sim
▲	5.0	86.0	2d sim
✕	2.5	8.60	2d sim
●	20.0	86.0	2d sim
□	35	23.0	exp

(a)   (b)

Figure 4. A steady-state molecular dynamics model of orthogonal metal cutting. (a) A single frame from our computer animated movie of the cutting process. The metal atoms are shaded according to their initial depth into the material. The shading pattern repeats every 16 layers. (b) The specific cutting energy (work per unit volume of material removed), calculated from our average cutting forces, as a function of depth of cut normalized by the tool edge radius. The open squares are the experimental measurements of Moriwaki and Okuda[13] on single crystal copper.

One observable that characterizes the mechanical properties of surfaces is the material hardness. By hardness, we mean the effort required to deform a surface, whether elastically or plastically. As our measure, we define hardness to be the load divided by the real area of contact[10]. It has been observed for a long time that this quantity is approximately three times the ultimate tensile yield strength (Y) of the bulk material. Because our tool is composed of atoms, we can define the real area of contact (actually this is a length of contact in 2D) to be the distance between tool atoms on which there is a non-zero force. The resulting contact length is shown in Figure 3(b) for elastic indentations of 0.5, 1, 2 and 3 layers (circles), initial plastic indentations of 4 and 5 layers (triangles), and full plastic indentations of 6, 7 and 8 layers (squares). The smooth curve is the 2D Hertzian contact mechanics result using our calculated elastic moduli for the Holian potential model[11]. The hardness, calculated from this data, is shown in Figure 3(c) as a function of load. We have normalized the hardness by our calculated theoretical yield strength (Y_t). Remarkably, the hardness is about $3Y_t$ for the fully plastic indentations. Unlike macroscopic indentation experiments, where plastic flow initiates at about $1.1Y$[10], we do not observe plastic flow until $H \approx 3Y_t$.

Shown in Figure 4(a) is a single frame from a computer animated movie based on our MD simulations of the orthogonal cutting process[12]. The calculation is performed in the reference frame of the tool with the material atoms flowing from left to right. We have performed the simulations for cutting speeds ranging from 8.6-86 m/s, tool edge radii ranging from 1.25 to 20 nm, and depths of cut as large as 10 nm. Some of the results are summarized in Figure 4(b). In this figure, we plot specific cutting energy (the work per unit volume of material remove) versus the depth of cut normalized by the tool edge radius. For comparison, we also show some recent experimental measurements of Moriwaki and Okuda[13]on single crystal copper. The rational for normalizing the depth of cut by the tool edge radius was to determine the dependence of specific energy upon the tool edge radius. However, we observe that our normalization does not remove all tool edge radius dependence from the MD data. Dull tools require less specific energy for the same normalized depth of cut as do sharp tools. Further comparisons are made with experimental data in a recent paper[14]. The dependence of the specific energy upon depth of cut is known as the size effect[15]. It has been observed during macroscopic machining that the slope of this log-log plot is about -0.2. For both the simulations and the micro diamond cutting experiments, we find a slope of about -0.6. The transition occurs at a length scale comparable to the material grain size and is probably a result of a change in the nature of plastic deformation, from intergranular at the macro scale to intragranular on the micro scale. We note that both the calculated and experimental specific energy exceed the cohesive energy per unit volume for copper (4.8×10^4 N/mm^2) at the smallest depths of cut.

3.2 Three Dimensional Simulations

Some results from our 3D simulation of the indentation of the copper {111} surface are shown in Figure 5. Figure 5(a) displays the pristine close-packed surface. The tip of the hard triangular asperity (tip of the tool) is very sharp. We estimate the effective tip radius as about 1 nm. As in the 2D simulations, the computer experiment consists of moving the tip into the surface at a constant velocity of 1 m/s. This speed was much slower than the propagation speed for dislocations in 2D. Unfortunately, we have not as yet verified this in 3D, but expect it to be true. Microindentation experiments are performed in the laboratory at a much slower rate, typically at about 10^{-6} m/s. Rates as slow as this are beyond the current capabilities of MD simulations and we are unable to study as yet plastic deformation due to phenomena such as creep and solid state diffusion. We also neglect the presence of an atmosphere (adsorbed fluids and oxide layers) and electrostatic interactions due to charge transfer. For now, we leave the inclusion of such details as future work.

Shown in Figure 5(b) is the instantaneous load on the tool as a function of indentation. The upper portion of the curve is the load during indenting and the lower portion is the load during removal. After the initial van der Waals attraction, the load rises linearly as the surface responds elastically. After an indentation of about 1.5 layers, the surface yields plastically, through the creation of a small dislocation loop. Critical yielding of this type has been observed in the laboratory[16]. We note that the release of load during the first yield in 3D is much greater than we found in 2D. During further indentation, the system undergoes several of these loading-unloading events. After an indentation of about 6.5 layers, we reverse the direction of the tool and observe the load to quickly drop to zero. However, upon further removal, the load suddenly rises again. From our computer animated movie of this process, we observe that not only elastic recovery but also plastic recovery occurs during unloading. Presumably when one plastic event anneals out at the surface, the material returns to more intimate contact with the tool and the load rises.

Shown in Figure 5(c) is a cross-sectional slice through the center of the tool at an indentation of about 6.5 layers. Remarkably, very few atoms have bulged out around the tool. The surface appears to have accommodated most of the volume of the tool. In this figure, we have shaded the atoms according to their initial depth into the material. The shading pattern repeats every 8 layers. We have found in our 2D studies that the passage of a dislocation causes a mismatch between successive layers. The shading highlights the dislocation path so that it is easily discernible to the eye. We have made many cross-sectional plots like this. They all indicate that the range of plastic deformation is limited to at most a few lattice spacings surrounding the tool tip. We also show, in Figure 5(d), a snapshot taken after the tool was removed. The small step on the surface in front of the debris demonstrates the presence of a small dislocation loop extending from the left edge of the step, into the surface, and terminating at the bottom of the crater.

Results from our MD simulation of scraping the asperity across the clean metal surface are shown in Figure 6. Of the many possibilities for the orientation of the asperity with respect to the metal surface, we have chosen to study the orientation illustrated by the cartoon in Figure 6(c), in

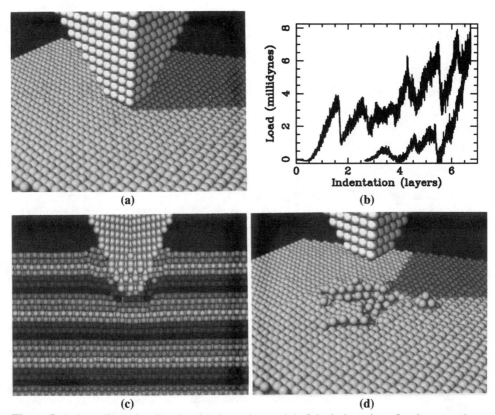

(a)

(b)

(c)

(d)

Figure 5. A three dimensional molecular dynamics model of the indentation of a clean metal surface by a sharp diamond tip. The work material contains 72,576 atoms in 36 layers. (a) The clean {111} surface of copper prior to indentation. (b) The load on the tool as a function of indentation. The upper curve is the load during indenting into the metal surface and the lower curve is the load during removal. (c) A cross section through the center of the tool at an indentation of about 6.5 layers. As in Figure 4(a), the copper atoms are shaded according to their initial depth into the material. The shading pattern repeats every 8 layers. (d) A snapshot taken at the end of the computer experiment. The step on the surface in front of the debris demonstrates the presence of a small dislocation loop between the near edge of the step and the bottom of the crater.

which the material flows towards the blunt edge of the asperity at a speed of 100 m/s. The results presented here are preliminary and an exhaustive study of different orientations and speeds is forthcoming. The snapshot in Figure 6(a) illustrates the geometry of our simulation. The indentation proceeds as before with the surface underneath flowing past the asperity. The indentation is turned off after 2000 time steps with the tip of the asperity at about 5 layers beneath the surface. Our animation of this simulation displays a peculiar phenomena of surface waves. The entire surface lifts up and down several times during the simulation. The time-scale for these events correlates well with the time required for an acoustic wave to propagate from the surface, reflect off the

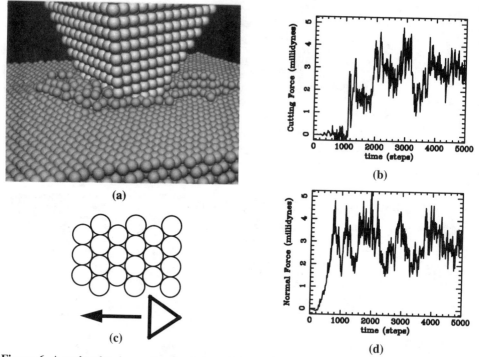

Figure 6. A molecular dynamics simulation of the scraping of a sharp triangular asperity across the clean metal surface. The computer experiment is performed by indenting into a moving surface. (a) A snapshot taken during the simulation. (b) The horizontal (cutting) force on the tool as a function of time. The indentation stops at about 5 layers after 2000 time steps. (c) A cartoon illustrating the direction of scraping the triangular asperity across the close-packed metal surface. (d) The vertical (normal) force on the asperity as a function of time.

lower rigid boundary, and return to the surface again. We did not observe these waves in our 2D simulations and suspect that the dispersion time in much faster in 2D than in 3D.

Shown in Figures 6(b) and 6(d) are the horizontal cutting force and the normal force (load) on the tool, respectively. The normal force displays a behavior similar to the load during the indentation simulation. This force rises rapidly until a critical yielding occurs, after which we observe several loading-unloading events. The cutting force displays a much different behavior. During the initial elastic loading of the surface, the cutting force remains nearly zero and rises sharply only after the surface has yielded plastically. It will be instructive to study this dry frictionless sliding with a more realistic description of adhesion between the tool and the surface. Finally, we note that the coefficient of friction observed in the simulation is about $\mu \approx 1$, as expected for a conical asperity with small adhesive forces[17].

4. Discussion

Both our 2D indentation simulations and experiments[7,16] indicate that surface hardness increases with decreasing indentation size and hence proximity to the surface. This hardness approaches that calculated from the theoretical yield strength for nanometer size indents. We find this dependence of yield stress upon length scale to also manifest itself in our cutting simulations. The specific cutting energy increases dramatically with decreasing depth of cut. Furthermore, there is a qualitative difference in behavior occurring at a length scale of a few micrometers. We interpret this difference as a change in the nature of the plastic deformation. At larger length scales, deformation occurs primarily between grains and at grain boundaries (intergranular). While at submicron length scales, all of the deformation occurs within a single grain (intragranular) where we might expect the yield stress to be determined by the force required to create dislocations. To see how the change in yield stress effects the specific cutting energy, we estimate the shearing contribution to the specific energy by assuming that the stress along the shear plane is equal to the yield strength (Y). We find $W_S \approx Y/\sin\phi$, where ϕ is the shear angle[18]. Taking $Y \approx 300$ N/mm^2 for copper[10] and $\phi \approx 25°$ as a typical shear angle for micron size depths of cut[13], we find $W_S \approx 700$ N/mm^2, somewhat smaller than the large depth of cut data shown in Figure 4(b). Taking $Y_t \sim G/10$ as an estimate of the theoretical yield strength, we find $W_S \approx 2 \times 10^4$ N/mm^2, comparable to our smaller depth of cut data. Unfortunately, this is not the whole story. As the depth of cut decreases, the cutting ratio (and hence the shear angle) decreases dramatically. This also leads to an increase in specific energy at small depths of cut.

As with our 2D simulations, our 3D simulations of indentation show many features that resemble experiments. We observe a critical yielding phenomenon. In our simulations, the surface yields through the creation of a single dislocation loop. This dislocation rapidly moves to equilibrium and the load drops considerably. This region of decreasing load is not observable in experimental studies of surface indentation. The experiments are performed at constant load, in which the tool suddenly jumps forward when the critical yielding occurs[16]. Probably the most startling result of our 3D studies is the small range of plastic deformation surrounding the asperity tip. We find this range to be limited to at most a few lattice spacings. In contrast, the dislocations in our 2D studies readily propagate many hundred lattice spacings. There appears to be an energy barrier associated with a minimum dislocation loop size that must be overcome before the dislocation can propagate on the slip plane[19]. This minimum loop size is probably about 1nm. Furthermore, the work performed in moving the dislocation in the nonuniform stress field is significantly greater for the dislocation loop in 3D than for the single dislocation edge in 2D. We can only conclude that on this nanometer length scale dislocations are not a very efficient mechanism of accommodating strain in three dimensions.

5. Acknowledgments

We gratefully acknowledge Mike Allison of the Engineering Graphics Lab at Lawrence Livermore National Laboratory for providing his graphics routines which have made the visualization of the dynamics of these molecular dynamics simulations possible. We also acknowledge Eugene Brooks and the Massively Parallel Computing Initiate at LLNL for both generous support and computing time on the 128 processor BBN-TC2000 on which most of these simulations have been performed.

6. References

1. Donaldson, R.R., Syn, C.K., Taylor, J.S. and Riddle, R.A. (1987) "Chip Science: A Basic Study of the Single-Point Diamond Turning Process," Lawrence Livermore National Laboratory Report UCRL-53868-87.

2. Belak, J. and Stowers, I.F. (1990) "A Molecular Dynamics Model of the Orthogonal Cutting Process," Proceedings of the 1990 Annual Meeting of the American Society of Precision Engineers, Rochester, NY, pp76-79.

3. Landman, U. (1991) "Computer Simulations," These Proceedings.

4. Daw, M.S. and Baskes, M.I. (1984) "Embedded Atom Method: Derivation and Application to Impurities, Surfaces, and other Defects in Metals," Phys. Rev. B **29**, 6443-6453.

5. Holian, B.L., Voter, A.F., Wagner, N.J., Ravelo, R.J., Chen, S.P., Hoover, W.G., Hoover, C.G., Hammerberg, J.E. and Dontje, T.D. (1990) Los Alamos National Laboratory Report LA-UR-90-3566.

6. Oh, D.J. and Johnson, R.A. (1989) "Embedded Atom Method Model for Close-Packed Metals," in V. Vitek and D.J. Srolovitz (eds.) **Atomistic Simulation of Materials: Beyond Pair Potentials**, Plenum Press, New York, pp233-238.

7. Pharr, G.M. and Oliver, W.C. (1989) "Nanoindentation of Silver—Relations between Hardness and Dislocation Structure," J. Mater. Res. **4**, 95-101.

8. McClintock, F.A. and Argon, A.S. (1966) **Mechanical Behavior of Materials**, Addison-Wesley Publishing, Reading.

9. Johnson, K.L. (1985) **Contact Mechanics**, Cambridge University Press, Cambridge.

10. Bowden, F.P. and Tabor, D. (1950) **The Friction and Lubrication of Solids**, Clarendon Press, Oxford.

11. Belak, J. (1990) Unpublished calculations. Our energy scale differs by a factor of three from that used by Holian et. al, because we take the standard Lennard-Jones model as our reference two-body potential model while Holian et. al use one third of this model. We find that both the Young's modulus (E) and Poisson's ratio (ν) are independent of direction for the 2D triangular lattice. At $T = T_{melting} / 4$, $E \approx 57.5 \pm 2.5$ ε/σ^2 and $\nu \approx 0.43 \pm 0.02$. The shear modulus (G) depends upon the direction of shear within the crystal. However, we can define an effective isotropic shear modulus as $G = E / 2(1 + \nu) \approx 20.1 \pm 0.5$ ε/σ^2. The theoretical yield strength (Y_t) is calculated by performing constant tension computer experiments on an initially perfect crystal. We find the system to yield at a strain of about 8 percent under an applied tension of $Y_t \approx 2.4 \pm 0.2$ ε/σ^2.

12. Belak, J, Hoover, W.G., DeGroot, A.J., and Stowers, I.F (1990) "Molecular Dynamics Modeling Applied to Indentation and Metal Cutting," Lawrence Livermore National Laboratory release PR-16039.

13. Moriwaki, T. and Okuda, K. (1989) "Machinability of Copper in Ultra-Precision Micro Diamond Cutting," Annals of CIRP **38**, 115-118.

14. Stowers, I.F., Belak, J., Lucca, D.A., Komanduri, R., Rhorer, R.L., Moriwaki, T., Okuda, K., Ikawa, N., Shimada, S., Tanaka, H., Dow, T.A. and Drescher, J.D. (1991) "Molecular Dynamics Simulation of the Chip Formation Process in Single Crystal Copper and Comparison with Experimental Data," Proceedings of the 1991 Annual Meeting of the American Society of Precision Engineers, Santa Fe, NM, pp100-103.

15. Shaw, M.C. (1984) **Metal Cutting Principles**, Clarendon Press, Oxford; Backer, W.R., Marshall, E.R., and Shaw, M.C. (1952) "The Size Effect in Metal Cutting," Trans. of ASME **74**, 61.

16. Gane, N. and Bowden, F.P. (1968) "Microdeformation of Solids," J. Appl. Phys. **39**, 1432-1435; see also Pethica, J.B. (1991) "Mechanical Responses of Materials on the Nanometer Scale," These Proceedings.

17. Suh, N.P. (1986) **Tribophysics**, Prentice-Hall, Englewood Cliffs, p96 and references therein.

18. Childs, T.H.C. (1991) Private communication; see also reference 15.

19. Hull, D. and Bacon, D.J. (1984) **Introduction to Dislocations**, Pergamon Press, Oxford.

MACHINES
AND
MEASUREMENTS

..... There has been little interest in achieving low friction in applications involving steady sustained sliding. It is precisely in these situations, of course, that friction dissipates such large amounts of energy (overall perhaps 5% of all the energy generated by mankind) and one might have anticipated enormous concern in this area.

E RABINOWICZ, in "Fundamentals of tribology" (N P Suh and N Saka, eds), MIT Press, 1980, p 351.

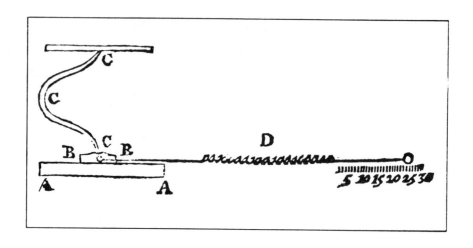

Amontons' sketch of his apparatus for friction experiments (Amontons, 1699). Test material A-A against B-B; spring C-C-C provides normal loading; spring balance with scale, D, for friction measurements. [From D. Dowson, <u>History of Tribology</u> (Longman, London, 1979) p. 155, with permission.]

SCALE EFFECTS IN SLIDING FRICTION: AN EXPERIMENTAL STUDY

PETER J. BLAU
Metals and Ceramics Division
Oak Ridge National Laboratory
P. O. Box 2008
Oak Ridge, TN 37831-6063
United States of America

ABSTRACT. Solid friction is considered by some to be a fundamental property of two contacting materials, while others consider it to be a property of the larger tribosystem in which the materials are contained. A set of sliding friction experiments were designed to investigate the hypothesis that the unlubricated sliding friction between two materials is indeed a tribosystems-related property and that the degree to which the materials or the machine and its environment will affect the measured friction is also system-dependent. Three tribometers were used: a friction microprobe (FMP), a typical laboratory-scale reciprocating pin-on-flat device, and a heavier, commercial wear tester. The slider material was stainless steel (AISI 440C) and the flat specimen material was an ordered alloy of Ni_3Al (IC-50). A sphere-on-flat geometry was used at ambient conditions and at normal forces ranging from 0.01 N to 100 N and average sliding velocities of 0.01 to 100.0 mm/s. The nominal, steady-state sliding friction coefficient tended to decrease with increases in normal force for each of the three tribometers, and the steady state value of sliding friction tended to increase as the size of the machine increased. The mechanisms for this behavior concern the relative role of oxide layers, roughness generation, and debris particle-trapping. The variation of the friction force during sliding was a characteristic of the stiffness of the test system. These studies support the idea that the frictional behavior of both laboratory and engineering tribosystems should be characterized by more than a single numerical value for friction coefficient at steady-state and that friction models should predict variations in frictional behavior more explicitly by considering system properties. Mechanistically, the present results underscore how the competition between frictional contributions can change under different testing conditions.

1.0 Introduction

The development of fundamental models for friction has depended on the viewpoint of the individual investigator and is largely based on his or her scientific or technical training and accumulated experience. It is also based upon the concept of what a tribosurface is, how it reacts, and what simplifying or boundary conditions the modeler chooses to impose. A mechanics-oriented investigator is likely to lump materials properties into such bulk properties as shear strength and elastic modulus, a materials scientist may often opt for

I. L. Singer and H. M. Pollock (eds.), Fundamentals of Friction: Macroscopic and Microscopic Processes, 523–534.
© 1992 *Kluwer Academic Publishers. Printed in the Netherlands.*

qualitative or descriptive models which include dislocation interactions, grain sizes, and crystal structure-related parameters, and a chemist or surface physicist may impose chemical or atomic bonding characteristics in modeling friction.

Techno-cultural differences are significant in a field as broad as tribology. Individuals with a mechanical engineering or mechanics background have no trouble recognizing the importance of machine stiffness or vibrations in frictional behavior, but only recently, have materials scientists come to acknowledge the possibility that friction cannot be modelled exclusively in terms of materials deformation and fracture under the influence of a uniformly-applied normal force and constant interfacial velocity. Atomic-scale friction studies based on surface physics approaches may take the mechanics of the machine fixtures into account during the construction of their instruments, but may not always consider such effects in interpreting experimental results. Given adequate precautions, however, the distances travelled by fine tips may be so small that the effects of the machine compliance may be negligible. The question then arises: At what scale, and under what conditions do the characteristics of the tribometer begin to affect frictional behavior?

Considering the myriad possible phenomena which can occur simultaneously at various size scales during solid sliding, all approaches to modeling friction of the general case are incomplete. It is true, however, that in specific cases of very well-characterized and controlled tribosystems, a large number of the possible influences on friction can be considered inconsequential, and reasonably accurate, semi-empirical models can be developed.

Our evolving understanding of physical surfaces parallels the history of instrumentation available to measure and image surfaces. Surface concepts based mainly on profiling techniques may produce quite different friction models than concepts based on microscopy of contact damage at various magnifications or views of cross-sections below contact surfaces. As microscopy and surface imaging techniques evolve, an evolution of friction models is likely to occur in conjunction with them.

A more detailed discussion of the subject of scale effects on friction has been published earlier [1]. Reference was made to the excellent earlier review of friction by Bowden [2] and to the discussion of the statistical nature of friction by Rabinowicz [3]. The conclusions of the recent work [1] were that successful, predictive models for friction must in general consider systems characteristics as well as materials properties. When dealing with unlubricated or poorly lubricated systems, there may not be a single value for friction, but rather a set of characteristic values, and therefore, modelers should strive to define the boundaries and distributions of values rather than to try to arrive at a single value for friction. Particularly in the case of running-in [4], a system can pass through a number of frictional states, and other transitions are also possible [5]. The present investigation of scale effects considers only the characteristics of steady-state, post-running-in behavior of tribosystems.

Under poor or inadequate lubrication, solid contact can result in a hierarchy of damage effects in materials, depending on the manner in which the sliding partners accommodate the imposed surface shear. Shear-induced damage in metals can result in increased dislocation density [6], the formation of adiabatic shear bands or deformation twins [7], recrystallization of the grain structure [8], and crystallographic texturing [9]. In ceramics, dislocation structures and microfractures commonly occur [10]. In polymers, visco-elastic behavior leads to periodic changes in materials response to surface tractions [11]. These kinds of evolving changes in the microstructure alter the mechanical properties of surfaces and their morphology, hence, their frictional characteristics can evolve as well.

In the present experimental work, sliding friction tests were conducted to investigate the hypothesis that the unlubricated sliding friction characteristics between two fixtured materials is a tribosystem-related property and that the degree to which the materials or the machine and its environment will affect the measured friction is also system-dependent.

2.0 Experimental Details

Three machines were used in this investigation: a low-load, friction microprobe (FMP), a typical laboratory-scale reciprocating pin-on-flat device, and a heavier, commercial wear tester (Cameron-Plint TE-77).

A schematic illustration of the friction microprobe (FMP) is given in Figure 1. It consists of a small "boat" suspended between two elastic webs. A stationary stylus containing a 1.0 mm diameter sphere of stainless steel AISI 440C was the slider. As the computed-controlled, position-encoded traverse stage moves, the slider friction restrains the stage and deflects the supporting webs. This deflection is measured by a capacitive transducer system. By calibrating the force and deflection characteristics of the webs, the instantaneous stylus position with respect to the flat specimen, as well as the tangential (friction) force, can be determined. In the current experiments, the stage traverse velocity was set at 10.0 μm/s and the track length was 100 μm. Normal forces ranged from 9.8 to 98.1 mN. Friction force

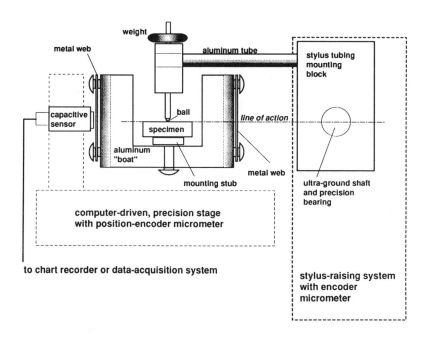

Figure 1) Diagram of the friction microprobe (FNP). The unit is contained in a plastic box to shield it against air currents, and the base stands on an air-supported anti-vibration table.

readings were taken when the stylus friction had become relatively level so that the sliding velocity of the stylus matched that of the traversing stage. All tests were performed in air. Further details of the FMP have been described elsewhere [12].

Figure 2 shows a schematic diagram of the pin-on-flat reciprocating wear machine. The "pin" specimen was a 9.52 mm diameter bearing ball of AISI 440C stainless steel. Metallographically-etched cross-sections of both the 1.0 mm diameter and 9.52 mm diameter balls revealed similar microstructures. Normal forces ranged from 0.98 N to 15.0 N. The reciprocating velocity averaged 0.1 m/s over a 5.0 mm track length. The commercial wear testing machine used the same diameter ball as the pin-on-flat tester, the same track length, and the same average velocity. The normal force ranged from 10.0 to 100.0 N. Table 1 summarizes the various tests performed in the course of this study.

Table 1. Summary of the Testing Conditions

Testing Variable	FMP*	P-O-F*	C-P*
slider diameter (mm)	1.0	9.52	9.52
stroke length (mm)	0.1	10.0	10.0
average velocity (mm/s)	0.01	100.0	100.0
duration of test (cycles)	10.	1000.	1000.
minimum normal force, F_{min}(N)	0.01	1.0	10.0
Hertz stress at F_{min} (GPa)	0.42	0.44	0.94
maximum normal force, F_{max}(N)	0.98	15.0	100.0
Hertz stress at F_{max} (GPa)	0.90	1.08	2.02
testing environment	air	air	air
relative humidity range (% RH)	50 ± 15	55 ± 10	65 ± 12
testing temperature	room T	room T	room T
tests per condition	3	3	3

* FMP = friction microprobe
P-O-F = laboratory-scale pin-on-flat machine,
C-P = heavy-duty, Cameron-Plint TE-77 machine

The flat specimen material was a mirror-polished surface of Ni_3Al alloy IC-50 of nominal composition: 11.3 wt% Al, 0.6 wt% Zr, 0.02 wt% B and the balance Ni. It was a single-phase, polycrystalline material produced by strip rolling with a grain size of about 10-20 μm. The tribological behavior of this material against a variety of slider materials, unlubricated and lubricated, was recently reported elsewhere [13]. The surfaces of all contacting materials were swabbed with acetone and ethanol within ten minutes of testing. Unworn surfaces were used for each test.

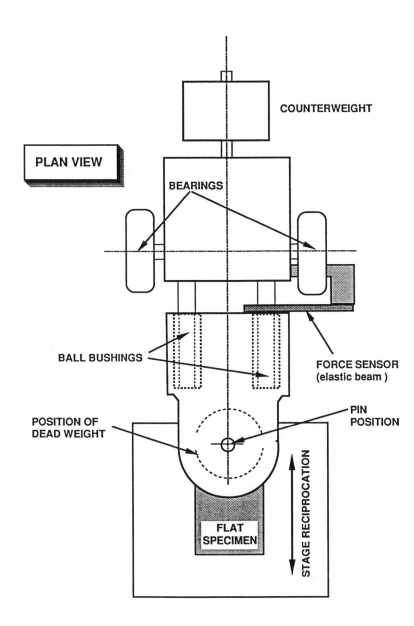

Figure 2) Diagram of the pin-on-disk machine.

3.0 Results

The nominal friction coefficient and the range in values were estimated from analog friction traces (see Fig. 3). The friction coefficient variability represents the typical maximum and minimum value observed during a single pass of the slider during the steady state period.

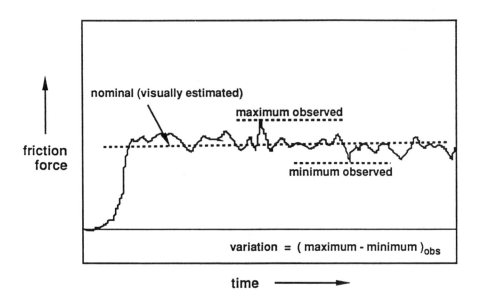

Figure 3) Definition of the nominal friction coefficient and the range in friction coefficients.

The nominal friction coefficient is plotted as a function of normal force for all three machines in Fig. 4 (a)-(c). All three machines displayed a trend of decreasing friction coefficient with increasing load.

The variation in friction coefficient is plotted against normal force in Fig. 5 (a)-(c). The FMP (a) and the commercial heavy-load machine (c) both exhibited a decreasing variation with normal force, but the mid-range, pin-on-flat machine exhibited a wide scatter band of frictional variability (b) with no obvious trends.

Examination, by optical microscope, of the tip of the slider and the contact surface of specimens tested in the FMP showed no evidence for transfer or wear after friction testing. However, the flat specimens tested in the larger machines, all exhibited the types of surfaces normally associated with severe metallic wear. The wear features included deep grooving, surface roughening, and the generation of debris which collected in dispersed deposits within the apparent contact area. Spherical sliders were flattened and occasional heterogeneous deposits of transferred material were observed.

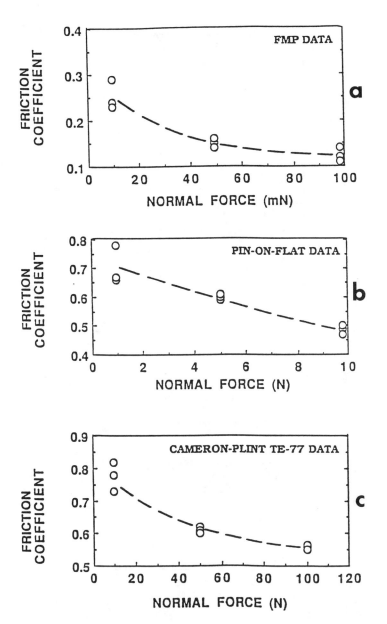

Figure 4) Nominal friction force as a function of the normal force for (a) the friction microprobe, (b) the pin-on-flat laboratory tribometer, and (c) the Cameron-Plint TE-77 commercial wear testing machine. Each datum represents a single run.

530

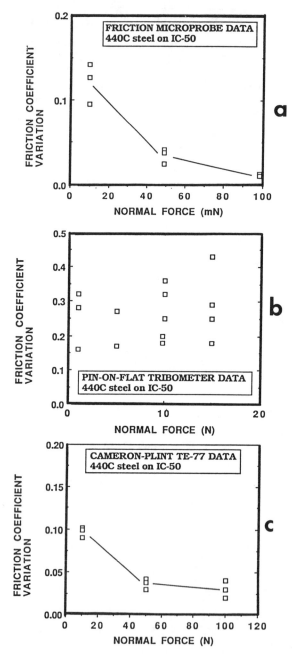

Figure 5) Variation in the "steady state" friction force for runs on three different machines: (a the friction microprobe, (b) the pin-on-flat laboratory tribometer, and (c) the Cameron-Plint TE 77 commercial wear testing machine. Each datum represents a single run.

4.0 Discussion

The friction coefficient declined as a function of rising normal load for all three testing machines, exhibiting a departure from Amontons' law. Because data from all three machines behaved similarly in this respect, it is concluded that the observed behavior was independent of the machine. However, the existence of a common behavioral trend does not necessarily imply that the same explanation is valid in all three cases.

Bowden and Tabor [14] reported a declining friction coefficient with increasing load when testing indium films on steel and extended their interpretation to explain experiments with oxide-covered surfaces. Their explanation was that when a thin film is present, increased deformation of the substrate under the film, due to an imposed increase in load, produces only a small increase in the real contact area and therefore, the resulting friction force is not linearly proportional to the normal force. This explanation may hold for FMP results where friction was likely to have been governed by the shear of thin ambient oxide films on both the nickel aluminide [15] and the stainless steel ball. In support of the role of thin oxides, no evidence for metallic wear was observed by light microscopy on the FMP ball or flat specimen.

Wear tracks on the pin-on-flat machine and on the commercial machine, however, showed extensive evidence for severe metallic wear behavior indicating that the initial oxides had been disrupted and their effects overwhelmed by a major contribution from plowing, debris, and subsurface plastic deformation. In light of the physical evidence, strictly elastic, Hertzian analyses would not apply in these case. Instead, the role of surface roughening and wear debris particle layers must be addressed further.

As load on the sliding contacts was increased, their wear rates increased. The increase in wear rate produced three results: (1) a wider nominal contact area was created on the ball and flat specimens, (2) the surface of both pin and flat became rougher, and (3) a larger quantity of particulate debris (third-bodies) was generated during sliding. These effects, whose presence was confirmed by optical microscopy, are believed to be the causes of the reduction in friction force with increasing load.

The role of third-bodies in mitigating friction has been discussed at length by Godet [16] and later by Berthier et al [17]. In essence, up to twenty different mechanisms can operate within the sliding interface to accommodate the relative velocity between two solid surfaces if particles are present between them. Quantitative formulations of particle effects on wear and friction have been more recently developed by Heshmat [18] who treated the behavior of interposed layers using rheological and hydrodynamic arguments. In the present case, however, the particle layers were not continuous to the extent that they behaved as a continuous fluid.

The presence of wear debris can raise the friction force in an interface (e.g., [19-20]). But, it has also been demonstrated that the effect of the debris can be reduced if one or both contact surfaces contain depressions or hole patterns [19]. As the sliding surfaces of the nickel aluminide flat specimens became more highly worn, as they would with higher loads, the greater surface roughness was believed to have provided additional facility for debris trapping to reduce the overall friction.

To describe this behavior, it is possible to use a transitions model proposed earlier by the author [21]. The instantaneous friction coefficient (μ) arising from contact of the solids (S) is reduced by a factor (L), whose value ranges from 0.0 to 1.0. The factor L represents the lubricating effectiveness of any interfacial friction-reducing process (such as an oxide film, removal of debris, or presence of solid lubricating films). Thus,

$$\mu = L\,S \tag{1}$$

L can be expressed as

$$L = [\,1\,/\,(1 + c\,f)\,] \tag{2}$$

where constant c is the measure of the effect on friction arising from the given lubricating mechanism if it operated over the entire, real area of contact. The variable f is the fraction of the real contact area over which the given mechanism is operating. For example, if $c = 1$, the friction of the contacting solids (S) would be reduced by a factor of 0.5 provided that friction-reducing process operated over the entire area of contact. Less than full-area coverage would reduce the friction of the system by a smaller amount. The decrease in friction with increasing load, as observed in the two larger reciprocating test machines, is consistent with the behavior of Eqn. (2) when f increases due to increasing amounts of debris particle trapping in rough surfaces. The determination of the value of c for various friction-reducing mechanisms is the subject for further research.

The variation in friction coefficient is affected by the stiffness of the testing machine. The more compliant the machine, the more likely that relatively small frictional perturbations will be amplified by the machine and that they will persist long enough to be detected by the force-measuring system. The natural frequency of the FMP and the pin-on-flat machine were similar (about 24-26 Hz), therefore the greater frictional variation of the latter machine cannot be explained in terms of its difference in natural frequency. However, the pin-on-flat machine measured friction by a relatively compliant strip of strain-gaged steel. The FMP, with its twin webs and sensitive capacitive displacement system, is relatively stiff in comparison, as is the commercial machine which uses a stiff, piezoelectric load cell. The compliance of the pin-on-flat machine therefore tended to mask the effect of lowering frictional variation with increasing load because the contact oscillated forward and backward more freely in that case. The location of the contact area in compliant systems is a function of the friction force and frictional variations may produce large accelerations or decelerations in the localized contact zone. The microscopic sliding velocity at any instant will in general be different than that imposed macroscopically. In a stiff system, the off-set in the position of the contact is not so significant and the sliding velocity remains more constant.

The role of system stiffness and vibrations on friction has been widely discussed elsewhere [22-25], and the various models which incorporate these effects will not be elaborated here. Unfortunately, few of the vibration-based models consider the level of detail of surface interactions discussed in this volume. Considering present results, it is interesting to note the observation of Chiou et al [25] that it was possible to dramatically change the severity of wear by altering the stiffness of the testing system.

Overall, interpretation and understanding of the friction results in the present study were complicated by: (1) a change in sliding interfacial conditions due to the transition between oxide-dominated effects and effects from increasing surface deformation and wear, (2) the introduction of third-bodies and roughness which mitigate friction when wear becomes significant, and (3) the influence of the machine stiffness on the stability of the steady-state friction coefficient. As a result, to effectively model and understand solid friction, including its transitions and variability, a hierarchy of scale-effects-based friction models is required.

These models must depend on both the balance of microscopic processes dominating the macroscopic frictional response and on the characteristics of the system under consideration.

5.0 Summary

A series of unlubricated sliding friction experiments were conducted with three different machines to examine the hypothesis that the friction coefficient between two materials can be different depending upon the tribosystem involved. The findings and conclusions were:

1. In all three machines, the nominal friction coefficient decreased as the normal force increased, in exception to Amontons' law, and friction coefficients did not agree when comparing results from one machine to those from another.

2. The build-up and trapping of third-bodies in the sliding interface resulted in the decrease in friction with increasing load in the two larger testing machines. In the friction microprobe, the role of thin oxide layers, as explained by Bowden and Tabor, caused the decrease in friction with increasing load.

3. The steady-state variability in the friction coefficient between one machine and another can be explained in terms of differences in the stiffness of the fixturing. If the frictional behavior of engineering systems is to be modeled accurately, stiffness as well as material properties must be taken into account.

4. The nominal friction coefficient and its variation at steady state are both materials and system variables. The extent that each of these contributes to the macroscopic friction force characteristics must be analyzed on a case-by-case basis.

ACKNOWLEDGEMENTS

The assistance of Ms. Fang-Lei Wang, Mechanical Engineering, Princeton University, with these experiments is appreciated, and the continuing support of the U. S. Department of Energy, Office of Transportation Materials, Tribology Project, is also acknowledged. The author thanks the reviewers of his draft for their helpful suggestions.

REFERENCES

1) Blau, P. J. (1991) "Scale Effects in Steady State Friction," *Trib. Trans.*, Vol. **34**, pp. 335-342.
2) Bowden, F. P. (1957/58) "A Review of the Friction of Solids," *Wear*, Vol. **1**, pp. 333-346.
3) Rabinowicz, E. (1957) "Investigation of Size Effects in Sliding by Means of Statistical Techniques," Proc. of the Conf. on Lubr. and Wear, Instit. of Mech. Engr., London, pp. 276-280.
4) Blau, P. J. (1991) "Running-in: Art or Engineering?", *J. of Mater. Engrg.*, Vol. **13**, pp. 47-53.
5) Blau, P. J. (1989) <u>Friction and Wear Transitions of Materials,</u> Noyes Publications, Park Ridge, NJ, pp. 197-267.
6) Ohmae, N. (1980) in <u>Fundamentals of Tribology</u>, ed. N. P. Suh and N. Saka, MIT Press,

pp. 201-222.

7) Samuels, L. E. (1982) <u>Metallographic Polishing by Mechanical Methods</u>, 3rd ed., ASM, Metals Park, Ohio, pp. 123-126.

8) Ruff, A. W., Ives, L. K. and Glaeser, W. A. (1981) "Characterization of Wear Surfaces and Wear Debris," in <u>Fundamentals of Friction and Wear of Materials</u>, ed. D. A. Rigney, ASM, Metals Park, Ohio, pp. 235-289.

9) Blau, P. J. (1979) "A Study of the Interrelationships Among Wear, Friction and Microstructure in the Unlubricated Sliding of Copper and Several Single-Phase Binary Copper Alloys," Ph.D. dissertation, The Ohio State University, 341 pp.

10) Evans, A. C. and D. B. Marshall, D. B. (1981) "Wear Mechanisms in Ceramics, in <u>Fundamentals of Friction and Wear of Materials</u>, ed. D. A. Rigney, ASM, Metals Park, Ohio, pp. 439-452.

11) Bartenev, G. M. and Lavrentev, V. V. (1981) <u>Friction and Wear of Polymers</u>, Elsevier, Chapters 3-5.

12) Blau, P. J. (1990) "Friction Microprobe Studies of Composite Surfaces," in <u>The Tribology of Composite Materials</u>, ed. P. K. Rohatgi, P. J. Blau, and C. S. Yust, ASM International, Materials Park, Ohio, pp. 59-68.

13) Blau, P. J. and DeVore, C. E. (1990) "Sliding friction and wear behavior of several nickel aluminide alloys under dry and lubricated conditions," *Trib. Int'l.*, Vol. **23** (4), pp. 226-234.

14) Bowden, F. P. and Tabor, D. (1986) "The Friction and Lubrication of Solids," Oxford Science Pub., pp. 119-120.

15) Wood, G. C. and Chattopadhyay, B. (1980) "Transient Oxidation of Nickel-Base Alloys," Corrosion Sci., Vol. **10**, p. 271.

16) Godet, M. (1984) "The Third-Body Approach: A Mechanical View of Wear," Wear, Vol. **100**, pp. 437-454.

17) Berthier, Y., Godet, M., and Brendle, M. (1986) "Velocity Accomodation in Friction," STLE Preprint No. 88-TC-3A-2.

18) Heshmat, H. (1991) "The Rheology and Hydrodynamics of Dry Powder Lubrication," Trib. Trans., Vol. **34**, pp. 433-439.

19) Suh, N. P. (1986), <u>Tribophysics</u>, Prentice-Hall, Englewood Cliffs, New Jersey, pp. 416-442.

20) Sheasby, J. S. (1981), "Attainment of Debris-Free Dry Wear Conditions," Proc. ASME Wear of Materials, ASME, New York, pp. 75-81.

21) Blau, P. J. (1987), "A Model for Run-in and Other Transitions in Sliding Friction," J. of Trib., Vol. **109**, pp. 537-544.

22) Aronov, V., D'Sousa, A. F., Kalpakjian, S., and Shareef, I. (1984) "Interactions Among Friction, Wear, and System Stiffness - Part 1: Effect of Normal Load and System Stiffness," J. of Trib., Vol. **106**, pp. 54-58.

23) Aronov, V., D'Sousa, A. F., Kalpakjian, S., and Shareef, I. (1984) "Interactions Among Friction, Wear, and System Stiffness - Part 2: Vibrations Induced by Dry Friction," J. of Trib., Vol. **106**, pp. 59-64 (1984).

24) Kato, K., Iwabuchi, A. and Kayaba, T. (1982) "The Effects of Friction-Induced Vibration on Friction and Wear," Wear, Vol. **80**, pp. 307-320.

25) Chiou, Y. C., Kato, K. and Kayaba, T. (1985) "Effect of Normal Stiffness in Loading System on Wear of Carbon Steel - Part 1: Severe-Mild Wear Transition," J. of Trib., Vol. **107**, pp. 491-495.

UNSTEADY FRICTION IN THE PRESENCE OF VIBRATIONS

D. P. HESS
Department of Mechanical Engineering
University of South Florida
4202 East Fowler Avenue
Tampa, Florida 33620-5350
U.S.A.

A. SOOM
Department of Mechanical and
Aerospace Engineering
University at Buffalo
Buffalo, New York 14260
U.S.A.

ABSTRACT. In this paper, results from the authors' recent work on the vibrations of sliding contacts in the presence of dry friction are summarized. We examine some idealized models of smooth and rough contacts, in which the assumed sliding conditions, the kinematic constraints and the mechanism of friction, i.e., the adhesion theory of friction, in this case, are well-defined. Instantaneous and average normal and frictional forces are computed. The results are compared with experiments. It appears that when contacts are in continuous sliding, quasi-static friction models can be used to describe friction behavior, even during large, high-frequency fluctuations in the normal load. However, the dynamics of typical sliding contacts, with their inherently nonlinear stiffness characteristics, can be quite rich and complex, even when the sliding system is very simple. A three degree-of-freedom vibratory model of a rough block in sliding contact with a planar moving countersurface, from which some preliminary results have been obtained, is used to illustrate some of these complexities.

1. Introduction

All friction phenomena that take place between bodies in relative motion are inherently dynamic. Unsteady friction includes not only the usual interplay of mechanics, physics and chemistry that is associated with friction, but also introduces the added dimension of mechanical vibrations. In experimental studies of unsteady friction, proper dynamic testing practices must be followed. Regarding unsteady friction, one can ask how contact vibrations are influenced by the presence of different types of friction or one can seek to determine the extent to which vibrations can alter the mechanisms of friction itself.

In this paper we will give an overview of our recent work on dry friction in the presence of contact vibrations. The reader is referred to other papers [1-4] for details. The existing literature on this subject is extensive and has been recently reviewed in considerable detail [5-7]. Nevertheless, we will make some general observations regarding the nature of unsteady friction and the interpretation of friction coefficients

535

I. L. Singer and H. M. Pollock (eds.), Fundamentals of Friction: Macroscopic and Microscopic Processes, 535–552.
© 1992 *Kluwer Academic Publishers. Printed in the Netherlands.*

under unsteady conditions.

We consider friction to be unsteady when the normal or the friction (actually the tangential) force at a sliding contact varies with time. By this definition, unsteady friction can occur in the presence of normal oscillations even when the coefficient of friction remains constant. At some level, e.g., during individual asperity interactions, all sliding is unsteady. Quite often, however, the contact forces remain nearly constant so that, for practical purposes, steady (or smooth) sliding takes place. In the presence of significant normal or tangential vibrations, sliding becomes unsteady. There is not necessarily a clear line of demarcation between the two.

The most common occurrence of unsteady friction is during the acceleration or deceleration of machine elements in sliding contact, especially during starting or stopping of motion. These conditions can be of great interest to the motion control engineer [7] where smooth movement of components is required. Also, in many devices, starts, stops and reversals of motion represent the times during which the most wear occurs. Vibrations may be undesirable from the point of view of the stresses that are induced or noise that is generated and may need to be controlled. Furthermore, vibrations can affect the outcome of friction and wear tests.

In discussing unsteady friction, one should distinguish between continuous and intermittent sliding. Under some conditions, sliding between two surfaces may be interrupted by intervals of complete loss of contact or by "sticking," complete or partial. At low sliding speeds, momentary reversals of the sliding direction can occur.

Intermittency of sliding can be quite subtle. For example, momentary loss of contact may last less than a millisecond and occur only a few percent of the time [8,9]. The loss and subsequent resumption of contact may take place quite smoothly. Such contact loss may not be audible or immediately noticeable from an average friction force measurement. Intermittent loss of contact will generally reduce the average "friction" force, since during loss of contact there is no friction force. The situation may be altered if strong impacts occur.

Stick-slip oscillations have been studied for more than fifty years (the reader is referred to some classic works [10-12] and recent summaries [5-7]). Yet there are many aspects of the transitions between stick and slip that are not widely appreciated or fully understood. For example, it has been recently shown, using a simple lumped single degree-of-freedom model, and verified by experiment [13] that stick-slip can occur under a constant normal load with friction described by a single global sliding friction coefficient, with no distinction between static and kinetic friction coefficients. In such cases, the magnitude of the tangential force will be lower (possibly even negative) during the stick phases and the average "friction" force is reduced compared to continuous sliding [5,13]. Patterns of stick and slip will change when normal vibrations accompany sliding. Such is the case with vibratory parts feeders where a number of modes of behavior can be found when a single constant friction coefficient is used to model the problem [14,15]. Also, Martins et al. [5] describe a number of interesting "apparent friction" phenomena that can occur when contact oscillations during sliding with dry friction are accompanied by various patterns of stick, slip and loss of contact.

When one includes the possibility of stiction, significant tangential contact compliance, velocity-dependent friction, normal adhesive forces, or the simultaneous interaction of more than a single mechanism of friction, an unambiguous interpretation of transitions between stick and slip can become difficult indeed. The dynamic model of the sliding system is usually different during stick from that during slip, just as the dynamic model of a system changes when contact is lost. Multi-dimensional high frequency transient motions during start-up or stopping can be quite complex.

Plastic deformations, different contact geometries and mechanical constraints, changes in surface topography with running-in, wear particles, thermoelastic effects, the presence of lubricants and non-rigid body vibrations (e.g., bending or torsion) of the sliding bodies can complicate matters during both continuous sliding and transitions accompanying intermittent sliding.

2. Overview

In the models that we will discuss below, we limit ourselves to continuous sliding, although extensions to loss of contact or sticking could be made. The unsteadiness arises due to forced normal contact vibrations of a rigid rider mass, supported by smooth Hertzian or randomly rough planar compliant contacts undergoing elastic deformation. Initially the rider is constrained to move only along a line normal to the sliding direction. The vibration problem is solved for the normal motions. To allow a well-defined mechanism of friction to be explicitly inserted into the dynamic model, the instantaneous friction force is related to the normal motion through the adhesion theory of friction. Accordingly, the instantaneous friction force is taken to be proportional to the instantaneous real area of contact. While we recognize the limitations of the adhesion theory, it is selected due to its simplicity and its ability to describe many situations of practical interest [16]. This approach allows the model of the friction force to initially be decoupled from the normal motions.

A general feature of the results is that as the normal oscillations increase, the average separation of the surfaces increases. This is due to the nonlinear character of the contact stiffness which increases (hardens) as the instantaneous normal load increases from its mean value and decreases (softens) as the load is reduced. This increase in average separation is, under the assumptions stated above, sometimes, but not always, accompanied by a decrease in the average friction force.

A more interesting, yet still simple, model is that of a rough block in planar contact that is allowed to translate and rotate with respect to the countersurface against which it slides. We have developed a modification of the Greenwood-Williamson [17] rough surface model for this purpose. The basic equations are given and general features of the problem are discussed. Some comparisons are made with our experiments and with part of the work of Martins et al. [5], in which they examine a similar problem using a phenomenological constitutive contact model.

Before proceeding, we comment on the interpretation of the coefficient of friction under unsteady conditions. If both the load and the friction force at a contact vary with time, the instantaneous friction coefficient, $\mu(t)$, is

$$\mu(t) = \frac{F(t)}{P(t)} \tag{1}$$

Of particular interest is the interpretation of average friction. One interpretation of average friction is to take the time average of $\mu(t)$, denoted by $\langle \mu(t) \rangle$. Alternatively, one could define an average friction coefficient, μ_{av}, as the average friction force divided by the average normal load, so that

$$\mu_{av} = \frac{\langle F(t) \rangle}{\langle P(t) \rangle} \qquad (2)$$

If the normal load remains constant or the instantaneous friction coefficient does not change with time, the two interpretations are equivalent. Otherwise they are not.

This is readily demonstrated by considering the example of a smooth, massless, circular Hertzian contact to which an oscillating load $P_o(1 + \cos\Omega t)$ is applied. This amount of load fluctuation is just enough to give impending contact loss at one extreme of the motion. The friction coefficient is μ_o when the load is at its mean value, P_o. For illustration purposes, the instantaneous friction force is assumed to be proportional to the instantaneous real area of contact. It is easy to show [1] that, in this case, $\frac{\mu_{av}}{\mu_o} = 0.92$

whereas $\frac{\langle\mu\rangle}{\mu_o} = 1.84$. This is illustrated in Fig. 1. The time average of the friction coefficient, $\langle\mu(t)\rangle$, increases while the average friction force decreases! When F, P and μ all vary with time, the coefficient of friction seems to be of limited value. Particular difficulties arise when $P(t) \approx 0$. For defining average friction, we prefer the definition of equation (2).

Sometimes, in friction testing, only the instantaneous friction force is measured. Even this requires a measurement system with sufficient frequency bandwidth to accurately measure the fluctuating forces. The normal load is not monitored. If one incorrectly assumes that the normal load remains constant, when it does not, one obtains what Martins et al. [5] call the "apparent friction" coefficient which can be quite different from the actual friction. Apparent friction sometimes includes stick or loss of contact which do not represent friction in the usual sense.

3. Average Friction Under Unsteady Sliding

3.1. HERTZIAN CONTACT, NORMAL MOTION

As the first and simplest example, we examine the dynamic behavior of a circular Hertzian contact under dynamic excitation. The system is shown in Fig. 2.

The rider has mass, m, and is in contact with a flat surface through a nonlinear stiffness and a viscous damper. The lower flat surface moves from left to right at a constant speed, V. The friction force, F, acts on the rider in the direction of sliding. The rider is constrained to motion normal to the direction of sliding. The model accommodates the primary normal contact resonance. The contact is loaded by its weight, mg, and by an external load, $P = P_o(1 + \alpha\cos\Omega t)$, which includes both a mean and a simple harmonic component. The normal displacement, y, of the mass is measured upward from its static equilibrium position, y_o. The equation of motion during contact, obtained from summing forces on the mass is

$$m\ddot{y} + c\dot{y} - f(\delta) = -P_o(1 + \alpha\cos\Omega t) - mg \qquad for \quad \delta > 0 \qquad (3)$$

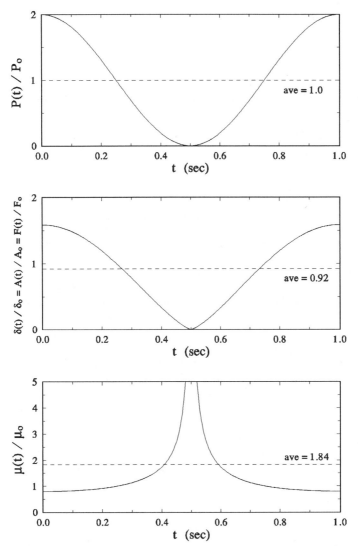

Figure 1. Instantaneous and average load, area, and friction (force and coefficient) for a smooth massless Hertzian contact.

where δ is the contact deflection and $f(\delta)$ is the restoring force given by

$$f(\delta) = \frac{4}{3} E' R^{\frac{1}{2}} \delta^{\frac{3}{2}} = K_1 (y_o - y)^{\frac{3}{2}} \quad , \quad y_o = \left(\frac{P_o + mg}{K_1} \right)^{\frac{2}{3}} \tag{4}$$

An approximate steady-state solution to this nonlinear system has been obtained [1] using the perturbation technique known as the method of multiple scales.

The contact area, A, is proportional to the contact deflection, $(y_o - y)$. Based on the adhesion theory of friction, we assume that the instantaneous friction force is proportional to the area of the contact. Therefore,

$$\frac{F}{F_o} = \frac{A}{A_o} = 1 - \frac{y}{y_o} \tag{5}$$

The normal oscillations, $y(t)$, are asymmetrical due to the nonlinear contact stiffness, and give rise to a decrease in average contact deflection, $(y_o - \langle y \rangle)$, (i.e., an increase in separation of the sliding bodies) by an amount $\langle y \rangle$, where $\langle y \rangle$ is the average of $y(t)$. Since equation (5) is linear, we can also write

$$\frac{\langle F \rangle}{F_o} = \frac{\langle A \rangle}{A_o} = 1 - \frac{\langle y \rangle}{y_o} \tag{6}$$

As oscillations increase, the average contact area, and, by implication, the average friction are reduced. A reduction in average friction force of up to ten percent was shown to occur [1] prior to loss of contact. This is not greatly different from the result obtained without considering inertia forces or damping and illustrated in Fig. 1.

Godfrey [18] conducted experiments to determine the effect of normal vibration on friction. His apparatus consisted of three steel balls fixed to a block that slid along a steel beam and was loaded by the weight of the block. The beam was vibrated by a speaker coil at various frequencies. The normal acceleration of the rider and the friction at the interface were measured. His measurements, under dry contact conditions, are illustrated in Fig. 3. If one assumes that occasional contact loss begins to occur when the normal acceleration reaches an amplitude of one g, one can superimpose the friction reduction predicted by our model as indicated by the heavy line. Reasonably good agreement is obtained. At higher normal accelerations, where there is progressively more intermittent contact loss, a larger reduction in friction occurs.

The dynamic behavior of continuously sliding Hertzian contacts under random roughness-induced base excitation has also been examined [3]. At sufficiently high loads, such contacts can be represented by a smooth Hertzian contact [19]. The role of the surface roughness is only to provide a base excitation as it is swept through the contact region.

By restricting the effective surface roughness input displacement to stationary random processes defined by the spectral density function $S_{y_i y_i}(k) = L\pi^{-1}k^{-4}$ (where L is a constant and k is the surface wavenumber), the Fokker-Planck equation can be used to obtain the exact stationary solution.

Again, one finds a decrease in the mean contact compression under dynamic loading. This also leads to a reduction in the mean contact area and the average friction force, under the assumption that the instantaneous friction force is proportional to the instantaneous area of contact. Based on the analysis the reduction in average friction force when vibration amplitudes approached the limit of contact loss was around nine percent.

A pin-on-disk system with a steel against steel Hertzian contact, excited by surface irregularities, was used to obtain measurements of average friction at various sliding speeds. The normal vibrations increased with sliding speed. The analysis was

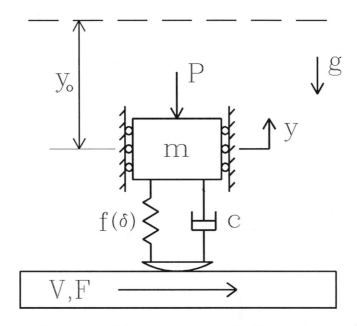

Figure 2. *Dynamic model of normal motion for Hertzian contact.*

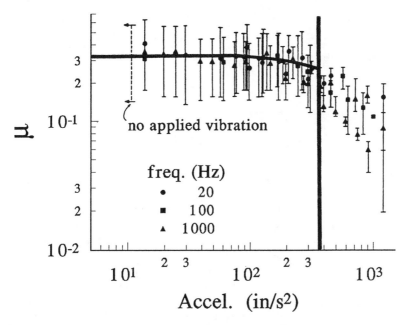

Figure 3. *Measurements from Godfrey (1967) showing the effect of vibration on friction;* ▬▬▬ *, present theory.*

compared with the experiments by adjusting the parameter, L, so that the analytical model gave initial loss of contact at the same speed (i.e., at 50 cm/s) as observed during the tests. The computed results are shown together with the measurements in Fig. 4.

The measurements show a decrease in friction with increasing sliding speed. Considering that the load criterion of Greenwood and Tripp [19] is not satisfied at all times during the motion, the agreement with the theoretical model is quite good. The measurements illustrate that, only at speeds well above those associated with initial loss of contact, can one obtain large reductions in average friction, of as much as thirty percent.

3.2. PLANAR CONTACT, NORMAL MOTION

The Greenwood and Williamson [17] statistical formulation of the elastic contact of randomly rough surfaces is still the best known and most widely-used model. The normal vibrations of such a contact can be cast in the same form as equation (3) with the real contact area and the normal elastic restoring force expressed by

$$A = \pi \eta \tilde{A} \beta \sigma \int_{h}^{\infty} (\varepsilon - h)\, \Phi^{*}(\varepsilon)\, d\varepsilon \tag{7a}$$

$$f(\delta) = \frac{4}{3}\eta \tilde{A} E' \beta^{\frac{1}{2}} \sigma^{\frac{3}{2}} \int_{h}^{\infty} (\varepsilon - h)^{\frac{3}{2}}\, \Phi^{*}(\varepsilon)\, d\varepsilon \tag{7b}$$

where

$\beta \equiv$ asperity radius

$\sigma \equiv$ standard deviation of asperity height distribution

$h = \dfrac{d}{\sigma} = \dfrac{y + d_o}{\sigma} \equiv$ normalized separation

$d_o \equiv$ static separation

$\varepsilon = \dfrac{z}{\sigma} \equiv$ normalized asperity heights

$\eta \equiv$ surface density of asperities

$\tilde{A} \equiv$ nominal contact area

$\Phi^{*}(\varepsilon) \equiv$ normalized asperity height distribution

The nonlinear vibration problem is solved in [2]. The contact stiffness nonlinearity is stronger than that of the Hertzian model. Again one finds that, on average, the sliding surfaces move apart during sliding. The change in average separation is typically around thirty percent of the vibration amplitude, $|y|$. Although the surfaces on average, move apart, the average friction force obtained by taking the time average of the contact area, remains unchanged in the presence of normal vibrations! This seemingly paradoxical result is not unexpected when one recognizes that the Greenwood-Williamson model leads to a direct proportionality between the normal load and the real contact area at all separations, i.e., a constant instantaneous friction

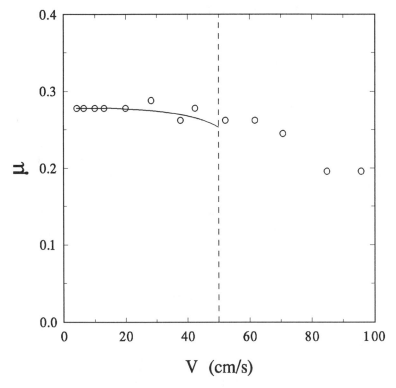

Figure 4. Average coefficient of friction at various sliding speeds: o, measurements; —— , theoretical.

coefficient. While the nonlinear contact vibrations can be complicated, and the instantaneous friction force may change considerably, the friction coefficient is not expected to change. Other rough surface models, may give somewhat different results.

Linearized equations for the normal vibration problem have also been developed [4]. One rather remarkable result of the linearized analysis is that the small amplitude normal natural frequency of a weight-loaded rigid block supported by a Greenwood-Williamson type rough surface is $\omega_{oy} = \sqrt{\frac{g}{\sigma}}$. The natural frequency is independent of the block and countersurface materials. The natural frequency is independent of the block dimensions and, at least on earth, depends only on the standard deviation of the asperity heights, σ. The acceleration spectra of a steel block (a 4.4 cm cube) obtained during sliding against a large steel base at a speed of 3 cm/s are shown in Fig. 5. One finds the normal natural frequency at around 1300 Hz which is in general agreement with the block roughness that was measured($R_a \approx 0.2\,\mu m$). Angular motions, with a resonant frequency of 1070 Hz are also observed and shown in Fig. 5. It is clear that the possibility of angular motions must be included in a model of the problem.

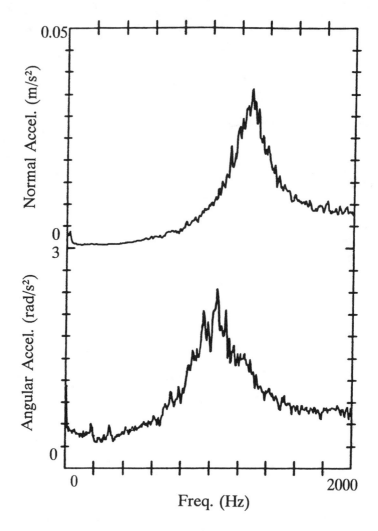

Figure 5. Average rms spectra of measured normal and angular accelerations from sliding block.

3.3. PLANAR CONTACT, NORMAL AND ANGULAR MOTIONS

A model that allows for both normal and angular motions of a nominally stationary block pressed against a moving countersurface is shown in Fig. 6a in its frictionless equilibrium position and in Fig. 6b in its steady sliding equilibrium position. Some angular displacement, θ_o, and offset, c, of the normal reaction force are necessary to maintain moment equilibrium of the block.

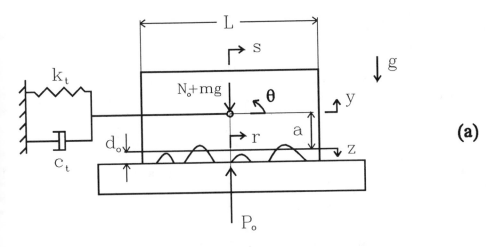

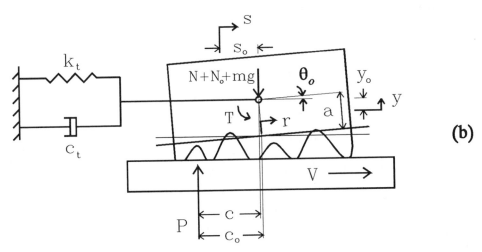

*Figure 6. Three degree-of-freedom system model: (a) frictionless equilibrium position,
(b) steady sliding equilibrium position with friction.*

We have extended [4] the Greenwood-Williamson model to account for angular as well
as normal motions. For an exponential distribution of asperity heights, we find that

$$\frac{A(y,\theta)}{A_o} = \frac{P(y,\theta)}{P_o} = \frac{\sigma}{L\theta}\, e^{-\frac{y}{\sigma}}\,[\, e^{\frac{L\theta}{2\sigma}} - e^{-\frac{L\theta}{2\sigma}}\,] \tag{8}$$

where A_o and P_o denote the real contact area and the normal load at the frictionless equilibrium position. y and θ are measured from that position. Both the contact area and normal load depend on the normal displacement, y, and the angular displacement, θ. However, in the presence of both angular and normal motions, the area of contact remains proportional to the normal load. Therefore, within the assumptions of the analysis, we do not expect the coefficient of friction to change with normal or angular vibrations during continuous sliding, since $A(y,\theta)/A_o = F(y,\theta)/F_o = P(y,\theta)/P_o$.

For other surface topologies, the above result may not hold. For example, it has been shown [20] that for a periodic surface consisting of a regular pattern of hemispherical asperities of equal height, it is the friction force rather than the friction coefficient that remains constant when relative angular motions occur at a given normal load.

The dynamic equations of the three degree-of-freedom sliding system can now be written directly:

$$m\ddot{s} + c_t\dot{s} + k_t s = F(y,\theta) \tag{9a}$$

$$m\ddot{y} + b\dot{y} - P(y,\theta) = -N(t) - N_o - mg \tag{9b}$$

$$J\ddot{\theta} + B\dot{\theta} - c(\theta)P(y,\theta) - (a + d_o)F(y,\theta) = T(t) \tag{9c}$$

Viscous damping terms b and B have been introduced to account for some damping of the motions. Introducing the changes in variables, $q = \dfrac{y}{\sigma}$ and $\phi = \dfrac{L\theta}{2\sigma}$, the equations become (see [4])

$$\ddot{s} + \frac{c_t}{m}\dot{s} + \frac{k_t}{m}s = \frac{F_o}{m}f_1(q)\,f_2(\phi) \tag{10a}$$

$$\ddot{q} + \frac{b}{m}\dot{q} - \frac{P_o}{m\sigma}f_1(q)\,f_2(\phi) = -\frac{1}{m\sigma}(N(t) + P_o) \tag{10b}$$

$$\ddot{\phi} + \frac{B}{J}\dot{\phi} - \frac{L}{2J\sigma}f_1(q)\,f_3(\phi) = \frac{LT(t)}{2J\sigma} \tag{10c}$$

where

$$f_1(q) = e^{-q} \tag{11a}$$

$$f_2(\phi) = \frac{1}{2\phi}\left(e^\phi - e^{-\phi}\right) \tag{11b}$$

$$f_3(\phi) = \frac{1}{2\phi}\left(e^\phi - e^{-\phi}\right)\left[P_o\left(\frac{\frac{L}{2\phi}\left[(1-\phi)e^\phi - (1+\phi)e^{-\phi}\right]}{\left(e^\phi - e^{-\phi}\right)} + \frac{2\sigma a\phi}{L}\right) + (a+d_o)F_o\right] \tag{11c}$$

Equilibrium of the rider under steady sliding requires that

$$k_t s_o = F_o\,f_1(q_o)\,f_2(\phi_o) \tag{12a}$$

$$f_1(q_o)\,f_2(\phi_o) = 1 \tag{12b}$$

$$f_3(\phi_o) = 0 \tag{12c}$$

The equilibrium position of the rider under steady sliding can be determined by solving equation (12c) for ϕ_o, equation (12b) for q_o, and finally equation (12a) for s_o.

The equations of motion (10,11) clearly reveal nonlinear coupling among the translational and angular motions. Subharmonic, superharmonic, and combination resonances may occur when the system is in forced oscillation. Chaotic motions may also take place. In problems of this type, system stability, i.e., stability of sliding, may be of concern and was studied by Martins et al. [5]. In fact, our equations are similar to those of Martins et al.. For example, equations (10) and (12) of the present paper can be compared to equations (5.8) and (5.6) in their paper. An essential difference between the two approaches is that the normal and angular contact stiffnesses have different forms. Martins et al. used a power law form that has also been used by Back et al. [21] and Kragelskii and Mikhin [22].

4. Stability of Steady Sliding

A linear stability analysis of the three degree-of-freedom contact model developed above reveals some interesting aspects of sliding systems. The linearized equations of motion for small perturbations about the steady sliding equilibrium position without forcing terms are (see [4])

$$\ddot{s}_s + \frac{c_t}{m}\dot{s}_s + \frac{k_t}{m}s_s + \frac{F_o}{m}q_s - \frac{F_o c_2}{m}\phi_s = 0 \tag{13a}$$

$$\ddot{q}_s + \frac{b}{m}\dot{q}_s + \frac{P_o}{m\sigma}q_s - \frac{P_o c_2}{m\sigma}\phi_s = 0 \tag{13b}$$

$$\ddot{\phi}_s + \frac{B}{J}\dot{\phi}_s - \frac{L c_4}{2J\sigma}\phi_s = 0 \tag{13c}$$

where

$$c_2 = \frac{e^{-q_o}}{2\phi_o^2}\left[\phi_o\left(e^{\phi_o} + e^{-\phi_o}\right) - \left(e^{\phi_o} - e^{-\phi_o}\right)\right] \tag{14a}$$

$$c_4 = e^{-q_o} \left\{ \frac{P_o L}{4}\left[\frac{1}{\phi_o}\left(-e^{\phi_o} + e^{-\phi_o}\right) - \frac{2}{\phi_o^3}\left[(1-\phi_o)e^{\phi_o} - (1+\phi_o)e^{-\phi_o}\right]\right] \right.$$
$$\left. + \frac{\sigma a P_o}{L}\left(e^{\phi_o} + e^{-\phi_o}\right) + \frac{(a+d_o)F_o}{2}\left[\frac{1}{\phi_o}\left(e^{\phi_o} + e^{-\phi_o}\right) - \frac{1}{\phi_o^2}\left(e^{\phi_o} - e^{-\phi_o}\right)\right] \right\} \tag{14b}$$

These linearized equations, having an asymmetric stiffness matrix, describe a circulatory system. These equations are similar to equation (5.15) of Martins et al. [5], which they found to exhibit a high frequency flutter instability at friction values well below that which would result in the block tumbling, when $\mu_o = \frac{L}{2a}$, which is a divergence instability.

This flutter instability does not seem to occur with the model of equation (13). The eigenvalues that we have computed with the damping set to zero are always purely imaginary, never exhibiting positive real parts. In Martins et al. [5], the flutter instability occurs when the two eigenvalues associated with the angular and quasi-normal natural frequencies take on the same value. In the present model, the eigenvalue ratio (angular divided by quasi-normal) always remains less than unity, approaching this value only when the block is very long, i.e., when $\frac{2a}{L} < 1$. The

548

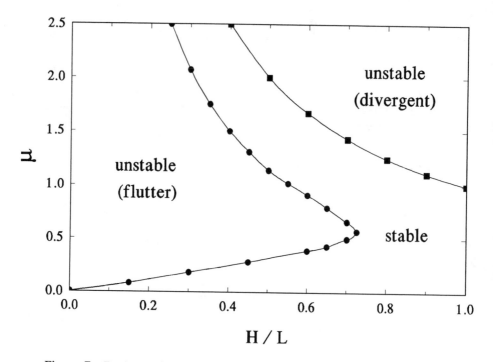

Figure 7. Regions of instability based on Martins et al. (1990). (H = 2a)

differences in the qualitative behavior of the two models seems to be due to the details of the normal and angular restoring forces. These contact stiffnesses are very sensitive to the details of the surface texture. This may largely explain the elusive nature of many high frequency instabilities and squeal phenomena which can occur in sliding systems and can change and appear or disappear as surfaces run-in or wear.

Fig. 7 shows the instability regions found by Martins et al. [5] in the absence of damping. In their analysis, and with weight loading, the stability depends only on the height to length ratio of the block, the sliding friction coefficient, and a damping parameter. These results indicate instability over a fairly broad range of aspect ratio (H/L). The addition of normal damping causes the flutter instability boundary to shift to higher friction values for small aspect ratios.

Interestingly, the addition of a linear torsional stiffness, k_ϕ, to our sliding block model can lead to instability. The resulting linearized equations of motion for this case are

$$\ddot{s}_s + \frac{c_t}{m}\dot{s}_s + \frac{k_t}{m}s_s + \frac{F_o}{m}q_s - \frac{F_o c_2}{m}\phi_s = 0 \qquad (15a)$$

$$\ddot{q}_s + \frac{b}{m}\dot{q}_s + \frac{P_o}{m\sigma}q_s - \frac{P_o c_2}{m\sigma}\phi_s = 0 \qquad (15b)$$

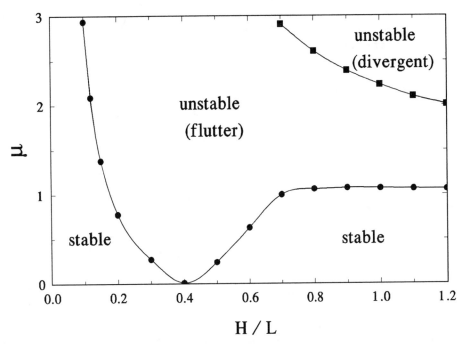

Figure 8. Regions of instability when a torsional spring is added. (H = 2a)

$$\ddot{\phi}_s + \frac{B}{J}\dot{\phi}_s + \left(\frac{k_\phi}{J} - \frac{Lc_4}{2J\sigma}\right)\phi_s + \frac{Lc_3}{2J\sigma}q_s = 0 \tag{15c}$$

where c_2 and c_4 are defined in equations (14) and

$$c_3 = e^{-q_o}\left[\frac{P_oL}{4\phi_o^2}\left[(1 - \phi_o)e^{\phi_o} - (1 + \phi_o)e^{-\phi_o}\right] + \frac{P_o\sigma a}{L}\left(e^{\phi_o} - e^{-\phi_o}\right) + \frac{(a + d_o)F_o}{2\phi_o}\left(e^{\phi_o} - e^{-\phi_o}\right)\right] \tag{16}$$

From the nonlinear equations of motion (10), equilibrium of the rider under steady sliding for this case requires that

$$k_t s_o = F_o\, f_1(q_o)\, f_2(\phi_o) \tag{17a}$$

$$f_1(q_o)\, f_2(\phi_o) = 1 \tag{17b}$$

$$-\frac{L}{2\sigma} f_1(q_o) f_3(\phi_o) + k_\phi\phi_o = 0 \tag{17c}$$

From these equations the equilibrium position during sliding can be computed.

The stability of the system defined by equation (15) can be assessed by computing the eigenvalues. Fig. 8 shows the stability regions for a particular value of linear torsional stiffness and no viscous damping. The instability region of particular interest is the flutter instability between aspect ratio values of 0.2 and 0.6. As the torsional stiffness is decreased, this region becomes narrower and shifts to lower values of aspect ratio.

550

The effect of damping on this region is currently being investigated. An alternate means of adding torsional stiffness to the system is to offset the horizontal suspension from the center of mass.

We are in the process of carrying out experimental studies on the stability of such systems. To date, we have not been able to duplicate these flutter instabilities in the laboratory. It may be that the instabilities are suppressed by damping, stabilized by low level surface-roughness induced vibrations, or affected by other dynamic characteristics of the apparatus.

5. Conclusions

We have shown that, under certain well-defined assumptions, unsteady dry friction phenomena at continuously sliding contacts can be defined with respect to steady conditions. It seems that average friction forces are not greatly affected by even large fluctuations in the contact forces, unless some intermittency, i.e., loss of contact or sticking occurs. Quasi-static friction mechanisms, which may be formulated to relate the instantaneous relative approach and orientation of the sliding bodies to the instantaneous friction force, can also be applied to unsteady sliding. Extension to intermittent conditions would not be difficult.

This does not mean that unsteady friction problems are trivial. The complex nonlinear dynamic behavior of seemingly simple sliding systems such as blocks on flat surfaces become even more challenging when such elements are embedded in larger dynamic systems. Furthermore, the details of the surface topographies, contact compliances, constraints, and dynamic inputs are generally not known. Yet we have shown at least analytically, that high frequency flutter instabilities, which can cause squeal noise and affect apparent friction are very sensitive to these parameters.

Since the present approach specifically includes individual asperity contacts, it may be viewed, in a sense, to relate microscopic to macroscopic friction. It would be interesting to know if an approach that statistically combined complete time histories of many asperity interactions (or whatever other micro events are present) would yield significantly different interpretations of steady or unsteady friction.

In any case, there are numerous open questions to be answered and many interesting phenomena to be explained. However, due to both the theoretical and experimental uncertainties always associated with friction phenomena, real progress will only be made when clearly formulated analyses are closely coupled to careful experimental studies.

Acknowledgement

The authors would like to acknowledge the support of the National Science Foundation under Grant No. MSM 85-14220 and Grant No. MSS 9100458.

References

1. Hess, D. P. and Soom, A., 1991, "Normal Vibrations and Friction Under Harmonic Loads: Part I -Hertzian Contacts," *ASME Journal of Tribology*, Vol. 113, pp. 80-86.

2. Hess, D. P. and Soom, A., 1991, "Normal Vibrations and Friction Under Harmonic Loads: Part II -Rough Planar Contacts," *ASME Journal of Tribology*, Vol. 113, pp. 87-92.

3. Hess, D. P., Soom, A. and Kim, C. H., "Normal Vibrations and Friction at a Hertzian Contact Under Random Excitation: Theory and Experiments," *Journal of Sound and Vibration*, in press.

4. Hess, D. P. and Soom, A., "Normal and Angular Motions at Rough Planar Contacts During Sliding with Friction," *ASME Journal of Tribology*, in press.

5. Martins, J. A. C., Oden, J. T. and Simoes, F. M. F., 1990, "A Study of Static and Kinetic Friction," *International Journal of Engineering Science*, Vol. 28, pp. 29-92.

6. Oden, J. T. and Martins, J. A. C., 1985, "Models and Computational Methods for Dynamic Friction Phenomena," *Computer Methods in Applied Mechanics and Engineering*, Vol. 52, pp. 527-634.

7. Armstrong-Helouvry, B., 1991, *Control of Machines with Friction*, Kluwer Academic Publishers, Boston.

8. Soom, A. and Kim, C. H., 1983, "Interactions Between Dynamic Normal and Frictional Forces During Unlubricated Sliding," *ASME Journal of Lubrication Technology*, Vol. 105, pp. 221-229.

9. Soom, A. and Kim, C. H., 1983, "Roughness-Induced Dynamic Loading at Dry and Boundary-Lubricated Sliding Contacts," *ASME Journal of Lubrication Technology*, Vol. 105, pp. 514-517.

10. Bowden, F. P. and Leben, L., 1939, "The Nature of Sliding and the Analysis of Friction," *Proceedings of the Royal Society of London*, Vol. A169, pp. 371-391.

11. Derjaguin, B. V., Push, V. E. and Tolstoi, D. M., 1957, "A Theory of Stick-Slip Sliding of Solids," *Proceedings of the Conference of Lubrication and Wear* (Institution of Mechanical Engineers), pp. 257-268.

12. Rabinowicz, E., 1965, *Friction and Wear of Materials*, Wiley, New York.

13. Marui, E. and Kato, S., 1984, "Forced Vibration of a Base-Excited Single-Degree-of-Freedom System With Coulomb Friction," *ASME Journal of Dynamic Systems, Measurements, and Control*, Vol. 106, pp. 280-285.

14. Taniguchi, O., Sakata, M., Suzuki, Y., and Osanai, Y., 1963, "Studies on Vibratory Feeder," *Bulletin of JSME*, Vol. 6, pp. 37-43.

15. Sakaguchi, K. and Taniguchi, O., 1970, "Studies on Vibratory Feeder (2nd Report)," *Bulletin of JSME*, Vol. 13, pp. 881-887.

16. Tabor, D., 1981, "Friction - The Present State of Our Understanding," *ASME Journal of Lubrication Technology*, Vol. 103, pp. 169-179.

17. Greenwood, J. A. and Williamson, J. B. P., 1966, "Contact of Nominally Flat Surfaces," *Proceedings of the Royal Society of London.*, Vol. A295, pp. 300-319.

18. Godfrey, D., 1967, "Vibration Reduces Metal to Metal Contact and Causes an Apparent Reduction in Friction," *ASLE Transactions,* Vol. 10, pp. 183-192.

19. Greenwood, J. A. and Tripp, J. H., 1967, "The Elastic Contact of Spheres," *Journal of Applied Mechanics*, Vol. 34, pp. 153-159.

20. Hess, D. P., 1991, *Nonlinear Contact Vibrations and Dry Friction at Concentrated and Extended Contacts*, Ph.D. Dissertation, University at Buffalo.

21. Back, N., Burdekin, M. and Cowley, A., 1973, "Review of the Research on Fixed and Sliding Joints," *Proceedings of the 13th International Machine Tool Design and Research Conference* (edited by S. A. Tobias and F. Koenigsberger), Macmillan, London.

22. Kragelskii, I. V. and Mikhin, N. M., 1988, *Handbook of Friction Units of Machines*, ASME Press, New York.

INTERACTION AND STABILITY OF FRICTION AND VIBRATIONS

M. T. BENGISU and A. AKAY
Department of Mechanical Engineering
Wayne State University
Detroit, Michigan 48202
USA

ABSTRACT. A stability analysis of stick-slip vibrations is presented, considering single- and multi-degree-of-freedom systems to demonstrate the significance of the dimensions of a system on friction-induced vibrations. First, a stability analysis is given for a single-degree-of-freedom system, based on the linearized model, where the existence of limit cycles is demonstrated. The analysis is then extended to a multi-degree-of-freedom system. The increase in the dimensions of the system is shown to change the response characteristics from those of a single-degree-of-freedom system; now the number of bifurcations is more than one. The spectrum of the response is composed of a combination of fundamental frequencies corresponding to each bifurcation and their harmonics. These fundamental frequencies may become synchronized depending on the coupling between the modes of the system. In cases of higher dimensions and several independent frequencies, it is also possible to have a chaotic response.

1. Introduction

Vibrations are an undesirable by-product of friction between sliding surfaces. Friction-induced vibrations, often referred to as stick-slip, are an inhibiting factor for increased accuracy and efficiency in mechanical systems. In the case of friction measurements, vibrations, either of the test setup or of the components in contact, can significantly influence the measured values. Similar effects on wear tests are to be expected. Friction-induced sounds, such as brake and bearing squeals, and chatter in machine tools, have plagued some industries for years without a universally accepted explanation or solution.

Although until recently vibrations have been treated purely as a consequence of friction, there is strong evidence of a vital interaction between friction and vibrations. The friction force generated between two bodies in sliding contact is very much dependent on the resulting vibrations, and in most cases there is significant *feedback* between friction and the vibratory response of the system. The existence of a feedback mechanism suggests that friction forces may be modified by judicious design of the components in contact, or even their attachments, to have particular vibration characteristics, not unlike the adjustment of tension in the strings of a musical instrument.

I. L. Singer and H. M. Pollock (eds.), Fundamentals of Friction: Macroscopic and Microscopic Processes, 553–566.
© 1992 *Kluwer Academic Publishers. Printed in the Netherlands.*

The first notable study of the effects of change of dry friction forces with relative velocity was performed by Thomas [1] in 1930. Variation of frictional force with velocity as the fundamental cause of friction-induced vibrations was recognized and studied with the now-classic instrument developed by Bowden and Leben [2] in 1939 at Cambridge. Their demonstration, which consisted of a block connected to an anchored spring and set on a slab in motion, led to recognition of the now well-known stick-slip phenomenon, which was later analyzed systematically in greater detail by Blok [3]. In 1946, Bristow [4] argued that "... stick-slips are dependent on the speed of sliding, the dynamics of the instruments and the nature of the dependence of the kinetic friction on velocity." He further suggested that failure to consider the dynamics of the instrument used to measure friction is bound to give erroneous results. These developments in stick-slip were later followed by a large number of studies which employed mass-spring-damper models using velocity-dependent friction forces, considering oscillations in the tangential direction only. The recognition that friction causes vibrations in a direction normal to the interface [5] further complicated models of both the systems and the friction force.

During sliding the force developed in the interface has both tangential and normal components. Even for nominally flat surfaces, we know that these components of the contact force at an interface do not necessarily have uniform distributions. Nonuniformity of the interface force further exacerbates the vibrations by providing coupling between the oscillations among angular, tangential and normal directions. Accurate modelling of the effects of such asymmetry of the excitation forces on the response of a system requires more than one degree of freedom. Recently, significant studies were performed on self-excited vibrations due to friction by considering the normal and rotational degrees of freedom in addition to the sliding motion of a rigid mass, for example [6,7], as explained in more detail in the paper by Hess and Soom in this section.

Just as friction forces cause vibrations, the resulting vibrations in turn affect the friction forces. Most of the early studies were concerned with the effects of vibration on static friction and on its measurement. Later, studies by Tolstoi [5] and Godfrey [8] showed a decrease in friction force as a result of vibration, with the general conclusion that when the vibration acceleration exceeds the gravitational acceleration, friction force decreases. The decrease is attributed to a possible reduction in the contact area, thus allowing an eventual increase in the hydrodynamic lifting power.

The investigations cited above deal almost exclusively with the interaction of friction and vibrations, using rigid-body dynamics to illuminate the difficulties involved in the modelling and understanding of the problems encountered. Many problems of friction and vibration, however, require consideration of the elastic waves within the components of the system.

When two extended surfaces slide against each other, stresses develop at the actual contact areas, generating elastic waves that emanate into each body, as depicted in Fig. 1. The nature of these elastic waves and their range are determined by both macroscopic and microscopic parameters related to the objects that embody these surfaces. In addition to the external normal and tangential forces that cause them to slide against each other, their

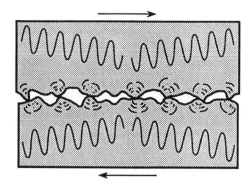

Fig. 1. Schematic of waves emanating from contact areas.

elastodynamic properties (material, geometry, boundary conditions, etc.) and interfacial properties (surface composition, debris, asperities, and humidity) are key factors.

The contact forces will induce waves and internal vibrations of the bodies. When the interface forces have frequencies or spatial distributions that coincide with those of one of the natural modes of a body, we can expect different behavior from the friction-excited system. Elastic objects in sliding friction often exhibit *self-tuning*, where the combined system response *locks in* to a particular frequency very close to one of the natural frequencies of either object. This paper deals with this particular aspect of the friction-vibration interaction problem by considering the response of an elastic system subject to a tangential friction force. First, an overview of the classical stick-slip problem will be given, with some new insights into the problem, and then the results will be compared with those for a multi-degree-of-freedom (mdof) model representing a continuous elastic system.

2. Dynamic Model

For the purposes of the present analysis, simply to demonstrate the stability aspects of stick slip in single- and multi-degree-of-freedom systems, the problems considered here will be restricted to having only the horizontal components of the contact force and motion. The friction force, as illustrated in Fig. 2, will be expressed by the following equation

$$F_f = F_n sgn(V - \dot{x}) \left(1 - e^{-\beta|V-\dot{x}|}\right) \left[1 + (F_r - 1)e^{-\alpha|V-\dot{x}|}\right] \tag{1}$$

where F_n is the instantaneous normal contact force, α and β are parameters related to surface roughness or the bonding ability of the surfaces and F_r is the ratio of the asymptotic value of the curve to its maximum. In a sense, F_r gives the ratio of the static and kinetic friction forces. The parameter β is related to the "stick" and α is related to "slip" state of the motion. The shape of the curve, of course, depends on the relative *instantaneous* interface speed $(V - \dot{x})$. Where V and $\dot{x}$ are the absolute velocities of the interacting flexible

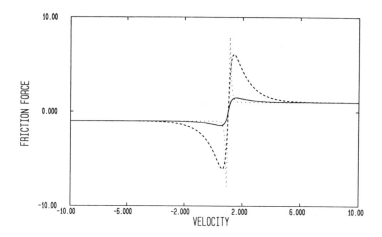

Fig. 2. Friction force as a function of relative interface velocity.

bodies at the contact point. Most of the friction force expressions utilized in the literature can be recovered from Eq. (1) by appropriate combinations of parameters α, β and F_r.

In order to demonstrate the stick-slip phenomenon using the friction force expressed in Eq. (1), a simple, single-degree-of-freedom system is formulated. In this simple case, the normal motion of the system is neglected and the normal contact force is assumed constant. The system considered is a rigid mass, horizontally connected to a linear stiffness and a damping element. The mass is placed on another flat rigid surface which moves at a constant velocity V. The interface of the rigid bodies is macroscopically smooth and the linear (horizontal) motion of the lower mass is coupled to the linear motion of the upper mass by the friction force given above.

The equation of motion of the system is the familiar second-order differential equation with the friction force on the right-hand side.

$$x''(\tau) + 2\xi x'(\tau) + x(\tau) = F_{n0} sgn\left[V - x'(\tau)\right]\left(1 - e^{-\beta_0|V - x'(\tau)|}\right) \left[1 + (F_r - 1)e^{-\alpha_0|V - x'(\tau)|}\right] \tag{2}$$

where the independent variable $\tau = \omega_n t$ is the nondimensional time. The other nondimensional variables are $\xi = c/2m\omega_n$ is the damping ratio, $F_{n0} = F_n/mg$ is the normal force, $\beta_0 = \beta g/\omega_n$, $\alpha_0 = \alpha g/\omega_n$, $V = v\omega_n/g$ is the driving velocity, and F_r is the ratio of the static and kinetic friction forces. ω_n is the natural frequency of the mass-damper-spring system and g is the gravitational constant. The $'$ and $''$ in Eq. (2) denote differentiation with respect to nondimensional time τ.

Similarly, equations of motion can be written for a mdof system. Here, we consider the case of a simple three-degree-of-freedom dynamic system with a friction force acting at a single point, as shown schematically in Fig. 3. The linear part of the equations of the

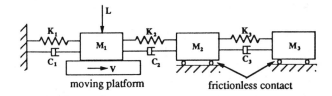

Fig. 3. Schematic of a three-degree-of-freedom system excited by a friction force.

system, given in Eq. (3), are decoupled; however, there will still be coupling between the modes due to the nonlinear friction force.

$$
\begin{bmatrix} 1 & 0 & 0 \\ 0 & \lambda_1^2 & 0 \\ 0 & 0 & \lambda_2^2 \end{bmatrix} \begin{Bmatrix} x_1'' \\ x_2'' \\ x_3'' \end{Bmatrix} + \begin{bmatrix} 2\xi_1 & 0 & 0 \\ 0 & 2\xi_2\lambda_1 & 0 \\ 0 & 0 & 2\xi_3\lambda_2 \end{bmatrix} \begin{Bmatrix} x_1' \\ x_2' \\ x_3' \end{Bmatrix} + \begin{Bmatrix} x_1 \\ x_2 \\ x_3 \end{Bmatrix} = \begin{Bmatrix} a_{11} \\ a_{12} \\ a_{13} \end{Bmatrix} F_f \quad (3)
$$

where the velocity is

$$
x' = \left[a_{11}x_1' + a_{12}\lambda_1^2 x_2' + a_{13}\lambda_2^2 x_3' \right]
$$

The second-order nondimensional equation of motion, Eq. (2), can be reduced to two first-order equations by the following transformation

$$
x = X_1
$$
$$
x' = X_2
$$
$$
x'' = X_2'
$$

where the state variables $X_1(\tau)$ and $X_2(\tau)$ denote the displacement and the velocity of the sliding rigid mass. Hence, two first-order equations in state-space are obtained as,

$$
X_1'(\tau) = X_2(\tau)
$$
$$
X_2'(\tau) = -2\xi X_2(\tau) - X_1(\tau)
$$
$$
+ F_{n0} sgn[V - X_2(\tau)] \left(1 - e^{-\beta_0|V - X_2(\tau)|} \right) \left[1 + (F_r - 1)e^{-\alpha_0|V - X_2(\tau)|} \right]
$$

$$(4)$$

Similar transformations can easily be shown for Eq. (3).

3. Stability Analysis

Considering the sdof case, the stability of the nonlinear system, represented by Eq. (4), can be investigated by a stability analysis of the corresponding linearized equations. If the roots of the linearized system have negative real parts, the nonlinear system is asymptotically stable, as in the linear case. If they have positive real parts, the nonlinear system is not asymptotically stable and, in general, it either has a sustained oscillatory response or an unbounded response. If the real part of the roots is zero, the stability of the nonlinear system cannot be determined by stability analysis of the linearized systems.

The equilibrium point of the dynamic system is found from Eq. (4) as

$$X_{1e} = F_{n0} sgn(V) \left(1 - e^{-\beta_0 V}\right) \left[1 + (F_r - 1)e^{-\alpha_0 V}\right]$$
$$X_{2e} = 0 \tag{5}$$

and the linearized equations of motion of the system about this point are,

$$\left\{ \begin{array}{c} X_1' \\ X_2' \end{array} \right\} = \left[\begin{array}{cc} 0 & 1 \\ -1 & -2\xi + S_f \end{array} \right] \left\{ \begin{array}{c} X_1 \\ X_2 \end{array} \right\} \tag{6}$$

where,

$$S_f = F_{n0} \left[\alpha_0 e^{-\alpha_0 V}(F_r - 1)(1 - e^{-\beta_0 V}) - \beta_0 e^{-\beta_0 V}\left(1 + (F_r - 1)e^{-\alpha_0 V}\right)\right] \tag{7}$$

The root locus of the linearized system described by Eq. (6) is given in Fig. 4. The parameter S_f is actually the slope of the friction force with respect to nondimensional velocity X_2 at the equilibrium point, and describes the damping effect of the friction force. Its value determines the stability of the nonlinear system as illustrated in Fig. 5.

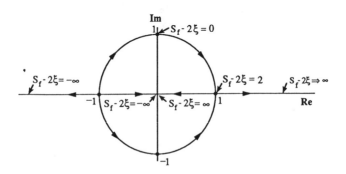

Fig. 4. The root locus of the dynamic system governed by equations of motion given by Eq. (6).

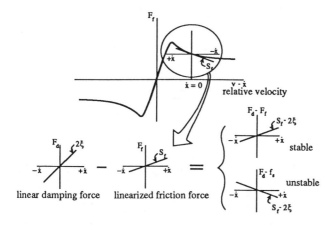

Fig. 5. Illustration of the linearized friction force about the equilibrium point.

In accordance with this criterion, the nonlinear dynamic system governed by Eqs. (4) is asymptotically stable if

$$S_f - 2\xi < 0 \tag{8a}$$

not asymptotically stable if

$$S_f - 2\xi > 0 \tag{8b}$$

and its stability is undetermined by this analysis if

$$S_f - 2\xi = 0 \tag{8c}$$

The frequency $\Omega = f/f_n$ of the oscillations of the nonlinear system can be approximated by that of the linearized system and is given by

$$\Omega = \frac{f}{f_n} \cong \sqrt{1 - \frac{1}{4}(S_f - 2\xi)^2} \tag{9}$$

provided that

$$1 - \frac{1}{4}(2\xi - S_f)^2 > 0$$

When the system is not asymptotically stable it may still be stable in the Lyapunov sense, having a limit cycle.

4. Stability and Stick-Slip

Generation of stick-slip vibrations and the relation of the fundamental frequency of the oscillations to the natural frequency of the dynamic system can be characterized by the stability analysis presented here. The onset of stick-slip depends on the reduction of the friction force at the end of the "stick" state. If α is zero, there would be a steady increase in the friction force as relative interface speed increases. In such a case the dynamic system is asymptotically stable and would reach its equilibrium position following transient oscillations. As stated in Eq. (8a), the condition for stability is $S_f - 2\xi < 0$. Substitution of $\alpha = 0$ in this inequality yields

$$-F_{n0}F_r\beta_0 e^{-\beta_0 V} - 2\xi < 0 \tag{14}$$

since F_{n0}, F_r and β_0 are all positive numbers, the inequality is always satisfied. Hence for $\alpha = 0$, the nonlinear dynamic system is asymptotically stable.

In cases when the linearized system is unstable, i.e., $S_f - 2\xi > 0$, stick-slip oscillations are generated, and the frequency of such oscillations can be approximated by Eq. (9) in the range $0 < S_f - 2\xi < 2$. As the value of the term $(S_f - 2\xi)$ increases, the fundamental frequency of stick-slip is reduced. As it approaches and exceeds 2, the approximation given in Eq. (9) is no longer valid, and unlike an unstable linear system which would show an exponential growth, the present nonlinear system still exhibits oscillations at frequencies decreasing with increasing values of $(S_f - 2\xi)$.

On the other hand, as the term $(S_f - 2\xi)$ approaches zero, the fundamental frequency of the stick-slip oscillations approaches the natural frequency of the linear part of the dynamic system. Therefore, the fundamental frequency of the stick-slip oscillations for the present single-degree-of-freedom system, when not asymptotically stable, is always less than the natural frequency of the system but higher than zero. For a multiple-degree-of-freedom system this behavior is somewhat different and will be discussed in the next section.

The most significant parameter that influences the frequency of stick-slip is the normal load. Under light normal loads the system response displays a frequency closer to the natural frequency of the system. Under heavy normal loads, the response frequency is closer to zero. In general, as the normal contact force increases, the fundamental frequency of stick-slip vibrations will be shifted to lower frequencies, which indicates the potentially significant influence of normal vibrations on the interaction of friction and vibrations.

5. Numerical Results

5.1 SINGLE-DEGREE OF FREEDOM

Stability analysis of the linearized system provides a good insight into the general behavior of the actual nonlinear system. In this respect, stability analysis can be used for guidance in selecting the parameters of the friction force for detailed numerical analyses. To demonstrate the conditions under which stick-slip can or cannot exist and the conditions which cause "stronger" stick-slip to occur, four different cases are presented here. The first two cases correspond to an asymptotically stable dynamic system where stick-slip vibrations cannot be sustained under steady state. The other two cases show different levels of stick-slip vibrations as a function of the fundamental frequency of stick-slip.

5.1.1. *Asymptotically Stable Cases.* The parameters of the friction force are so selected that the friction force is not reduced at the end of the so-called stretch period. This corresponds to choosing $\alpha = 0$, which causes the dynamic system to be asymptotically stable, as discussed earlier. When $\alpha = 0$, the other parameters of the friction force cannot change the behavior of the system. The dynamic response of the system diminishes after several oscillations, and the system becomes stationary at its equilibrium point. In this position the spring is stretched and the mass is acted on by a constant friction force, as indicated by the phase plot (a) in Fig. 6. It is not always necessary to have the parameter $\alpha = 0$ to have an asymptotically stable system. Certain other combinations of the friction force parameters can cause the system to be asymptotically stable, as shown in plot (b) of Fig. 6.

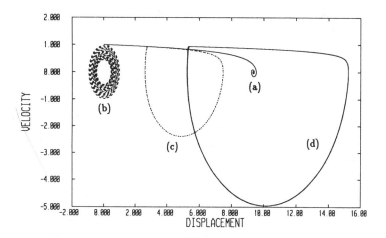

Fig. 6. Phase plots for a single-degree-of-freedom system for different system parameters. $V = 1, \beta_0 = 5, F_r = 2, \xi = 0.01$. (a) asymptotically stable; $\alpha_0 = 0$, $F_{n0} = 4.7$, (b) asymptotically stable (partial response shown); $\alpha_0 = 1, F_{n0} = 0.005$, (c) stick-slip; $\alpha_0 = 1, F_{n0} = 4.7$, (d) stronger stick-slip than in (c); $\alpha_0 = 1, F_{n0} = 10$

5.1.2. *Stick-Slip Response.* The dynamic system with $\alpha = 0$, whose response is shown in Fig. 6a, can be made to generate stick-slip vibrations simply by increasing the parameter α from zero to one while keeping all other parameters constant. By making $\alpha \neq 0$, the linearized system becomes unstable and hence, as mentioned earlier, the nonlinear system is stable in the Lyapunov sense. The steady-state response exhibits stick-slip vibrations as indicated by the partial phase plane trajectory (c) in Fig. 6.

An increase in the normal force F_{n0} while keeping all the other parameters constant causes stronger stick-slip vibrations to take place. The stick period is longer as compared to lower normal forces, i.e., the fundamental frequency of stick-slip is reduced. The response of the system is shown in plot (d) of Fig 6.

5.2 MULTI-DEGREE OF FREEDOM

As an example of mdof system, a three-degree-of-freedom system is analyzed. Approximate solutions are obtained by solving Eq. (3) using numerical integration for several different values of the friction force parameters. The purpose here is to demonstrate the bifurcations which take place as the values of the nonlinear friction force parameters are changed. If the system parameters are such that the roots of the linearized system are on the left-hand side of the complex plane, the system is asymptotically stable and oscillations cannot be sustained. When the friction force parameters are changed so that the first pair of linearized system roots cross over to the right-hand complex plane, as illustrated in Fig. 7, the first bifurcation occurs. The response trajectory of the system in phase space closes on itself and a limit cycle is attained. The projection of the phase space trajectory onto the velocity-displacement plane of the driving point is shown by plot (a)in Fig. 8. Further changes in the parameters, in this case an increase in normal contact force, cause a second bifurcation which roughly corresponds to the crossover of the second pair of the linearized system roots to the right-hand complex plane. This bifurcation is not a period-doubling bifurcation. Instead its period is increased several times. The response is a combination of two periodic functions with synchronized fundamental frequencies. The corresponding projection of the phase-space trajectory onto the velocity-displacement plane of the driving point is given in plot (b) of Fig. 8.

Changing the system parameters such that the third pair of poles also crosses over to the right-hand complex plane produces another bifurcation. The response of the system is such that all of the fundamental frequencies are synchronized. The projection of the corresponding phase space trajectory onto the velocity-displacement plane of the driving point shows four loops, as shown in plot (c) of Fig. 8.

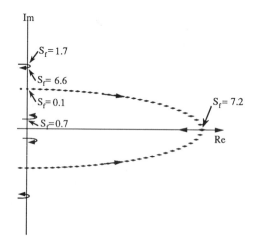

Fig. 7. The root locus of the three-degree-of-freedom system described by Eq. (3).

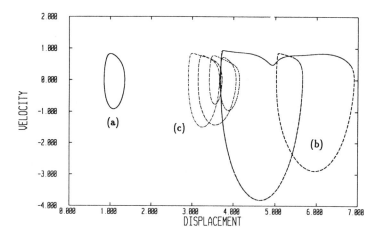

Fig. 8. Projection of the phase-space trajectories onto the velocity-displacement plane of the driving point response for different sets of parameters. $V = 1, \alpha_0 = 1.11, F_r = 2.3$, (a) one bifurcation; $\beta_0 = 3.33, F_{n0} = 2.8$, (b) two bifurcations; $\beta_0 = 3.33, F_{n0} = 15.5$, (c) three bifurcations; $\beta_0 = 3.33, F_{n0} = 10.0$.

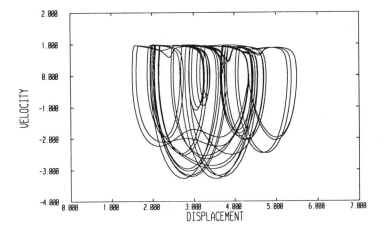

Fig. 9. Projection of the phase-space trajectory of the driving-point response onto the velocity-displacement plane exhibiting chaotic behavior. $V = 1, \alpha_0 = 1.1, F_r = 2.3, \beta_0 = 11.1, F_{n0} = 10.0$

All three bifurcations, obtained by changing the system parameters, produce periodic system response. Their occurrence is roughly related to the location of the linearized system roots in the complex plane. The fundamental frequencies of the periodic response

564

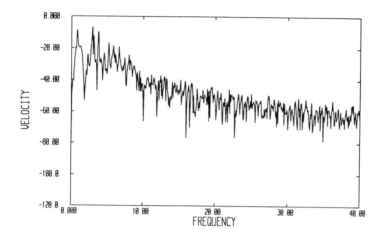

Fig. 10. Spectrum of the response for the phase-plane given in Fig. 9.

produced by the bifurcations is always lower then the frequency of the corresponding pair of roots.

It should be noted that the same root locations of the linearized system on the complex plane may give different responses of the actual nonlinear system. For the last example with three bifurcations, a new response is obtained while keeping the root locations the same by only changing the ratio α_0/β_0 from 3 to 10. The motion of the system is now chaotic. The projection of its phase space trajectory onto the velocity-displacement plane of the driving point is shown in Fig. 9 with, the corresponding spectrum of the driving point velocity in Fig. 10.

6. Discussion

The behavior of lumped-parameter linear systems acted upon by a nonlinear friction force is considerably more complicated than in sdof systems. Such dynamic systems, although linear, respond nonlinearly due to the nonlinear nature of the friction force at a contact interface. In the case of single-degree-of-freedom systems there is only one bifurcation point which separates the asymptotically stable state from the state of limit cycles. This bifurcation point corresponds to the crossing of the linearized system poles over the imaginary axis on the complex plane, which is a property of sdof systems that makes analysis considerably simpler. The mdof systems behave essentially in the same manner, and the first bifurcation takes place at the crossover of the most unstable pair of linearized system poles; see Fig. 7. The following bifurcations, however, are only roughly defined by the crossover of the poles. In certain cases, additional pole pairs crossing over may not immediately affect the system response. However, travel of the poles further right in the complex plane leads to subsequent bifurcations.

Each bifurcation manifests itself as an independent fundamental frequency in the response of the system. The term "fundamental frequency" is used here to describe the first frequency of the Fourier series of a periodic time function. Then each bifurcation contributes to the response a new periodic function in the time domain. In addition, the response has the harmonics of each of these "fundamental" frequencies. The relative values of the "fundamental" frequencies can have a significant influence on the system behavior. Specifically, whether these frequencies are multiple integers of each other, i.e., commensurate, or not becomes important. Of course, the system parameters determine such conditions. In the case of several periodic time functions where the fundamental frequencies are notintegralmultiples of each other, their combination results in a nonperiodic system response. When each fundamental frequency of the response is someintegralmultiple of the other, this leads to a periodic overall system response. The latter condition is caused by the synchronization of the fundamental frequencies. The spectrum of the system response looks as if all the frequency components are harmonics of a new frequency lower than those generated by the bifurcations. Synchronization of the fundamental frequencies is ascribed to the coupling between the modes of the system as a result of the nonlinear friction force. Of course, other parameters such as the eigenvectors of the linear system and the points of application of the friction force also determine the strength of the coupling between the modes of the system. A system whose modes are weakly coupled may produce asynchronous vibrations, whereas a system whose modes are heavily coupled will produce synchronous vibrations.

In summary, the phase space trajectories of mdof systems depend on the number of bifurcations present. If the poles of the linearized system are all on the left-hand side of the complex plane, the system will be asymptotically stable. If one pair of poles is on the the right-hand complex plane, then the response of the system is periodic with one fundamental frequency and possibly its harmonics, and the phase-space trajectory is a limit cycle. Two pairs of poles on the right-hand side of the complex plane present several possibilities. The system may have had only one bifurcation and the trajectory would be a limit cycle, or there may have been two bifurcations and the trajectory would be on the surface of a three-dimensional torus. If the trajectory is on a three-dimensional torus, there are only two independent frequencies in the response. These frequencies may be syncronous or asyncronous, depending upon the coupling between the modes. Similarly, three bifurcations will produce three independent frequencies, and the phase-space motion would be on the surface of a four-dimensional torus, and may or may not be syncronous. As the number of bifurcations grows, the dimension of the motion in the phase-space grows accordingly. Finally, for N bifurcations the motion will take place on the surface of an N+1 dimensional torus.

There is, however, one exception to all has been said about the motion of such a system. When three independent frequencies present in the response, the motion in the phase-space may not take place in integer number of dimensions. This is the case of chaos, where the motion has a fractal dimension.

7. Conclusions

Stability analyses of oscillatory systems excited by friction indicate that significant differences exist between the responses of single- and multi-degree-of-freedom systems. As the dimension of the system increases, the number of bifurcations increases, in some cases exhibiting chaotic response. Thus, in modelling friction-induced vibrations it is necessary to consider multiple degrees of freedom to represent the system response more accurately. Although the effects of normal motion were not explicitly explored in this paper, oscillations in the normal direction always exist. Normal oscillations, when coupled with tangential oscillations, can also exhibit behaviors similar to those shown here.

Stability analysis of nonlinear systems, such as those considered here, based on the linearized equations of motion, can be a useful approach in predicting the interaction of friction and vibrations.

8. Acknowledgement

The authors would like to acknowledge support for this research by the National Science Foundation and the Wayne State University Institute for Manufacturing Research.

9. References

1. Thomas, S. (1930) "Vibrations damped by solid friction," *Philosophical Magazine*, **9**, 329 – 345.
2. Bowden, F. P. and Leben, L. (1939) "The nature of sliding and the analysis of friction," *Proc. Roy. Soc. London,* **A169**, 391 – 413.
3. Blok, H. (1940) "Fundamental mechanical aspects of boundary lubrication," *SAE Journal,* **46**, 54 – 68.
4. Bristow, J. R. (1947) "Kinetic boundary friction," *Proc. Roy. Soc. London,* **A189**, 88 – 102.
5. Tolstoi, D. (1967) "Significance of the normal degree of freedom and natural normal vibrations in contact friction," *Wear*, **10**, 199 – 213.
6. Soom, A. and Kim, C. (1983) "Interactions between dynamic normal and frictional forces during unlubricated sliding," *J. Lubrication,* **105**, 221 – 228.
7. Martins, J.A.T., Oden, J.T. and Simoes, F.M.S. (1990) "A study of static and kinetic friction," *Int. J. Engineering Science,* **28**, 29 – 92.
8. Godfrey, D. (1967) "Vibration reduces metal to metal contact and causes an apparent reduction in friction," *ASLE Transactions,* **10**, 183 – 192.

APPENDIX

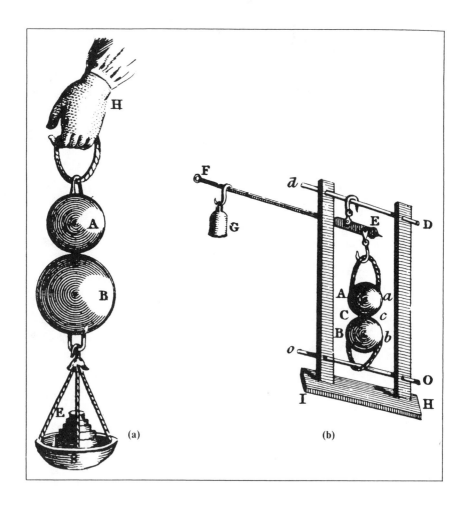

The Rev. J.T. Desaguliers' demonstration before the Royal Society of the cohesion of lead (Desaguliers, 1725) [From D. Dowson, <u>History of Tribology</u> (Longman, London, 1979) p. 160, with permission.]

EPILOGUE TO THE NATO ASI ON FUNDAMENTALS OF FRICTION

1. Introduction

The NATO ASI at Braunlage was a fruitful exploration of the current state of knowledge and new approaches to friction processes (and other tribology-related matters). In addition to formal presentations, most of which are included in this book, the ASI devoted much of the last two days to discussing what areas of friction are well-understood and what issues need further clarification. The epilogue, divided into 5 parts, is a summary of those discussions as well as more recent inputs. The last day of the meeting was devoted to questions, answers and commentaries on "Future Issues in Friction at the Microscopic and Macroscopic Level." One of the commentaries, by David Tabor, nicely captures the atmosphere of the meeting and provides a clear and concise summary of many of the issues addressed. A transcribed version of his comments is given in part 2, below.

In preparation for the "Future Issues" session, the penultimate day was devoted to topical discussions. Eleven groups were organized, and each was asked to identify:
- O   What concepts of friction have been made clear on the macroscopic and microscopic level?
- O   What is missing (concepts, theoretical or experimental approaches)?
- O   What topics were not treated (fully) that need to be dealt with in the future?

Groups were kept smaller than fifteen, and participants were asked to wander from group to group and contribute to topics of interest. At the end of the session, the participants reconvened to hear 5-minute summaries (later transcribed and printed) from each group leader. The summaries and transcribed comments from the "Future Issues" session were distilled by the editors into a list of ISSUES and RECOMMENDATIONS. These lists **(printed in bold letters)**, along with editorial comments and selected commentaries (*in italics*), are given in part 3. Two of the topics that generated much discussion, "Energy dissipation by friction processes" and "New ways of probing friction processes," are given extended attention.

Finally, everyone loved the molecular dynamic simulation videos of Uzi Landman, Jim Belak and Judith Harrison and the BBC science program with Thomas Mathia on "tribology of skiing." While there is no substitute for a good movie, a good diagram is sometimes second best. Lots of creative effort was spent generating diagrams depicting "Hierarchy of Friction Models" that span the microscopic to macroscopic. Three of these are presented in part 4. We end the epilogue in part 5 with the final comments of a lubrication engineer, Duncan Dowson, who wonders out loud how this knowledge of "microscopic friction processes" might influence engineers.

2.   Commentary by David Tabor.

"This Conference I think has been for me the most exciting and worthwhile Conference on

I. L. Singer and H. M. Pollock (eds.), Fundamentals of Friction: Macroscopic and Microscopic Processes, 569–588.
© 1992 *Kluwer Academic Publishers. Printed in the Netherlands.*

Friction that I have ever attended. I have been greatly stimulated, and I think all of us have been, by the lectures, by the attention of the attendees, by the discussions, by the posters, by the participation of the audience as well as the lecturers and by the flexibility in approach shown by all our specialist participants. I would like to thank the organisers and the organising committee on my behalf and on behalf of all of you for arranging such a successful event.

The continuum approach. In attempting to assess major conclusions it seems to me that there is a consensus that continuum mechanics, contact mechanics and fluid mechanics will continue to supply the practicing engineer with the tools for designing and producing effective and viable machinery. But he or she will need to know more about the properties of materials and the properties of fluids, and this remains an ongoing problem for the tribologist. From the practical point of view bulk properties will be sufficient but when, for example, the EHL film reaches thicknesses of the order of a few hundred angstroms (as indicated by Duncan Dowson) we may have to consider in greater detail the molecular structure of the liquid. I shall refer to this later.

Surface topography. In this continuum approach to friction and lubrication we still have to struggle with the problem of surface topography. Probably the design engineer will continue to use the elegant profilometry techniques described twenty five years ago by Greenwood and Williamson - in spite of what Jim Greenwood tells us about some fundamental unresolved conceptual difficulties. I feel in the discussion of this issue like the happily married couple who after 25 years of married felicity suddenly learn that their marriage certificate is faulty and may be legally invalid. Nothing in their relationship is changed but they are not quite sure where they are. My impression is that engineering tribologists are well and truly wedded to the established ideas of surface profiles, distribution of asperity heights, radii of curvature of asperity tips and sampling distances. For the foreseeable future the old relationship will remain as a fundamental constituent of tribological design. It is up to those who have discovered a flaw in the marriage certificate to rectify the situation. I am sure we all wish Jim Greenwood and his colleagues every success in this serious task of preventing a divorce.

Energy dissipation. When we talk about friction we nearly always think of it as a force. It is very rare that workers talk of energy dissipation and when they do they refer to it in terms of bulk properties e.g. plastic flow, viscoelastic losses, viscous flow. And I think it is at this point that I notice a change in attitude at least amongst those who are interested in the mechanisms of friction. We are beginning to recognise that maybe we should be thinking of energy dissipation in terms of atomistic processes. For example the talk I gave here is not a lecture I took out of my file of lectures: it was one I had never delivered before because I had not been thinking specifically about energy dissipation, except in terms of bulk properties. It was stimulated by a questionnaire sent out by the Conference organisers. My approach seems very simple and obvious. I started with the friction of metals where it is known that plastic deformation usually occurs at the regions of real contact. We know that plastic deformation involves unstable displacements by unit atomic spacings along the slip plane, the atoms vibrating as they jump from one position to another. I came to see that all frictional processes - or most of them - involve a distortion of the atomic arrangement, an instability, a flicking back to a new position of equilibrium while the distorted material vibrates and the vibrations ultimately degrade into heat.

Of course with metals we may still think in terms of plastic deformation and simply assert that there is no problem here since it obviously involves the expenditure of energy. Similarly with viscoelastic solids we may explain energy loss in terms of springs and dashpots. But polymers are not made of springs and dashpots. They are made of molecules which in themselves have intrinsic elastic properties of rate-dependent modulus and they are impeded in their movement by *molecular interactions or entanglements which involve viscous losses.*

Surface films. You can see from what I have just said that I have been thinking of friction in terms of bulk properties interpreted in molecular terms. I have not said a word about surface films which in the real world play such a significant role in the frictional process. Sometimes these films are applied directly to the surface or are incorporated into the surface during manufacture. But in many cases they are formed by chemical reaction between the surface and the environment. For this reason I welcome, as one of the important aspects of our lectures and discussions, our exposure to the chemistry of film formation. Of course with engineering systems many of these surface reactions are facilitated by the temperatures developed by frictional heating and I hope that at some stage we shall have a more detailed understanding of the temperatures developed between sliding surfaces and their role in chemical reactivity. A day or two ago we heard that, according to molecular simulation calculations, surface temperatures of three or four thousand degrees can be achieved and we need to look more deeply at this (see later).

Apart from the way surface films are formed we need to know how well they are attached to the substrate, whether they are ductile or brittle and how we can quantify these properties. So far we can only refer to them in descriptive terms.

Surface proximity devices. In referring to surfaces and surface films this brings me naturally to that part of the Conference which has dealt with the surface-force microscope (SFM) and I am glad that some effort is now being made (as we heard from previous speakers) to enlarge the area of surface that can be usefully examined by a modification of the SFM technique. This may bridge the gap between atom-by-atom studies and the somewhat more macroscopic aspects of friction which have been studied so elegantly by Jacob Israelachvili and Steve Granick.

Molecular dynamics: atomic simulations. Here we come I think to what I regard as one of the fundamental problems involved in our understanding of friction: that is we do not have a way of seeing what is actually taking place at the interface while sliding is taking place. We can look at the surface before and after sliding has occurred. But what took place at the interface is inaccessible to direct examination. There are of course some exceptions. If one surface is transparent we can study the hot spot temperatures from the radiation emitted but the details of the contact regions remain unknown. And that is why the beautiful atomic simulations by Uzi Landman and Judith Harrison present such an exciting prospect. I was in fact familiar with some of the earlier work in molecular dynamics especially the classical pioneering studies of Alder but at that time it did not seem to me that this could offer a useful approach to friction simply because so few atoms (or molecules) were involved whereas the individual junctions in friction experiments, even at loads of micronewtons, contain thousands or millions of atoms. Indeed it has always been my view, in the past, that atomic models of friction are "unrealistic" and in some ways I am surprised at my own lecture with which I opened this Conference. But we now hear that modern computations make it possible to study the behaviour of 10,000 atoms, say a cube of 20 x 20 x 20 atoms and that these simulate, to a remarkable degree, the behaviour of bulk matter.

Seeing that modern molecular simulation has been so successful I would like to hope that further work along the lines of Evans in Australia will be carried out by tribologists on the viscous properties of liquids, in particular on the effect of temperature, pressure and shear rate. Perhaps too such studies will clarify the boundary conditions which determine whether there is any slip between the liquid and a solid surface.

Surface temperatures and surface films. In connection with molecular simulation I would like to repeat two items that I mentioned earlier in relation to the frictional process. We can calculate surface temperatures in friction from a knowledge of the frictional work, the speed, the *thermal conductivity of the solids and from a reasonable estimate of the asperity size. But Uzi Landman has suggested that very much higher temperatures (not yet thermodynamically equilibrated) may be generated in the brief instant that transpires as atoms slide over one another. It is important for us to know if these are real or significant in affecting surface properties and surface reactivity. This certainly merits further study. The second concerns the behaviour of surface films during sliding whether they are organic lubricants or solid films. Can molecular simulation help us to understand their mechanical properties and how they break down? In practical affairs these films are of great importance.*

A new springtime? This is the first time I have been exposed to a detailed account of the power of molecular (or atomic) simulation in the study of surface interactions. I do not know if the details depend crucially on the assumed interatomic potentials but I think this technique offers great promise for further work in this field. When as a physicist I began my research on friction over 50 years ago our surface techniques were limited to the optical microscope and a little later to electron microscopy and diffraction. At that time we could at best interpret our results in terms of bulk properties. Almost every experiment that we did was being done for the first time and I suppose I could say that the first 15 or 20 years of this work constituted the spring time of our work in terms of our macroscopic approach. I think we are now approaching another spring time (as Ernest Rabinowicz indicated in his opening talk) and that this will involve a critical and constructive application of molecular simulation to the frictional process. It is not possible to tell if such an approach will lead to better tribological practice but I am sure it will shed new light on the mechanisms involved.

There is of course a great danger. I can imagine that if in three or four years' time we have another Conference on Friction we shall have nothing but papers on the atomic force microscope and molecular simulation. Meanwhile in the background the practical people will be studying the wear properties of surface films (especially plasma deposited diamond), the degradation of organic lubricants and other mundane matters - while our engineering tribologists, like Duncan Dowson and Ken Johnson will be busily engaged designing more efficient and more viable machinery on the basis of continuum properties. And Ernest Rabinowicz will be measuring the slipperiness of floors and earning the just rewards that flow from the American obsession with litigation.

This has been a most exciting Conference. I hope that you will not feel more tired than I am and that you will return to your laboratories stimulated by the papers, the discussions and the camaraderie of the last ten days.

3. Issues and Recommendations.

3.1 ENGINEERING MACHINE DESIGN

1. How can engineers make quantitative predictions of machine behavior from present-day models of friction behavior (continuum mechanics, lubrication, contact vibrations)?
2. Can we characterize the boundary/interface conditions of machines well enough to apply friction models?
3. What friction-related parameters can be used to monitor vibration and vibration instabilities?

4. How can engineers distinguish and cure or control the different forms of stick-slip ?

5. Can machine motion be tailored to optimize transfer films for low wear and stable friction?

6. Microscopic engineering: How might fine-tipped proximal probes influence the design of future engineering devices, from the submicrometer to sub millimeter scale? They can be used to manipulate materials, cut patterns, punch holes, polish asperities,...

7. How will engineers cope with the intrinsic "high friction" of micromachines, given that they will likely be fabricated from "clean" materials (e.g. Si) and will likely be operated in "clean" environments and at relatively low loads?

8. Can engineers take advantage of the "new found" properties of confined fluids (<10 nm thick) to control friction?

9. How can properties of ultra-thin lubricant films be incorporated into lubrication models to predict behavior of ultra-smooth engineering surfaces?

3.2 DEFICIENCIES IN PRESENT-DAY MACROSCOPIC MODELS OF FRICTION.

1. Models are relatively well developed for idealized materials subjected to elastic, visco-elastic or simple two-dimensional plastic deformation.

2. Models are needed to account for: strongly layered thin films, work hardening, and brittle crack growth under compression in idealized materials and in real materials (crystal structure, grain size, dislocation behavior,...).

3. How does friction evolve with sliding? Can "dynamic friction maps" be constructed analogous to "static friction maps?" [see chapter by Childs and extended discussion by Johnson.]

4. At what separation do continuum models (e.g. elastic/plastic deformation, roughness,...) break down?

5. What are the mechanics of interfacial wave propagation?

6. How does friction enhance the stress intensity field in mode II fracture?

7. What role does fracture mechanics play in sliding motion and attendant frictional resistance? Is there mixed-mode failure? This question follows from the presentation of Johnson (p. 227) in which he showed how certain materials fit onto a friction "map" whose different regions represent steady sliding, brittle stick-slip and ductile plastic behavior. During the discussion that followed, Arvin Savkoor made the point that the possibility of mixed-mode failure is still an open question. The relevant issues here are:

   a. what are the alternative "rival" processes (e.g. mode I-type Schallamach waves for elastomers, constrained plastic failure for metals,...);

   b. given that the normal and transverse stress distributions determine K_I and K_{II} respectively, how do these combine to give the total work of adhesion? Once this is known, it should be possible to predict the condition for instability (slip);

   c. if mode II fracture is operating, do we assume that fracture energy is *not* recovered through healing of the crack? [This question is quite separate from the question of energy loss mechanisms!];

    d. is it necessary to postulate an intrinsic shear stress of the interface, τ_o? Can this be related to K_{IIc}? If not, must we always postulate the existence of a surface film whenever we need τ_o? [see chapters by Briscoe and Savkoor].

8. How can the models be modified to account for the "intermediate scale discreteness" of real materials including dislocation behavior, crystal structure, grain size,..?

9. Can material-dependent models be formulated that account for "intermediate scale discreteness?" Clearly, in some cases we can bridge this very large gap between the continuum models used by engineers and the models involving discrete atoms. But, as pointed out by many speakers, microstructure gives rise to another type of discreteness. In this connection, Johnson asked how we can integrate the activities of people who work on the atomic scale with those on the engineering scale:

"These remarks have not been carefully thought out and should be regarded as a spontaneous reaction to the meeting which is coming to an end. They relate to length scales and the relation between those who work on the small scale and those on the engineering scale. First of all it has been a splendid occasion for people who work in these two areas to talk to each other and to discover what we are each trying to do: in many cases in different departments.

I am an engineer: probably most of the people in the room are physicists and we do not meet as often as we should. So that has made it a splendid meeting. On the other hand, although we now have this feeling of esprit de corps, I think we somehow kid ourselves if we think that we can start with forces between atoms and finish by calculating answers to problems such as shoes gripping floors that Ernie Rabinowicz was talking about. I don't think that one does that sort of thing.

One of the things that has struck me is that there is an intermediate length scale which we have not heard much about at this meeting. It lies between the atomic/molecular behavior that we have heard a lot about and the continuum approach. In metals, for example, it is the scale of dislocations and other defects in crystals. It is the scale where crystal structure matters.

I think back to when I was a beginner in the tribology game, just about the time that dislocations were 'invented' or at least the time when they were beginning to seriously affect material scientists' thinking. I was terribly worried whether I would have to learn all this very complicated dislocation mechanics in order to perform calculations of the plastic deformation of surfaces. But that isn't the way it is: one doesn't solve plastic deformation problems of an engineering sort by starting off with forces between dislocations.

Thanks to this meeting, we engineers now have a whole new area of activity to get used to, one which will surely be important and feed into tribology. But we kid ourselves if we think we can start off with forces between atoms and solve engineering problems on a continuous line carried out by a single person."

This lead on to the question of asperities:

3.3 ROLE OF ASPERITIES IN SLIDING CONTACT.

1. Have friction models properly accounted for material properties at the asperity scale? In addition to the general points made above, it was pointed out that a serious attempt to correlate observations of frictional behaviour with grain boundary behaviour could be worthwhile. It appears that on the whole, single-asperity fibre/fibre sliding experiments are consistent with macroscopic models – the gap here is between the asperity scale and the atomic simulations.

2. How does the contact in static loading evolve as sliding commences?
3. How does the surface roughness evolve during dry sliding contact?
4. At what length scale does the flow stress approach the theoretical stress?
5. How do fracture and plastic effects compete at the asperity scale?
6. Does the wear mode change as the contact size gets smaller?
7. Do AFM and fine-point friction studies depict friction at the "asperity" level? [see chapter by Pollock.] If so, why are the results inconsistent with results inferred from powder flow friction studies? [see chapter by Adams.]
   a. Perhaps these proximal probe experiments are in the regime that should not be considered the single-asperity scale?
   b. What other experiments can be done to understand friction at this scale?
8. Does surface energy play an important role in friction processes at the asperity scale? Brian Briscoe argued that surface energy plays an "intermediate" role - to hold surfaces together long enough to allow strain energy to be transferred to the regions adjacent to the contact. However, the strain energy itself is dissipated into the bulk.

3.4 ROLE OF INTERFACIAL SLIDING AND RHEOLOGY.

1. Where does sliding occur? Technically, where is the "velocity accommodation locus?"
   a. How does this locus change with separation between the counterfaces?
   b. Where is the microstrain accommodated? At a counterface (by sliding) or within the interfacial material (by shear)? We must identify the velocity accommodation locus: 1) to advise the microscopic modelers what occurs in practice, and 2) to understand wear, transfer and traction cracking.
2. What mechanism governs the magnitude of the "velocity accommodation parameter" or "interfacial rheology parameter," commonly called the interfacial shear strength or contamination factor and labeled τ, q, or s by various authors?
   a. What is the significance of the velocity accommodation parameter and how is it related to bulk properties?
   b. How do gas-surface reactions alter the velocity accommodation parameter (and wear behavior) of sliding contacts?
   c. How is this parameter related to energy dissipated in solids and fluids?
3. Can the magnitude of the velocity accommodation parameter ("shear strength") be predicted from properties of interfacial films or their bulk material counterparts?
4. How does the "third body" evolve?
   a. What are the rheological changes and dissipation characteristics?

b. Are the processes isothermal, because of small volumes involved?

c. What are the processes (mechanical and chemical) by which frictional energy transforms solid/liquid/gas interactions into surface films and third bodies?

5. What is the relationship between friction and the generation of wear debris?

   a. Do frictional stresses control the size, shape, structure or microstructure of wear debris?

   b. If so, what can this analysis tell us about the friction processes that generated the particles?

   c. Can this friction-wear synergism be used to control wear?

6. Can wear models be devised that can predict "lubricating" third body products that reduce friction and minimize wear?

7. Are friction processes at the macroscopic and microscopic scales governed by different mechanisms? Interfacial sliding studies in both traditional friction tests and in the SFA show similar friction behavior - either F $\propto$ load (Amontons' Law) or F $\propto$ area. However, different interpretations of apparently similar behaviors have been given. Amontons' Law is accounted for traditionally, e.g. in the Bowden-Tabor model, by shearing of junctions accompanied by *permanent* plastic deformation. Israelachvili, however, [see p. 371] derives the same behavior using the cobblestone model, in which heat is dissipated with *no permanent* deformation. Moreover, the friction coefficient in the two cases is determined by very different materials properties [see p. 372]. The second behavior (F $\propto$ area) leads to two different explanations of the interfacial shear strength, $\tau = \tau_0 + \alpha$. Briscoe [see p. 167] interprets τ (generically) as a dissipation parameter and α as the pressure dependence of the shear strength (of the "third body" or the interfacial itself); Israelachvili [see p. 371] suggests τ_0 is due to attractive van der Waals forces and the α term is the "bumpiness" of the contact.

3.5 SURFACE FILMS AND ATOMIC LEVEL TRIBOCHEMISTRY.

1. If we want to control friction and we have the ability to precisely modify the chemistry and structure of surfaces, what chemical and physical properties would we give to the first few hundred angstroms of a metal or ceramic?

2. How do surface films accommodate stresses or transmit them to the subsurface (quantitative models)?.

3. Is frictional energy dissipated in surface films? If so, what is the mechanism?

4. How do friction processes activate surface chemistry? Traugott Fischer offered several possible mechanisms:

   a. By direct mechanical excitation of a chemical reaction, e.g., by straining surface bonds;

   b. By thermal activation of a local volume of material;

   c. By removal of reaction products;

   d. By acceleration of diffusion processes (mass transport);

   e. By exoemission;

a. By triboelectricity.
7. If friction produces the rapid local temperature rises suggested by molecular dynamics simulations, ca. rates of 10^{10} K/sec, how should one describe the chemistry?
   a. Would it be vastly different than what one would get from a surface held at a constant temperature of that value?
   b. What kinds of chemistry e.g. the modes of energy dissipation (vibrational, rotational, electronic,...) takes place under these conditions? Perhaps, chemical dynamic simulations will be necessary to identify the chemistry. This point pertains to issues discussed following the lecture by Fischer, on page 310.
8. How does crystallinity (lattice imperfections, surface registry, amorphous vs crystalline,...) affect tribochemical reactions?
9. How do tribochemical films that form during sliding or rolling contact affect tribological behavior?
10. What are the chemical and mechanical contributions to the breakdown of tribochemical films?
11. What are the important properties of films, e.g. surface energy, that should be examined experimentally?

3.6 HOW CAN ATOMISTIC MODELING MAKE AN IMPACT ON UNDERSTANDING FRICTION?

Approaches:
1. Identify the microscopic phenomena underlying macroscopic behavior. (topics listed below)
2. Describe properties of small material aggregates.
3. Assess and evaluate the range (particularly spatial) of validity of continuum descriptions.
4. Topics that can be treated include:
   a. Material transfer,
   b. Surface and subsurface diffusion,
   c. Atomistics of plastic response,
   d. Energy dissipation mechanisms such as transient heat generation or excitations (phonons),
   e. Schallamach and Stoneley waves -- generation and propagation,
   f. Phase transformation (amorphization, glassification, solid-liquid, etc.),
   g. Boundary conditions (stick-slip and partial slip),
   h. Non-equilibrium phenomena: Energy pathways and mode coupling,
   i. Mechanical response to transient heat and pressure generation, including shock conditions,
   j. Chemistry under temperature jump (flash) conditions,
   k. Interfacial shear strength, its variation with depth and the tribological consequences,
   l. Tribological consequences of surface modifications (e.g. ion implantation,

alloying).

Problems:

1. How critically does the deduced friction process depend on the form of the *assumed* interatomic potential?

2. Can algorithms (e.g. hybrid methods) be developed to simulate friction processes at time and length scales longer than can be treated in molecular dynamics calculations alone? e.g. that extend computational simulations from the nm/femtosec scale to the μm/μsec scale. It seems that molecular dynamics is unlikely to ever be able to incorporate behavior over all relevant time scales into a single simulation. In a given engineering application, it will always be necessary to couple the molecular dynamics approach with the continuum mechanics approach at the appropriate length scale.

3. Can atomic models treat particular materials or are they only able to qualitatively distinguish between classes of materials e.g. ionic vs covalent vs metallic?

4. In practical machines, the surface films that sustain sliding -- whether organic lubricants, oxides or other solid films -- eventually break down. Can molecular simulations help us to understand the mechanical properties of these films and how the films break down?

Fischer's group of wandering scholars analyzed how one constructs an atomic theory of sliding friction. Assuming that shear takes place at a solid-solid interface and considering only those mechanisms by which translational movement is transformed into heat, they suggest six building blocks for such a theory:

1. The interatomic force potentials inside the sliding body;
2. The atomic force potential between atoms of the two opposing bodies;
3. The relation between these potentials, to determine whether adiabatic (non-dissipative) sliding is possible or if the interbody bonds 'snap' and cause vibrations of the atoms;
4. The crystal structure of the two solids must be considered, including questions of registry between the two structures;
5. Dynamics (how fast is vibrational energy emitted into body?) and viscoelasticity;
6. Collective effects (sliding by dislocation motion, shear melting, creation of defects).

Finally, Belak's group came up with a strategy for calculating engineering friction coefficients by atomistic methods; their strategy is presented in part 4.2 below.

3.7 SQUEEZED LIQUID LAYERS AND BOUNDARY CONDITIONS.

1. Can lubricants be tailored to take advantage of the dynamic properties of certain fluids [see "chemical hysteresis" in chapter by Israelachvili, p. 351, and extreme non-Newtonian fluid behavior (e.g. enhanced critical shear stress, viscosity, elasticity...) [see chapter by Granick, p. 387].

2. What is the origin of the time-dependent static friction of confined fluids?

Granick, on p. 387, suggests two explanations: growth of the area of contact and slow molecular rearrangements at the boundary.
3. **What atomic scale behavior influences liquid lubrication?**
   a. At the boundary, e.g. degree of stick or slip at wall,
   b. In the "squeezed fluid," changes in phase or chemical composition, changes in rheology.
4. **How do the above changes influence cracking and wear?**

3.8 HOW CAN ATOMISTIC MODELING MAKE AN IMPACT ON UNDERSTANDING LUBRICATION?

Computer simulations can be used to study:
1. **Effect of molecular structure on phase and rheology,**
2. **Effect of temperature, pressure and strain rate on phase and rheology,**
3. **Changes induced by additives or mixtures,**
4. **Effect of loading rate,**
5. **Modes of energy dissipation,**
6. **Non-Newtonian flow,**
   a. Where does shear occur? (at surfaces or in lubricant?)
   b. How do the molecules move (what is the activation volume)?
   c. What are the relevant time scales?
   d. What determines the large shear-rate limit of the shear stress?
7. **Artificial quasi-solids as a fast technique for solving continuum equations.**

3.9 STICK-SLIP PHENOMENA.

1. **How should we categorize the many contributions to stick-slip? Should we distinguish engineering phenomena (e.g. thermal softening in brakes, roughness effects on rolling surfaces, incipient seizure on clean surfaces) from friction oscillations or ratcheting at the atomic scale? Is the latter stick-slip?**
2. **What criteria can be used to distinguish stick-slip from ratcheting?**
   a. **length scales?** Mark Robbins suggested we call an effect stick-slip if the length of the slip increases as you decrease the spring constant. If you have ratcheting motion, the length of slip is tied to a length scale intrinsic to the underlying substrate. This suggestion could provide a needed criterion for differentiating between true stick-slip and ratcheting over atomic potentials.
   b. **spring stiffness?**
      i. Robbins also indicated that varying the spring stiffness can produce qualitative changes in the relationship between force and velocity.
      ii. spring stiffness also imposes a fundamental limitation on friction measurements with an AFM: for high sensitivity, the spring must be so weak that it moves almost the characteristic length of the atomic potential variations.
      iii. an important point is whether or not the kT energy of the sensing spring used to measure frictional force is as large as the frictional energy that you

are trying to detect.

3.10 COMMENTARIES ON ENERGY DISSIPATION MODES IN FRICTION PROCESSES.

Tabor's lecture and the discussions that followed (see pp. 3-23) set the scene for subsequent contributions to the understanding of this central topic. On the last day of the meeting, he added the following remarks on the subject of surface energy:

"Before I attempt to express my main impressions of this Conference I feel I must interpose a few words as a penitent and apologise for the mistake I made at the end of my original presentation. On that occasion I presented a model which I thought demonstrated that in pulling two surfaces apart the process was reversible and only the surface energy was involved. I now realise that this is not so and that as the bonds go "pop" across the interface elastic energy is lost in the bodies. I do not know if this is true in peeling. But my original model involving the separation across a plane appears to be wrong.

As a practising scientist I recognise three levels of achievement. The first and best is to get it right and to have the results accepted by others. The second is to get it wrong or to be controversial and to evoke discussion which finally resolves the issue. The third and least satisfactory is to have your work ignored. My involvement in surface energy is in the second category."

Other commentaries on energy dissipation:

*(John Yates) "It seems to me that the field of friction is just beginning to meld together macroscopic and microscopic pictures. A similar transformation happened at an earlier time in the field of chemical kinetics, where for the previous 75 years, investigators had focused on measuring and explaining rate constants for chemical reactions. Then the dynamicists came in and began to measure the individual trajectories of processes that occurred on the atomic level. The objective of that work **never** was to calculate the overall rate constants from all the microscopic measurements, but instead to do the microscopic physical measurements and the modeling in a way that allows one to see entirely new concepts developing. Such concepts could never come out of studies of rate constants which are simply the global averages of the effects of all of these processes. It's likely, therefore, that the goal of the microscopic approach to friction is to introduce entirely new concepts of friction processes, rather than simply finding a way of calculating friction coefficients or other engineering parameters from more fundamental processes."*

(Gary McClelland) "We must develop experimental techniques to learn how energy is dissipated in frictional processes. Following the path taken in chemical physics, instead of simply looking at the rate at which chemical energy is dissipated, we should look at the modes of energy dissipation. In tribology, we should look at the spectrum of energy dissipated in frictional processes. For example, what events generate high frequency phonons and what events generate low frequency sound waves? High frequency phonons, for example, may be generated by short, prompt events. Since these events are very short, you would have to use a spectroscopic technique which captures these rapid events and likely do experiments at low temperature where the lifetimes are much longer."

(John Pethica) "We recognize that friction processes result in heat dissipation. But that doesn't mean that heat was the initial energy loss process. We know, for example, from triboemission studies, that there are high energy processes that occur during rubbing. Therefore, following the chemical kinetics/dynamics physics analogy, perhaps if we sort out what all these

spectral responses are initially, we may come to recognize the dominant processes controlling friction and the mechanics of energy dissipation. "

(Jacob Israelachvili) *"With friction, just as with adhesion, what you always have is a loading followed by an unloading. This may be viewed in terms of an approach then a separation, or an advancing then a receding contact angle. And, regardless of the size of the interaction zone, the front part of a sliding asperity is in the loading (adhering or entangling) state and the back part in the unloading (de-adhering or disentangling) state. The loading and unloading events are, therefore, separated by some characteristic length and time, which clearly depends on the sliding velocity.*

*Now let's consider energy dissipation. Dissipation will be governed by two **competing** processes occurring at the front and back of the sliding contact zone. Consider the two extremes of fast and slow sliding velocities. At very high speeds there is no time for entangled or adhesive junctions to form (at the front), while at very low speeds there is plenty of time for separation or disentanglement of junctions to occur at the back. In both cases, there is little departure from equilibrium and therefore minimum energy dissipation. At some intermediate velocity (time scale), there will be maximum dissipation, often with pronounced stick-slip.*

This argument should be quite general and will apply at various length and time scales. McClelland's Tomlinson-type model (see page 406) may be analyzed in this way. In other cases the length scale may be determined by molecular or asperity dimensions, or may be macroscopic, as in the case of a squeaking door. "

Several other issues on energy dissipation and time and length scales are listed below:

1. What are the energy dissipation mechanisms in friction processes? some suggestions are: lattice excitations (phonons) and heat generation; Schallamach and Stoneley waves-generation and propagation; phase transformations (amorphization, glassification, solid-liquid, etc.); internal friction (e.g., in polymers)...

2. At the atomic scale, can some of the vibration energy be used to overcome the potential barrier of the next friction event or is all of it dissipated into the bulk?

3. Are there different characteristic length and time scales associated with static friction and kinetic friction?

4. While time and length scales are certainly important, what other scales may be pertinent to friction processes, e.g., film thickness/contact length or molecular vibration period/sliding speed?

5. Is energy dissipation a necessary consequence of discontinuous motion during friction?

3.11 NEW WAYS OF PROBING FRICTION PROCESSES.

Why do we know so little about friction processes? Stated simply by Tabor, *"we do not have a way of seeing what is actually taking place at the interface while sliding is taking place. "* Hence, one of the experimental difficulties in investigating friction is that friction occurs at a buried interface. A second difficulty is that friction events can take place rapidly: less than nanoseconds for atomic events and microseconds for micrometer-sized asperity events. New approaches are therefore needed to investigate, in real time, the prompt loss processes buried in the interface of a sliding junction.

Pethica, Tabor, Robbins and others have suggested several spectroscopies to investigate friction in the buried interface:

1. Optical spectroscopy To gain access to the interface, make one sliding face sufficiently thin or "transparent" to the radiation. An example would be the real-time optical spectroscopies of elastohydrodynamic (EHD) fluids being performed at Shell Research Ltd. and at Imperial College (by Spikes et al, see end of part 4). The work partially bridges the gap between macroscopic and molecular processes in EHD lubrication. Interference gives the film thickness and shape down to vanishing film thickness; the contact zone is scanned by a laser and the Raman shift is used to determine the pressure distribution in the film; then optical birefringence gives a measure of molecular orientation in the film during flow.

   To resolve tribochemical processes at the picosecond time scale, time-resolved pulsed laser spectroscopy was recommended to probe the reactions, possibly in conjunction with pulse energy to trigger events. Non-linear spectroscopies of buried interfaces were also described by Dick Polizzotti.

2. Electromagnetic wave probes using e.g. microwaves.

3. Phonon spectroscopy. Perform experiments at liquid He temperatures, where phonons have longer lifetimes, and use high quality (low phonon scattering) substrates to transport the prompt phonons ballistically into a detector. Determine from the phonon energy spectrum whether the energy at the interface appears in vibrational, translational or rotational form.

Pethica then recommended measurements of friction at small time scales, since many chemical processes and the molecular dynamics simulations of these processes take place in a time domain below a nanosecond. But, he cautioned, it may be exceedingly difficult to resolve forces on a time scale less than 10 μsec, because of the mechanics of an apparatus. To follow the mechanical response e.g. stress/strain behavior in real time, an apparatus would have to have a sub-nanosecond response time. However, calculations show that such small response times require that the mass of the apparatus be impractically small e.g. only a few atomic masses. Even present day atomic force measurements, coupled through a fine tip, find it hard to achieve a time constant less than 10 μsec. As a result, the slower responding apparatus cannot see the true behavior, but only a response curve dictated by the time constant and load-line of the apparatus. Therefore, there is a great incentive to make these measurements **indirectly**, and the scientist who comes up with the method will certainly be breaking new ground. [Several "unorthodox" friction probes are mentioned at the end of part 4.]

Uzi Landman suggested an approach for determining the role of chemisorbed films in friction processes, an important issue to understanding boundary lubrication. The approach combines theoretical modeling of phonons interactions with states of chemisorbed molecules and experimental infrared spectroscopy that gives information on lifetimes and transitions for stretched or distorted molecules on a surface. If we know the molecular distortions produced by friction and the distorted states of the molecule, then we can calculate how energy is dissipated by friction and how dissociation occurs on the surface.

In addition, there are still many investigations that can be made using more conventional friction probes:

1. **Investigate complex motions (load-axis spin, periodic loading) using a traditional friction tester.**
2. **Proximal probes: friction at controlled microstrain, at ultralow loads, ...**

4. Hierarchy of Friction Models.

4.1 FRICTION OF MACHINES.

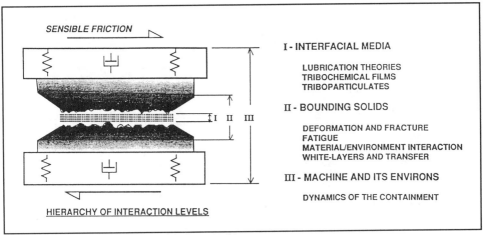

Fig. 1. Interaction levels in a friction machine.

Peter Blau pointed out that the entire tribosystem must be considered when modeling friction. He suggested[1] that we develop a hierarchical model based on where the shear forces are concentrated, from the interface out to the machine itself. Friction in a machine is often influenced by physical processes operating at more than one location and size scale in the tribosystem -- from the atoms and molecules interacting in the lubricant at the interface to the bolts holding the machine to the floor. Start with an interface between two moving surfaces and work outward (see Fig. 1). If the surfaces never make contact, we can probably model friction accurately knowing only the properties of the media (I) between the surfaces. If however there is solid-solid contact, the topography, chemistry and mechanical properties of the bounding solids and third body particulates must be considered. Finally, if shear stress cannot be accommodated within those layers, the shear forces will be transmitted beyond the near-surface zone (II) to the constraining machine and fixtures (III) and result, perhaps, in vibrations which may feedback from the machine to the interface.

584

4.2 ATOMISTIC APPROACHES FOR ENGINEERING SOLUTIONS.

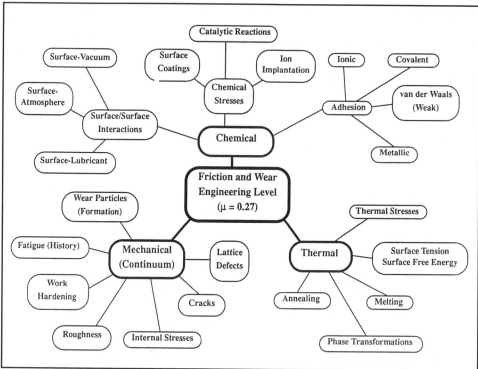

Fig. 2. Hierarchy of inputs required to calculate friction coefficient at an engineering level.

Much has been said at this Institute about atomistic experiments. Hence it would be valuable to consider how such experiments could be used in the reliable and efficient manufacturing of sensible products, i.e., how can atomistic methods make an impact at the engineering level? Jim Belak and his group of wandering scholars, including Kristian Glejbol, Dieter Klaffke and Arvin Savkoor, came up with a chart that symbolizes the transition from the atomistic approach to the engineering level (Fig. 2).

They started at the engineering level where it was agreed that a single number, namely the coefficient of friction, is adequate for designing a machine and assign it a reasonable friction coefficient: 0.27. The various levels of understanding needed to apply an atomistic approach are represented in the tree diagram. The Friction and Wear Engineering Level at the center of the tree is characterized by $\mu = 0.27$ and represents the least amount of fundamental understanding. At the outermost branches, they put atomistic descriptions of the materials and their interactions, about which much is known, both experimentally and theoretically.

Although three areas of tribological phenomena (Thermal, Chemical and Mechanical) are represented, the mechanical branch will be discussed in order to illustrate the hierarchy of levels of understanding. Moving away from the center, one first encounters a continuum elastic/plastic description with a single yield criterion (Y = constant). At the next level, there are polycrystalline grains which can produce inhomogeneous and anisotropic behavior in a material. Then, within each grain, deformation is controlled by dislocations and other defects. These defects are ultimately controlled by the interatomic interactions at the outermost perimeter of the branches.

4.3 THEORETICAL MODELS AND EXPERIMENTAL APPROACHES.

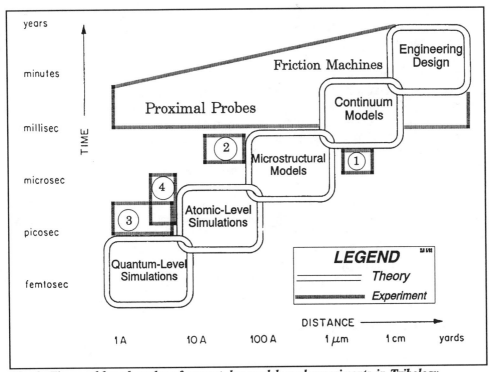

Fig. 3. Time and length scales of present-day models and experiments in Tribology.

How can we use the power of microscopic modeling to gain new insights into macroscopic friction processes and, ultimately, to solve technological problems? Bill Goddard[2] suggests that this can be done by progressing along the "chain-linked" ladder, illustrated in Fig. 3, from quantum-level studies to engineering design. His "hierarchy of modeling tribological behavior" unites atomistic models, which operate in very short length-time scales, with engineering models, which describe tribological behavior in length-time scales perceivable by more traditional measuring equipment. This approach "...allows consideration of larger systems with longer time scales, albeit with a loss of detailed atomic-level information. At each level, the precise parameters (including chemistry and thermochemistry) of the deeper level get lumped into those of the next.

The overlap between each level is used to establish these connections. This hierarchy allows motion up and down as new experiments and theory lead to new understanding of the higher levels, and new problems demand new information from the lower levels."

But where are the experimental approaches for investigating the "lower (short scale) levels?" As illustrated in Fig. 3, most "friction machines," including the proximal probe devices, are operated at long time scales. An abbreviated search of the recent literature by one of the editors, Irwin Singer, found only three "tribology" tests and a fourth proximal probe method that come close to investigating friction behavior at short time and length scales:

1. Bair et al.[3] have used fast IR detectors to measure flash temperatures during high speed frictional contacts of asperities of length 10 μm and greater, with time resolution of about 20 μsec.

2. Spikes et al. have developed *real time* optical techniques for investigating the physical behavior of EHL films down to 5 nm thick[4] and chemical processes occurring in contacts 10 μm wide by 80 nm thick[5].

3. Krim et al.[6,7] have used a quartz crystal microbalance to measure the sliding friction of molecularly thin films (both solid and liquid) condensed onto surfaces, probing atomic vibrations amplitudes between 0.1 to 10 nm and time scales from 10^{-12} to 10^{-8} sec.

4. Hamers and Markert[8] have shown that STM images are sensitive to the recombination of photo-excited carriers whose lifetimes are in the picosecond range.

Clearly, innovative experimental approaches for measuring friction processes at short and intermediate time-length scales (like those outlined by Pethica earlier) are needed to assist the modelers who are already there.

Finally, there is another time scale to consider, and that is "the question of time to translating this (fundamental) knowledge into engineering practice." Duncan Dowson tells us how this can be accomplished in the next and final part of the Epilogue.

5. Commentary by Duncan Dowson.

"I think that before making a few observations on the meeting itself, I would like to take this personal opportunity to thank the organizers for providing me with the opportunity to attend. It has been a remarkable experience for me. I would echo David Tabor's comment that this has been the most valuable meeting in the subject of friction that I have ever attended.

We have heard quite a lot about the subject of friction from the atomic scale up to the macroscopic scale. I suppose that we are all reflecting this morning on the issues we clarified and the points that we will take back in our minds to our laboratories or to our computers and to determine how the information that we have received will affect the direction of our future research. The answer will be different depending upon where you come from, what your background is, whether you are a physicist concerned with the atomic scale or whether you are an engineer and trying to design the next generation of jet engines.

In some ways we could take a pessimistic view of the outcome this fortnight. What have we heard that we did know about before we came to this meeting? I think we should all ask ourselves that question. If the answer in some cases is negative, I think we should reflect further on the point already made this morning that the new bits of information that have come in strengthen our confidence and understanding of the subject as a whole. A lot of loose ends have been brought together and clarified very effectively by the participants, the speakers and the discussers, and I have found that very valuable.

There has, of course been a great deal of discussion on size effects, a subject of great importance to both solids and fluids. In general terms, we have been talking about length scales -- angstroms and nanometers up to micrometers. Little has been said this morning about fluids and indeed little has been said at the meeting about friction associated with fluids. But we have heard a few very revealing observations about the behavior of fluids adjacent to solid boundaries. In a sense, as I mentioned earlier, I take great comfort in the fact that these remarkable changes in fluid behavior do seem to be limited to very small distances from the solid boundary. And since most engineers are talking about effective fluid films which are a few orders of magnitude bigger than those distances, you might argue that the boundary effects as such may not be very meaningful. While they certainly provide details of boundary behavior under no-slip conditions, whether dry or not, and indicate unusual viscosities of molecularly-thin layers adjacent to the boundary, do they really affect the outcome of our overall calculation? The answer is no as far as I can see. We can confidently apply continuum mechanics at the present scale.

But the present scales **are** related to effective lubricant film thicknesses. Jim Greenwood and others have demonstrated that we are now starting to talk about nominally smooth surfaces in which the effective film thickness is on the order of the roughness of protuberances on engineered surfaces. Will it be difficult to separate smooth surfaces by coherent films of a lubricant? These surfaces are going to come very close together and touch on asperities either in steady state motion or certainly while starting and stopping machinery. And so we are concerned with the effects we have heard about here, both in relation to solid friction at such interactions and of course in relation to the effects on the lubricant properties and behavior within two molecular layers from the surface.

There are other aspects of scale I think we should reflect on. One is the question of time in terms of translating this knowledge into engineering practice. Let me assume that you have heard something during the last fortnight that might have an impact on the design and operation of machinery. You may be impatient to wait to see how long it takes for that to show itself in practice. The time scales are generally **enormous**. We are not talking about next month or next year. There are engineers out there who are already designing the motorcar engines for the year 2000 and jet engines for the next century. Before there can be a translation of some of the concepts, and that assumes you have new concepts, I think you should take note that it is going to be about a decade if not a generation before that impact will be seen fully in engineering. This makes it all the more important for meetings like this to bring together physicists, chemists and engineers, so that the engineers can absorb by osmosis the concepts that you are revealing. Don't shut us out just because it will take us 10 years or 20 years. It is important that these ideas feed into our consciousness so that we apply them sensibly in future developments.

Little has been said about the subject of wear at the meeting. Perhaps I was slightly surprised that there was not greater emphasis and greater attention paid to the results of friction, and whether friction processes are intimately related to production of wear particles. What is the morphology of the wear particles? Is there anything to be learned from them that we could feed back into the understanding of the friction process? For example, there are many revealing studies in the field of wear about the size and the nature of wear particles. About a year ago, I heard Professor Kato from Japan give an illuminating talk. He would have us believe that as

I walked across this wooden floor this morning, with every step I took, between 1000 and 10000 wear particles were produced between my shoe and the floor. Maybe the size and the shape of the wear particles have something to say about the processes involved, the interactions of the solids associated with friction and wear. Perhaps we should have devoted a little more time to that interface than we had time to do. (Maybe Irwin you should start thinking now about the subject for your next ASI.)

In concluding, I would like to say that there is also another aspect that I think is important. As I listen to physicists and chemists discussing in intimate detail the atomic interactions, I think at times there tends to be a feeling of competition between science and engineering. That I think is totally wrong and misleading. Certainly from the engineers' point of view, in my experience, we love the opportunity to try to hang on for those physical concepts, to the understanding of basic interactions in solids and in liquids and all phases of matter. We may not fully understand all of them, but it is most important that we try to keep the communication as strong as possible. I hope equally that scientists will not be too frustrated by the fact we take 10 to 20 years to incorporate their bright new ideas in designing better skis or whatever it might be. Let us perpetuate this interaction between groups of people who all have one objective and that is to understand the laws of physics in order to apply them as effectively as we possibly can for the good of society through the manufacture of reliable, efficient and sensible products. "

REFERENCES

1. P.J. Blau, STLE Trib. Trans., <u>34</u> (1991) 335.
2. "Tribology of Ceramics" National Materials Advisory Board Report 435, National Academy Press, 1988, Washington DC, p. 88.
3. S. Bair, I. Green and B. Bhushan, J. Tribology <u>113</u> (1991) 547.
4. G.J. Johnston, R. Wayte and H.A. Spikes, Tribology Transactions, <u>34</u> (1991) 187.
5. P.M. Cann and H.A. Spikes, Tribology Transactions, <u>34</u> (1991) 248.
6. J. Krim and A. Widom, Phys. Rev., <u>B38</u> (1988) 12184.
7. J. Krim, D.H. Solina and R. Chiarello, Phys. Rev. Lett., <u>66</u> (1991) 181.
8. R.J. Hamers and K. Markert, Phys. Rev. Lett., <u>64</u> (1990) 1051.

Irwin Singer and Hubert Pollock
May 1992

Introduction to Contact Mechanics:

a summary of the principal formulae

K L JOHNSON
Department of Engineering
University of Cambridge
Trumpington Street
Cambridge, England

Reference 1: K L Johnson, "Contact Mechanics", Cambridge University Press (1985).

1. Geometry of non-conforming surfaces in contact

1.1 UNDEFORMED BODIES

Figure 1 illustrates the necessary description of the geometry. The principal radii of curvature are R_1', R_1'' for the first surface and R_2', R_2'' for the second surface: they are the maximum and minimum values of the radius of curvature of all possible cross-sections of the profile.

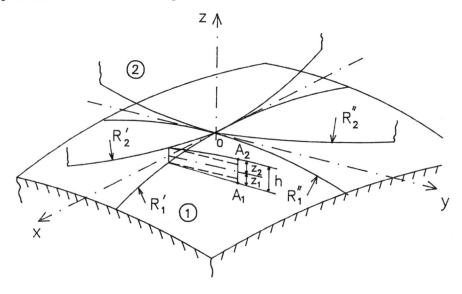

Fig. 1

I. L. Singer and H. M. Pollock (eds.), Fundamentals of Friction: Macroscopic and Microscopic Processes, 589–603.
© 1992 *Kluwer Academic Publishers. Printed in the Netherlands.*

The profiles are expressed as:

$$\text{Body 1:} \qquad z_1 \simeq \frac{x^2}{2R_1'} + \frac{y^2}{2R_1''} \qquad\qquad x, y \ll R_1', R_1''$$

$$\text{Body 2:} \qquad z_2 \simeq \frac{x^2}{2R_2'} + \frac{y^2}{2R_2''} \qquad\qquad x, y \ll R_2', R_2''$$

Where the bodies are in contact at 0 with their principal axes of curvature aligned, the separation of corresponding points A_1 and A_2 is:

$$h(x,y) = z_1 + z_2 = \frac{1}{2}\left[\frac{1}{R_1'} + \frac{1}{R_2'}\right]x^2 + \frac{1}{2}\left[\frac{1}{R_1''} + \frac{1}{R_2''}\right]y^2$$

i.e.

$$h(x,y) = \frac{1}{2R'}x^2 + \frac{1}{2R''}y^2 \qquad\qquad (1)$$

where $R' = R_1' R_2'/(R_1' + R_2')$ and $R'' = R_1'' R_2''/(R_1'' + R_2'')$ are the *relative* radii of curvature of the two surfaces. For $h = $ constant, equation (1) represents an ellipse:

$$\frac{x^2}{a^2} + \frac{y^2}{b^2} = 1 \qquad\qquad (2)$$

where $a/b = \sqrt{R'/R''}$

It follows that contours of equal separation are similar *ellipses*.

1.2 DEFORMED BODIES

Before deformation the bodies touch at 0, and the separation of points A_1 and A_2 is

$$h = \frac{1}{2R'}x^2 + \frac{1}{2R''}y^2$$

from eq (1).

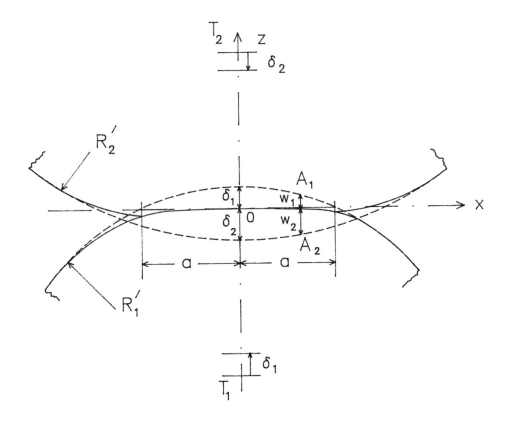

Fig. 2

A normal load is now applied and the bodies are compressed so that distant points T_1 and T_2 approach the origin by δ_1 and δ_2. The final separation of A_1 and A_2 is now

$$h' = h - (\delta_1 + \delta_2) + (w_1 + w_2)$$

where w_1 and w_2 are the normal elastic displacements of the surface at A_1 and A_2.

If the points A_1 and A_2 lie within the contact area, $h' = 0$

therefore

$$w_1 + w_2 = (\delta_1 + \delta_2) - h$$

i.e.

$$w_1 + w_2 = \delta - \frac{x^2}{2R'} - \frac{y^2}{2R''} \tag{3}$$

where $\delta \equiv \delta_1 + \delta_2 = w_1(0) + w_2(0)$

$\quad\quad\quad$ = the approach of two distant points T_1 and T_2.

$\quad$ If A_1 and A_2 lie outside the contact area, $h' > 0$

i.e.

$$w_1 + w_2 > \delta - \frac{x^2}{2R'} - \frac{x^2}{2R''}. \tag{4}$$

1.3 TRANSMITTED FORCES AND MOMENTS

$\quad$ The resultant force transmitted from one surface to another through a contact area S is resolved into a *normal load* P acting along the common normal, and components Q_x and Q_y of *tangential force* in the tangent plane sustained by friction. In terms of the normal traction (pressure) p and the *tangential tractions* q_x, q_y we have:

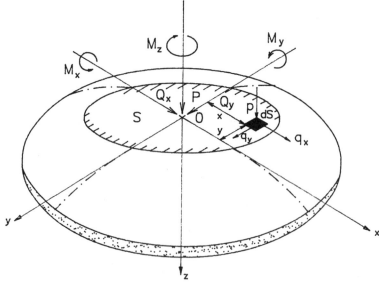

Fig 3.

$$P = \int_S p.dS \tag{5}$$

$$Q_x = \int_S q_x dS \tag{6}$$

$$Q_y = \int_S q_y dS \tag{7}$$

where $Q_x = -\dfrac{\Delta v_x}{|\Delta v|}\mu P$; $Q_y = -\dfrac{\Delta v_y}{|\Delta v|}\mu P$, since in a purely sliding contact the tangential force reaches its limiting value in a direction opposed to the sliding velocity v.

Since S is finite, the contact is able to transmit *rolling moments:*

$$M_x = \int_S p.ydS \tag{8}$$

$$M_y = \int_S -p.xdS \tag{9}$$

Acting about the common normal, and arising from friction, we also have the *Spin moment:*

$$M_z = \int_S (q_y x - q_x y)dS \tag{10}$$

2. Hertz theory of elastic contact

Hertz analysed the stresses set up at the contact of two elastic solids. Denoting the significant dimensions of the bodies both laterally and in depth by l, we may summarise the assumptions made in the Hertz theory as follows:

2.1 ASSUMPTIONS

(1) the strains are small, and within the elastic limit: $a \ll R$;

(2) the surfaces are continuous and non-conforming: $a \ll R$;

(3) Elastic deformation can be calculated by assuming that each solid is an elastic half-space:

$$a \ll R_{1,2}, \qquad a \ll l;$$

(4) The surfaces are frictionless: $Q_x = Q_y = 0$.

2.2 POINT FORCE ON AN ELASTIC HALF-SPACE

The displacement $w(r)$ at a distance r from a point force P is:

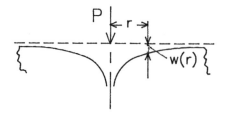

$$w(r) = \frac{1 - \nu^2}{\pi E} \frac{P}{r} \qquad (11)$$

where E is Young's modulus and ν is Poisson's ratio.

Fig. 4

2.3 DISTRIBUTED PRESSURE

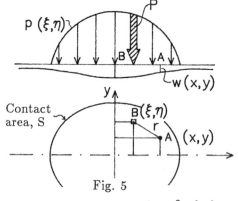

Consider an elemental force at B, given by $P = p(\xi, \eta).d\xi.d\eta$. At another point A, such that $r = BA = \sqrt{(x - \xi)^2 + (y - \eta)^2}$, the displacement due to P is:

Fig. 5

$$w(x, y) = \frac{1 - \nu^2}{\pi E} \int \int_S \frac{p(\xi, \eta) d\xi d\eta}{\sqrt{(x - \xi)^2 + (y - \eta^2)}}$$

$$(12)$$

2.4 CONTACT EQUATION

Inside the area S, from geometry (Eq 3), we have:

$$w_1(x,y) + w_2(x,y) = \delta - \frac{x^2}{2R'} - \frac{y^2}{2R''} \tag{13}$$

Substituting (2) into (3), since p acts equally on both surfaces,

$$\frac{1}{\pi}\left[\frac{1-\nu_1^2}{E_1} + \frac{1-\nu_2^2}{E_2}\right] \int \int_S \frac{p(\xi,\eta)d\xi d\eta}{\sqrt{(x-\xi)^2+(y-\eta^2)}} = \delta - \frac{x^2}{2R'} - \frac{y^2}{2R''} \tag{14}$$

This is a (singular) integral equation for $p(\xi,\eta)$. By analogy with electrostatic potential Hertz recognised that this equation would be satisfied by an ellipsoidal distribution of pressure:

$$p(x,y) = p_0\sqrt{1 - x^2/a^2 - y^2/b^2} \tag{15}$$

acting on an elliptical area of semi-axes a and b.

2.5 CIRCULAR POINT CONTACT

For spheres or crossed cylinders: $R' = R'' = R$ (say),[1] $a = b$; so that equation (15) becomes $p(r) = p_0\sqrt{1 - r^2/a^2}$, where $r^2 = x^2 + y^2$. $\qquad$ (16)

From (12), $w(r) = \dfrac{1-\nu^2}{E}\dfrac{\pi p_0}{4a}(2a^2 - r^2)$. $\qquad\qquad$ (17)

Substituting into (13) and performing the integration gives:

$$\left[\frac{1-\nu_1^2}{E_1} + \frac{1-\nu_2^2}{E_2}\right]\frac{\pi p_0}{4a}(2a^2 - r^2) = \delta - r^2/2R \tag{18}$$

For this to be identically true for all r,

$$\delta = \pi a p_0/2E^* \tag{19}$$

and

$$a = \pi p_0 R/2E^* \tag{20}$$

where

---

[1] or, for example: $R_1' = R_1'' = R_1; R_2' = R_2'' = R_2$; relative curvature $1/R \equiv 1/R_1 + 1/R_2$.

$$\frac{1}{E*} = \frac{1-\nu_1^2}{E_1} + \frac{1-\nu_2^2}{E_2}$$

The pressure given by (16) is in equilibrium with the load P, i.e.

$$P = \int_0^a p_o \sqrt{1-r^2/a^2} \ \ 2\pi r dr = \frac{2}{3}p_o\pi a^2 \tag{21}$$

Combining (19), (20) and (21):

$$a = \left[\frac{3PR}{4E^*}\right]^{\frac{1}{3}} \tag{22}$$

$$\delta = \frac{a^2}{R} = \left[\frac{9P^2}{16RE^{*2}}\right]^{\frac{1}{3}} \tag{23}$$

$$p_0 = \frac{3P}{2\pi a^2} = \frac{3}{2}\bar{p} = \left[\frac{6PE^{*2}}{\pi^3 R^2}\right]^{\frac{1}{3}} \tag{24}$$

2.6 LINE CONTACT

For two cylinders with their axes both parallel to the y-axis, making contact over a long strip of width 2a, we have:

$$p(x) = p_0\sqrt{1-x^2/a^2} \qquad \text{(Fig.7)} \tag{25}$$

Per unit length, the force is $P = \int_{-a}^{+a} p_0\sqrt{1-x^2/a^2}\,dx = \pi a p_0/2$ (26) giving:

$$a = \left[\frac{4PR}{\pi E^*}\right]^{\frac{1}{2}} \tag{27}$$

$$p_0 = \frac{4}{\pi}\bar{p} = \left[\frac{PE^*}{\pi R}\right]^{\frac{1}{2}} \tag{28}$$

In this case, the compression δ depends on the choice of datum for displacements. In physical terms, δ cannot be found by consideration of the local contact stresses alone; it is also necessary to consider the stress distribution with the bulk of each body.

2.7 APPROXIMATION FOR MILDLY ELLIPTICAL CONTACTS

If $R'/R'' < 5$, say, it may be shown that

$$b/a \simeq (R''/R')^{\frac{2}{3}} \tag{29}$$

and $\quad c \equiv \sqrt{ab} \simeq \left[\dfrac{3PR_e}{4E^*} \right]^{\frac{1}{3}} \quad$ where $\quad R_e \equiv \sqrt{R'R''} \tag{30}$

δ and p_0 are given by equations (23) and (24) with R replaced by R_e.

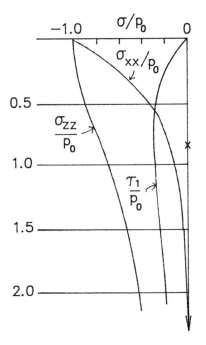

Fig. 6

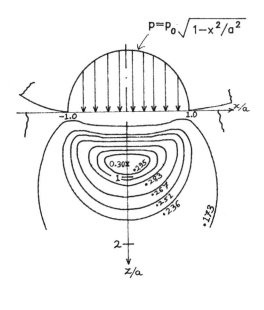

Fig. 7

3. The elastic stress field induced by Hertz contact

3.1 STRESSES WITHIN THE SOLIDS: LINE CONTACT OF CYLINDERS

σ_{xx} and σ_{zz} are principal stresses along the z-axis, as shown in figure 6. The principal shear stress τ_1 (radius of Mohr's circle) is given by $\tau_1 = \frac{1}{2}(\sigma_{xx} - \sigma_{zz}) = p_0 a\{z - z^2(a^2 - x^2)^{\frac{1}{2}}\}$;

$$\text{(31)}$$

its contours are shown in figure 7.

At the surface, $z = 0$, $\sigma_{xx} = \sigma_{zz} = -p$, $\tau_1 = 0$, and $(\tau_1)_{max} = 0.30p_o$, at depth $z = 0.78a$.

3.2 ELLIPTICAL CONTACTS (BETWEEN BODIES WITH GENERAL PROFILES)

The maximum shear stress occurs on the z-axis at a point beneath the surface whose depth depends upon the eccentricity of the ellipse:

	LINE					CIRCULAR
a/b	0	0.2	0.4	0.6	0.8	1.0
z/a	0.785	0.745	0.665	0.590	0.530	0.480
$(\tau_1)_{max}/p_0$	0.300	0.322	0.325	0.323	0.317	0.310

4. Plastic contact stresses

4.1 ONSET OF PLASTIC YIELD

Yield criteria:

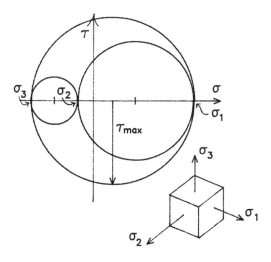

(i) *Tresca (maximum shear stress):* let σ_1, σ_2 and σ_3 be principal stresses where $\sigma_1 \geq \sigma_2 \geq \sigma_3$

$$\tau_{max} = \tfrac{1}{2}|\sigma_1 - \sigma_3| = k = \tfrac{1}{2}Y \qquad (32)$$

where k = yield stress in simple shear and Y = yield stress in tension.

Fig.8

(ii) *Von Mises (shear strain-energy; maximum octahedral stress):*

$$(\sigma_1 - \sigma_2)^2 + (\sigma_2 - \sigma_3)^2 + (\sigma_3 - \sigma_1)^2 = 6k^2 = 2Y^2 \equiv 6J_2 \qquad (33)$$

<u>N.B.</u> The maximum difference between (i) and (ii) is 15%.

4.1.1. *Line contact of cylinders*

On the axis of symmetry $(x = 0), \sigma_{xx}, \sigma_{yy}, \sigma_{zz}$ are principal stresses.

By Tresca: $\tau_{max} = \tfrac{1}{2}(\sigma_{xx} - \sigma_{zz}) = 0.30p_o = k$, for yield,

$$(p_o)_Y = 3.3k = 1.67Y. \qquad (34)$$

By von Mises:

$$(p_0)_Y = 3.1k = 1.79Y. \tag{35}$$

From Eq (28), the load for initial yield is thus

$$P_Y \simeq \frac{8.8RY^2}{E^*} \tag{35}$$

4.1.2 Circular point contact

From §3.2, $\tau_{max} = 0.31 p_o$ at $z = 0.48a.$

By Tresca: $\tau_{max} = k$, and

$$(p_0)_Y = 3.2k = 1.60Y \tag{37}$$

By von Mises:

$$(p_0)_Y = 2.8k = 1.60Y \tag{38}$$

From Eq (28) the load for initial yield is:

$$P_Y = 21\frac{R^2Y^3}{E^{*2}} \tag{39}$$

Note how this resistance to yield varies as the *cube* of the yield strength (or of the hardness).

4.1.3. Effect of frictional traction

Reference 1 (above) discusses the effects of a combination of normal and tangential forces which do not cause actual sliding (normal and tangential surface tractions and compliance); the Cattaneo-Mindlin solution to the problem of partial slip; and the Hamilton-Goodman analysis of the stress distribution within the solid: see also chapter by A R Savkoor, page 113.

As discussed in the chapter by T H C Childs, the point of maximum shear stress moves towards the surface as the coefficient of friction is increased. At the same time, the contact pressure for first yield, i.e. the maximum Hertz pressure, decreases. The frictional traction also introduces shear stresses into the contact surface itself, and these can reach yield for large enough values of traction coefficient Q/P.

4.2 PLASTIC INDENTATION: CONTACT STRESSES ABOVE THE YIELD POINT

We consider three regimes: (1) Elastic, (2) Contained plastic (subsurface), (3) Fully plastic.

4.2.1. *Elastic:*

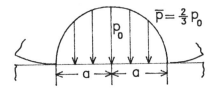

$$\bar{P} = \tfrac{2}{3}P_0$$

Fig. 9

From the Hertz equations for circular point contact, the mean pressure $\bar{p}$ is given by

$$\frac{\bar{p}}{E^*} = \frac{\tfrac{2}{3}p_0}{E^*} = \frac{4}{3\pi}\frac{a}{R},$$

so

$$\frac{\bar{p}}{Y} = \frac{4}{3\pi}\frac{a}{R}\frac{E^*}{Y} \tag{40}$$

The load P is related to the contact area A by $P = \pi a^2 \bar{p} = \frac{4}{3}\frac{E^*}{R}a^3 = \frac{4}{3}\frac{E^*}{R}(A/\pi)^{\frac{3}{2}}$, so

$$\frac{P}{R}\frac{E^{*2}}{Y^3} = \frac{4}{3\pi^{3/2}}\left[\frac{AE^{*2}}{R^2Y^2}\right]^{\frac{3}{2}} \tag{41}$$

4.2.2. *Contained plastic*

Elastic-plastic cavity model:

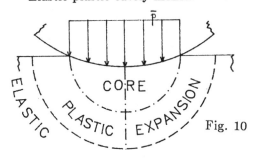

Fig. 10

For a spherical indentation,

$$\frac{\bar{p}}{Y} \simeq \frac{2}{3}\left\{1.7 + ln\left[\frac{1}{3}\frac{aE^*}{RY}\right]\right\} \tag{42}$$

4.2.3. *Fully Plastic*

Slip-line field models:

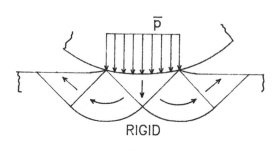

With modest friction ($\mu <$ 0.15), a cap of undeforming material adheres to the indentor.

$$\bar{p} \simeq H(hardness) \simeq 3Y;$$

Fig. 11

$$\frac{\bar{p}}{Y} \simeq 3, \qquad (43)$$

$$P = 3AY; \qquad \frac{PE^{*2}}{RY^2} = 3\left[\frac{AE^{*2}}{RY^2}\right] \qquad (44)$$

5. Non-dimensional plots (contact pressure; load)

5.1 ELASTIC-PLASTIC CONTACT: MASTER CURVES (Figs. 12 and 13)

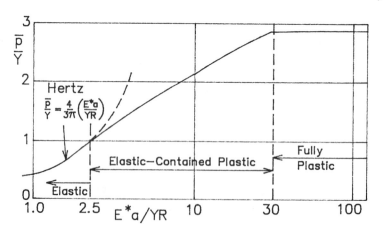

Fig. 12

In accordance with §4.1.3, we see that if a tangential force is present also, an increase in friction coefficient shifts the elastic/elastic-plastic boundary to lower values of $E^* a/(Y R)$: see chapter by T H C Childs, fig 12.

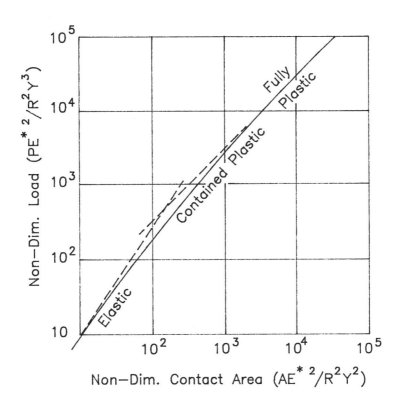

Fig. 13

5.2 EFFECT OF STRAIN HARDENING

Experiment and finite element analysis have shown that the above results can be used if Y is replaced by a representative yield stress Y_r, measured at a representative strain $\epsilon_r = 0.02\, a/R$.

NATO Advanced Study Institute on the
FUNDAMENTALS OF FRICTION
Braunlage, Germany, 29 July to 9 August 1992:
LIST OF PARTICIPANTS

M J Adams, Unilever Research, Port Sunlight Laboratory, Quarry Road East, Bebington, Wirral, Merseyside L63 3JW, UK

M Aderin, Dept of Mechanical Engineering, Imperial College, Exhibition Road, London SW7 2B, UK

N Agrait, Dept Fisica de la Materia Condensada (C-111), Universidad Autonoma de Madrid, 28049 Madrid, Spain

A Akay, Wayne State University, Dept of Mechanical Engineering, Detroit MI 48202, USA

M Akkurt, Istanbul Technical University, Fac. of Mechanical Engineering, Gumussuyu 80191 Istanbul, Turkey

S Akkurt, Istanbul Technical University, Fac. of Mechanical Engineering, Gumussuyu 80191 Istanbul, Turkey

P Arteaga, University of Surrey, Dept of Chemical & Process Engineering, Guildford, Surrey GU2 5H, UK

J F Belak, Department of Physics, Lawrence Livermore National Lab., P O Box 808, Livermore CA 94550, USA

P J Blau, Oak Ridge National Laboratory (MS 6063), P O Box 2008, Oak Ridge, TN 37831-6063, USA

M Brendlé, Centre de Recherches sur la Physico-Chimie des Surfaces Solides, 24 Avenue du Président Kennedy, 68200 Mulhouse, France

B J Briscoe, Dept of Chemical Engineering, Imperial College, Prince Consort Road, London SW7 2BY, UK

N A Burnham, KFA-ISI, Postfach 1913, D-5170 Jülich 1, Germany

L Caravia, Dept of Mechanical Engineering, University of Leeds, Leeds LS2 9JT, UK

G Carson, Dept of Materials Science & Engineering, University of Illinois, 1304 West Green Street, Urbana IL 61801, USA

Y L Chen, Dept of Chemical Engineering, University of California, Santa Barbara, CA 93106, USA

T H C Childs, Dept of Mechanical Engineering, University of Leeds, Leeds LS2 9JT, UK

R Christoph, CSEM, Rue Breguet 2, 2007 Neuchatel, Switzerland

R J Colton, Code 6177, Naval Research Laboratory, Washington DC 20375-5000, USA

L R Denton, Naval Weapons Support Centre, Mail Code 6051, Crane, Indiana 47522, USA

D Dowson, Dept of Mechanical Engineering, University of Leeds, Leeds LS2 9JT, UK

R Dwyer-Joyce, Dept. of Mechanical Engineering, Imperial College, Exhibition Road, London SW7 2BY, UK

A Erdemir, Tribology Section (MCT/212), Argonne National Laboratory, 9700 South Cass Avenue, Argonne, IL 60439, USA

J Ferrante (MS 5-9), NASA Lewis Research Center, 21000 Brookpark Road, Cleveland, OH 44135, USA

T E Fischer, Dept. of Materials Science and Engineering, Stevens Inst. of Technology, Hoboken NJ 07030, USA

B Gans (Code 6177), Naval Research Laboratory, Washington DC 20375-5000, USA

A J Gellman, Department of Chemistry, University of Illinois, 105 South Goodwin Ave, Urbana IL 61801, USA

R Wolf, JAVE Institut fur Angewandte Verschleibforschung, Hansastrasse 26, 7500 Karlsruhe 21, Germany

J-M Georges, Ecole Centrale De Lyon, B.P. 163, 36 Rue Guy de Collongues, 69130 Ecully-Cédex, France

M Ghadiri, Dept of Chemical & Process Engineering, University of Surrey, Guildford GU2 5H, UK

K Glejbol, Lab. of Applied Physics, Technical University of Denmark, DK-2800 Lyngby, Denmark

S Granick, Dept of Materials Science and Engineering, University of Illinois, 1304 West Green St., Urbana IL 61801, USA

J A Greenwood, University Engineering Department, Trumpington Street, Cambridge CB2 1PZ, UK

S Hainsworth, Materials Divison, The University, Newcastle-upon-Tyne NE1 7RU, UK

J A Harrison (Code 6110), Naval Research Laboratory, Washington DC 20375-5000, USA

D Hess, Dept of Mechanical Engineering, University of South Florida, Tampa, Florida 33620-5350, USA

J Holland, Institut für Reibungstechnik und Maschinenkinetik, Technische Universität Clausthal, Leibnizstr. 32, 3392 Clausthal-Zellerfeld, Germany

S Ikonomou, Department of Metallurgy & Materials Engineering, KU Leuven de Croylaan 2, B-3001 Heverlee, Belgium

J Israelachvili, Dept. of Chemical & Nuclear Engineering and Materials Department, University of California, Santa Barbara CA 93106, USA

S Jacobson, Dept of Technology, Uppsala University, Box 534, S-751 21 Uppsala, Sweden

K L Johnson, University Engineering Department, Trumpington Street, Cambridge CB2 1PZ, UK

G J Johnston, Mechanical Engineering Dept., Imperial College, Exhibition Road, London SW7 2B, UK

K Jung, Fachbereich Physik, Universität Kaiserlautern, Postfach 3049, D-6750

Kaiserlautern, Germany

A Kehrel, Institut für Metallforschung Metallphysik, Technische Universität Berlin, Hardenbergstr. 36, 1000 Berlin 12, Germany

D Klaffke, BAM Lab 5.22, Unter den Eichen 87, W-1000 Berlin 45, Germany

J Krim, Physics Dept., Northeastern University, Boston, MA 02115, USA

U Landman, School of Physics, Georgia Institute of Technology, Atlanta GA 30332-0430, USA

J Larsen-Basse, National Science Foundation, Surface Engineering & Tribology Program, 1800 G St NW #1108, Washington DC 20550, USA

J Y C Law, Dept of Chemical Engineering, Imperial College, Prince Consort Road, London SW7 2AZ, UK

B R Lawn, Ceramics Division, National Institute of Standards and Technology, Gaithersburg, MD 20899, USA

J Lepage, Laboratoire de Science et Genie des Surfaces, Ecole des Mines de Nancy, Parc de Saurupt, 54042 Nancy-Cédex, France

D P Leta, Exxon Research and Engineering Co., Rt 22 East Clinton Township, Annandale NJ 08801, USA

A Linnenbrügger, Inst. für Reibungstechnik u. Maschinenkinetik, TU Clausthal, Leibnitzstrasse 32, D-3392 Clausthal-Zellerfeld, Germany

W D Luedtke, School of Physics, Georgia Institute of Technology, Atlanta GA 30332-0430, USA

T Mathia, Lab. de Technologie des Surfaces, Ecole Centrale de Lyon, B.P. 163, 69131 Ecully-Cédex, France

D Maugis, CNRS/LCPC, 58 Bd Lefèbvre, 75732 Paris-Cédex 15, France

D Mazuyer, Laboratorie de Technologie des Surfaces, Ecole Centrale de Lyon, B P 163, 69131 Ecully-Cédex, France

G M McClelland, IBM Almaden Research Center, Dept K33, 650 Harry Road, San José CA 95120-6099, USA

E Meyer, Institute of Physics, Klingelbergstrasse 82, CH-4056 Basel, Switzerland

J W Mintmire, Theoretical Chemistry Section (Code 6119), U S Naval Research Laboratory, Washington DC 20375-5000, USA

J Moser, Institut de Physique Appliquée, Ecole Polytechnique Fédérale, CH-1015 Lausanne, Switzerland

J A Nieminen, Oxford University, Department of Materials, Parks Road, Oxford O1 3PH, UK

F D Ogletree, Material Science Division, Lawrence Berkeley Laboratory, CA 94720, USA

M Olsson, Uppsala University, Department of Technology, Box 534, S-75121 Uppsala, Sweden

R Overney, Institute of Physics, Klingelbergstrasse 82, CH-4056 Basel, Switzerland

B K Peterson, Mobil Research & Development Corporation, Central Research Lab., P O Box 1025, Princeton NJ 08543-1025, USA

J S Pethica, Department of Materials, Parks Road, Oxford O1 3PH, UK

R S Polizotti, Exxon Research and Engineering Laboratory, Corporate Research Laboratory, Route 22 East Annandale, New Jersey 08801, USA

H M Pollock, School of Physics and Materials, Lancaster University, Lancaster LA1 4YB, UK

E Rabinowicz, Rm 35-010 M I T, Cambridge MA 02139, USA

W A Rakowski, Academy of Mining & Metallurgy, Ul. Mickelwicz 30, IZPBM, Krakow, Poland

M O Robbins, The Johns Hopkins University, Department of Physics & Astronomy, 3400 North Charles Street, Baltimore MD 21218, USA

A R Savkoor, Delft University of Technology, Faculty of Mechanical Engineering and Marine Technology, Mekelweg 2, 2628 CD DELFT, The Netherlands

I L Singer, Code 6170 Naval Research Laboratory, Washington DC 20375-5000, USA

C R Slaughterbeck, Dept of Physics FM-15, University of Washington, Seattle WA 98195, USA

H M Stanley, Computer Mechanics Laboratory, University of California, 5149 Etcheverry Hall, Berkeley CA 94720, USA

B D Strom, Computer Mechanics Laboratory, 5129 Etcheverry Hall, U. C. Berkeley, Berkeley CA 94720, USA

B Stuart, Dept. of Chemical Engineering, Imperial College, Prince Consort Road, London SW7 2BY, UK

D Tabor, P.C.S., Cavendish Laboratory, Madingley Road, Cambridge CB3 0HE, UK

A Thölén, Lab. of Applied Physics, Building 307, Technical University of Denmark, 2800 Lyngby, Denmark

P S Thomas, Dept. of Chemical Engineering, Imperial College, Prince Consort Road, London SW7 2BY, UK

R Timsit, Kingston R&D Center, Alcan International Ltd., P O Box 8400, Kingston, Ontario K7L 5L9, Canada

M Tychsen, Inst. für Reibungstechnik und Maschinenkinetik TU Clausthal, Leibnitzstrasse 32, D-3392 Clausthal-Zellerfeld, Germany

E Vancoille, Dept. MTM, Katholieke Universiteit Leuven, de Croylaan 2, B-3001 Heverlee, Belgium

D Williams, Dept. of Chemical Engineering, Imperial College, Prince Consort Rd., London SW7 2BY, UK

T Wright, Materials Division, Herschel Building, The University, Newcastle-upon-Tyne NE1 7RU, UK

J T Yates, Surface Science Center, Department of Chemistry, University of Pittsburgh, Pittsburgh PA 15260, USA

Z Zhang, Dept. of Chemical & Process Engineering, University of Surrey, Guildford, Surrey GU2 5H, UK

SUBJECT INDEX

616